D0086781

ASUA BOOKSTORE
University of Arizona

DEC 23 1988

4251 LM

USED | 29.25
BOOK | PRICE

THIRD EDITION

Economic Geography

Truman A. Hartshorn
Georgia State University

John W. Alexander
Formerly University of Wisconsin, Madison

Prentice Hall, Englewood Cliffs, New Jersey 07632

102B Chris McDavid

Library of Congress Cataloging-in-Publication Data

Hartshorn, Truman A.
    Economic geography.

    Rev. ed. of: Economic geography/John W. Alexander,
Lay James Gibson. 2nd ed. 1979.
        Bibliography: p.
        Includes index.
        1. Geography, Economic.   I. Alexander, John W. (John
Wesley), 1918–   . II. Alexander, John W. (John
Wesley), 1918–   . Economic geography.   III. Title.
HF1025.H296   1987   330.9   87–17407
ISBN 0-13-225160-4

Editorial/production supervision and
    interior design: Fay Ahuja
Cover design: Photo Plus, Inc.
Photo research: Tobi Zausner
Photo editor: Lorinda Morris-Nantz
Manufacturing buyer: Paula Benevento
Editoral assistance: Julian Dangerfield

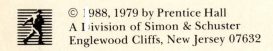

© 1988, 1979 by Prentice Hall
A Division of Simon & Schuster
Englewood Cliffs, New Jersey 07632

All rights reserved. No part of this book may be
reproduced, in any form or by any means,
without permission in writing from the publisher.

Printed in the United States of America

10 9 8 7 6 5 4 3 2 1

ISBN 0-13-225160-4 01

PRENTICE-HALL INTERNATIONAL (UK) LIMITED, *London*
PRENTICE-HALL OF AUSTRALIA PTY. LIMITED, *Sydney*
PRENTICE-HALL CANADA INC., *Toronto*
PRENTICE-HALL HISPANOAMERICANA, S.A., *Mexico*
PRENTICE-HALL OF INDIA PRIVATE LIMITED, *New Delhi*
PRENTICE-HALL OF JAPAN, INC., *Tokyo*
PRENTICE-HALL OF SOUTHEAST ASIA PTE. LTD., *Singapore*
EDITORA PRENTICE-HALL DO BRASIL, LTDA., *Rio de Janeiro*

To my parents Gailan and Carolyn Hartshorn and to the budding geographers in the family: Alan, Mary Karen, Sil, Louise, and Lois.

Contents

Preface

The first edition of the Alexander *Economic Geography* textbook became my first text as a rookie instructor of the Economic Geography course at Western Illinois University in 1966. Some 20 years later I still have a fondness for the book. It has always represented the mainstream of economic geography to me. I hope this third edition, drastically revised to keep abreast of the times, can maintain that stature.

ORGANIZATION OF BOOK

As with earlier editions, this book begins with a discussion of low-technology subsistence activities, followed by an examination of advanced commercial agricultural systems. Sections on energy, mineral resources, and manufacturing activity follow. In the final cluster of chapters sophisticated high-technology research and development activity and the structure of the modern metropolitan economy are discussed. The growing role of retail, office, and other service functions in the emergence of the multicentered metropolis is examined in this final section focusing on the urban economy.

This format provided a successful framework for the first edition of the book and is maintained here even though many chapters have been consolidated or eliminated. We do not cover as many commodities or functions as did earlier editions but explore in greater detail the contribution of various sectors of the economy, with in-depth discussions of representative activity at an international level. Highlighting this change is a dramatic scaling back of the earlier emphasis on the United States. The global system perspective followed promotes a more complete understanding of the interdependence exhibited by the contemporary functioning of the world economy.

Whereas most readers are familiar with the lifestyles and problems associated with modern, highly developed economies such as those in the United States, Canada, western Europe, and Japan, the situation facing third-world economies remains considerably different and much less familiar. Since all areas of the world economic system are discussed here, we felt that it would be useful to discuss in the first chapter the nature of poverty and development issues facing the third world as a prelude to later discussions of economic development trends.

WORLD-INTERDEPENDENCE THEME

The economic geography of the world has changed as much in the 25 years since the first edition of this book appeared in the early 1960s as it did in the previous century. In this 25-year period, the world economy mushroomed in size and complexity. At the same time, greater interdependence among nations added new dimensions to the world system. Major new work forms emerged as the postindustrial economy revolutionized the job market. The propelling force in economic growth increasingly became information and technology in the place of traditional raw materials and smokestack industries.

New forms of management and organization developed to shape and lead these changes. The world became particularly aware of Japanese business practices in the 1980s. The most visible and influential institution associated with business activity remained the multinational corporation, albeit much larger than before. Governments became more actively involved in promoting economic development. World inflation rates accelerated in the 1970s, an energy crisis emerged, and a crisis of finance gained momentum as the disparities between the developed and developing countries increased in the mid-1980s. Several newly influential groups of countries became important actors in the global marketplace. These included the Organization of Petroleum Exporting Countries bloc (OPEC), the newly industrializing countries (NICs), the Organization of Economic Cooperation and Development group (OECD), the Council for Mutual Economic Aid group (CMEA) in the Soviet bloc, and the European Economic Community (EEC).

World trade became a crucial factor in the development process, one that affected all parties. Gone were the days when raw materials flowed one way and finished goods the other. More and more goods were "international" in the sense that complex combinations of management, raw materials, technology, and semiprocessed goods, from many countries, interacted to create them. As less developed countries climbed the technology ladder, they began producing products at home to substitute for previously imported items and eventually began exporting more sophisticated products as well. In turn, the more developed nations moved to knowledge-intensive activities such as electronics, integrated circuits, robots, aerospace, telecommunications, and biogenetics.

To finance this process, nations bought and sold one another's products at an accelerating and unprecedented pace. Unfortunately, aberrations in the process have occurred, such as protectionism in the form of import restrictions or tariffs to insulate declining activities in some nations. The more developed nations generally financed the expansion of activity in less developed areas by extending credit, leading to growing dependency on the major powers by the third-world countries.

The growing economic interdependence among nations inspired the use of the term *global village* to refer to the scale and functions of the interworkings of the world economy today. The geographical consequences of this arrangement become apparent. It reminds one of a closed system in which any and all changes in one portion or place directly affect all others. Of course, there is spillover into the political arena because of this linkage. We discuss several political manifestations of this process in Chapter 20.

In the first chapter we define economic geography and discuss the particular plight of developing areas today, focusing specifically on poverty. Before delving into a discussion of particular economic activities in Chapter 3, the process of economic development and the evolution of the world system is reviewed in Chapter 2. Only from a better understanding of the differences in the setting and historical circumstances facing the third world in relation to the developed world can one appreciate how the third world posesses a completely separate and unique setting from the developed world's experience. Rather than following in the footsteps of the western world in terms of stages of economic development over time, that area appears to be forging its own destiny, even though ties to the developed world remain strong. The reasons for this difference are obscure and complex, but they relate to the widespread presence of poverty, the historical experience of the area (discussed in Chapter 2), and the growing gap in affluence between the developed and developing areas following recent inflationary price increases, high interest rates, and skyrocketing energy price hikes, to name a few.

A RECOMMENDATION

The names of many nations and of other places are mentioned in the chapters to follow. Therefore, every reader should have ready access to an atlas, an indispensable tool for mastering the material of geography as well as for comprehending world affairs. Among the best atlases for economic geography are *The Oxford Economic Atlas of the World; The Oxford Regional Economic Atlas, the United States and Canada; The Oxford Regional Economic Atlas, Western Europe; Man's Domain, a Thematic Atlas of the World; The National Atlas of the United States of America;* and *Goode's World Atlas.* In addition, it is helpful to purchase country or regional outline maps at a bookstore to make notations about places and areas discussed in the text.

Truman A. Hartshorn
Georgia State University

Acknowledgments

A worthy manuscript for the third edition of *Economic Geography* required five years of developmental effort. I have been ably served by capable secretaries, editors, reviewers, and patient students and colleagues during this time. I would not have attempted the project at all if it were not for the thoughtful counsel and encouragement from Betsy Perry, the first of four Geography editors with whom I worked at Prentice Hall. She was ably followed by Nancy Forsyth, Curt Yehnert, and Dan Joraanstad. Their assistants, Joan-Ellen Messina, and Jennifer Schmunk have been the unsung heroines during these transitions. Executive editor Bob Sickles supported the project throughout.

Reviewers also deserve plaudits for their critiques of at least two versions of the entire manuscript. For this effort, I will be indebted always. This group included the following: Geoffrey Hewings, and Howard Roepke of the University of Illinois; Murray Austin, University of Northern Iowa; David Goldstein, College of DuPage; Peter Muller, University of Miami; and James McConnell of the State University of New York at Buffalo. I am saddened to report that Professors Roepke and Goldstein did not live to see the fruits of their work on this project, but may we rejoice that *Economic Geography* is stronger for their contribution. Other, anonymous reviewers, have also played a helpful role in the manuscript development.

At the local level, I was the beneficiary of many enlightening discussions about the emerging world system with a former colleague, Nanda Shrestha, now of the University of Wisconsin, Whitewater. He authored Chapter 2 and entered practically all the manuscript into the word processor. My colleague Sanford Bederman penned Chapters 3 and 8 in superb fashion. During this time, I was also ably served by Barbara Denton, Georgia Nixon, Cynthia Fox, Steven Fievet, and Richard Sheets at Georgia State University. Several graduate students also provided helpful services including Steven Johnson, Sara Yurman, and Malissa Carling. No long-term project such as this could be accomplished without the support of the Dean, and I am most grateful for the positive commitment extended by Clyde Faulkner in this regard. Finally, my colleagues deserve thanks for their good nature during this extended period.

This edition includes a more balanced selection of maps at the world, regional, and local scales. Even those retained from earlier editions have been redesigned, providing a uniform product. Appreciation is extended to Jeff Mellander and Mark Smith of Precision Graphics in Champaign, Illinois who handled the drafting.

On the production side I would like to thank Barbara Zeiders for her copy-editing prowess, Jeanne Hoeting, managing editor of the math/science group and Fay Ahuja, production editor.

Courtesies are extended to co-author John Alexander whose legacy I have tried to uphold here. Lay Gibson similarly carried the day with the second edition, and is commended for his encouragement.

Economic Geography Today

1

Economic geography: What is it? Everyone knows that the term "economic" refers in one way or another to business activity, jobs, and/or money, and most people also have some notion of what geography means based on their experience in learning about the world around them. But what is economic geography?[1] *Economic geography* refers to the field of study focused on the location of economic activity at the local, national, and world scale (Figure 1-1). Economic geographers study not only highly developed areas such as the United States and Canada, but also developing areas and centrally planned economies. *Developing areas* lack a modern urban-industrial structure and are sometimes referred to as *third-world nations. Centrally planned economies* include China and countries in the Soviet bloc, which are also known as the *second world*. The growing interdependence among activities in all these areas has intensified in recent years. We now talk of global interdependence, and the geography of international business in recognition of this situation. This internationalizing process, which affects all world economies, is given particular attention in this book.

CLASSIFYING ECONOMIC ACTIVITY

A useful way to classify the various ways in which goods and services can be produced is to think in terms of a continuum from simple to complex, from the harvesting of fruits and nuts from nature's storehouse to creating them purposefully using modern agricultural practices (see Table 1-1). *Primary production* includes age-old activities such as hunting animals and gathering wild berries and nuts; extracting minerals from the earth's crust; fishing from rivers, lakes, and oceans; and the harvesting of trees. Primary producers might be labeled *red-collar workers* due to the outdoor nature of their work.

Secondary production increases the value (or usefulness) of a previously existing item by changing its *form*. Such activities include manufacturing and commercial agriculture. The farmer, for instance, applies hybrid seeds, fertilizers, and modern technology in the form of cultivating and harvesting equipment to increase the yields of crops. Steelmakers turn iron into a more durable metal in blast furnaces and steel mills. We often think of this group collectively as the *blue-collar labor force*.

Tertiary production involves the service sector rather than tangible goods. This work refers to a range of personal and business services involving a rapidly growing share of the labor force in highly developed areas. In a colloquial sense, persons engaged in personal service occupations "take in one another's washing." Retail clerks, barbers, beauticians, and secretaries all fall into the personal and business service

---

[1]The word *economic* pertains to all the activities in which people engage, the world over, in the production, exchange (or distribution), and consumption of goods and services. Anything people buy, barter, or work to produce, consume, or exchange is an economic item.

1

Figure 1-1 Retail Shops in Tokyo. *The majority of retail sales in Japanese cities continues to occur in small "mom and pop" retail outlets. Operators often live upstairs over the store in neighborhoods with narrow streets, colorful signs and sidewalk plants. (T.A.H)*

Table 1-1

Classification of Economic Activities

A. Production
1. Primary: harvesting commodities from nature (subsistence agriculture, forestry, fishing, mining) *red*
2. Secondary *blue*
   (a) Purposeful tending of crops and livestock (commercial agriculture)
   (b) Increasing the value of commodities by changing their form (manufacturing)
3. Tertiary: services (clerical, personal, business) *pink*
4. Quaternary: financial, health, entertainment, education, information, and data-processing services; middle-management administrative services; government bureaucrats *white*
5. Quinary *gold*
   (a) High-level managerial and executive administrative positions (public and private)
   (b) Scientific research and development services

B. Exchange
1. Transportation and distribution services
   (a) Increasing the value of commodities by changing their location (freight transportation)
   (b) Exchanging services and ideas by telecommunication or face-to-face contact
   (c) Satisfying the needs of people by changing their location (passenger transportation)
   (d) Warehousing and distribution function
   (e) Wholesale trade
   (f) Retail trade

C. Consumption: use of commodities and services by human beings to satisfy needs and wants

---

categories as a group and have been described as *pink-collar workers.*

Quaternary services represent a special type of service work, focusing on professional and administrative services, including financial and health service work, information processing, teaching, and government service as well as entertainment activity. Specialized technical, communication, and/or motivation and leadership skills provide the common thread linking these activities. Practically all quarternary activity occurs in office building environments or specialized environments provided by schools, theaters, hotels and hospitals, and we think of this group as the *white-collar work force* (see Figure 1-2).

The final grouping, *quinary activities,* remains more restricted in size in comparison to the other groups of activities just reviewed. The most visible persons in this group include chief executive officers and other top-management executives in both government and private service. Research scientists, legal authorities, financial advisers, and professional consultants who

provide strategic planning and problem-solving services belong to this cluster. Most of these high-order analytical and managerial activities occur in larger urban centers or in close proximity to large university, medical, and/or research centers. New York, London, and Tokyo, for example, being the primary world financial centers, possess a large number of specialized banking and other financial executives, giving them a very large cadre of quinary workers. An appropriate label for this group is the *gold-collar worker.*

Exchange Services

In addition to the "goods and services" occupations just discussed, people engaged in the exchange of items, whether it involves handling freight, wholesaling, storage, telecommunications, or passenger movement, also play important roles in modern economies. Most such exchanges increase the value of an item because of the services provided. This type of exchange is the purpose of wholesale trade and retail trade. A radio, for example, is worth more when it leaves a distributor than when it leaves the manufacturer, and its value continues to increase as it passes from distributor to retailer and from retailer to consumer. This additional value is created by the specialized services provided at each level of handling, including packaging, promotion, financing, and merchandizing of the product.

Figure 1-2 Downtown and Mid-town Manhattan. *The two downtowns of Manhattan have distinct histories. Downtown Manhattan, the "original" New York that developed on the waterfront, has evolved into the world's premier financial district, while mid-town has become the corporate office center and retail district. (Courtesy of Port Authority of New York and New Jersey)*

Consumption

A third aspect of all economic activity involves the consumption of goods and services. Until recently, the geography of consumption has largely been ignored by geographers. Today there is a small but growing literature that deals with both the patterns of consumption and the spatial aspects of consumer behavior.

The term *consumption* refers to the final or direct use of goods and services to satisfy the wants and needs of human beings. Some forms of consumption devour goods quickly, as is the case with nondurable goods produced for final consumption (textiles, food, etc.). Other forms of consumption engulf a commodity slowly, bringing about its gradual depreciation, as in the case of producing goods to make other goods, such as occurs with the machine tool industry. Still other forms of consumption, such as tourism and travel, may or may not diminish the quality or quantity of a commodity at all. For example, gazing at the Alps or skiing down the snowy slopes may not diminish the value of the product, especially if environmental concerns are not evaluated. Occasionally, consumption may actually increase the worth of an object, as in enjoying an antique table or a Rembrandt painting.

This gamut of economic pursuits, ranging from hunting and gathering in primitive societies to collecting antiques in highly developed areas, provides a topic for study by economists, historians, and several other scientists, including geographers. Just how does geography differ from these other disciplines in the study of economic activity? A partial answer can be found in the meaning of *geography* itself. The geographer's perspective deals with location and space, and even though the subject matter is the same as that of other disciplines, the approach or viewpoint of the geographer in dealing with this material provides the field with an identity and uniqueness.

THE MEANING OF GEOGRAPHY

Two widely held but false notions of geography may assist with the definition of geography. A good many people seem to think that geography is simply a matter of place-name recognition; to them, a geographer is a person who knows the location of county seats, state capitals, rivers, and seas. When a contestant on a quiz program chooses the "geography" category, for example, questions asked invariably involve place-name recall.

Other persons have the idea that geography is solely the study of the natural or physical environment. To them, the geography of Illinois, for example, would deal with its climate, topography, drainage, natural vegetation, soil, and minerals. In this view, geography represents a medley of excerpts from geology, meteorology, and biology.

The word *geography* comes from two Greek roots: *geo,* which means "earth," and *graphos,* which means "description." The meaning would seem to be simple and clear. But many scholarly disciplines "describe the earth," for instance, geology, pedology, botany, zoology, and meteorology. Surely, geography cannot claim to be the sum total of all earth sciences. In fact, the hallmark of geography is not so much *what* it studies as *how* it studies. Geography is unique because of the perspective from which its practitioners study the earth from a spatial perspective.

Analogy with History

The approach used in the field of geography shares at least one common perspective with that of the field of history. There would be no history if human events never changed and were invariable from day to day; it is because of the variation through time that the discipline of history exists. Because of these temporal variations, the historian can identify periods, such as the Elizabethan period, the Middle Ages, or the cold war era, on which to focus attention. If the main concern is with the manner in which people express themselves, one can identify various cultural periods, such as the Renaissance or the Victorian or the Mayan era, when classifying historical events. Regardless of the historian's predilection, then, the fundamental fact that phenomena differ from one *time* to another enables the researcher to distinguish chronological periods.

But this is only the beginning. The historian's main objective is to understand relationships between events. The scholar, for example, may want to know how an incident in 1914 relates to other events that took place in that year, in subsequent years, and decisions in earlier years that led to World War I.

The geographer is concerned primarily with variations from *place to place* rather than from time to time. Understanding relationships among places provides the integrating theme (Figure 1-3). There would be no geography if physical and human phenomena were distributed uniformly over the face of the earth. But rainfall, elevation, temperature, population, farming, mining, manufacturing, and cities themselves do vary

Figure 1-3 The Site and Situation of St. Paul, Minnesota. *St. Paul, located on the east side of the Mississippi River, but in an area where the river actually flows east-west, was settled due to its favorable site for crossing the river.*

markedly from one location to another. Any phenomenon whose distribution differs from place to place is termed a *spatial variable* and qualifies as an element of geography.

Definition of Economic Geography

It is at present impossible to formulate a definition of geography that is both complete and simple—and that all geographers will accept. Nevertheless, beginning students are entitled to some declaration of position on the field. To that end, the following definition is proposed: *Geography* is the study of spatial variation on the earth's surface. "Earth's surface" is construed rather broadly here to refer to the milieu in which human life exists—the lower portion of the atmosphere, which people breathe; the outer part of the lithosphere, upon which they walk and from which minerals are extracted; and the hydrosphere, where people fish and sail.

Spatial variation, like temporal variation, has profound significance for human life and underlies many of the problems facing nations, states, cities, farms, factories, families, and individuals today. The geographer deals with and searches for relationships among variables over space. Thus a more complete definition of *geography* would be: the discipline that analyzes and explains variations in activities over space. In this sense a better name for the discipline might be "spatial science."

Now we are in a position to answer our original question more fully. What is economic geography? By blending our definitions of the two constituent terms, we can derive this statement: *Economic geography* is the study of the spatial variation on the earth's surface of activities related to producing, exchanging, and consuming goods and services. Geographers rely heavily on maps, analytical methods, and models in their search for explanations. Whenever possible the goal is to develop generalizations and theories to account for these spatial variations.

How Do You Explain Activity Patterns?

An economic geographer might be asked, for example, to explain the location of shopping centers in a metropolitan area (Figure 1-4). A knowledge of central place theory, familiarity with the hierarchy of shopping centers typically found in an urban market, and an understanding of consumer travel behavior would all be useful in the analysis. In other words, one calls upon all relevant theories and generalizations when making the assessment. Theory suggests that the size of the market (population distribution), its buying power (income), and the transportation network all play a vital part in the explanation.

In this book we use this classical geographical approach to explain economic activity patterns. In some instances the explanations appear simple and straightforward, as in the shopping center example just reviewed. In other cases, they are more elusive and complex, as is the case with world poverty, which is discussed later in the chapter. Often, historical, political, technological, and/or cultural factors play important roles in these explanations, together with locational principles. Before proceeding with this discussion, it will be useful to discuss a framework that accounts for the organization of this book, together with a discussion of several useful concepts.

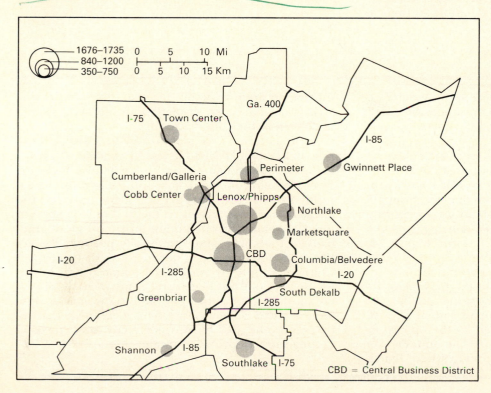

Figure 1-4 Location of Regional Malls in Metropolitan Atlanta. *Regional malls blanket the metropolitan area, but are more frequent in higher income neighborhoods, such as the north side of Atlanta. Proportional circles reflect retail sales volume in 1982.*

Locational Analysis

Locational analysis in economic geography involves not only an explanation of activities already present on the landscape but may also involve the selection of a future location for an activity such as a restaurant or shopping mall. Geographers use theories insofar as possible to explain why activities are located as they are. For example, in this book we are concerned primarily with agriculture, manufacturing, and service activities and the theories that have been developed to assist understanding the locations of these activities: (1) the von Thünen model in the case of agriculture (Chapter 5), (2) the Weber model for manufacturing (Chapter 13), and (3) Christaller's central place theory for tertiary, quaternary, and quinary activities, including retail location (Chapter 17). Geographers also rely on concepts in their analytical work, such as distance, interaction, and region. The concepts of interaction and region are explained more fully in the accompanying boxes.

In addition to calling on relevant theories to help account for patterns observed in this book, we will also use an overarching theme to integrate the many disparate events that are shaping the economic geography of the modern world. The *global system* framework is such a theme, and it will be expanded on to demonstrate more clearly the dynamics affecting economic activities and the variations in the level and interdependencies that exist in international economic development. Even those persons in the most remote areas of the world now participate in an economic system that is less local and regional and more national and international in scope. The world economy today increasingly generates levels of spatial and social interaction that transcend national space economies. The *political economy* of a nation, as well as

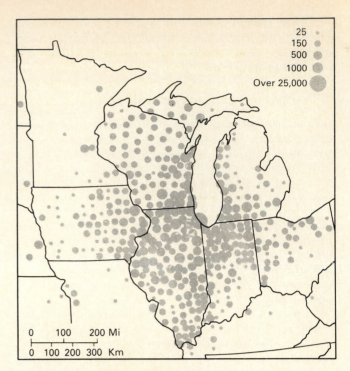

Figure 1-5 Nodal Region of Chicago Tribune Circulation. *A map of the circulation provides a good example of a nodal region. Notice the wide penetration of the Chicago Tribune throughout Illinois, northern Indiana, southern Wisconsin, and eastern Iowa. The movement of the paper into Missouri is restricted due to competitive strength of the St. Louis paper. Note that the intensity of the circulation decreases with distance from the city of origin. The increase in the circulation intensity in the northern peninsula of Michigan represents an exception explained by the presence of retirement and second home communities owned by Chicagoans.*

REGION

Geographers use the *region* concept to generalize about areas under study. Regions can be subdivided into two types, uniform and nodal. A *uniform region* is a homogeneous area defined on the basis of common characteristics that serve to make areas within the region more internally similar than areas outside its boundaries. It is an arbitrary area, often without a clear delineation on the earth's surface. An example would be an agricultural production area such as a dairy farming region or a tobacco-growing area. An example of a uniform region occurs with Figure 5-3. A *nodal region* is an area focused on a node or point away from which movement to or from the node occurs. Often, the node is a city and the movement involves the flow of goods or services to or from the city, for example, newspaper circulation or commuting. All areas around the node dominated by the flow to or from it are included in the region. The boundary of the nodal region is drawn at the point where the attraction to another node becomes stronger. In this way, a variation in the strength of the attraction to the node is built into the definition.

The pattern of circulation for the Chicago Tribune, illustrating the nodal region concept, is shown in Figure 1-5. Note that the region extends southward from Chicago into downstate Illinois and that it abruptly stops at the Missouri boundary, owing to the allegiance of Missourians to the St. Louis paper. On the other hand the circulation tapers off more gradually in Iowa. The map graphically portrays the notion of dominance which declines with distance from Chicago. Another type of nodal region can be drawn on the basis of flows from rural to urban areas, as with the movement of milk from the farm to market, creating a *milkshed* (see Figure 5-6). A *laborshed* that incorporates the area from which a city draws its commuting workers provides another example.

The concept *spatial interaction* refers to the movement process demonstrated by the flows of goods and services and people over space. The *principle of least effort* generally accounts for the length and intensity of this movement.* This principle is based on the notion that one minimizes distance and selects the shortest path when moving between two points. Underlying this conceptualization is the notion of *friction of distance,* which refers to the resistance to movement over space. In its simplest form this inverse relationship between movement and distance can be expressed as

$$I = \frac{M}{d}$$

where I = interaction
M = mass or attraction
d = distance

The greater the mass or attraction, the greater the level of interaction that will occur. Conversely, the greater the distance, the less the interaction. In other words, the amount of movement or interaction depends simultaneously on both the attraction or mass and the distance between the two points. The mass or attraction variable is typically based on the number of people in an area. The attraction between two large cities, for example, is greater than that between two smaller ones, even if they are farther apart. Distance can be measured in many ways in addition to the traditional length measurement (e.g., miles). It can be modified by travel time, cost, or social distance factors. In any case, the greater the distance between two points, the less movement or interaction that will occur between them. The diagram of a trade area around a city shown in Figure 17-10 provides an example of this relationship.

Trade as Spatial Interaction

The geographer Ullman suggested that for spatial interaction to occur, the impact of three interrelated conditions must be satisfied: (1) complementarity, (2)

*G. K. Zipf, *Human Behavior and the Principle of Least Effort* (Reading, Mass.: Addison-Wesley, 1949).

intervening opportunity, and (3) transferability.† *Complementarity* refers to the need for a supply-and-demand relationship to exist prior to any movement occurring between places. The *intervening opportunity* concept assists in an understanding of the source of supply for a particular item. When more than one source of supply exists, the nearest source to the final destination (the intervening opportunity) will be chosen. When additional sources of supply occur, closer to the destination, the new source area is substituted. *Transferability* reflects the time and cost factor of movement. As distance or costs of movement increase and reach a critical threshold, movement may decrease or cease altogether as costs and time factors create an economically unjustifiable situation.

When this movement stops, another item will be substituted for the one in demand. The transferability of an item generally reflects its value per unit of weight and its cost of transportation. Bulky products are the least transportable over long distances, as they cannot withstand expensive transportation charges which would increase the final product price dramatically. Product substitution becomes commonplace with such goods. High-value-added products, such as electronic goods, by contrast, can withstand large transportation charges because they are so valuable per unit of weight, and product substitution is rarely required.

In the field of international trade we observe that raw materials experience considerable sensitivity to intervening opportunities and transferability factors, leading to a substitution of sources and products, respectively. High-technology goods and highly fabricated manufactured products are less sensitive to these pressures. One could therefore use the spatial interaction concept to explain the growing tendency for initial processing of goods to occur in low-cost environments favoring less developed, third-world locations, whereas high-value goods remain competitive in highly developed areas. As this differentiation process has intensified in recent years, so has the volume of international trade.

†Edward Ullman, "The Role of Transportation and the Bases for Interaction," in William Thomas, Jr., ed., *Man's Role in Changing the Face of the Earth* (Chicago: University of Chicago Press, 1956), pp. 862–880.

macro forces associated with the transition of the world economy from a manufacturing to a postindustrial base, the international monetary system, and multinational corporations, among others, also affect these activities. Political economy refers to the impact of public policy and government programs on economic activity.

In terms of the level of economic development, the standard of living of a large share of the world's population has increased substantially in recent decades. Multinational corporations, based both in the developed and less developed world, now operate worldwide and the level of international trade has accelerated, greatly assisting this development process. But tensions have also increased, and political considerations have increasingly directed economic decision making—whether it be investment priorities or trade policy.

To be sure, enormous spatial contradictions and disparities still exist between and among countries at the world scale—particularly those between developed and developing countries. Rapid population growth prevents many areas from improving their standard of living, drought and famine afflict other regions, and warring factions disrupt yet other countries.

In many cases the roots of these spatial contradictions and disparities, particularly pronounced in today's developing world, run much deeper than those created by current conditions. As we shall see in Chapter 2, the political and social forces that emerged at the beginning of the industrial revolution and before account for many contemporary world economic problems. Before proceeding with a review of these historical-political forces, let us examine the present situation in the most disadvantaged area of the world today, the developing countries.

POPULATION AND DEVELOPMENT ISSUES IN THE DEVELOPING WORLD

The well-being of approximately three-fourths of the world's population—the poverty-ridden peoples in the underdeveloped countries of Asia, Africa, and Latin America—remains a significant and growing world problem. "It is indeed sobering that, two centuries after the Industrial Revolution, most of the world remains poor, still suffering from inadequate standards of living."[2]

Issues of population and development in the developing-world context are interrelated mainly because of the circular relationship between population and development. Population is both the subject and the basis of development. That is, development is directed at improving the socioeconomic well-being of population and at the same time requires people for its technical and manual labor power. It should be kept in mind that population, although necessary to get the process of development moving in the forward direction, can also stifle development meant to help itself.

Regardless of the nature of the relationship, the overriding issues of population and development in the developing world involve poverty, unemployment, underemployment, inequality, malnutrition, poor sanitation, and many other social miseries. Dudley Seers has asserted that a country cannot be considered developed or developing unless it has been able to reduce the levels of poverty, unemployment, and socioeconomic inequality affecting its population.[3] In the following subsection we discuss a specific problem—poverty—which is the most obvious issue and the most representative manifestation of all other problems.

[2]G. M. Meier, *Leading Issues in Economic Development*, 3rd ed. (New York: Oxford University Press, 1976), p. 1.
[3]Dudley Seers, "Meaning of Development," in C. K. Wilber, ed., *The Political Economy of Development and Underdevelopment* (New York: Random House, 1973).

Poverty

The persistently high level of poverty facing most of the world's underdeveloped countries severely handicaps economic growth and development planning. It was widely believed until recently that rapid growth of a country's *gross national product* (GNP) would solve this critical problem. The gross national product of a country equals the total value of goods and services produced during a specific period, typically one year. A prominent development economist with the World Bank has stated: "We were taught to take care of our GNP as this will take care of poverty."[4] But it has become only too painfully apparent to most development economists and policymakers in recent years that a rapid growth rate of GNP has not reduced the level of poverty, due to rapid inflation and the fact that benefits are not equally shared. In some cases, poverty has actually increased.

No matter how policymakers and planners view poverty, it is normally perceived as a social burden by well-off persons and by national government leaders, who view it as a political embarrassment. But for those who suffer chronic poverty, who survive at the margin of bare subsistence and starvation, it is also a human tragedy. The question of poverty is also so politically charged that often, ideological views and rhetoric overshadow the real problem. Because of these political overtones the issue of poverty provokes plenty of national and international attention, concern, and discussion, but effective solutions are far and few between.

Generally, poverty is measured in terms of per capita income (per capita GNP figures). Per capita GNP figures are often criticized as being a poor measure of poverty because of their inability to account for various types of household-level informal economic activities which help to sustain millions of families in underdeveloped countries and do not officially enter the accounting system. Since there are very few better alternative measures, GNP is still commonly used to analyze poverty.

Relative Poverty. The issue of poverty can be analyzed from a number of perspectives, but the focus of the present discussion relies on a simple classification of poverty into *relative poverty* and *absolute poverty* groupings. The conventional approach to poverty involves defining it in relative terms. Not every country (or population) has the same standard of living and per capita income. In comparison to other countries, a country is generally poorer or richer. Most underdeveloped countries are considered poor because their per capita GNP is much lower than that of most developed countries (see Table 1-2). Even among underdeveloped countries (and developed countries as well) there is a striking disparity in per capita GNP figures. For example, Brazil has a per capita income

[4]Mahbub ul Haq, in Meier, op. cit., p. 9.

Table 1-2

Per Capita Income Distribution: Comparison
between Underdeveloped and Developed Nations, 1983

Nation	Population (millions)	Area (thousands of square kilometers)	GNP per Capita (dollars)
Underdeveloped			
India	733.2	3,288	260
Indonesia	155.7	1,919	560
Egypt	45.2	1,001	700
Nigeria	93.6	924	770
South Korea	40.0	98	2,010
Mexico	75.0	1,973	2,240
Brazil	129.7	8,512	1,880
Developed			
Great Britain	56.3	245	9,200
Japan	119.3	372	10,120
West Germany	61.4	249	11,430
United States	234.5	9,363	14,110

Source: The World Bank, *World Development Report, 1985.*

of $1570, whereas the comparable statistic for India is only $180.

The problem of the international disparity in poverty is further aggravated by the presence of large internal disparities in poverty levels within a country, regardless of whether it is poor or rich in the aggregate. The internal disparity in underdeveloped countries appears greater than that in developed countries primarily because of the underdevelopment of their space economy (e.g., the lack of well-developed transportation networks).

The internal disparity generally has two dimensions: (1) a *spatial disparity* between regions, and (2) a *social* or *class disparity* between people. The spatial disparity in poverty refers to the fact that within a country some regions are relatively poor and some are relatively rich in terms of income and resource distribution. On the other hand, the social disparity refers to the fact that some people are rich and others are poor. Whereas regional disparity in resource distribution is generally determined by physical and historical factors, class disparity is sustained by social and institutional factors.

The case of Nepal, an underdeveloped country sandwiched between China and India, provides a clear example of the degree of spatial disparity between regions which exists in most underdeveloped countries (Figure 1-6). The distribution of cultivated land area between ecological regions in Nepal is very unequal.

Table 1-3 shows that the hill region has only 26 percent of the total area under cultivation in Nepal, but supports over 53 percent of the population. The tarai (plains) region, on the other hand, has the majority of the cultivated area (almost 72 percent), but supports only about 40 percent of the total population of Nepal. Because of such a disparity in popu-

Table 1-3

Spatial Disparity in the Regional Distribution of Population
and Cultivated Land Resources in Nepal, 1971

Region	Population	Percent of Regional Total	Cultivated Area (000s hectares)	Percent
Hill	6,167	53	510	26
Tarai (plain)	4,770	41	1,422	72
Kathmandu Valley	619	5	48	2
Regional total	11,556	100	1,980	100

Source: Adapted from Nanda Shrestha. "A Preliminary Report on Population Pressure and Land Resources in Nepal, *Journal of Developing Areas,* 16 (1982), p. 198.

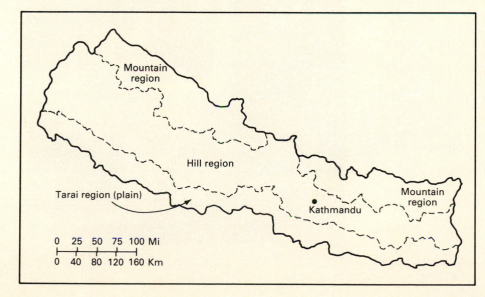

Figure 1-6 Regions of Nepal. *The distribution of cultivated land area between regions in Nepal is very unequal. Over one-half the population lives in the hill region, but that area has only about one-quarter of the cultivated land. The tarai (plains) area, on the other hand, contains about 40 percent of the population and over 70 percent of the cultivated land. (Courtesy of Nanda Shrestha)*

lation and land resource distribution, the hill region is much poorer than the tarai region.

The case of class disparity can easily be observed by looking at the figures of income distribution between classes presented in Table 1-3. Since the issue of class disparity is detailed in the discussion of absolute poverty in the following subsection, the discussion here is limited. The figures in Table 1-4 show that income distribution in underdeveloped countries is highly skewed in favor of the upper classes (higher quintiles). The income share of the highest quintile in Table 1-4 averages 52 percent, while that of the lowest quintile (class) is 5 percent for the underdeveloped countries listed. Class disparity is quite evident. The population in the lower-class bracket is poor relative to the upper-class population. Note also that the relative poverty ranges are quite large in developed countries as well, the primary difference being that the share of income controlled by the highest 20 percent of the population is generally much smaller.

Absolute Poverty. "The concept of relative poverty is important and reveals the problem of inequality associated with skewed income and uneven resource distribution. But it tells us little about the extent of *absolute poverty* in underdeveloped countries. Yet much of the current interest in relative poverty is not due simply to the existence of socioeconomic inequality. It is more often a concern with absolute poverty, which refers to being unable to meet minimum human needs, such as adequate food, clothing, health care, education, and shelter, even by developing world standards. The incidence of absolute poverty in underdeveloped countries has a powerful appeal for dramatizing the need for policy action in both domestic and international spheres."[5]

In spite of its widespread existence, estimates of absolute poverty are difficult to obtain, for a number of reasons. The difficulty arises not only from the lack of data, but also from the involvement of a large share of the population in informal economic activities involving few or no monetary transactions (Figure 1-7). A major difficulty also stems from the arbitrary nature of the definition of the poverty line itself.

The question is: What is the standard poverty line below which absolute poverty exists? This is a problematic question in the context of the third world, where the level of people's tolerance of poverty and survivability under difficult circumstances is consid-

Figure 1-7 Poverty in the Third World. *This barefoot peasant girl in Nepal is carrying fodder for animals. (Courtesy of Nanda Shrestha)*

Table 1-4

Percentage Share of Household Income by Percentile Groups of Households for Selected Nations

Nation	Quintile (20%)				
	Lowest	Second	Third	Fourth	Highest
Underdeveloped					
India (1975–1976)	7.0	9.2	13.9	20.5	49.4
South Korea (1976)	5.7	11.2	15.4	22.4	45.3
Mexico (1977)	2.9	7.0	12.0	20.4	57.7
Taiwan (1971)	8.7	13.2	16.6	22.3	39.2
Brazil (1972)	2.0	5.0	9.4	17.0	66.6
Average	5.3	9.1	13.5	20.5	51.7
Developed					
Great Britain (1979)	7.0	11.5	17.0	24.8	39.7
Japan (1979)	8.7	13.2	17.5	23.1	37.5
West Germany (1978)	7.9	12.5	17.0	23.1	39.5
United States (1980)	5.3	11.9	17.9	25.0	39.9
Average	7.2	12.3	17.4	24.0	39.0

Source: The World Bank, *World Development Report, 1985.*

[5]Meier, op. cit., p. 21.

Table 1-5

Estimates of Population below $75 Annual Income
in Developing Nations, 1969 (millions of people)

Continent	Absolute Number	Percent of Total Population
Asia	499	57
Africa	37	44
Latin America	43	17

Source: G. M. Meier, *Leading Issues in Economic Development*, 3rd ed. (New York: Oxford University Press, 1976), pp. 22–23.

ered to be remarkably high. Attempts have been made, however, to establish an absolute poverty line. Annual family income levels below $75 will be used here as the absolute-poverty-level indicator for developing countries.

In Table 1-5 we present absolute poverty estimates at the aggregate continental level for Asia, Africa, and Latin America. Note that these figures are estimates and that a wide margin of error is possible. Among the three continental areas, absolute poverty is highest in Asia and lowest in Latin America. In terms of absolute numbers, Latin America has almost 6 million more absolute poor than Africa, but both areas together do not approach the 500 million absolute poor estimated in Asia. As shown in Table 1-6, over 350 million poor are located in India, where in 1969, two-thirds of the population existed below the absolute poverty level. In Brazil and Mexico, about one-fifth of the population experiences absolute poverty. These estimates lead us to an important question: What causes poverty?

Causes of Poverty

A brief survey of the literature on poverty provides a wide range of explanations of poverty, depending on one's perception, understanding, and experience. The estimates presented in Table 1-4 provide some indi-

Table 1-6

Estimates of Population below $75 Annual Income
for Selected Nations, 1969 (millions of people)

Nation	Absolute Number	Percent of Total Population
India	359	67
Thailand	15	44
South Korea	2	17
Taiwan	2	14
Brazil	18	20
Mexico	9	18
Uganda	4	50
Senegal	1	35

Source: G. M. Meier, *Leading Issues in Economic Development*, 3rd ed. (New York: Oxford University Press, 1976), pp. 22–23.

cation that the scale of absolute poverty in underdeveloped countries is directly related to per capita income.

Latin America, for example, has the highest average per capita income and the lowest incidence of absolute poverty of the areas studied here. On the other hand, Asia has the lowest average per capita income and the highest incidence of absolute poverty. The average per capita income of Asia is almost four times lower than that of Latin America. Its poverty level is almost three times higher. A note of precaution when using absolute poverty figures to measure poverty is nevertheless in order. Explanations of poverty based solely on low per capita incomes appear too simplistic, because more overriding and comprehensive issues, such as overpopulation and lack of access to resources, may affect both income levels and poverty.

There are, in fact, many causes of poverty. These causes are invariably interrelated and complicated. Often, they are colored with ideological overtones. For example, the Malthusians, following the argument propounded by Malthus (see the accompanying box), believe that overpopulation is the principal cause of poverty. Although the Malthusian position has a certain validity in the case of a few Asian countries, it lacks a sound logical basis. It cannot explain why so much affluence exists in the midst of abject poverty. Nor can it explain why so much poverty exists in countries with a tremendous amount of resources but with a relatively small population.

On the other hand, Marxists, following the theoretical discourse of Karl Marx and Frederick Engels, argue that poverty exists not because of overpopulation, but because of the unequal distribution of productive resources such as land and capital available in the society. Unequal distribution excludes certain population groups from having secure access to productive resources. As a result, they cannot be productive and are forced into poverty. The situation gets worse in a socioeconomic environment in which employment opportunities are extremely irregular and limited, which is the case in most underdeveloped countries. Marxists do not deny that excess population growth adds to the problem of poverty, affecting those who are deprived and resourceless.

Rapid Population Growth. Excessive population growth aggravates the problem of poverty in societies where resources are limited, land productivity is stagnating, technological development is at a rudimentary stage, and resource distribution is quite uneven. High population growth rates often deplete available resources before they are fully developed and made more productive than what exists at the present state of technological development. Population growth also tends to aggravate the situation of unequal resource distribution because it leads to a greater degree of competition for limited resources among the population. But in this competition, not

An Englishman, Thomas Malthus, published his book *Essays on the Principle of Population* in 1798 at the beginning of the industrial revolution. The principal tenet of the theory holds that population growth accelerates in geometric fashion over time, whereas the food supply grows at a more modest arithmetic rate. This disparity creates an unstable situation that can lead to starvation and marginal living standards for the world's population. As food supplies increase, greater expansion rates of population outstrip those gains, creating an environment of helplessness.

Fortunately, the Malthusian interpretation of the linkage between food supply and population growth has not accurately explained conditions in many parts of the world in the past 200 years. Countries that successfully industrialized introduced new technology to increase the productivity of workers, developed modern medical practices, and lowered their birthrates. Food supply increases also outstripped population gains, such that higher standards of living also occurred.

Nevertheless, in many developing areas the Malthusian doctrine of population growth outstripping the food supply appears to be valid up to the present time. Not only do birthrates remain high in these areas, but the introduction of modern medical practices, including the use of antibiotics, has reduced infant mortality. Life-expectancy levels have also increased, such that explosive population growth rates threaten to outpace the ability of people to feed themselves. Africa's population, for example, could double in the next 25 years, while that in many developed countries, such as those in western Europe, may remain stable or even decline. Greater use of birth control measures may reduce overpopulation levels in many developing areas in the future. India, for example, has developed an attractive economic incentive program to encourage family planning. Unfortunately, the traditional rural outlook that favors large families to "do the work" and support aging parents prevails in many areas of India and, as elsewhere, is a difficult barrier to overcome.

---

everybody has the same chance (probability) of winning. It is almost certain that the wealthy and well-to-do class will win the competition because they can outbid the poor and deprived.

Population growth rates of most underdeveloped countries are very high compared to those of developed countries (see Table 1-7). Population growth rates in underdeveloped countries appear to have declined to some extent from the 1960–1970 period to the 1970–1978 period. This decline is attributed partly to the adoption of family planning policies which have emphasized the need for population control. Despite this declining trend, most of these countries have a

long way to go before they achieve the current growth rate of developed countries.

The *demographic transition* process helps account for the variable population growth rate in various countries. As countries modernize, they typically experience a change from slow to rapid growth and then back again to slow population growth as a result of changing birth- and death-rate relationships (see the accompanying box).

Agricultural Development and Food Production. Another significant cause of persistent poverty in most underdeveloped countries is the low level of land productivity (i.e., crop yield per unit of land). Land productivity in these countries has remained stag-

Table 1-7

Average Annual Growth Rates of Population (Percent)

Nation	1965–1973	1973–1983	Change
Underdeveloped			
India	2.3	2.3	0
Indonesia	2.1	2.3	+0.2
Egypt	2.3	2.5	+0.2
Nigeria	2.5	2.7	+0.2
South Korea	2.2	1.6	−0.6
Mexico	3.3	2.9	−0.4
Brazil	2.5	2.3	−0.2
Developed			
Great Britain	0.4	0.0	−0.4
Japan	1.2	0.9	−0.3
West Germany	0.7	−0.1	−0.8
United States	1.1	1.0	−0.1

Source: The World Bank, *World Development Report, 1985.*

Figure 1-8 Demographic Transition. *As nations modernize, they typically experience a sequence of changes in birth and death rates. This four-stage model shows how population growth relates to changes in these rates. The highest growth rates occur in stages II and III, while stages I and IV identify with slow growth.*

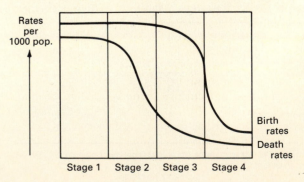

In most underdeveloped areas today, and in the remainder of the world prior to 1800, stage 1 of the demographic transition process reflects the presence of both high birthrates and high death rates (30 to 40 or more per thousand population per year) (Figure 1-8). Slow population growth levels occur in areas experiencing these conditions. As countries modernize they typically enter a second stage in demographic structure, retaining high birthrates while death rates decline significantly. Life-expectancy rates are extended by the introduction of improved health and sanitary practices, especially preventive medicine. This leads to a decline in mortality rates. Corresponding fertility (birth) rates remain high and a rapid rate of increase in total population occurs. Western Europe and the United States entered this stage after 1800. Many African, South American, Middle Eastern, and southeast Asian countries experience this situation today.

The third stage of change unfolds as dramatic declines in birthrates occur, approaching the earlier declines in death rates. This situation substantially decreases the rate of population growth. Death rates typically approximate 10 to 20 per thousand population and birthrates 20 to 30 per thousand in this stage. Modern urban industrial nations experienced this transition in the late nineteenth and early twentieth centuries. In Japan this transition took place after World War II. Parts of Latin America and Africa, including Chile, Bolivia, and Angola, are experiencing this transformation today.

Traditional thinking about the origins of the declining birthrate phenomenon associated with the demographic transition linked the change with the industrial revolution with the declining death rate. Although it is true that large families became more of a burden for nonfarm families in the industrial age, current thinking suggests that the Age of Enlightenment, which began in France in the eighteenth century, was more specifically the catalyst for this change. The emancipation of women began at that time. In the place of predestination and fatalism, more emphasis was placed on individual freedom in the Age of Enlightenment. "At the outset, then, fertility decline in Europe and North America probably was more a function of women seeking to gain greater control over their lives than it was a direct response to the industrial revolution and declining mortality."*

In recent history fertility declines have occurred as economic development proceeded, but other factors have also intervened. Whereas in South Korea, Taiwan, and Singapore economic advances have been associated with this change, in many parts of Asia government family planning programs have also been influential. Government programs, for example, have been particularly influential in reducing the birthrate in China, but less so in India. Couples in China are urged to defer marriage, and once married, to have only one child. Declining birthrates have been conspicuously absent in Africa and the Middle East. Part of this resistance to change can be associated with a renewed commitment to Islamic fundamentalism, characterized by intense individualism and a patriarchal family structure. In Egypt, for example, fertility rates actually increased in the 1970s, as shown in Table 1-7.

A fourth and final stage in the demographic transformation occurs as low growth or no growth unfolds, when birthrates and death rates both drop to 10 to 15 per thousand per year. The United States, Canada, Europe, the Soviet Union, Australia, and New Zealand all fit this circumstance today. A growing number of countries, including East and West Germany and Great Britain, had stationary populations in the mid-1980s. Some observers anticipate that the birthrates will not even maintain a steady population size in many of these areas in the future and that they may decline in population if net in-migration does not occur.

*James L. Newman, "Fertility in Transition: A World of Fewer and Fewer Children," *Focus* (Spring 1986), 5–6.

nant, and in a number of cases it has even declined. Land productivity in underdeveloped countries is several times lower than that in developed countries such as Japan.

The inability of these countries to increase land productivity in the face of a rapidly growing population has severely hampered attempts to reduce poverty. This does not mean there have been no attempts made to increase agricultural production. In the mid-1960s, the Green Revolution was introduced. The term *Green Revolution* refers to major biogenetic advances in agricultural technology. The strengths and weaknesses of this approach to improving agricultural output are discussed in Chapter 4.

Resource Distribution. No doubt, high rates of population growth and low agricultural productivity, together with the lack of resources and technological advancement, have contributed significantly to the perpetuation of poverty in underdeveloped countries. However, no explanation of poverty is more obvious and significant than the existing inequality in the social distribution of productive resources, including both land and capital.

In most underdeveloped countries, land is the basic source of livelihood, income, and social status. In spite of the fact that industrial and service activity has expanded in these countries, the agricultural sector (land) remains the most important source of employment and livelihood. Control of land generally means control of capital, other productive resources, and ready access to market distribution networks and outlets (Figure 1-9).

Given such absolute importance attributed to land in people's lives, the social distribution of land resources becomes a crucial element in the discussion of poverty in underdeveloped countries. It has been reported that inequality in resource distribution leads to (1) the exclusion of a large number of people from useful activities, thus making them much less productive than they could be with secure access to resources; (2) the underutilization and low productivity of resources available in a given society; and (3) the lack of incentives among those who possess minimal or no resources of their own and depend on others for their subsistence. None of these situations helps to increase productivity and thus reduce poverty.

The unequal distribution of productive resources is analogous to a zero-sum game or gambling situation. Gains to one can be obtained only through losses by others. Whereas those with control over resources enjoy a luxurious life, those who lack resources live in perpetual poverty—at the margin of bare subsistence and starvation. The extent of inequality in resource distribution in underdeveloped countries can be observed from Table 1-4. Since the data on resource distribution are lacking, Table 1-4 uses its proxy variable—income distribution.

These income distribution figures clearly show that the lowest 20 percent of the population shares only a tiny fraction of national income. It is as low as 2 percent in the case of Brazil. In no case does it exceed 8.7 percent on the income distribution scale. The highest 20 percent of the population, on the other hand, enjoys the largest portion of national income—from a minimum of 38.8 percent in Great Britain (a highly developed country) to a maximum of 66.6 per-

cent for Brazil. As mentioned earlier, the disparity is quite large in the highest-income category between underdeveloped and developed countries.

SUMMARY AND CONCLUSION

In this chapter we have defined economic geography and provided a discussion of the framework on which the following chapters are hung. Generally speaking, we will discuss activities in the primitive world first, followed by a discussion of resource extraction, manufacturing, and finally service activity. In all cases we emphasize the tendency for a global world system to emerge.

An appreciation of economic geography provides the student with a conceptual understanding of the changing fortunes of the world economy. The economic geographer can explain patterns of economic activities using a locational analysis approach that incorporates relevant theories and generalizations. In turn, these theories account for the variable level of economic development in various world regions.

The sharpest contrasts in economic development levels in the world today occur between the prevailing poverty of the developing or third world and more widespread affluence present in highly developed areas. It is not clear that these third-world areas can expect to experience the same sequential economic development process as highly developed areas have in the past. Unequal resource distribution and inadequate agricultural development initiatives contribute to the problem, as does rapid population growth. These situations create much more visible spatial and class disparities in developing countries than those found in developed areas, but the problem is much more complex and is inextricably related to the developmental history of the present developed-world countries, which we discuss in Chapter 2.

Figure 1-9 Wealth and the Control of Land in the Third World. *In most third world countries status and wealth is associated with land ownership. Finding suitable land in the central hill region of Nepal, shown here, presents a difficulty due to the rugged topography. Terraces carved in the hillside therefore become very valuable property. (Courtesy of Nanda Shrestha)*

SUGGESTIONS FOR FURTHER READING

BROWN, LESTER, *The Global Economic Prospect: New Sources of Economic Stress.* Washington, D.C.: Worldwatch Institute, 1978.

BROWN, LESTER, *In the Human Interest.* New York: W. W. Norton, 1973.

BROWN, LESTER, *State of the World, 1986.* Washington, D.C.: Worldwatch Institute, 1986.

DE SOUZA, ANTHONY, and J. BRADY FOUST, *World Space-Economy.* Columbus, Ohio: Charles E. Merrill, 1979.

FLINN, MICHAEL W., *The European Demographic System.* Baltimore: Johns Hopkins University Press, 1981.

GOULD, PETER, *The Geographer at Work.* Boston: Routledge and Keegan Paul, Inc., 1985.

JUMPER, SIDNEY, ET AL., *Economic Growth and Disparities: A World Review.* Englewood Cliffs, N.J.: Prentice-Hall, 1980.

LLOYD, PETER, and PETER DICKEN, *Location in Space: A Theoretical Approach to Economic Geography,* 2nd ed. New York: Harper & Row, 1972.

MCCARTY, H. H., and JAMES B. LINDBERG, *A Preface to Economic Geography.* Englewood Cliffs, N.J.: Prentice-Hall, 1966.

MEIER, G. M., *Leading Issues in Economic Development,* 3rd and 4th eds. New York: Oxford University Press, 1976, 1984.

NESS, GAYL D., and HIROFUMI ANDO, *The Land Is Shrinking: Population Planning in Asia.* Baltimore, Md.: Johns Hopkins University Press, 1984.

NEWMAN, JAMES O., *Population Patterns, Dynamics, and Prospects.* Englewood Cliffs, N.J.: Prentice-Hall, 1984.

REED, JAMES, *From Private Vice to Public Virtue: The Birth Control Movement and American Society since 1830.* New York: Basic Books, 1978.

SHRESTHA, NANDA, "A Preliminary Report on Population Pressure and Land Resources in Nepal," *Journal of Developing Areas,* 16(1982), 197–212.

WHEELER, JAMES O., and PETER O. MULLER, *Economic Geography,* 2nd ed. New York: Wiley, 1986.

WORLD BANK, *World Development Report 1980.* New York: Oxford University Press, 1980.

WORLD BANK, *World Development Report 1985.* New York: Oxford University Press, 1985.

Historical Evolution of the World System

2

In this chapter we present a broad overview of the evolution of the world system. The world system perspective builds on the network of linkages that tie together the various countries of the world. The origin and nature of these political and economic connections is reviewed in succeeding sections together with the mechanisms that have sustained them over time.

The world system perspective is quite broad and general, yet insightful and relevant to an economic geographer studying changing locational relationships at the international scale. It is based on an historical and political economic analysis of relationships, particularly those between developed and underdeveloped countries. Explaining the process of development and underdevelopment in the world space economy provides the major focus.

The analysis can be extended to show (1) how the historical relationships between developed and underdeveloped countries are sustained, and (2) why most underdeveloped countries have failed to develop their economies in a manner that alleviates such basic problems as poverty, unemployment, underemployment, and spatial and social inequalities.

One may find several interpretations and criticisms of the world system perspective. Some may even question its lack of clear articulation and a well-defined framework within which relationships between specific countries can be examined. However, no one can deny that today's space economy, whether that of a small country such as Liberia or a large one such as the United States, is a part of the international space economy.

When discussing the issues of economic development, no space economy can be analyzed in isolation. Authors such as Immanuel Wallerstein and Samir Amin argue that the spatial integration within the world system innately generates a dependency relationship.[1] They argue that the economic underdevelopment of most of today's third-world countries derives from dependency relationships with developed countries, dating back to the precolonial mercantilist period, when extensive trade relations first emerged.

Developed countries also compete with and depend on one another. The particularly acute nature of this situation becomes most visible in the areas of energy resources, basic raw materials, and mineral resources. The economic vulnerability of developed countries, for example, became evident during the 1973 Arab oil embargo, which greatly affected the supply and price of energy resources.

Within the world system framework the concept of *center-periphery* relationships has been utilized to analyze the economic positions of countries. Some authors have extended this division to include what they call "semiperiphery." The classification of *center, semiperiphery*, and *periphery* countries is based on different

---

Note: Chapter authored by Nanda Shrestha, Assistant Professor of Geography, University of Wisconsin at Whitewater.

[1]Samir Amin, *Unequal Development* (New York: Monthly Review Press, 1976).

criteria, such as access to, or control of, technology.

Center nations exhibit strong economies and robust economic activity. The United States, Japan, the western European countries, and Australia are prominent examples of this group. On the other hand, such countries as Pakistan, Thailand, Indonesia, Kenya, Nigeria, Mexico, Chile, and Jamaica are considered peripheral. Peripheral areas generally have "weak" states, low wages, and simple technology economies. Semiperipheral areas mix these features. In comparison to the center, they appear peripheral, while in comparison to the periphery, they appear centerlike, examples being Israel, Taiwan, South Africa, and even India. This center-periphery relationship at the global scale can also be extended to the national scale, and the national space economy can similarly be subdivided into center, periphery, and semiperiphery regions.

MEDIEVAL FEUDAL ECONOMIES

In the following subsections we provide a brief historical account of how the present world system evolved and continues to sustain a network of intricate relationships among nations. No international space economy in the modern sense existed in medieval times. Very few long-distance spatial economic relationships and interactions existed between and among nations. Prevailing relationships were regional and limited in scope. An inward-looking economy characterized European society in medieval times and it functioned under a feudalistic structure (Figure 2-1). Lekachman writes: "Change was slow in towns and slower elsewhere partly because of the conception of life which emanated from the period's dominant social institution, the Church."[2] The Church formed the apex of the socioeconomic structure as it frowned on materialistic values and the development of commercial and industrial activities.

Sustained spatial economic relationships were generally nonexistent in other parts of the world in medieval times. Long-distance interactions generally occurred during tribal feuds and territorial expansionary wars. Most societies were self-content with a production structure characterized by subsistence and self-sufficiency in almost every respect. Such situations are discussed in subsequent chapters because they still exist in some parts of the world.

In terms of its interactions and production structure, the space economy was essentially domestic, local, and regional at most. Very little spatial manipulation, planning, or selection in terms of the location of economic activities occurred. Physical environmental factors primarily affected such choices. Advanced technological elements played an insignificant role because the development of technology remained at a rudimentary state.

[2]Robert Lekachman, *A History of Economic Ideas* (New York: Harper & Row, 1959).

Figure 2-1 Agriculture in Medieval Europe. *An inward-looking feudal economy characterized European society in medieval times. A self-sufficient agriculture system as shown in this photograph of a feudal manor supported most of the population. (The New York Public Library Picture Collection)*

THE RISE OF MERCANTILISM

By the middle of the fifteenth century the medieval socioeconomic order began crumbling at the threshold of Europe's new era—the Age of Discovery and Exploration. The Age of Exploration proceeded with full speed in the sixteenth century as Europeans set out on voyages in search of wealth, to gain personal glory, and to spread the Christian gospel. This process ultimately led to the rise of mercantilism as Europeans began to dominate the worldwide import–export trade in Africa, Asia, and the new world of North and South America.

As a result of the worldwide mercantilist penetration by Europeans following the Age of Exploration in the fifteenth and sixteenth centuries, long-distance economic interactions between various nations grew and expanded spatially. Certainly, Arab merchants and traders conducted long-distance trade prior to the European penetration. They tied east Africa with India and the rest of Asia. But the scale of operation was smaller compared with the more comprehensive and systematic organization of the European mercantilist venture. European mercantilism demonstrated true international dimensions in almost every sense and gave birth to what we now call international trade.

This trade network, for the first time, systematically tied all major continents together into one world system dominated by the countries of Europe: Spain, Portugal, England, France, and the Netherlands. Walter Rodney stated that in the seventeenth century the European traders carried most of the east African ivory to markets in India, where "they engaged in buying [Indian] cotton cloth to exchange for slaves in Africa to mine gold in Central and South America. Part of the gold in the Americas would then be used to purchase spices and silks from the Far East."[3] These intercontinental trade links created a complete circle of trading relationships. Western Europe occupied a pivotal position at the apex of these relations and benefited tremendously in both political and economic terms. A diagram showing these complex trading ties appears in Figure 2-2. Note that western Europeans served as the merchants and brokers in this trade process, the profit from which was expropriated to Europe, thus laying the foundation for the industrial revolution.

ECONOMIC BENEFITS FROM MERCANTILISM

Lekachman writes that "mercantilists believed that the gain of one country was inevitably the loss of another." If this is truly the case, some would argue that the tremendous amount of economic benefit Europe derived from mercantilist trade came at the expense of many of today's underdeveloped countries in Africa, Asia, and Latin America. No matter how one interprets this phase of world economic history, the fact remains that mercantilist international trade contributed tremendously to thrust western Europe into the process of dynamic and accelerated economic development by the end of 1750 (Figure 2-3). This process helped western Europe emerge as the most dominant world power, a position it retained until World War II.

Mercantilism's contribution to Europe's dynamic development came in other forms as well. First, it defied the reigning socioeconomic order of the Church-dominated feudal production structure, which declined in importance with the transition to mercantilism. The commercial production of goods surfaced as an important economic activity. This activity occurred in existing towns and small cities and began to yield surpluses. Buying and selling land, labor, and

[3]Walter Rodney, *How Europe Underdeveloped Africa* (Washington, D.C.: Howard University Press, 1974).

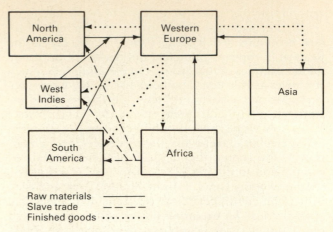

Raw materials ——————
Slave trade – – – – –
Finished goods ·········

Figure 2-2 Generalized Mercantilistic Trade Relationships. *Triangular trade relationships dominated in the mercantilist era involving the movement of slaves from Africa to North and South America and the West Indies; raw materials flowed from these areas to Western Europe for manufacture and then returned to the colonies as finished goods, completing the triangular flow.*

capital began, as each of these factors entered the market as a commodity. Economic activities expanded rapidly. Finally, the growth of cities proceeded at an accelerated pace in number as well as size, as they became centers of production and trade and the economic hinge of mercantilism.

These domestic economic activities became an important basis for the process of capital accumulation, which is essential for dynamic economic growth. The whole process was greatly aided and promoted by the profits from overseas mercantilist trade, including transatlantic slave trade, all of which in turn provided the groundwork for the industrial revolution in England in the late eighteenth century. Before we discuss the role of this powerful industrial revolution in the world system, we examine briefly a critical feature of the mercantilist period.

SLAVE TRADE

One important feature of the mercantilist era was the transatlantic African slave trade carried out by European merchants. This trade involved a massive transfer to the new world of able-bodied young Africans, the large majority of whom were males between approximately 15 and 35 years of age. In some cases, slave traders began to transfer younger persons—under 15 years of age, but rarely older persons, whose productive potential was limited.

Although the European involvement in the transatlantic slave trade began much earlier, the most intensive period occurred between 1700 and the very early 1800s, when such trade was abolished by the British Parliament. Even after its abolition in Britain, the depletion of young Africans continued for several more decades, as slave merchants from the United

Figure 2-3 Old Dutch Tobacco and Spice Warehouses. *It is not uncommon to still find today well-maintained buildings that once served as warehouses for the storage of imported goods in the Mercantilist era. They are usually located in small Dutch towns with their distinctive architecture, cobble stone-paved streets, and canals. (T.A.H.)*

States and other parts of the new world carried on this very profitable business. Altogether, almost 30 million people were removed from Africa during the slave trade era. The slave trade movement not only created a systematic linkage between Africa and the new world of South and North America and the West Indies, but it made possible the European-based exploitation of resources in the new world. Walter Rodney has written:

> When Europeans reached the Americas, they recognized its enormous potential in gold and silver and tropical produce. But that potential could not be made a reality without adequate labor supplies. The indigenous Indian population could not withstand new European diseases such as small pox, nor could they bear the organized toil of slave plantations and slave mines, having barely emerged from the hunting stage. . . . At the same time, Europe itself had a very small population and could not afford to release the labor required to tap the wealth of the Americas. Therefore, they turned to the nearest continent, Africa, which incidentally had a population accustomed to settled agriculture and disciplined labor in many spheres. Those were the objective conditions lying behind the start of the European slave trade, and those are the reasons why the capitalist class in Europe used their control of international trade to ensure that Africa specialized in exporting captives.[4]

The repercussions of slave trade at home and abroad were immense. The impact on the African economies, for example, was disastrous. The removal of able-bodied young Africans created a severe shortage of labor in Africa. The scarcity of labor greatly hampered the agricultural process and disrupted all other domestic production activities, inhibiting Africa's development.

[4]Ibid.

On the other hand, the economic impact of the slave trade on the European economy was extremely positive. Eric Williams in his book *Capitalism and Slavery* describes the many benefits derived from trading and exploiting slaves. For example, two British businessmen, David and Alexander Barclay, used the profit from slave trade to establish Barclay's Bank. Lloyd's of London—one of the world's largest banking and insurance companies—is also an outgrowth of the investment of profits derived from slave trade. Walter Rodney mentions that James Watt expressed "eternal gratitude to the West Indian slave owners who directly financed his famous steam engine"—one of the vehicles powering the industrial revolution.

THE INDUSTRIAL REVOLUTION

The industrial revolution, which began in England in the late eighteenth century, succeeded the mercantilist era. The capitalist mode of production, laid out and nourished during the mercantilist era, gained maturity with the coming of the industrial revolution, which started with the mechanical harnessing of steam power. These new advances changed production technology dramatically, and in the process the role of capital shifted to a central position. The textile industry—probably the most important industrial sector in England at that time—was greatly affected by this production technology change (Figure 2-4).

Transportation innovations, paced by steamships and railroads, also characterized the industrial revolution. These new modes helped conquer the friction of distance and reduce the cost of moving goods and people. The range of product distribution and market areas increased accordingly. Raw materials and natural resources from distant places now became easily accessible to English and other European factories on a regular, sustained basis.

Figure 2-4 Early Textile Mill in England. *This nineteenth-century needle mill in Ridditch, England shows how water power was harnessed to drive belts and pulleys in the mill. Women dominated the mill labor force. (The New York Public Library Picture Collection)*

Expansion of the industrial production capacity occurred first in England and later in other European countries and the United States. To utilize this increased production capacity effectively, a greater supply of raw materials and expanded markets for finished products became necessary, neither of which Britain or other European countries could easily provide at home. Further, the industrial revolution led to the production of capital at such a rapid pace that a *capital glut* occurred.

John Stuart Mill, a leading nineteenth-century English economist, discussed the problem of substantial capital glut resulting from the creation of far more industrial capital (e.g., machines and tools) than could be utilized domestically. This situation decreased the productivity of capital, as a substantial amount remained idle. Consequently, the margin of profit began to decline.[5]

After 1870, the British economy encountered growing difficulties with regard to international trade. Sustained economic growth became a bigger problem due to excess capital on the domestic front. On the international front, greater competition had emerged with Germany, France, and the United States, as these countries became industrialized.

[5]John Stuart Mill, *Principles of Political Economy* (London: Longman, 1909).

THE EMERGENCE OF COLONIALISM

The newly industrializing powers—France, Germany, and the United States—began to compete for markets and natural resources, forcing Britain to protect its foreign market and resource bases. This situation led Britain to pursue colonization aggressively to ensure continued economic growth so successfully propelled earlier by mercantilist trade and later by the industrial revolution. Using its powerful naval forces, Britain rapidly annexed foreign territories (colonies) all over the world. Colonialism allowed Britain to monopolize the markets and raw materials of its colonies, thereby neutralizing competition. Cecil Rhodes, a leading advocate of the colonial policy, expressed his view this way:

> I was in the East End of London and attended a meeting of the unemployed. I listened to the wild speeches, which were just a cry for 'bread, bread' and on my way home I pondered over the scene and I became more than ever convinced of the importance of imperialism. . . . My cherished idea is a solution for the social problem, i.e., in order to save the 40,000,000 inhabitants of the United Kingdom from a bloody civil war, we colonial statesmen must acquire new lands to settle the surplus population, to provide new markets for goods produced in the factories and mines. The

Table 2-1

Colonial Possessions of Foreign Territories
(millions of square miles and millions of people)

| Year | Great Britain | | France | | Germany | |
	Area	Population	Area	Population	Area	Population
1815–1830	—	126.4	0.02	0.5	—	—
1860	2.5	145.1	0.2	3.4	—	—
1880	7.7	267.9	0.7	7.5	—	—
1899	9.3	309.0	3.7	56.4	1.0	14.7

Source: V. I. Lenin, *Imperialism, the Highest Stage of Capitalism* (Peking: Foreign Language Press, 1969).

Empire, as I have always said, is a bread and butter question. If you want to avoid civil war, you must become imperialist.[6]

Britain's colonial drive did not, however, go uninterrupted. Other European nations, including Germany and France, began to expand their colonial involvement. As the colonial powers of western Europe intensified their "scramble for foreign territories or colonies," particularly in Africa, they became embroiled in a tug-of-war. When they realized that the intrapower conflict would not serve their long-range interests, they convened a conference which later came to be known as the *Berlin Conference*. This conference, led by Otto von Bismarck, the imperial chancellor of Germany, took place in November 1884 and included the representatives of 14 powerful nations, including the United States. They gathered primarily to settle the issue of the territorial partitioning of Africa into several colonies controlled by the British and French, among others. The extent of the colonial control of foreign territories by western European colonial powers is shown in Table 2-1, and a breakdown of their holdings is presented in Table 2-2.

ECONOMIC BENEFITS FROM COLONIALISM

As Mill pointed out, the colonization of resource-rich territories in Africa, Asia, Latin America, and elsewhere guaranteed a productive outlet for the use of excess capital generated in industrial countries.[7] In this context, Lenin stated: "In these backward countries [colonies] profits are usually high, for capital is scarce, the price of land is relatively low, wages are low, raw materials are cheap."[8]

In addition, Mill discussed many other benefits that the colonial powers of western Europe derived from formal colonization. First, the employment of capital in the colonies helped increase Europe's profits by maintaining the productivity level of capital that

remained behind in Europe. Second, colonization provided guaranteed access to raw materials and markets for Europe's industries. This was particularly important to effectively sustain the process of industrial development. This is an important point in two respects. First, Europe lacked many of the resources required to keep industries and factories operating at full capacity and to sustain the industrial drive; therefore, it had to rely on foreign territories for raw materials. Second, the absence of a large population base within the continent did not provide Europe with a market large enough to consume the growing industrial output. Europe therefore needed large foreign markets to absorb its finished products.

Third, aristocratic consumers in Europe had come to savor exotic, high-quality foreign products, for example, silk and porcelain from China, tea and spices from India, and ivory from Africa, not to mention gold and diamonds. At that time imported sugar from the West Indies and Latin America was also regarded as a special commodity in Europe. Colonization allowed Europeans ready access to all these products at minimal prices.

Fourth, colonization provided a crucial safety valve releasing an increasing number of domestic unemployed and underemployed people for work in the colonies. Finally, colonies became a cheap source of food products for the colonial powers. By shipping cheap agricultural products back home to feed the industrial labor force, inflation was controlled and wages could be kept lower.

Table 2-2

Share of Territory Ruled by European Colonial Powers, by Continent, 1876–1900 (percent)

Colonies in:	1876	1900	Change
Africa	10.8	90.4	+79.6
Polynesia	56.8	98.9	+42.1
Asia	51.5	56.6	+5.1
Australia	100.0	100.0	—
America	27.5	27.2	−0.3

Source: V. I. Lenin, *Imperialism, the Highest Stage of Capitalism* (Peking: Foreign Language Press, 1969).

[6]Quoted in V. I. Lenin, *Imperialism, the Highest Stage of Capitalism* (Peking: Foreign Language Press, 1969).

[7]Mill, op. cit.

[8]Lenin, op. cit.

The transnational relationships established in colonial times still affect the modern international political and economic environment. Colonial rule invariably destroyed a growing local industrial base and technological processes in the colonies by flooding markets with European products. In this way, colonial powers avoided competition for their finished products from small-scale domestic producers in the colonies. In addition, colonial governments rarely trained the indigenous population to handle managerial or highly technical tasks, which served to psychologically dampen their confidence and innovative abilities and to institutionalize the dependency situation found in the modern world system (Figure 2-5).

MECHANISMS OF THE MODERN WORLD SYSTEM

We can identify several mechanisms of the modern world system that maintain the dependency relationships established in a previous era. These include international trade, multinational corporations, international labor migration, foreign aid (both economic and military), and technology transfer. However, in this section we discuss only one major mechanism: the operation of multinational corporations in underdeveloped countries, because of their overarching impact. Other themes are developed in subsequent chapters and explored in more detail in Chapter 20, where we evaluate the geography of international business.

Multinational Corporations

Multinational corporations (MNCs), also known as *transnational corporations,* are interrelated with international trade, foreign aid, and technology transfer in many ways. In a sense, MNCs are a modern version of the mercantilist trade discussed earlier, but the mode of operation is very different.

MNCs are primarily private firms and companies that have legitimately, and with the consent of host governments, established branch operations in foreign countries. These companies command vast

Figure 2-5 British Officer in India, circa 1870. *Colonial governments rarely trained the indigenous population to handle managerial or highly technical tasks which psychologically dampened their confidence and innovative abilities not to mention the institutionalization of dependency relationships. (The New York Public Library Picture Collection).*

amounts of resources in the form of capital, technology, managerial expertise, and information. Many of these firms are on the forefront of product innovation, research, and development.

Direct foreign investment, and the trade of goods and services, provide examples of how multinational firms operate across national boundaries. From the locational perspective of economic geography, multinationals have truly internationalized the space economy. They have become a textbook example of how global locational operations work in terms of investment decision making (Figure 2-6).

The fundamental concern in the locational decision of MNCs is to select a location or locations that minimize the total cost of production, including the cost of transportation, and thus maximize the total volume of profit. In this regard, multinationals provide a classical example of efficient and effective locational decision making concerning the establishment of their activities at the international scale.

MNCs often separate the various operations of the firm locationally in such a way that they can take advantage of inexpensive raw materials, a cheap labor force, and/or the large markets of the third world. While these firms typically maintain a headquarters location in their home country, they establish multiple processing, fabricating or manufacturing, and distribution operations in various parts of the world, depending on the product and its market. Thus each firm maintains a global operating network.

Limits present in the domestic market in terms of sales potential and production costs often lead MNCs to branch out from their home countries. They invest excess capital in underdeveloped countries where this resource is scarce and labor and raw materials are readily available. This overseas investment of capital allows them to increase profit margins over what would be possible domestically. From a locational perspective, a list of specific reasons that large companies headquartered in developed countries, such as Nestlé, Goodyear, General Motors, IBM, Mitsubishi, United Fruit Company, and many others, have expanded production and market operations widely in the underdeveloped countries follows:

1. *Raw materials and natural resources:* Accessibility to these basic production inputs often provides a lifeline for the effective production operation of MNCs. These resources become readily available to MNCs through their branches, affiliates, and subsidiaries in the third world. Furthermore, the establishment of production units in these countries helps MNCs reduce the cost of transportation of raw materials. In most cases, the transportation of unfinished bulky raw materials and natural resources is much more expensive than the transportation of higher-value finished products. Furthermore, once a final product is fabricated, it can be marketed in underdeveloped countries, further minimizing transportation time and cost.

2. *Cheap labor force:* Another basic production input is labor, which can, in turn, be subdivided into manual (blue-collar) labor, and managerial (or technical) labor, often called white-collar labor. MNCs are concerned primarily with the availability of cheap manual labor, which most underdeveloped countries have in abundance. The labor cost or the wage rate in these countries is typically several times cheaper than that in developed countries such as the United States, Japan, West Germany, France, Great Britain, and others, where the vast majority of MNCs are based. Often, there is less indirect cost attached to labor in underdeveloped countries, such as insurance or retirement benefits. Rarely do pension plans or health and life insurance benefits exist for workers. By avoiding such overhead costs,

Figure 2-6 Container Ship in Savannah, Georgia. *The container ship, carrying every conceivable product from bulk raw materials to frozen french fries, and electronics products, has routinized and intensified the flow of goods among nations. Goods can be transported by several different modes of transportation without repacking using container technology, significantly reducing the time and cost of transit. (T.A.H.)*

a substantial savings occurs, which means that the total labor cost of production operations is drastically reduced.

3. *Markets:* The size of market and the accessibility to market are crucial considerations in the locational decision. Although the standards of living in most underdeveloped countries are low, the countries do represent large markets for MNCs' products, owing to their large and rapidly growing populations. These markets also expand as western-oriented consumerism and life-styles become incorporated into their cultures. By locating production, distribution, and exchange operations in these countries, multinationals can readily capture these large and expanding markets.

4. *Labor unions:* There are very few labor union organizations in underdeveloped countries. As a result, MNCs do not have to be overly concerned about the restrictive working environments often associated with organized labor. This labor union–free environment again reduces production costs.

5. *Environmental laws:* Most underdeveloped countries do not have strict environmental protection and safety and preservation regulations. This situation frees MNCs from having to abide by costly pollution-abatement measures and other restrictions that often prevail in their home countries and raise production costs.

6. *Tax benefits:* To attract MNCs, many underdeveloped countries have policies that provide tax breaks for these companies, which assist firms by enhancing profit margins. Additionally, these companies are not typically required to reinvest their profits in host countries. They are allowed to expropriate profits back to their home countries or wherever they choose to reinvest. As a result, they have gained the status of what some call the *footloose* industries—free to move in or out, as they wish, without a long-term commitment to any country.

7. *Political power structure:* The political power structure is generally controlled by the elites in underdeveloped countries, who have close ties to MNCs. By devising and maintaining industrial, commercial, and trade policies that are favorable to MNCs, national elites have much to gain from their close relationships and collaboration with MNCs. Local political leaders can offer political support and in turn receive financial and material benefits. The focus here, however, is not on which party stands to benefit most. The point is that a corrupt political power structure in underdeveloped countries sometimes facilitates the operation of MNCs in these countries. On the other hand, such a political structure, if it is weak, unstable, or lacks powerful military support from within, can act to deter MNCs from operating in these countries. A weak and unstable political situation signals social upheavals and economic uncertainties—an environment that MNCs avoid.

Economic Benefits from Multinational Activities

For the reasons just discussed, the locational decisions of MNCs headquartered in the United States, Japan, West Germany, Great Britain, and many other developed countries frequently favor expanded production and distribution operations in underdeveloped countries. This arrangement enhances accessibility to necessary production inputs; reduces total production costs, including the cost of transportation; enlarges markets; and thus increases the margin of profit for multinationals.

Certainly, the transnational economic penetration by MNCs sustains the operation of the world economic system, but not without significant repercussions for underdeveloped countries and even for the home countries themselves. Some argue that MNCs play an important role in the development process of underdeveloped countries. According to this argument, MNCs make direct investments in these countries, transfer advanced technologies, create jobs for their unemployed, and help them harness natural resources. Some counter such arguments by asserting that the projected beneficial results of MNCs' economic activities appear shallow when examined critically.

Both of these viewpoints have some validity. Although some MNCs do little to help the countries in which they operate, others have positive effects on local economies. Thus, when weighing the benefits, it is useful to keep in mind various types of MNCs and their modes of operation. Some MNCs are involved in primary and extractive activities, such as agriculture, forestry, and mining. Others are engaged in manufacturing activities, such as textiles, apparel, automobiles or electronics (Figure 2-7). Still others are found in tertiary and quaternary sectors, including banking and trade. In some cases the involvement of foreign companies in manufacturing activities has been found to be quite beneficial. For example, MNCs' textile and apparel manufacturing operations in underdeveloped countries produce large numbers of nonskilled jobs, because these industries are labor intensive rather than capital intensive. In addition, the production technology used in these industries is suitable and appropriate locally, contributing to the overall development of domestic economies. Hong Kong's initial industrialization, for example, was strongly tied to the development of its textile and apparel industries. Their success, in turn, attracted more foreign investment in other industries, leading to a more diversified industrial structure. These stages of industrial change are discussed in Chapter 14.

Multinational investments, however, do not routinely produce long-term benefits for the general population or local economies of host countries. This situation occurs predominantly in underdeveloped countries that lack a broad economic base. Typically, investments in these areas are selective in terms of both the types of economic activities and locational choices made, as we discuss in later chapters. This investment occurs primarily in large urban areas already experiencing rapid development. Since corporations typically invest in those economic activities and harness those natural resources which they want and need, not necessarily those which fulfill the needs of the host country and its people, benefits may or may not occur.

In addition, the capital and technology that MNCs transfer to host countries are very specialized and industry specific with little general application elsewhere because their use is limited to a particular firm. MNC production operations tend to be capital intensive, so they typically create far fewer local employment opportunities than anticipated. For this reason the argument that third-world countries derive considerable economic benefit from MNCs is often exaggerated.

The increasing expansion of MNCs into the third world also affects the employment situation in home countries and tends to hamper their further economic growth. For example, when the General Electric Company decided to close down its iron production unit in Ontario, California, and relocate in Singapore, GE workers in Ontario were laid off permanently.

When multinationals flood domestic markets with products manufactured in Taiwan, Korea, Hong Kong, Singapore, and other less developed areas, the demand for products produced domestically by non-multinationals decreases. This situation weakens the demand for locally produced goods and inhibits the expansion of these industries, reducing employment opportunities in these fields in MNCs' home coun-

tries. Such a loss may, however, be overcome by expanding new activities and investing in high-technology fields or introducing improved technological processes in more traditional industries, as we will see in later chapters.

SUMMARY AND CONCLUSION

In this chapter we have reviewed the historical evolution of the world economy, giving emphasis to the precolonial feudal system in medieval times, the rise of mercantilism, slavery, the colonial era, and the modern global multinational environment. The global operation of MNCs represents a major force in sustaining the modern world system in which the space economy has been internationalized. In many ways the present dependency relationships maintained by multinational corporations represent extensions of earlier control mechanisms utilized by European colonial powers. This dependency perspective provides a useful framework from which to evaluate agricultural, industrial, and service activities in both developed and developing economies.

In the next chapter we begin our survey of world economies by examining primitive subsistence activities that still occur in many less developed areas, much as they did throughout the world in medieval times. By understanding these contemporary economies we will gain a better understanding as to why some observers believe that the present less developed world may not experience the same footsteps as the western world did before it in terms of stages of economic development. Rather, these areas appear to be forging their destiny in their own way.

Figure 2-7 McDonald's in Downtown Tokyo. *The golden arches symbol of McDonald's fast food restaurant chain has become a universal phenomena with 9,000 operations in over 40 countries, McDonald's dominates the commercial fast food industry worldwide, serving more than 19 million customers a day. An outlet in the Mitsukoshi department store, reputed to be located on the most valuable property of any McDonald's shown here. (T.A.H.)*

BOORSTIN, J. DANIEL, *The Discoverers.* New York: Random House, 1983, pp. 79–289.

EICHER, CARL, and JOHN M. STATZ, *Agricultural Development in the Third World.* Baltimore, Md.: The Johns Hopkins University Press, 1984.

GALENSON, DAVID W., *Traders, Planters, and Slaves.* New York: Cambridge University Press, 1986.

GHATAK, SUBRATA, and KEN INGERSENT, *Agriculture and Economic Development.* Baltimore, Md.: The Johns Hopkins University Press, 1984.

GORDON, MARVIN F., *Agriculture and Population: World Perspectives and Problems.* Washington, D.C.: U.S. Department of Commerce, Bureau of Census, 1975.

HARVEY, DAVID, *Consciousness and the Urban Experience.* Baltimore, Md.: The Johns Hopkins University Press, 1985.

HOBSBAWM, E. J., *The Age of Capital, 1848–1875.* New York: Scribner, 1975.

LANE, PETER, *The Industrial Revolution: The Birth of the Modern Age.* New York: Barnes & Noble, 1978.

LEKACHMAN, ROBERT, *A History of Economic Ideas.* New York: Harper & Row, 1959.

MEINIG, DONALD W., *The Shaping of America: A Geographical Perspective on 500 Years of History, Vol. 1: Atlantic America, 1492–1800.* New Haven: Yale University Press, 1986.

PARRY, J. H., *The Age of Reconnaisance.* New York: Mentor Books, 1963.

RODNEY, WALTER, *How Europe Underdeveloped Africa.* Washington, D.C.: Howard University Press, 1974.

SMITH, NEIL, *Uneven Development.* New York: Basil Blackwell, 1984.

WANMALI, SUDHIR, *Periodic Markets and Rural Development in India.* Delhi: B. R. Publishing, 1981.

WILLIAMS, ERIC, *Capitalism and Slavery: The Caribbean.* London: André Deutsch, 1964.

WILLIAMS, ERIC, *Negro in the Caribbean.* Brooklyn, N.Y.: Haskell House, 1970.

3 Primitive Economic Activity

At one time everyone on earth was engaged in what are now referred to as *primitive* economic activities. The term "primitive" is inherently subjective and reflects a necessary comparison between different modes of life. People can be considered primitive when their *material culture* is simple in form and function. Material culture refers to the artifacts or products associated with a society's technology. These artifacts include tools, containers, shelters, processed food, items of clothing, and all other material objects and devices used by society. Even if the material aspects of the culture are primitive, the nonmaterial aspects of life, including religion, language, and social organization, might be extremely complex and even sophisticated. Another way to regard primitive societies is to recognize them as being almost completely preoccupied with securing food and other necessities of survival, the fundamental economic problem of humankind.

Primitive economic activity is *subsistence activity.* Subsistence entails a quality of life at its most basic level and provides for only primary human needs: food, shelter, and propagation. In this chapter extensive subsistence economies are discussed. These activities include *primitive gathering, primitive hunting, primitive herding,* and *primitive agriculture.*

In Chapter 4 emphasis is placed on a more intensive form of subsistence activity. That form of agriculture, generally emphasizing rice cultivation, is a more advanced system of subsistence cultivation. It features the use of the plow, permanent fields and settlements, and domesticated animals, and it supports greater densities of population. The difference, then, between intensive and extensive activity refers to the degree to which land and labor are utilized to produce food. Extensive activity uses a lot of land in relation to the product yield, whereas intensive activity yields considerable output from a small quantity of land through greater labor and technology investments.

The scale of primitive subsistence activity in the late twentieth century is small but by no means insignificant. Very few people, a fraction of 1 percent of the world's population, rely solely on gathering, hunting, or herding for their livelihood. On the other hand, primitive agriculture (called generically, "shifting cultivation") is practiced widely throughout the tropical world (Figure 3-1).

In the world today, both primitive gathering and primitive agriculture are concentrated in the tropics. Primitive hunters and nomadic herdsmen can be found in the arid regions of Africa and Asia and in the northern areas of North America and Eurasia. The spread of modern technology has all but supplanted these activities in other parts of the world. Over 200 million people are engaged today in these primitive forms of economic activities, the overwhelming majority of them being subsistence farmers and herders.

Five hundred years ago, primitive economic activities were practiced virtually everywhere, the prin-

Note: Chapter revised by Sanford H. Bederman, Professor of Geography, Georgia State University, Atlanta, Ga.

Figure 3-1 Shifting Cultivation in Southwestern Cameroon. *Subsistence agriculture in Africa is traditionally undertaken by women. In this photograph a woman is preparing a small clearing for planting. (Courtesy of Sanford Bederman)*

cipal exceptions being Renaissance Europe, Ming China, and a tiny portion of the Middle East. Whereas only about 5 percent of the world's population today make their living through the most primitive of means, the share in A.D. 1500 was somewhere between one-half and two-thirds of humankind.

PRIMITIVE GATHERING

Primitive gathering, the lowest order of economic activity, experiences continual retreat as a way of life. Only a few thousand people currently practice this form of livelihood worldwide. A diversity of cultures engaged in this activity does exist, however, and there are various levels of specialization.

Primitive gathering persists primarily in isolated pockets in the low-latitudes, including the territories of some Indian tribes dispersed throughout the Amazon basin (Brazil, Peru, Ecuador, and Venezuela), together with a few stretches within tropical Africa, the northern fringe of Australia, the interior of New Guinea, and the interior portions of southeast Asia (Burma, Thailand, and China). Primitive gathering is the oldest of all economic activities and one in which human beings supported themselves almost exclusively by collecting a variety of products provided by nature. Even today, where primitive gathering is the predominant way of life, people subsist on the fruits, nuts, berries, roots, leaves, and fibers that they collect from trees and from shrubs and smaller plants. From all types of wildlife (plant and animal) on land, in the air, and in the waters, they extract enough to satisfy basic needs for nourishment, shelter, warmth, clothing, and tools. They plant few seeds and spend no time cultivating the soil. They exert no effort to breed, feed, or protect animals, nor do they try to improve and control their habitat.

Of all economic endeavors, gathering requires the least amount of capital investment and effort, but considerable space is required. It is an *extensive activity* requiring a large quantity of land to support each person. Yields per acre and yields per person are so low that surpluses are almost nonexistent. A very low *man-land ratio* occurs in such areas, typically no more than two persons per square mile.

Gathering economies often involve tribal societies in which individuals or single families possess a strong recognition of territoriality. For example, individual family units of the Semang, an extremely small group of primitive gatherers in Malaya, each control a traditional territory containing about 30 square kilometers. Furthermore, their claim over certain valuable trees and fruits is recognized by neighbors. This notion of territoriality plays an important stabilizing influence in these areas, where a delicate balance must be maintained to assure an adequate long-term food supply.

Primitive gatherers still live in the Stone Age, much like their ancestors of 100 centuries before. Indeed, some of them are not even that far advanced. The Chavante and Chamayura tribes in the wilds of the Matto Grosso in western Brazil still do not use stone points for their arrows. Written language is unknown. Their overall health is generally poor, and their life expectancy is short.

PRIMITIVE HUNTING

Primitive hunters share many characteristics with gatherers. Both groups know how to use fire, prepare food, manufacture tools and implements, and construct shelters. They are cognizant of local conditions and have the ability to exploit food resources. Hunters and gatherers generally do *not* have domesticated food plants, domesticated animals (except dogs), permanent settlements, or high population densities. Hunters differ from gatherers in that they employ more sophisticated methods to secure food and depend much more on animals.

Hunting is primarily a communal activity, often requiring planned, large-scale expeditions and a very well developed division of labor. Almost every hunting group recognizes this method of obtaining food as a cooperative venture and mobilizes most of its members to help capture the prey. Tracking down wild animals and protecting families against enemies can be conducted more efficiently in groups. The tools and implements utilized by hunters include a variety of traps, snares, and lethal weapons (i.e., bows and arrows and spears). Even though hunting requires a higher level of technology than that required by gathering, many of those who practice this economic activity today still have a relatively low food productivity level and often live precariously close to starvation.

As it exists today, hunting occurs primarily in high-latitude zones, particularly in the arctic (Figure 3-2). In the middle of the nineteenth century, however, specialized hunters could be found throughout the Americas, in southern Africa, and in the interior of Australia. With the arrival of Europeans in North and South America, the indigenous hunters who were not killed lost their lands and their means of livelihood. The Great Plains of North America and the pampas

of South America, for example, were both superb hunting grounds. Today, much of that land is planted in wheat or sustains large commercial herds of cattle. The hunters of southern Africa, the Bushmen, have experienced a similar loss of territory and are now limited to the barren, inhospitable Kalahari region.

Typical of the peoples today who make a living by hunting are the North American Eskimo. These very skilled hunters have refined their abilities into a fine art. Their mode of life proves that the more diverse the prey, the more complex the tools which hunters are likely to employ. The Eskimo are experts at fashioning kayaks from skins and harpoons from bones. They display remarkable ingenuity in adapting animal products to satisfy their physical needs. Farther south, in a few parts of Canada, occasional Indian tribes remain essentially at the hunting stage. These Indians are landsmen whose main targets are deer. Another high-latitude people who still subsist mainly on the harvest of wild animals are the Yukaghirs of Siberia.

The physical environment of the arctic features extremely low average temperatures and a very short summer period. The barren land, called *tundra,* is comprised of meager vegetation (mosses, lichens, and shrubs). This tundra vegetation provides an adequate diet for caribou and musk-ox, and these and other animals, in turn, furnish food for the humans who range over these large expanses in search of animal prey.

Figure 3-2 Subsistence Gathering and Hunting. *Shaded areas on the map indicate general regions where nomadic peoples depend on subsistence gathering or hunting. Hunters predominate in the Arctic while subsistence gathering tends to identify with the tropics.*

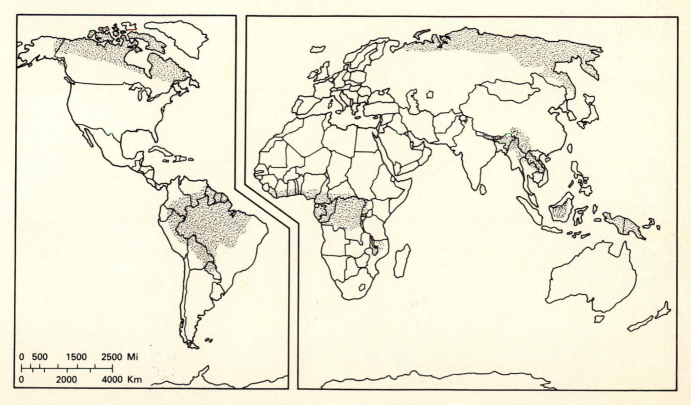

```
0  500   1500    2500 Mi
0      2000    4000 Km
```

Primitive hunters can be credited with maintaining a fine balance between the supply and the harvest of animal resources. This is in sharp contrast to the performance of some members of more advanced economies, who have seriously damaged the environment and have actually caused the extinction of many types of wildlife. One might argue that the technological abilities of these primitive peoples prevented them from becoming numerous enough to threaten the food supply. When it is realized that they killed only what they needed and wasted nothing (i.e., they used skins for shelter, tendons for cord, and bones for tools), they must be recognized for their conservation of natural resources. Primitive gatherers and hunters seem never to have destroyed their economic base by overharvesting the existing food supply, except when the techniques or tools for doing so (e.g., firearms) were introduced from other cultures.

Only a few thousand people still make their living exclusively by hunting. Many are under pressure to change their way of life and enter the commercial occupation of raising animals. The Eskimo never herded domesticated animals until they were taught to do so by Laplanders from Scandinavia. Indeed, the Eskimo during the past half-century have undergone considerable social and technological change, most all of it because of contacts with those who were searching for minerals or who were setting up early warning military defense systems. Those Eskimo who have abandoned hunting altogether find employment with private companies or with the Canadian or U.S. government, and often live in permanent coastal settlements.

PRIMITIVE HERDING (PASTORAL NOMADISM)

Primitive (or nomadic) herding is a more advanced economic activity than either gathering or hunting, since those who live by it make at least some investment to enhance natural production. The product is animal, and the investment is labor—not just the labor required to extract from the natural supply, but that necessary to nurture and increase that supply.

The domestication of animals marked a step upward, from animal gathering to primitive herding. Once this step was taken, people began to play a significant role in producing their commodities by bringing them to maturity as well as in harvesting them. No longer did human beings function as parasites living from nature's bounty, for now they made an investment of their own.

In the late twentieth century, one vast area of primitive herding can be demarcated (Figure 3-3). It extends all the way from the Atlantic shores of North Africa eastward across Africa through the Arabian peninsula, then deep into inner Asia, almost to the Pacific Ocean—a longitudinal extent of over 8000 miles. Latitudinally, this arid and semiarid region extends from 5 degrees south latitude (on the east coast of Africa) to 50 degrees north latitude in central Asia—a range of over 3500 miles.

Herding activity encompasses the single largest territory on earth. These pastoralists occupy some 10 million square miles, twice the area of land devoted to cultivation. In this core region of *pastoralism*, it is estimated that there are 3.5 million herders in the Sahara, Sahel, and Sudanic zones of Africa, over half a

Figure 3-3 Subsistence Herding. *Shaded areas on the map identify the major regions of subsistence herding in the world today, with the most noticeable concentrations occurring in Saharan Africa, the Middle East, Central Asia, particularly western China and Mongolia, and the Arctic coasts of Eurasia.*

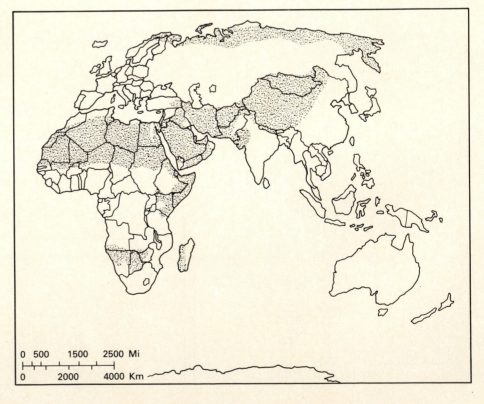

0   500   1500   2500 Mi

0   2000   4000 Km

million in the Middle East, and some 11 million in the northern fringes of the Middle East and the Indus River plain of Asia, in the Soviet Union, and in China. This makes for a total of approximately 15 million pastoralists in this huge area.

A lesser region of herding is found in northern Eurasia, extending into Alaska (Figure 3-3). In the southern hemisphere, too, there are small areas in southwest Africa and on the island of Madagascar. Primitive herding is absent from Australia, South America, and most of North America.

It should be noted that the existence of nomadic herding in a region does not preclude other types of economic activity. In the Persian Gulf area, for example, more people are concerned with oil or gas exploitation today than with herding animals.

The material culture of herders is characterized specifically by a dependence on domesticated animals. The economic needs of nomadic herders are met by animals that feed on wild plants rather than on cultivated crops. Animals supply food (milk, cheese, and meat), materials for clothing (fibers and skins), shelter (skins), fuel (excrement), and tools (bones). The animals that have proved most satisfactory in these roles are sheep, goats, cattle, camels, reindeer, and yaks. Horses are used in some areas, but they usually perform the special function of transporting the herdsmen as they tend their animals. Meat plays a small role in the diet of the true nomadic herder, because animals are rarely killed except on special ceremonial occasions.

As mentioned above, primitive herding is a regularized migratory undertaking. Migration (and the mobility that is required for successful herding) is a basic feature of the life of these people. Movement in search of pastures can be undertaken either over vast horizontal distances, or vertically from one elevation to another. The latter practice is known as *transhumance* and can be found in such places as the Andes, Himalayas, and east Africa. It must be noted that transhumance is also practiced by modern animal herders in the Alps, Pyrenees, Caucasus, and Rocky Mountains.

Physical Environments of Herding Areas

Most of the regions where nomadic herding occurs possess an arid or semiarid climate (usually less than 20 inches of precipitation is recorded). Under these physical circumstances, trees do not grow over broad areas, as grasses and shrubs comprise the natural vegetation (Figure 3-4). Wherever an adequate moisture supply is found, however, some form of agriculture normally is practiced. From Morocco to north China, the search for water and pastures is the critical fact of life. During the extended drought in the African Sahel (the grasslands immediately south of the Sahara Desert) from the late 1960s well into the mid-1980s, hundreds of thousands of animals could not find food and died. Pastoralists dependent on those animals either themselves starved or moved near population centers, where they received relief until environmental conditions permitted them to return to herding.

Most primitive herding occurs in the regions of shrubs, bunch grass, and short grass. Goats, which can

Figure 3-4 Herding Goats and Cattle in Ethiopia. *Several herdsmen with a herd of goats and a few cattle in the drought-stricken village of Bume, Ethiopia. The drought of the 1970s and 1980s, which affected much of northern Africa, killed about 85 percent of the cattle in Southern Ethiopia. Goats generally fare better in dry weather. (Courtesy of United Nations/Jerry Frank)*

endure considerable aridity, constitute the herds in the driest regions, whereas sheep and camels predominate in places with somewhat heavier rainfall. The wetter fringes of these drylands can support herds of cattle.

Not only the type but also the amount of natural vegetation is conditioned by the availability of precipitation. The vegetation cover in these areas is extremely restricted. The search for forage, therefore, is never ending. The herdsmen leave their animals in place as long as possible (i.e., until forage gives out). They then move on to another place where a meager supply of water and grasses suffices briefly—until the next move.

As noted above, the unending movement of subsistence herdsmen and their animals is both horizontal and vertical. They must move because supplies at any given time in any given place are limited and because the rains come at different times in different places. Because highlands tend to be wetter than lowlands, herdsmen practice transhumance by moving to higher altitudes to find grass for their animals. Those who practice transhumance recognize that there are seasonal variations in forage supplies at different elevations, but the factor that determines when they go up or come down is temperature, not precipitation. Pastoralists tend to winter in the lowlands and move to upland pastures during the summer. As autumn approaches and frost begins to occur, herds are returned to the lowlands.

The physical environment is different in the primitive herding region of northern Eurasia. Precipitation is low, but evaporation is even lower; thus the climate there is considered to be a humid one. However, the temperature drops so low that an arctic climate prevails and the natural vegetation is tundra; there are no trees, only a few shrubs, a little grass, and lichens and mosses. In addition, the ground is frozen much of the year. Forage is scanty, as in the steppe (grassland) and desert, but here the major handicap is the cold, not aridity. Plants that withstand the tundra climate are no more palatable to people than are the shrubs of the deserts or the grasses of the steppes. But certain animals—in this case, caribou and reindeer—can exist even on this limited vegetation. Because of limited forage, the herds migrate toward the pole in summer when tundra plants germinate and mature, and then move toward the equator in the winter when the intense cold drives the plants into dormancy.

Herding Cultures and Their Future

The development of herding as a distinct culture has been an issue of some controversy for anthropologists and for economic geographers. Some years ago the prevailing theory was that herding represented a lower form of economic activity than agriculture. Furthermore, this Darwinian notion asserted that groups of people evolved through stages of culture and that once they progressed to a certain level, they never regressed (or devolved).

In the case of primitive herding, evidence suggests that these groups, once engaged in cultivation, did devolve when environmental circumstances dictated. Some scholars, for example, the distinguished German geographer, Eduard Hahn, believe that herding became a way of life for many plow farmers in the ancient Near East who tried to cultivate land too marginal for successful crop growth. Although the land proved too dry for food crops, the farmers' domesticated draft animals could subsist on the limited natural vegetation. Rather than move, farmers adopted herding as their new means of support.

Two factors have combined in recent decades to challenge the prevailing life-style of the pastoralist. The imposition of (1) political boundaries by states, and (2) new, ambitious settlement plans in nations such as the Soviet Union and China have sharply limited the nomad's mobility. This situation exacerbated a traditional problem for the pastoral nomad: finding the right balance between herd size and the capacity of the land to support animals. By circumscribing the mobility of the nomad, governments have successfully limited the ability of the pastoralist to make a living.

When herders become sedentary (or settled), there is a tendency for the social fabric of tribes to tear apart. The tribal system of nomadic peoples is based on communal property and cooperation, coupled with intense loyalty. In some areas, governments have purchased the nomads' animals, and this, too, helps break down the system of communal ownership. This phenomenon has occurred extensively in the Middle East and north Africa. Even in Tibet and northern Siberia, where yak and reindeer herders have been extraordinarily isolated in the past, herders are now being influenced more and more by the modern economies that surround them.

On the margin of nomadic herding areas there frequently occurs a process that might be called *cultural osmosis,* whereby some nomads quietly shed their herding habits, settle down at the edge of agricultural areas, and adopt agricultural techniques from their neighbors. Areas of nomadic herding will thus shrink as these people are assimilated along the margins, or in some cases, the exigency of government policy demands that they change their long-standing way of life.

PRIMITIVE AGRICULTURE

Primitive cultivation represents the first endeavor of people to control static resources, that is, the bounty of the land. Agriculture, more than any of the economic activities described above, is influenced by technological innovations and applications of capital and energy. Primitive cultivation is called such because it manifests only rudimentary technical management of the land, and limited amounts of time, effort, and capital are devoted to this activity. It should be noted that unlike the other economic activities described, primitive forms of cultivation are still practiced widely in the modern world.

Different types of primitive cultivation are known by a bewildering variety of names, yet only a few reflect real differences between agricultural systems. For the purpose of this section, primitive agriculture is synonymous with the system of *shifting cultivation* and its more progressive counterpart, *rotational bush-fallow cultivation.*

Today, there are three broad regions where primitive agriculture can be found (Figure 3-5). The largest and most populous is in central Africa. Straddling the equator, nearly half of the continent lies in this zone. Most of west Africa's farmers practice rotational bush fallowing. The second major region lies in southeast Asia and the adjacent offshore islands, from Sumatra eastward through Borneo, Papua–New Guinea, the New Hebrides (now called Vanuatu), and numerous tropical islands of the Pacific Ocean. On the mainland of Asia, primitive cultivation tends to be confined to the interior of Burma, Thailand, Cambodia, and adjacent portions of India and China.

The third region where primitive agriculture is practiced embraces most of the Amazon basin, reaching from the Atlantic coast to the Andes Mountains, and from Bolivia to Venezuela. The practice of primitive agriculture is also found in Ecuador and Colombia and extends northward through Central America into southern Mexico. A small part of the West Indies is also included. It is generally recognized here, as well as in other major regions, that primitive cultivation is the prevailing activity, but it coexists with other forms of economic livelihood.

Recent estimates by the *Food and Agriculture Organization* (FAO) indicate that nearly 200 million people make their living by these systems of agriculture. The FAO also reports that shifting cultivators occupy 33 million square kilometers of land, which is nearly twice the total area of the world's permanent cropland. This statistic illustrates the low population densities of these areas and the extensive character of the economic activity. Of the 200 million shifting and rotational bush-fallow cultivators, it is estimated that perhaps 75 percent live in Africa, with most of the remainder in Asia. The region possessing the fewest primitive farmers is Latin America.

Primitive cultivation is clearly not a passing phenomenon as primitive hunting and gathering seem to be. From Table 3-1 it is clear that agriculture plays a critical role in the economy of most developing nations of tropical Africa and Asia. In virtually every third-world country, however, the vast majority of the economically active population are subsistence farmers.

Depending on the region, subsistence farmers rely on a variety of staple foods. The leading staples in the western hemisphere include maize and manioc (also

Figure 3-5 Primitive Subsistence Farming. *The shaded areas on the map indicate concentrations of subsistence farming. In these areas, primitive cultivation is the prevailing activity but other forms of economic livelihood also coexist. Almost all of this activity occurs in tropical environments.*

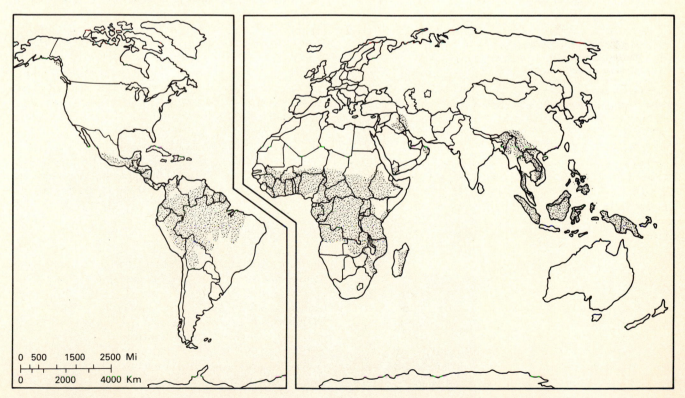

Table 3-1

Nations with 70 percent or more of the economically active population engaged in agriculture, 1984

Nation	Percent	Nation	Percent
Bhutan	93	Tanzania	79
Nepal	92	Guinea	78
Rwanda	88	Somalia	78
Central African		Botswana	77
Republic	85	Ivory Coast	77
Niger	85	Ethiopia	77
Mali	85	Gambia	76
Bangladesh	82	Afghanistan	76
Lesotho	81	Kenya	76
Burundi	81	Sudan	75
Malawi	81	Gabon	74
Mauritania	81	Yemen AR	73
Papua–		Thailand	73
New Guinea	81	Kampuchea	
Madagascar	80	(Cambodia)	72
Guinea-Bissau	80	Senegal	72
Chad	80	Zaire	72
Uganda	79	Equatorial Guinea	72
Burkino Faso	79	Laos	72
Cameroon	79		

Source: United Nations, *1984 FAO Production Yearbook,* 1985.

called cassava), together with beans, squash, plantains (the cooking banana), and sweet potatoes. In Africa, farmers subsist on manioc, millets, yams, cocoyams, cassava, plantains, and maize (Figure 3-6). Rice dominates all staple foods in Asia, whereas on the Pacific islands, such foods as yams and taro are preferred.

Characteristics of Subsistence and Bush-Fallow Agriculture

In both subsistence and rotational bush-fallow cultivation, a similar technology is employed to prepare fields for planting. Because these types of agriculture are practiced mainly in the tropics, farmers are continually confronted with the problem of clearing land from either dense forests or overgrown savanna grasses. Trees and underbrush are hacked out by using machetes, and the vegetation is burned. Sometimes, farmers dispense with the cutting and let fire alone do the job of clearing. This primitive technique is commonly called *slash/burn* agriculture. Methods of preparing the soil are equally primitive. By using a digging stick (dibble) or a hoe, farmers simply scratch the surface of the soil at the time of planting. No other attempts are made to prepare the soil beforehand. Plows are almost unknown.

Farmers have long known that fire is a simple method of clearing land of wild vegetation. Unfortunately, they also believe that ashes provide a good fertilizer for the soil. Ash helps a little, but because the organic matter in the soil is combustible, the fire also acts to partially destroy its fertility.

Tropical soils, because of constant high heat and humidity, do not contain significant amounts of organic material, the source of nutrients to living plants. Furthermore, for plants to assimilate soil nutrients, they must be soluble. Yet, because of heavy precipitation, soluble nutrients are leached from the soil. What remains in the soil are nonsoluble iron and aluminum oxides, both of which are extremely infertile and which give the soil its reddish color. The lack of

Figure 3-6 Land Clearing by Burning in Cameroon. *In the settlement shown here, farmers are clearing a plot for cultivation that has lain fallow for many years. In tropical Africa burning remains the preferred way of clearing vegetation for planting. (Courtesy of Sanford H. Bederman)*

organic material, excessive leaching, and the exacerbating effects of burning the soil cover combine to assure that cultivation on a piece of land is possible for only a very limited amount of time.

Of all forms of agriculture, primitive subsistence slash/burn agriculture supports the fewest people. Recent studies have calculated the average density in areas where it is practiced to be about 12 persons per square kilometer. On the other hand, rotational bush-fallow cultivation can support densities of over 100 people per square kilometer.

These differences in population density are accounted for by the fact that shifting cultivators abandon both their fields and dwellings every few years, moving to another area to practice shifting cultivation before moving again. In rotational bush fallowing, settlements remain stationary, but the fields are rotated throughout the environs. Abandoned fields quickly revert to bush vegetation cover and remain fallow to regenerate their fertility. After 10 or 15 years of fallowing, many fields can be cleared and cropped again. Because bush fallowing represents a more stable people–land relationship, population densities are considerably higher than those in slash/burn-shifting areas, and food surpluses are produced.

Even though subsistence farmers throughout the tropics consume most of the food they harvest, surpluses are taken to nearby villages where small markets are held at given time intervals. These *periodic markets*, usually found in places where a modern transportation network does not exist, allow farmers to barter or sell their limited produce in order to obtain other items they need (Figure 3-7). Depending on the region, periodic markets can be held in a specific village on cycles of anywhere from 2 to 12 days. This vital commercial institution can be found almost everywhere in west Africa, in much of Latin America, as well as in parts of interior south and southeast Asia. As much socializing as commerce takes place at these very colorful markets.

The Future of Primitive Subsistence Farming

Much criticism has been leveled at primitive agriculture, particularly as it manifests itself in shifting cultivation. The physical constraints of this system of cultivation have been amply described above. Nonetheless, it is obvious that rudimentary farmers have been able to sustain themselves, albeit marginally, for millenia. Before this century, the economies of areas that engaged in primitive farming were isolated and self-contained. In summary, shifting cultivation from the agricultural point of view is not considered objectionable as long as vegetation and the fertility of the soil are restored by nature. The opinion of those

Figure 3-7 Periodic Market in Nepal. *In these makeshift stalls, farmers bring their products to market, including vegetables, fruit, poultry, goats, and handicraft items. The Nepalese border town shown here lies near the India boundary. The market is open two times a week. (Courtesy of Nanda Shrestha)*

who regard it as a totally destructive agricultural system under *all* circumstances is largely unjustified.

In spite of the unjust criticisms, a very serious problem does exist today. The extremely rapid growth of population in Africa, Asia, and Latin America, coupled with a slow movement away from a subsistence livelihood to a money economy, has placed incredible strains on land that possesses only limited productive capability. Shifting cultivation is an extensive economic system, wherein 90 percent of all available land should be kept in fallow to make up for the low natural soil fertility. With the population explosion in developing areas, the demand for food has increased concomitantly. To produce more food, farmers cultivate plots that should be left fallow for many more years. The result is a continuing destruction of the soil, a condition that perpetuates low land productivity, or in some extreme cases, accounts for desertification of the land.

On the brighter side, in many parts of the world's tropics, those subsistence farmers who are advantaged by living near plantations and have the benefit of good transportation and communications systems are turning more and more to producing crops for sale. These *peasant farmers* are quite sedentary, and their personal wealth is frequently based on their entrepreneurial aggressiveness. (Peasant farmers, also called smallholders, are discussed in detail in Chapter 8.)

The fact that subsistence farmers can easily revert to *smallholding* indicates the existence of excess land and labor capacity to produce a surplus above the requirements needed for minimum subsistence. Smallholding peasant farmers have therefore been able to produce cash crops without reducing their output of foods for subsistence. This problem does not exist in Asia, because the major crop, rice, is both the subsistence and the export product.

In parts of west Africa, farmers often put their fallow plots to good use by nurturing tree products that are sold as cash crops. Subsistence crops are also interplanted with export crops. We have more to say on this topic in Chapter 8.

A process of modernization encouraged by the government in many countries with primitive subsistence economies has had an ambivalent effect. This process aims to improve the quality of life of people with the knowledge that primitive activities provide only a very precarious livelihood. Modernization has meant the introduction of new technologies (and western life-styles), together with the requirement that primitive peoples undertake to apply them. Through pressure, some primitive herders have become keepers of commercial livestock or farmers, and shifting cultivators have become permanently settled. Previously isolated territories are now being opened up, often through the vehicle of "intrusion," that is, when modern economic activities such as plantations, mines, or factories are established there. Subsistence cultures are thus turned outward, contacts with other groups made, and different social institutions developed.

This process can bring economic benefits. It can also bring profoundly upsetting social cultural changes. The primitive peoples described in this chapter have never been very receptive to change in their mode of living. Before this century, intrusion for primitive peoples often took the form of slavery or some other form of subjugation. Even today, the modernization of primitive peoples breeds unrest. In striving toward the goal of economic growth, developing nations must be careful to minimize such harmful social effects of unregulated modernization.

SUMMARY AND CONCLUSION

Primitive economic activities occupy a large share of the earth's surface but support a relatively small population. Most of these activities are practiced either in the tropics or in high-latitude regions where severe environmental constraints limit productive capacity. Several centuries ago, by contrast, nearly all the world population engaged in these types of subsistence activities. The retreat of primitive gathering, hunting, and herding activity continues today. Only primitive agriculture as practiced in nearly half of Africa, the Amazon basin, and southeast Asia remains a fairly stable type of agriculture among the forms discussed here. In fact, the area now covered by subsistence agriculture is twice as large as the area of permanent cropland in the world today and encompasses much of the third world.

SUGGESTIONS FOR FURTHER READING

AKINBODE, ADE, "Population Explosion in Africa and Its Implications for Economic Development," *The Journal of Geography*, 1977, pp. 28–36.

FORDE, C. DARYLL, *Habitat, Economy and Society: A Geographical Introduction to Ethnology.* London: Methuen, 1957.

HARRIS, DAVID R., "The Ecology of Swidden Agriculture in the Upper Orinoco Rain Forest, Venezuela," *Geographical Review*, 1971, pp. 475–495.

HUNTER, JOHN, and G. K. NTIRI, "Speculations on the Future of Shifting Agriculture in Africa," *Journal of Developing Areas*, 12, No. 2 (January 1978), 183–208.

LUSTIG-ARECCO, VERA, *Technology: Strategies for Survival.* New York: Holt, Rinehart and Winston, 1975.

MORGAN, W. B., *Agriculture in the Third World, a Spatial Analysis.* Boulder, Colo.: Westview Press, 1978.

MORGAN, W. B., "Peasant Agriculture in Africa," in M. F. Thomas and G. W. Whittington, eds., *Environment and Land Use in Africa.* London: Methuen, 1969, pp. 241–272.

SONNENFELD, JOSEPH, "Changes in an Eskimo Hunting Technology, an Introduction to Implement Geography," *Annals,* Association of American Geographers, 50 (1960), 172–186.

STEWART, NORMAN R., JIM BELOTE, and LINDA BELOTE, "Transhumance in the Central Andes," *Annals,* Association of American Geographers, 66 (1976), 377–397.

4 Intensive Subsistence Agriculture

Contemporary agricultural practices in east and south Asia contrast sharply with the situations in Africa and Latin America described in Chapter 3 (Figure 4-1). The classic forms of *intensive subsistence* agriculture, found today in China, India, southeast Asia, Korea, and Japan, involve high levels of output per unit of land. This intensive use of the land produces relatively large yields per acre, but frequently little surplus occurs because of the vast food needs of the tremendous domestic population that is supported by this agricultural system. In Japan, however, considerable surpluses do exist, owing to the impact of modern practices, including the use of hybrid seeds, mechanization, modern irrigation practices, and commercial fertilizers.

CROPPING TECHNIQUES

Cropping methods and the types of crops grown distinguish intensive agriculture from other types. Rice is typically the principal crop. The highest yields occur with *wet* rice, which is grown in *paddy* fields. These fields are typically small dug-out areas that lie about 3 feet below ground level, bounded by narrow dikes, dams, or roadways (Figure 4-2). The recessed fields permit flooding during the growing season. Another type of paddy is created by dammed-up terraces, which can also be irrigated (Figure 4-3). Upland rice grown in nonirrigated areas provides far lower yields. This form is called *dry* rice.

The Asian rice field typically occupies the space of a garden plot, being an acre or less in size. Elaborate irrigation, circulation, and drainage systems sustain the high yields. Floodplains are favored locations because of their proximity to water and the ease of setting up hydraulic systems on relatively flat land. In recent years the number of drilled wells and electric pumps used to supplement surface water from reservoirs and rivers in rice-growing areas has grown dramatically.

The highest concentrations of population in south and east Asia are associated with the floodplains and deltas of the major rivers, such as the Ganges, Irrawady, Mekong, and Yangtze, as a result of their ability to produce abundant rice crops. Where rainfall, irrigation systems, and growing-season conditions permit, double crops of rice are grown each year. This *multiple-cropping* technique requires considerable mechanization and/or hand labor. Typically, farmers attempt to replant a second crop, using transplanted seedlings, within a day or two of the first harvest. In places such as extreme southern China it is even possible to produce three crops a year.

Rice agriculture requires a warm (average temperature of 70°F) growing season 4 to 6 months in length. Water requirements are also high. In the Ganges valley and in extreme southern China, there is over 80 inches of annual rainfall. Areas with less than 50 inches of rainfall are generally unsuitable for rice cultivation. Rice plants grow best on thick alluvial deposits (2 to 3 feet thick) underlain by clay subsoil

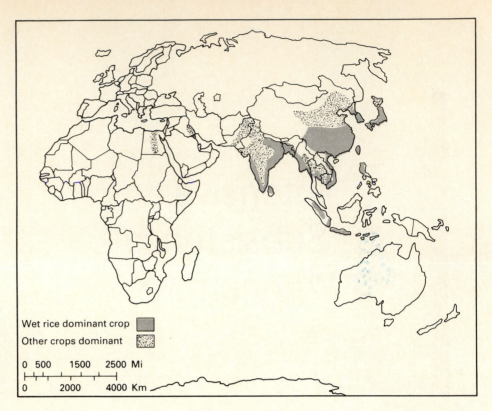

Figure 4-1 Intensive Subsistence *Rice cultivation is the primary agricultural activity in intensive subsistence areas. As indicated on the map, areas of particular significance include coastal and delta areas of India, Southeast Asia, southern China, and Japan.*

Wet rice dominant crop

Other crops dominant

0 500   1500   2500 Mi

0   2000   4000 Km

that prevents water seepage. Alluvial soil is a loose, friable material high in organic matter that promotes rapid root and stem growth.

When temperatures are cooler and growing seasons shorter, wheat is often substituted for rice as the second crop. In areas unfit for rice cultivation, two crops of wheat may be grown. Other grains, such as barley, millet, and sorghum, are also produced as food crops. Many farmers augment their cereal grain crops with corn, beans, peas, melons, and fruits to supplement their diets. Often, a practice called *interculture* permits the simultaneous growing of a second crop between the rows of the main crop or on dikes between paddy fields. Industrial cash crops are also grown, including cotton, tea, sugarcane, rapeseed, and jute.

GREEN REVOLUTION

The introduction of many cultivation and management techniques increases yields in this intensive agriculture environment. Several deal with modern technological applications, such as commercial fertilizer, pesticides, improved hybrid seeds, and machinery. These advances, collectively known as the *Green Revolution,* have been responsible for dramatic increases in yields in recent years. Greater local control and incentive arrangements, improved marketing assistance, the use of cooperatives, the expansion of credit arrangements, and the use of trained agricultural specialists are other factors promoting greater productivity.

The term "Green Revolution" refers to a major biogenetic advance in agricultural technology. In the mid-1960s, agricultural scientists developed high-yielding varieties (HYVs) of rice and wheat. These so-called "miracle seeds" offered an opportunity to increase agricultural output in many poverty-stricken third-world countries. We discuss these advances in subsequent sections. It is useful here to mention that problems also accompanied these advances. In many cases, for example, only the wealthy landed gentry benefited from the change, leaving the masses in poverty.

There are two major explanations for the failure of the Green Revolution to reduce the level of poverty in many areas. First, the Green Revolution is essentially class biased, in that only those agriculturists with sufficient land and other assets (mainly capital) could afford to adopt its innovations (Figure 4-4 on page 40). The Green Revolution required large quantities of chemical fertilizer and pesticides together with a well-developed system of irrigation. Worldwide increases in the price of petroleum products in the 1970s further distanced the poor from access to many of these products. Only wealthy and prosperous landowning farmers could afford to buy or obtain these products on credit. The poor and landless peasants, on the other hand, did not have access to these resources even on credit, owing to the lack of sufficient assets for collateral. Consequently, the benefits of the Green Revolution technology have accrued mostly to farmers with considerable land and wealth.

Second, once the wealthy and landed farmers realized that the Green Revolution together with agri-

Figure 4-2 Rice Paddy Field Complex in Japan. *This rice producing area in the Tokyo area near the Narita airport provides a classic example of the intensive cultivation of small fields in dug-out paddies. Taken at harvest time in late summer, this photograph shows rice straw drying on fences on the perimeter of the fields, even as the stubble is burned and plowed under for immediate replaintings. (T.A.H.)*

Figure 4-3 Terrace Agriculture in Nepal. *These flooded rice fields in foreground contrast with the hill terraces in background. Maize and millet are common crops on upland terraces in Nepal. (Courtesy of Nanda Shrestha)*

Figure 4-4 Wealthy Landholder's Mansion in Taral Region of Nepal. *The veranda architectural style of this lavish home promotes ventilation in the warm season. This farm produces rice in the summer and wheat in the winter. (Courtesy of Nanda Shrestha)*

cultural mechanization (e.g., use of tractors) provided great potential for self-cultivation, they began to withdraw land from the sharecropping and rental market. They began to cultivate the land themselves using hired laborers during the harvest and planting seasons when the labor need reached its peak. A large number of tenants and peasants who previously had access to this land on a rental or share-crop basis lost their jobs. They became more dependent on irregular seasonal labor. Thus, in many instances, the Green Revolution contributed to the aggravation of poverty rather than to its solution.

Another problem with the Green Revolution is that many targeted areas shifted a larger share of their fertile land to cash crops for export (e.g., coffee, tea, bananas, jute, cotton, cocoa, rubber production, etc.) rather than emphasizing food crops for domestic market consumption. Such cash crops provided important sources of hard currency necessary to pay for imported goods and to repay foreign debts. But the contribution these crops make to the reduction of poverty is negligible because the use of most productive land was diverted away from such basic crops as rice, wheat, corn, millet, and legumes. These problems are dealt with in more detail in Chapter 8, which examines plantation agriculture.

ANIMAL HUSBANDRY

Animal production has increased considerably in the past 20 years in east and south Asia but remains a secondary consideration. Waterfowl (e.g., ducks) and fish farming account for much of this expansion, which is discussed in Chapter 8. Widespread animal husbandry cannot be practiced, due to a shortage of land and feed. The land can be used more effectively to produce small grains for direct human consumption. In Hindu areas, religious practices prohibit the killing of cattle, and this has thwarted the use of this source of protein in the Indian diet. Cattle are used, however, as draft animals to pull plows and cultivators and for transportation. They also often serve the role as collateral or as a bridal dowry.

Sheep are raised in marginal grassland areas in the interior. Other small animals, such as swine, goats, and chickens, frequently thrive on waste as scavengers. In China, for example, it is not uncommon to find swine fed a fermented mash made of leaves, stems, pods, and bran from grain and vegetable waste. Animal excrement traditionally provided a very important fertilizer for croplands and a fuel source for heating and cooking. Human waste, or *night soil,* also has been an important organic fertilizer source. Cattle, especially water buffalo, also provide a source of milk, which offers protein in a diet otherwise predominantly vegetarian.

The economies of China and India, among the areas studied here, are particularly dependent on agriculture. Approximately 75 percent of the 1 billion people in China depend on agriculture for a livelihood, and much of the country is characterized by a rural village atmosphere. In India, 75 percent of the population of 600 million persons is engaged in agriculture. By contrast, only 7 percent of the population of Japan is engaged in agriculture. This statistic accurately suggests that the Japanese agricultural sector is more like that of other modern industrial nations than that of its Asian neighbors. Despite the fact that a modern commercial-industrial economy exists in Japan, many agricultural practices remain third world in nature. Japanese farmers still produce rice on traditional paddy fields and remain inefficient by international standards. In order to be more specific about the contemporary agricultural environment in India, China, and Japan, the remainder of the chapter is devoted to a separate discussion of each.

INDIA

Until the mid-1960s, the prophets of doom accumulated considerable evidence that the Indian economy could not survive, owing to rapid population growth and an ineffective agricultural sector that offered little hope of ever bringing self-sufficiency to the country. Illiteracy, the caste system, disease, and the lack of mechanization (see Figure 4-5) remain tremendous handicaps, but farming has made a remarkable turnaround in the last 20 years. The country is now self-sufficient in grain production, but the situation remains precarious because other sectors of the economy have not kept pace. Because of enormous population growth pressures, agricultural output must continue to expand. The ongoing Green Revolution has permitted this to happen. At the same time, care has been taken to ensure that mechanization does not displace agricultural workers, because it appears that at present, these workers cannot be effectively absorbed into other sectors of the economy.

Most farms in India are very small. In the 1970s, about half were no more than $2\frac{1}{2}$ acres in size. Such small holdings make it difficult to adapt to Green Revolution innovations. Moreover, as mentioned earlier, small farmers often do not have access to credit to buy seed, fertilizer, or mechanical aids.

The leading banks in India were nationalized in 1969 in order that more money could be directed to agricultural lending programs. A goal was set in 1980 that by 1985, 16 percent of all monetary advances from banks would go to agriculture. Small land holdings, those with less than 5 acres, receive particular attention. Economic assistance also encourages owner-operated rather than tenant-based operations. Restrictions against large holdings (over 15 acres) assure that large units do not displace family farms.

Advisory services, such as the training and visit (T&V) program of the World Bank, are helping small farmers modernize and increase output in India, but these small, inefficient operations limit progress. In 1971, 15 percent of the land-owning households in India controlled 45 percent of the land. Land reform, if needed, appears to be politically unrealistic. The problem may become worse if population growth rates do not fall. Fortunately, there is evidence that the population growth rate is indeed dropping.

Productivity increases accounted for the increased agricultural output in India in recent years. The growth is demonstrated most dramatically by the remarkable progress in Punjab, a state in northwest India bordering Pakistan. Three-fourths of the cultivated land is irrigated there, compared with one-fourth of India overall. All villages in Punjab had electricity by 1977, whereas only one-third of those in the

Figure 4-5 Harvesting Rice by Hand. *This woman is harvesting rice in the Solo River Basin of Indonesia by hand, cutting each stalk separately. (FAO photo by J. Dornbierer)*

entire country were electrified. Farm ownership rates are also higher in Punjab. A good marketing network is available to collect and distribute the grain harvest, primarily wheat, unlike many other areas, which remain isolated due to the poor transportation infrastructure.

Agricultural Patterns

Rice production is concentrated in eastern India, especially in the Ganges and Brahmaputra River delta regions. In no other rural areas in the world are population densities higher than they are in these areas, where they often exceed 2500 persons per square mile. Jute and sugarcane are also grown in the Ganges lowland area. Wheat production is widespread and the drier northwestern sections of the country, and millet and sorghum production flourish on the interior Deccan plateau. Coconuts and rice dominate in the extreme south.

CHINA

As in India, rice acts as the major food crop in China. Similarly, population growth exerts enormous pressure on the agricultural system in China. Until re-

cently domestic grain production shortfalls required large quantities of imported grains to meet food needs, but productivity increased dramatically following reforms in the late 1970s. Use of fertilizers, greater mechanization, and better hybrid seeds have all played a role in increased productivity, but a return to family-level decision making and better marketing techniques has been just as important.

In 1957, about 85 percent of the labor force was engaged in agriculture, whereas 77 percent are so employed today. A tremendous expansion in the size of the work force in this period means that about 100 million more persons work the land at present in comparison with the situation in 1957 (Figure 4-6). Shifts in the labor force are nevertheless occurring at a more rapid pace today as opportunities expand in small towns, in industrial, service, and in construction activity.

The communist revolution that culminated in 1949 when Mao Zedong assumed control of the government was a rural-based agrarian movement. Land reform was desperately needed at the time to break up large holdings and reallocate land among the peasants. It was not until the late 1950s, however, that the collectivization program achieved this reform. This program eliminated price incentives, free markets, and private plots. A goal of regional self-sufficiency

Figure 4-6 Harvesting and Thrashing Rice in Guangxi Province, China. *Note women with rice straw hats operating primitive rice thrashing machine in field. Note contrast with equipment shown in Figure 4-7. (Georg Gester/Photo Researchers)*

received emphasis in grain production. Greater priority was placed on increasing grain production rather than the cultivation of so-called "economic" crops such as cotton, oilseed, and sugar. These reforms were only partially successful, as production declined. A gradual return to free markets followed in the early 1960s.

The agricultural *collective* as implemented in 1958 consisted of a three-tiered organizational structure. At the top of the hierarchy is the *commune,* which averages about 3000 households or 10,000 persons under its supervision. The second-tier group is the production *brigade,* which incorporates about 300 households. Brigades act as local administrative units. In addition to agricultural interests, brigades often organize small-scale industrial plants to produce goods for local consumption and supervise off-season public works or construction projects, such as road building, reforestation, or land improvements. The third, grass-roots level unit is the production *team,* made up of 20 to 30 households or about 100 persons.

Traditionally, communes functioned as both a political/administrative management unit and as an economic institution. Reforms in the late 1970s and early 1980s reinstated a township-type government separate from the commune and placed more emphasis on private-sector initiatives.[1] Separated from its political responsibilities, the commune began pursuing broader economic interests including work with other communes on joint economic enterprises. This arrangement facilitated the development of agricultural-industrial-commercial complexes and the creation of companies that produced farm machinery or conducted seed research.

Other reforms expanded local decision-making flexibility and cut back the concentration of power at the managerial level. Several overlapping forms of responsibility provide incentives and room for adjustments to local needs and conditions in the commune system. Land is held publicly and production goals are negotiated. In the 1970s food allotments were given directly to individual households following harvests, and provisions made for additional remuneration in various ways in the form of "work points" and cash. Once output quotas were met, surpluses were shared among members.

By the early 1980s the complicated work point system was replaced by a simpler contract system called the *responsibility system.* Farmers now leased land from the collective, which continued to own it, but when the quota was met, any excess output could be marketed by the farmer as an individual and sold at a profit. Less reliance on self-sufficiency and more priority on producing "economic" specialty crops such as cotton, sugarcane, and oilseed accompanied this reform. By 1983, about 80 percent of farm families had joined the responsibility system, suggesting a

[1]Clifton Pannell, "Recent Chinese Agriculture," *The Geographical Review,* 75 (1985), 170–185.

strong trend toward privatization of Chinese agriculture. This trend also carried over to machinery ownership, fully one-third of which was owned by private families in the mid-1980s. Loans made directly to rural families also exceeded those made to collectives. Production levels increased with these reforms, but imports remained necessary due to population growth.

Each area is now encouraged to produce what it can do best, following the tenants of *comparative advantage* (see a discussion of this concept in Chapter 5 as it applies to agriculture in the United States). The Den Xiaoping administration has thus abandoned the regimentation and command policies of the past in favor of a more decentralized, pragmatic agricultural system for the country.

Crop yields in China are lower than in the United States and Europe, but output is considerably higher than in India. China has over one-third of its cropland under irrigation, whereas India has less than 20 percent. Two and one-half times as much fertilizer is used on the same amount of land in China as in India. As a result, yields are double those in India. China also has an advantage in that grain crops are grown on the best land, whereas India tends to grow industrial crops (jute and cotton) on the best land. Finally, there is more double cropping in China than in India. Strict guidelines control rural–urban migration in China to prevent excess urban population growth. The urban labor force cannot absorb additional workers without displacement. These restrictions were more rigidly enforced during the Cultural Revolution of the 1960s than they are today. At that time, city residents were actually deported to the countryside.

Agricultural Patterns

Although China produces the world's greatest output of cereal grain, only 10 percent of the land mass is cropped. Vast reaches of the country, including Sinkiang, Tibet, and Inner Mongolia, are desolate, and nomadic herding prevails. The interior grasslands are also too dry for agricultural use. In the warm and humid southern portions of the country, on the other hand, double-crop rice agriculture effectively increases the net cultivated area of the country by 40 percent. Upland dry rice cultivation occurs farther inland in the south. Rice and tea, as well as rice and winter wheat cultivation prevail farther north, although still in the southern half of the country.

The Chin-Ling mountains and Huai River valley form a boundary between rice cultivation to the south and cereal grain crops (primarily wheat) to the north. In the extreme northeastern region of the country, soybeans, sorghum, corn, and other coarse grains, including spring wheat, predominate. Cotton, millet, corn, and winter wheat are grown in the central reaches of China astride the North China Plain, in the Hwang Ho River valley, and in the Shantung peninsula.

JAPAN

Agriculture in Japan remains an enigma. In many ways it resembles that of other intensive subsistence economies—emphasis on paddy-field rice, labor-intensive cultivation, and small holdings. But there are significant differences as well. Advanced levels of mechanization, production surpluses, and shifts to commercial vegetable, meat, and fruit products all suggest a commercial market-based agricultural economy more like that of the mixed farming areas of western Europe and North America.

There is no doubt that Japanese farmers, unlike most of their Asian counterparts, are well-off economically. Government price supports heavily subsidize rice prices, quotas severely restrict the import of competing agricultural products, and many farmers work part-time in nonagricultural jobs. All these circumstances help keep the standard of living very high for Japanese farmers. Nevertheless, the younger generation is leaving the farm in unprecedented numbers. The average age of farmers rose from 41 to 51 from 1960 to 1980.

The side effects of farm subsidies and inefficiencies in the Japanese agricultural sector mean that Japanese consumers spend a far greater share of their disposable income on food (roughly one-third) than do western Europeans (one-fifth) or Americans (one-sixth). Some might argue that given population pressures and the limited amount of arable land, such high prices might be expected. It is true that 75 percent of Japan is forested and the population, roughly half that of the United States, lives in a fragmented land area roughly the size of the fifth largest state in the United States (between Montana and Nevada in size). But farmers have been slow to readjust to grow products that are in greater demand. In the meantime, consumers increasingly turn to western-style diets that require importing greater quantities of meat, resulting in greater surpluses of rice.

Wheat and livestock and other products in great demand in Japan could be grown more widely on fields presently producing rice. But tradition, red tape, and price supports favor the status quo. For example, the price support that Japanese farmers receive for rice is six times that available to American farmers. At the same time, import restrictions are levied on over 20 agricultural products, including beef, oranges, orange juice, peanuts, cheese, tobacco, and certain fish. The products that are imported are very expensive to buy. Major trading companies, *sogo sosha*, handle most of the imported food. See Chapter 13 for an explanation of the *sogo sosha*. "It suits them to buy cheap foodstuffs from efficient American, European, Australian, and Brazilian suppliers and sell them expensively alongside costly equivalences produced locally by inefficient Japanese farmers."[2] Moreover, the government in power in the post–World War II era has received a large share of its support from the

farmer, further decreasing the likelihood of reforms in current subsidy arrangements.

Over two-thirds of Japanese farms are less than $2\frac{1}{2}$ acres in size. Only 2 percent are larger than 7.5 acres in size. Fully 87 percent of the 47 million farm families depend on outside nonfarm jobs for support. As a result, farmers earn on the average over 12 percent higher incomes annually than do nonagricultural workers.

Agriculture Patterns

Despite inefficiencies, Japanese agriculture is changing and production of food crops has increased rapidly in recent years. Sophisticated specialized farm machines (tractors, combines, trucks, and self-propelled seeders and cultivators) are widely available (Figure 4-7). About 10,000 efficient cooperatives provide advice and buying services for seed and fertilizer as well as marketing assistance in selling products.

The tremendous scale of intensive greenhouse fruit and vegetable production accounts in large measure for rising production levels in Japan. Improved imported dairy cattle and beef stock breeding animals are raising livestock quality. Farm cooperatives have been particularly successful in developing networks of fish farms to grow freshwater fish. Rice fields are being converted to grape vineyards to support a rapidly expanding wine industry, but upland grapes appear to produce the best crops (Figure 4-8).

The country will generally remain self-sufficient in vegetables, pork, poultry, egg, and milk production in the future (Table 4-1). The largest deficit products appear to be soybeans and wheat, but if the figures in

Table 4-1

Japanese Self-sufficiency Ratios for Selected Farm Products, 1978–1990 (percent)

Product	1978	1990 (Projection)
Rice	111	100
Wheat	6	19
Barley	14	17
Soybeans	5	8
Vegetables	97	99
Fruits	78	83
Meat	80	83
Beef	73	71
Pork	90	95
Poultry	94	96
Eggs	97	99
Milk and milk products	89	89
Sugar	22	32
Summary figures		
Food	73	73
Foodstuffs	29	35
Grain	34	30

Source: John Lewis, "The Real Security Issue—Rice," *Far Eastern Economic Review,* June 19, 1981, p. 70.

[2]"Japanese Farmyard Follies," *The Economist,* July 17, 1982, p. 77.

Figure 4-7 Mechanized Rice Thrashing in Japan. *Modern specialized equipment, such as this self-propelled rice harvesting machine cuts, thrashes, cleans, and bags rice, as well as prepares bundles of straw tied with twine. (T.A.H.)*

Figure 4-8 Japanese Vineyard. *Since Japanese farmers produce a great surplus of rice, many have converted former rice paddies to grape vineyards. Note the screening over the grape crop shown here on the Kanto Plain northwest of Tokyo, which continues as a major rice-producing region. (T.A.H.)*

Figure 4-9 Japanese Greenhouse Farming. *Many vegetable, fruit, and flower items are grown in greenhouses located in rural and suburban settings in Japan. Fresh vegetables are in great demand as the Japanese diet typically includes them three meals a day. Note chickens scavenging outside greenhouse. (T.A.H.)*

Table 4-1 prove accurate, the self-sufficiency rate for these products will increase by 1990 as rice surpluses decline. Sugar self-sufficiency rates have also increased during the 1980s.

Greenhouse crop production began in Japan in the early twentieth century in shelters covered with oiled paper or glass. By the mid-1950s plastic vinyl film came into wide use (Figure 4-9). In the early 1970s over 125,000 acres of farmland were covered by vinyl-lined tunnels that grew a tremendous variety of fresh fruits and vegetable crops, including strawberries, cucumbers, tomatoes, lettuce, melons, and peas. Another 50,000 acres of greenhouses in suburban and urban settings provided additional produce.

Market gardens in Japan produce vegetables, fruit, and flowers in close proximity to all major urban markets. Corn, taro, sweet potatoes, and other garden crops can be seen all over the landscape interspersed with rice fields. Plastic sheets that hold in heat and restrict weed growth, placed between planted rows in fields, are another technique used to increase yields. Truck farming occurs on rural margins of major population centers and in upland hilly areas. Truck farms produce cabbage, Chinese cabbage, Japanese rad-ishes, eggplant, green paprika, and ginger, among other products. As in China and India, swine and hens are common farm animals, typically living on waste, as they do elsewhere in Asia.

SUMMARY AND CONCLUSION

Despite great success in producing high yields from intensive cultivation of rice in east and south Asia, considerable pressure exists to increase yields even more to support rapidly growing populations. Many agricultural practices, such as double cropping, inter-culture, and mechanization, assist this process, but the Green Revolution has been most successful in expanding output. The Green Revolution, however, has failed to reduce poverty levels in many countries because benefits accrue primarily to wealthy agriculturalists, not to the landless peasants. The shift to cash crop output for foreign markets rather than producing food for domestic consumption in some cases also contributes to a lingering food problem.

Very large portions of the population of China and India depend on agriculture as a livelihood, whereas only a small minority do so in Japan. In recent years, Indian and Chinese farmers have been very successful in expanding output. Reforms affecting the collective agricultural system in China have led to greater flexibility at the local level and have provided more incentives to increase productivity. China is, in fact, a net agricultural exporter today. Japanese farmers benefit from a generous price support system, but farms remain small and inefficient by international standards. The partial shift to western-style diets in Japan has encouraged the expansion of vegetable, egg, meat, and milk production, but rice output still dominates.

SUGGESTIONS FOR FURTHER READING

BARKER, RANDOLPH, and RADHA SINHA, *The Chinese Agricultural Economy.* Boulder, Colo.: Westview Press, 1982.

BARKER, RANDOLPH, ET AL., *The Rice Economy of Asia.* Washington, D.C.: Resources for the Future, 1985.

BAYLISS-SMITH, TIM P., and WANMALI SUDHIR, eds., *Understanding Green Revolutions: Agrarian Change and Development Planning in South Asia.* New York: Cambridge University Press, 1984.

HANKS, LUCIEN, *Rice and Man: Agricultural Ecology in Southeast Asia.* Arlington Heights, Ill.: Harlan Davidson, 1972.

HANSEN, GARY E., *Agricultural and Rural Development in Indonesia.* Boulder, Colo.: Westview Press, 1981.

HAYAMI, YUJIRO, and MASAO KIKUCHI, *Asian Village Economy at the Crossroads: An Economic Approach to Institutional Change.* Baltimore, Md.: Johns Hopkins University Press, 1982.

HAYAMI, YUJIRO, ET AL., eds., *Agricultural Growth in Japan, Taiwan, Korea, and the Philippines.* Honolulu: University of Hawaii Press, 1979.

ISLAM, M. NURL, ET AL., eds., *Rural Energy to Meet Development Needs: Asian Village Approaches.* Boulder, Colo.: Westview Press, 1984.

KOJIMA, KIYOSHI, and TERUTOMO OZAWA, *Japan's General Trading Companies: Merchants of Economic Development.* Paris: Organization for Economic Co-operation and Development, 1985.

LARDY, NICHOLAS R., *Agriculture in China's Modern Economic Development.* New York: Cambridge University Press, 1983.

PERRY, ELIZABETH I., and CHRISTINE WONG, eds., *The Political Economy of Reform in Post-Mao China.* Cambridge, Mass.: Harvard University Press, 1985.

SMIL, VACLAV, *The Bad Earth, Environmental Degradation in China.* Armonk, N.Y.: M. E. Sharpe, 1984.

WITTFOGEL, KARL, *Agriculture: A Key to the Understanding of Chinese Society, Past and Present.* Canberra, Australia: Australian National University Press, 1970.

WORLD BANK, *China, Long-Term Development Issues and Options: A World Bank Country Economic Report.* Baltimore: Johns Hopkins University Press, 1985.

5 Dairying and Mixed Farming

Commercial agriculture in western Europe and North America focuses on the production of the dairy, meat, and grain products preferred in western-style diets. Market conditions, physical constraints, and tradition largely explain the pattern. The von Thünen conceptual model of agricultural land use (discussed below) provides insight into the location of major production regions. In the Soviet Union, growing conditions are more restrictive, limiting production options. Although output focuses on similar products (grain, dairy products), significant differences occur because of the collective production system. Soviet agriculture is much more labor intensive and undercapitalized than that in western Europe and North America, and separate, distinctive producing regions for various products are less common.

CONCEPTUAL OVERVIEW

Commercial agricultural location patterns in the mid-latitudes in North America and western Europe conform remarkably well to the von Thünen agricultural land use model developed in Germany early in the nineteenth century. Von Thünen observed that rings of agricultural production occurred around major urban market centers (Figure 5-1). Perishable items in strong demand and those products with high transportation costs captured locations close to the city. These items could compete favorably on higher-priced land near the center because of higher market prices. Farther from the market, products less perishable in nature, with lower transport costs and lower market prices, predominated. In remote locations the most distant from the market, extensive agriculture, including grazing, replaced more intensive grain production and general farming.

Von Thünen argued that three factors influenced the type of production at any particular location: (1) distance to market, (2) selling price of product at the market, and (3) *land rent,* which is roughly equivalent to *economic rent* in classical economics. Economic rent is the revenue a farmer receives after deducting the cost of production of a commodity. The principles that von Thünen developed still hold, but the product produced at a given location has, in some cases, changed over time as a result of the different market demand and technology considerations. For example, wood production captured a more prominent close-in location in von Thünen's time because of strong dependence on firewood for domestic heating and cooking and high costs of transportation.

GRAPHING BID-RENT CURVES

The relationship between economic rent and distance from the market for one or several products can easily be demonstrated in graphic form using a *bid-rent curve.* The bid-rent curve is the line showing the economic return at varying distances from the market (see Fig-

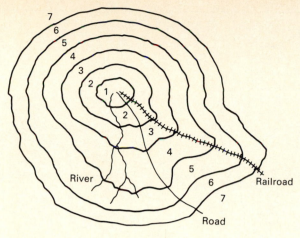

Zones of land use

1. Urban area
2. Market gardening
3. Dairying (fluid milk)
4. Dairy products (butter, cheese)
5. Grain production (wheat, corn)
6. Livestock and general farming
7. Grazing

Figure 5-1 Von Thünen Agricultural Land Use Model. *This model helps explain the rings of agricultural production that occur around major urban markets; it works remarkably well in accounting for the commercial agricultural patterns found in North America and Western Europe.*

In comparing the bid-rent curves of two products, the basis for rings of production around the market can be determined. In Figure 5-2B a bid-rent line is shown for two products (wheat and peas). The curve for peas starts out in a higher position than that of wheat because its market price is significantly higher than that for wheat ($150), more than offsetting the increased cost of production ($75). A return of $75 is indicated at the market. Transportation charges are also higher for peas than for wheat. Five miles from the market, $50 of transportation charges are incurred, leaving economic rent levels at $25. Ten miles from the market another $50 charge occurs, plunging the rent level below zero. Connecting the points with a straight line reveals that it is uneconomical to produce peas beyond 5 miles from the market because at greater distances wheat becomes more profitable. The curve in top position at any given distance from the market will give the highest return. In this example, peas will be produced closest to the market and wheat farther away.

In North America today we can observe agricultural regions roughly recreating the von Thünen model. Broad regional agricultural zones spread out from the urban core regions of Ontario, New York, New Jersey, and Pennsylvania (Figure 5-3). *Specialty farming* and *dairy farming* predominate in this zone. A

ure 5-2). The bid-rent curve slopes down to the right for each item because additional transportation costs occur with greater distance from the market. For example, production located at the market would incur no transportation costs. At any given distance away from the market, economic rent would decrease in direct proportion to transportation cost increases. The height of the curve at any distance can be determined by subtracting production and transportation costs from the price received for the product at the market. Figure 5-2A shows the economic rent return from a ton of wheat at the market as $50. This figure can be determined by subtracting the cost of production ($50) from the selling price at the market ($100), yielding an economic rent of $50. At any distance from the market the rent decreases in proportion to the level of transportation charges. Assuming that transportation costs are $5 per mile, 5 miles from the market, an additional $25 decrease in rent occurs, leaving a return of $25 (see Figure 5-2A). At 10 miles, with an added transportation charge of $25, the rent becomes zero, making it uneconomical to produce wheat at any greater distance from the market, where returns would become negative. Any increase in market prices would contract the production area. Changes in transportation charges would similarly alter production areas. As shipment costs increased, the curve would become steeper and contract the production area. Decreases in rates would flatten the curve, encouraging production at greater distances.

Figure 5-2 Agricultural Rent Gradients. *Using a bid-rent curve one can determine economical locations for the production of specific products. In the example provided here, the preferred production areas for two products can be determined. (Redrafted from Alonso, Papers and Proceedings, Regional Science Association, 6, 1960)*

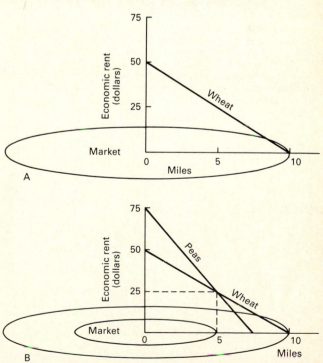

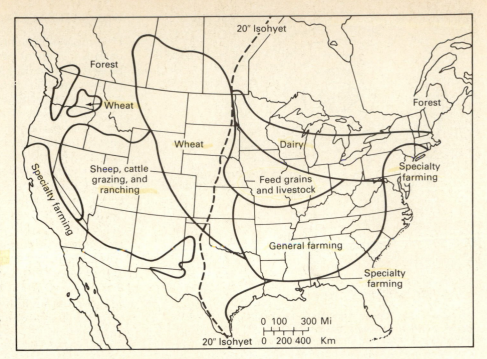

Figure 5-3 North American Agricultural Regions, after von Thünen. *The broad regional agricultural production areas shown here spread outward from the specialty farming and dairy areas in the East, through feed grain and livestock and wheat areas to grazing and ranching in the West.*

Figure 5-4 Comparative Agricultural Situations for North America and the Soviet Union. *Agricultural activity in the Soviet Union faces a more restrictive physical environment than in North America. Most of the productive land in the Soviet Union lies within a fertile triangle, the vertices of which include Leningrad in the north, Odessa in the south, and Novosibirsk to the east, shown on the map here by the diagonal lines. The shaded area on the map shows the Soviet Union superimposed on a map of North America. Note the relative locations of North American and Soviet cities. (Source: Central Intelligence Agency, U.S.S.R. Agricultural Atlas, 1974)*

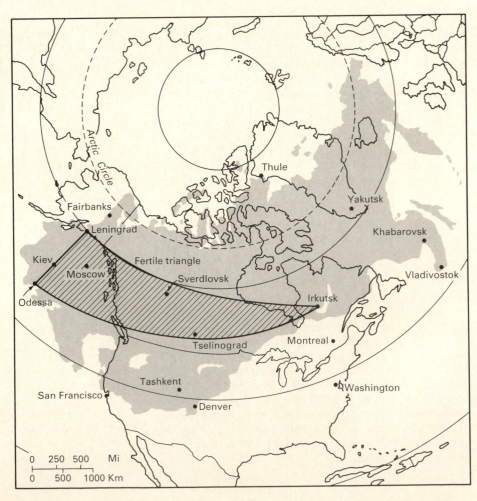

feed grain and livestock belt in the middle west lies adjacent to that zone. Corn and soybean crops provide feed for cattle and hogs, also produced on the farm, hence the label "feed grain and livestock region" for that area. Wheat and other small-grain production occupies an adjacent zone farther west, together with a broad general farming region to the south. An extensive sheep, cattle grazing, and ranching zone occupies the western intermountain zone. Specialty farming regions envelope the west and Gulf coasts.

Interruptions to these sweeping macro regions occur at the local scale due to the presence of other market centers. Market garden and dairy production zones, for example, occur around all major metropolitan markets, interrupting broad regional patterns. The intensity and form of agricultural operations also varies regionally according to population density, land value, and climatic considerations. Perhaps the greatest variety of activities occurs in the coastal specialty crop zone. Examples of specialty crops include citrus production in Florida, tobacco on the Atlantic coastal plain, and peanut output on the interior Gulf coastal plain. Specialty crops and poultry farming provide the focus for Chapter 7.

Similar broad regional patterns according to the von Thünen model prevail in western Europe, with a dairy zone focused on Great Britain, northern France, the Netherlands, Belgium, and Denmark; small grains extending eastward from France across the north German plain through eastern Europe; and grazing (especially sheep) occurring in upland areas of Spain and Great Britain. The growing season in western Europe is too short to allow corn to fully mature north of Italy, although it can be grown as a forage crop. Dry summers in southern Europe restrict output to small grains, sunflowers, grapes, and olives, among other crops in the Mediterranean region, unless irrigation is provided.

Agricultural activity in the Soviet Union also faces a more restrictive physical environment than does that found in North America. Most of the productive land lies within a fertile triangle the vertices of which include Leningrad in the north, Odessa in the south, and Novosibirsk in the east (see Figure 5-4). Outside this realm, conditions are not favorable for agriculture. To the north, the growing season is too short and temperatures too cold, and to the south, rainfall is too low. Comparing the setting of the Soviet Union with

Figure 5-5 Soviet Agriculture Vegetation Regions. *Only two of the five regions (C and D) on this map identify with prominent agriculture areas in the Soviet Union. They roughly correspond with the fertile triangle area. Areas A and B have short or non-existent growing seasons and Area E is too dry, except where it is irrigated. (Source: Central Intelligence Agency,* U.S.S.R. Agricultural Atlas, *1974)*

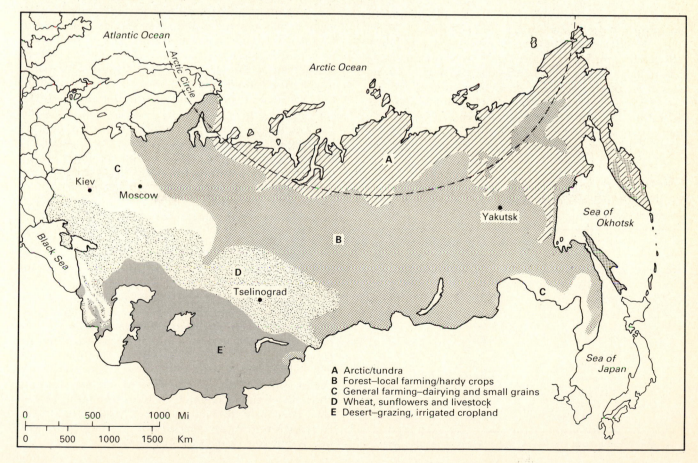

A Arctic/tundra
B Forest–local farming/hardy crops
C General farming–dairying and small grains
D Wheat, sunflowers and livestock
E Desert–grazing, irrigated cropland

that of the United States and Canada in terms of latitude shows how much farther north the core agricultural areas lie in the Soviet Union. Within the fertile triangle producing area, agricultural zones favor milk and meat livestock in the north, together with small grain and potato production (Figure 5-5 on page 51). Farther south, wheat begins to dominate, together with swine and cattle production, especially in the Ukraine. Toward the east, other small grains predominate, including oats, barley, and rye, and to the extreme south, outside the fertile triangle, cotton and other specialty crops are locally produced on irrigated land.

DAIRYING AND MIXED-FARMING INDUSTRY

In the remaining portion of this chapter, emphasis is placed on a closer examination of the dairying and mixed-farming regions of North America, western Europe, and the Soviet Union. Dairy-producing regions are more specialized in the United States and western Europe than in the Soviet Union, but the latter leads in world milk production. Mixed farming is much more productive in the United States than in western Europe and the Soviet Union because of more favorable climatic and soil conditions.

North American Dairy Belt

Dairy farming in North America extends from the St. Lawrence valley in the northeast westward 2000 miles to the Minnesota/Manitoba border, including much of Vermont, upstate New York, southern Ontario, Michigan, Wisconsin, Minnesota, Pennsylvania, and Ohio. The leading milk-producing states are shown in Table 5-1. In the eastern half of the dairy-producing zone and in secondary production zones around all the major metropolitan regions in this belt, fluid milk is produced for the market. In the western reaches of this

Table 5-1

U.S. Fluid Milk Production, by State, 1982
(millions of dollars)

Rank	State	Value of Shipments	Percent of U.S. Total
1	California	2,561	13
2	Ohio	1,321	7
3	Texas	1,102	6
4	New York	1,046	6
5	Pennsylvania	833	4
6	Florida	717	4
7	Michigan	697	4
8	Minnesota	653	3
9	Wisconsin	619	3
	Subtotal	9,549	50
	U.S. total	19,028	100

Source: U.S. Bureau of the Census, *1982 Census of Manufactures.*

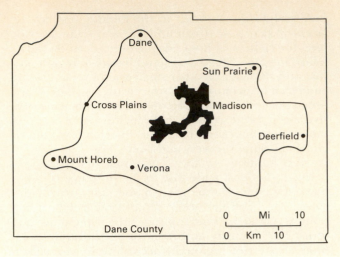

Figure 5-6 Madison, Wisconsin Milkshed. *This diagram delimits the region from which milk traditionally originated for the Madison market. More urban development in the urban fringe and improved refrigeration and transportation facilities have extended the region beyond these limits in recent years.*

belt, the fluid milk market share declines and milk for manufacture into various dairy products predominates. Surplus milk, even in fluid-production zones, can also be diverted into manufactured products. In fact, great surpluses of milk are produced today in the dairy belt, creating enormous problems for the processor, the government, and the farmer.

Secondary milk-producing areas form *milk sheds* around major metropolitan markets as discussed in Chapter 1. A *milk shed* is the local area around a city from which it receives its fluid milk (Figure 5-6). In the northeast, the large metropolitan market created by the megalopolis causes these areas to coalesce into one tremendous producing region, but elsewhere, local belts emerge.

The large size of the southern California market has propelled Los Angeles County to the first-ranking fluid milk–producing region in the country. Overall, California now ranks as the first milk-producing state, significantly outdistancing Ohio, Texas, and New York (Table 5-1). The relatively high ranking of Texas can be explained by the large size of its market.

Dairy Product Production Patterns. Among milk-producing states, further regional specializations occur with dairy product processing activities. Ice cream production generally occurs relatively close to the market, with cheese products occupying intermediate locations, and butter production located the farthest from the market (Table 5-2). The leading ice cream–producing state in 1982 was California, with Pennsylvania ranking second. Wisconsin dominated cheese production in the United States, shipping about three times as much as second-ranking Minnesota. Wisconsin also led in butter production, but in this case Minnesota was a far stronger competitor, with nearly one-fourth of total production.

Table 5-2

Leading Dairy Product–Producing States, 1982
(millions of dollars)

Product	Rank	State	Value of Shipments	Percent of U.S. Total
Ice cream	1	California	352	12
and frozen	2	Pennsylvania	244	9
desserts		Subtotal	596	21
industry		U.S. total	2,855	100
Cheese	1	Wisconsin	4,116	38
industry	2	Minnesota	1,385	13
	3	Missouri	1,057	10
		Subtotal	6,558	61
		U.S. total	10,763	100
Butter	1	Wisconsin	553	33
industry	2	Minnesota	375	22
		Subtotal	928	55
		U.S. total	1,687	100

Source: U.S. Bureau of the Census, *1982 Census of Manufactures.*

Production and transportation characteristics explain this patterning. It takes a pound of milk to make a pound of ice cream, whereas it might take about 10 pounds of milk to make a pound of cheese. A pound of butter, in turn, requires 20 pounds of milk. This makes butter the most economical and concentrated form for transportation purposes. Therefore, butter production thrives at the greatest distance from its market. The principle demonstrated by these statistics is that if a raw material is turned into a product higher in value per unit of weight, it can withstand the greater transport costs to market that the longer distances entail. Second, by converting the product (raw milk) into a more concentrated processed form, its market shelf life is extended, as deterioration is retarded.

Evolution of Dairy Farming. The highly specialized dairy cows now bred worldwide for milk production are all native to western Europe. The most popular cow, the Holstein-Friesian, originated in the coastal lowlands of the Netherlands and Germany. The first native-born black-and-white Holstein calf in the United States was born in Madison County, New York, in 1890. Other popular dairy cows include the Guernsey and Jersey from the Channel Islands, Ayrshires from Scotland, and Brown Swiss from Switzerland.

Dairy farming remains a labor-intensive activity overall because of the large amount of care that cows require. Not only do they not produce milk until they are over 2 years of age, but once in production, cows must be milked at least twice a day. Moreover, the dairy belt in the United States is situated in a severe winter region. Cows remain indoors at least 5 to 6 months a year, and the trend is for indoor housing throughout the year. Even though modern milking parlors and loafing barns typically have automated feeding and cleaning apparatus, constant monitoring is required (Figure 5-7).

Higher prices for milk consumed as fluid rather than in processed form encourage its production closer to the market. Federal milk marketing orders, government price supports, and the elaborate spatial-economic integration of feed, equipment, breeding, and health services industries that serve the dairy in-

Figure 5-7 Modern Diary Loafing Barns. *This dairy farm operation shows a classic mix of farm buildings including the 2-story main barn designed for hay storage on the upper level, several large silos for silage, and other storage buildings. Unlike the situation here where the Holsteins graze in pasture, many farms house cattle in loafing barns throughout the day to enhance milk production. (Grant Heilman Photography)*

dustry reinforce the status quo. The truck-based refrigerated collection and distribution network, typically operated by farmer-owned cooperatives, ties the system together.

Dairy farms have grown in size in recent years as mechanization allowed for the cultivation of more land per unit. They now average about 300 acres in the United States and Canada, making them comparable in size with feed grain units in the middle west, discussed later in this chapter. Dairy farms have decreased dramatically in number in recent decades as many units have gone out of production, and the size and output of the remaining units has increased accordingly through consolidation. The number of dairy farms in the United States exceeded 600,000 in 1950, whereas fewer than 200,000 exist today.

The typical dairy farm in North American dairy areas has 100 milk cows and nearly as many young female stock, called heifers, raised as replacement milkers. Each farm produces about 15,000 pounds of milk per cow per year. Milk prices received by farmers doubled in the 1970s and averaged about $13.00 per hundred pounds in the early 1980s, but prices have slipped in recent years and dairy farmers are losing money in greater numbers. Growing quantities of surplus milk add to the problem. Traditionally, farmers have increased production to offset price decreases. Federal milk subsidies also affect prices and production. Disincentives to increased production, including complete herd buyouts, are now being incorporated into federal programs.

Elaborate buildings and expensive machinery require large capital investments and give dairy farms very high market values (Figure 5-8). A typical dairy farm today has several feed silos and machinery storage sheds in addition to the dairy barn itself. Chopped forage for feed is stored in the silos as silage, usually made from corn, hay, or sorghum. The configuration of the dairy barn has changed dramatically in the past 20 years as the need for lofts to store baled hay has declined. Rather than large second-story hay lofts, modern barns are broad, sprawling one-story shed-type structures, often covering several acres, that combine milking parlors with loafing pens.

Machinery needs for the dairy farm go far beyond those required for feeding, milking, and cleaning responsibilities. The cost of bulk milk cooling and storage tanks alone can run into hundreds of thousands of dollars. In addition, dairy farms typically have several large tractors, plows, harvesters, wagons, and trucks. Dairy animals themselves represent a large investment. Mature cows sold for milk production have a value of about $1500 each. No wonder the indebtedness of dairy farmers has risen dramatically in recent years. The average equity of a U.S. dairy farmer today is about 60 percent on a total investment averaging $600,000.

The large array of equipment owned by dairy farmers is used to produce a variety of crops. If it were not for the dominance of milk-sales revenue receipts, dairy farmers would easily qualify as mixed or general farmers. The most important crops grown by farmers are hay, corn, oats, and wheat. Pastureland also remains a mainstay on dairy farms, especially for the rearing of young stock as replacement animals. Farmers have intensified their feeding operations in recent

Figure 5-8 Milking Parlor. *This automated milking parlor allows one person to milk several cows at one time. Many parlors include a computer controlled feeding during milking, based on the production level of the particular cow. (Grant Heilman Photography)*

years to become more efficient. The pastureland this strategy releases becomes cropland, where feasible, making operations more productive.

It should be pointed out that dairy farms typically do not occupy the best farmland. Glacial deposition left the rolling topography typically occupied by these farms very stony, and steep slopes not suitable for cropping can be pastured. Heavy use of lime and fertilizer are necessary to build up fertility levels on these marginal soils. Farmer-owned cooperatives and government extension agents have assisted farmers become much more efficient over the years through the introduction of improved agricultural practices, including the use of drainage tiles in fields, crop rotation, contour farming, improved seeds, and better cattle-feeding programs.

Unlike dairy operations in the northeast, milk production in southern California in the Los Angeles basin typically occurs as an intensive feedlot operation, relying on imported grain and alfalfa grown on irrigated fields in the Central and Imperial valleys of California or imported from Arizona and elsewhere in the southwest. Very large milking parlors produce enormous quantities of milk in a limited number of independent units. A few such operations have over 10,000 cows each. Similar high-intensity operations now occur in Florida and Arizona, serving growing markets in these areas.

Soviet Dairy Industry

The Soviet dairy industry cannot be as clearly delineated as that in the United States because the role of cattle is different and production areas are more widely distributed. The distinctions between beef and dairy cattle that exist in the United States do not occur in the Soviet Union. All cattle are milked and also used as beef animals in the Soviet Union. Only one-fourth of all cows are milked as dairy cattle in the United States. It is not surprising to learn, therefore, that milk production in the Soviet Union per cow is half that of the United States.

Cattle husbandry in the Soviet Union generally conforms to the population distribution, with virtually all agricultural areas having some production. This situation contrasts dramatically with that in the United States, where much more specialization occurs. Much of the cattle and milk output in the Soviet Union now occurs in small private plots farmed by both industrial workers and farmers themselves on land adjacent to their homes.

Notwithstanding the wide distribution of cattle rearing in the Soviet Union and the large concentration of cattle in the Ukraine, a belt of predominantly dairy farming activity exists in extreme northwestern Russia, encompassing Latvia, Lithuania, and Estonia on the west, extending to the east of Leningrad. This agricultural production area is comparable to that in the northeastern United States. In much of northern Europe and the Soviet Union, potatoes are widely grown as an industrial crop and used for both animal feed and human consumption. Cattle and hogs feed on potatoes in those areas, much as they depend on corn for feed in the United States.

Mixed Farming

Mixed farming encompasses much of the eastern United States, western Europe, and large portions of the fertile triangle in the Soviet Union. Crops and animals occur in various combinations throughout these regions, but the role of crops is particularly crucial in that they provide multiple roles as feed for animals, as a cash crop, and as a food supply for farm families.

U.S. Corn Belt. Corn is the premier feed crop in the United States. Corn production occurs extensively in the middle west, extending nearly 1000 miles from central Ohio westward through Indiana, Illinois, and Iowa and into southern Minnesota and eastern Nebraska—hence the label *corn belt* for this region. In the United States as a whole, corn occupies around one-fourth of all harvested cropland, but in the corn belt it often reaches or exceeds 50 percent. But more than corn grows in the corn belt, the reason the region is labeled a feed grain and livestock area in Figure 5-3.

Wherever possible, U.S. farmers prefer to grow corn both because of its effectiveness in fattening animals and its high yield per acre. The lofty status of corn is thus reflected in its ability to command locations ideal for its production. Corn commonly yields 120 to 130 bushels or more per acre in Iowa and Illinois, whereas soybeans and wheat produce half the yield per acre. On the other hand, soybeans are also grown widely because they command more than twice the price of a bushel of corn. A comparison of U.S. corn and soybean production areas shows similar distributional patterns of the two crops, the major difference being the greater emphasis on soybeans in the lower Mississippi valley (Figure 5-9). Other crops produced in the corn belt include oats, wheat, and hay. Often, these crops are rotated in a multiyear cycle, but with increasingly heavy applications of chemical fertilizers and pesticides, farmers can sustain corn production every year on the same field.

At the northern and eastern boundaries of the corn belt, a transition to dairy farming occurs. Poorer soils and a shorter growing season are important in defining this transition to the north. Proximity to northeastern corridor metropolitan markets (Boston–Washington, D.C.) and the more rugged topography of the Allegheny plateau make dairying more competitive to the east. On the western boundary, the major factor is precipitation. Corn requires a warm, humid climate during the growing season. The 20-inch average annual rainfall line (isohyet) generally defines this boundary. Although the location of the boundary varies from year to year, it generally runs

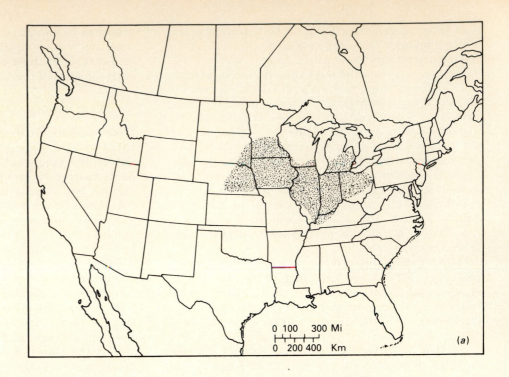

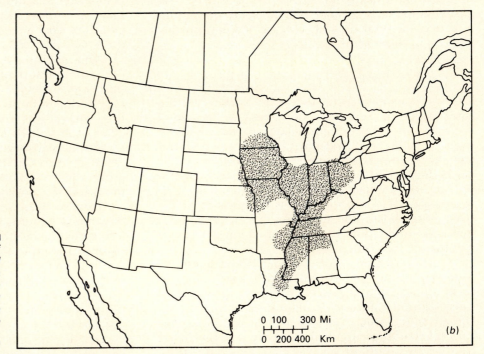

Figure 5-9 Corn and Soybean Farming in the United States. *A comparison of major U.S. corn production (a) and soybean production (b) areas shows a similar spatial distribution for each crop, with the major difference being the greater emphasis of soybeans in the lower Mississippi Valley and the general absence of corn in that area. (Source: U.S. Census of Agriculture)*

north and south through central Nebraska (Figure 5-3). Farther west, rainfall totals drop off dramatically; wheat production competes better on this drier land unless irrigation is available (see Figure 6-2 and the discussion in Chapter 6). To the south, a transition to general farming occurs, characterized by smaller units practicing less specialized cultivation and more grazing. Upland rolling-hill topography and poorer soils make corn less competitive in that direction.

The most significant cropping change in the U.S. corn belt region in recent decades has been the phe-

nomenal expansion of soybean production. The soybean provides an extremely high quality source of protein; therefore, the market for soybeans, both as a cash crop and an animal food, continues to expand. Separated into oil and meat, the soybean has hundreds of industrial and commercial uses, from paint and plastics to cosmetics and candy, in addition to its growing use as a cattle, hog, and pet food. In fact, 60 percent of the world soybean output now comes from the United States and it has become a major U.S. farm export.

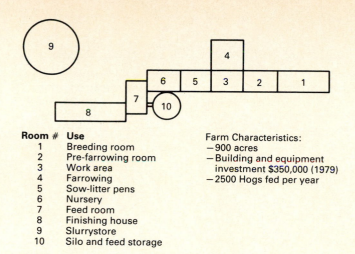

Room #	Use
1	Breeding room
2	Pre-farrowing room
3	Work area
4	Farrowing
5	Sow-litter pens
6	Nursery
7	Feed room
8	Finishing house
9	Slurrystore
10	Silo and feed storage

Farm Characteristics:
- 900 acres
- Building and equipment investment $350,000 (1979)
- 2500 Hogs fed per year

Figure 5-10 Hog Farm Production Facilities. *The specialized hog farm building facilities include several specialized and automated rearing and feeding rooms as shown on this schematic diagram. These operations are typical of a large farm in the feed grain belt as discussed in the accompanying Box.*

U.S. Feed Grain and Livestock Region. In the heart of the U.S. corn belt, sales of beef cattle and hogs bring in the most revenue for the farmer.[1] The importance of this activity often leads observers to label the area as a *feed grain and livestock region* rather than simply a corn belt, as mentioned earlier. The correlation between the locational pattern of hogs sold and the mixed-farming region, extending from eastern Nebraska to Ohio, is very strong. We therefore use the term "feed grain and livestock region" to refer to this area in the remainder of the chapter. Indeed, this area is best understood in the context of a system having multiple roles in which various crops and livestock play specific parts. The system analogy extends downward to the individual farmstead, which has evolved into an industrial feedlot in many instances (see the accompanying box).

Iowa farmers alone owned 14 million hogs in 1982, over twice as many as are raised in any other state (Table 5-3). Few other areas of the United States have large concentrations of hogs. Only one other state, Illinois, had over 5 million, but only 12 other states, mainly in the corn belt, had over 1 million hogs at that time. The pattern of swine production in the Soviet Union and western Europe is far less localized, owing to a broader range of feeds utilized and the fact that pigs are often fed garbage and reared as scavengers, greatly extending their range.

As mentioned earlier, the distribution of cattle in the United States displays several distinct concentrations of specialized animals. Milk animals predominate in the dairy region and mature beef cattle in the feed grain and livestock region (Table 5-4), but beef

[1]In the United States, the term "hogs" refers generally to mature pigs, whereas the term "swine" refers to the entire pig population.

cows are not limited to the latter area. The beef cattle distribution exhibits concentrations in three broad production regions that encompass the nine states that had more than 1 million beef animals each in 1982: (1) the feed grain and livestock area of the middle west, encompassing all or part of the states of Missouri, Nebraska, Iowa, and Kansas; (2) the Plains states bordering the feed grain belt (Oklahoma, Texas, South Dakota, and Montana), where young cattle are reared as range (grazing) animals prior to shipment eastward to the core producing area; and (3) production in general farming areas where cropping is not as competitive (Florida and Kentucky), serving the growing southern market.

A major feature of the beef cattle distribution is the large concentration in Texas, which accounts for 15 percent of the beef cows in the United States today. Most of these animals are grown on the range, and this mode of agriculture is discussed in Chapter 6. The other major anomoly is Florida, which also produces a significant quantity of beef. A plentiful year-round grazing environment makes Florida a suitable producing region; the industry is somewhat new to the state because of the need in this growing area for tick-resistant cattle, which were not developed until the 1920s.

The dynamics of cattle-producing regions are also influenced by economics at various stages in the life cycle of the animal. In some instances an animal will be reared in three distinct regions during its lifetime; many calves are born on the range, in Wyoming or Colorado; later transported to Nebraska, Texas, or Missouri for growth; and finally shipped to an Iowa or Illinois feedlot for a final 90-day program of fattening on a high-protein grain diet. In this way cattle are moved to successive food sources, not vice versa. The movement pattern also takes them closer to eastern markets at each stage prior to slaughter. It should be noted, however, that slaughtering itself has shown the reverse locational tendency in recent years. Whereas large-scale slaughtering formerly occurred in large cities on the eastern margins of producing areas, such as Chicago, it has now moved westward, nearer the farmstead, to ensure greater quality control of the

Table 5-3

Leading U.S. Hog-Producing States, 1982 (millions)

Rank	State	Number	Percent of U.S. Total
1	Iowa	14	25
2	Illinois	6	11
3	Minnesota	5	8
4	Indiana	4	8
5	Nebraska	4	7
6	Missouri	3	6
	Subtotal	36	65
	U.S. total	55	100

Source: U.S. Bureau of the Census, *1982 Census of Agriculture.*

Table 5-4

Leading U.S. Cow-Producing States, 1982* (thousands)

	Beef Cows				Milk Cows		
Rank	State	Number	Percent of U.S. Total	Rank	State	Number	Percent of U.S. Total
1	Texas	5,223	15	1	Wisconsin	1,853	17
2	Nebraska	2,024	6	2	California	946	9
3	Missouri	1,933	6	3	New York	875	8
4	Oklahoma	1,830	5	4	Minnesota	839	8
5	South Dakota	1,596	5	5	Pennsylvania	691	6
6	Iowa	1,536	4		Subtotal	5,204	48
7	Montana	1,528	4		U.S. total	10,850	100
8	Kansas	1,524	4				
9	Florida	1,098	3				
10	Kentucky	983	3				
	Subtotal	19,275	56				
	U.S. total	34,203	100				

Source: U.S. Bureau of the Census, *1982 Census of Agriculture.*
*Excluding calves, bulls, and steers.

INDUSTRIALIZED HOG FARMING: CONFINEMENT FEEDING

Hog farming has changed dramatically in the past 25 years in the U.S. feed grain belt. Industrial methods, involving close confinement of animals, use of automated feeding and watering systems, use of advanced drugs and antibiotics, and mechanized manure removal has allowed for increased specialization and concentration in the business. Producers selling over 1000 hogs per year now account for about half the total U.S. output, compared with less than 10 percent in the mid-1960s. Moreover, the share of units producing 5000 or more hogs per year is growing rapidly.

Floor plans of hog production facilities demonstrate the specialization and automation now overtaking the industry (Figure 5-10). The facilities in a typical hog confinement operation are discussed below for a hog operation on a 900-acre farm in Illinois feeding 2500 hogs per year.

1. Breeding room
   a. Six sow pens with 14 sows (females) in each
   b. Three boar pens with 4 boars (males) in each, all equipped with automatic feeders and waterers
2. Pre-farrowing room
   a. Individual crates for each sow, with
   b. Automatic feeders and waterers
   c. Hogs removed 1 week before farrowing (giving birth)
3. Work area
   a. Feed storage
   b. Scrubbing area
   c. Office
4. Farrowing
   a. Individual crates for each sow
   b. Concrete floor heated to 90°F
   c. Separate feeders and waterers for piglets
   d. Sows and piglets housed here for 2 weeks
5. Sow-litter pens
   a. Two sows and litters (young) in each pen
   b. Piglets fed protein ration in addition to milk from their mothers
   c. Sows and piglets stay here 3 weeks until weaning; piglets then moved to nursery and sows moved to breeding house
6. Nursery
   a. Ten pens with 25 to 30 pigs per pen
   b. Automatic feeding and watering
   c. Pigs stay here until they weigh 85 pounds
7. Feed room
   a. High-moisture corn, commercial corn, and antibiotics ground and mixed
8. Finishing house
   a. Nine pens with 35 pigs each
   b. Pigs enter at 85 pounds and are sold for slaughter at 220 pounds
9. Slurrystore
   a. Manure storage tank, which handles manure in liquid form
   b. Manure pumped into tank from central pit under the workroom once a month
   c. Manure scraped into pit daily
   d. Pit holds 30,000 gallons
   e. Tank holds 300,000 gallons
   f. Manure spread on fields twice yearly

finished product and to save transportation costs. Live cattle are much more difficult and bulkier to ship than is processed meat. Greater use of refrigerated transport has also benefited this process.

Beef cattle breeding, like dairy cow breeding, began in Europe, especially in England and Scotland, to provide higher-quality meat for the English market. Shorthorn beef animals were first imported into the United States in the 1780s. Shorthorns are red, white, or roan (red and white) in color. In the early 1800s, Hereford imports came to the United States. Herefords are red animals with white faces. A third major breed is the Aberdeen Angus, introduced in the United States in the 1870s. These animals are black and constitute the second most important beef animal in the United States, after Herefords. Brahmans from India have been popular in the south because of their heat- and disease-resistant characteristics. Santa Gertrudis animals, an American breed, are a cross of Brahmans with Shorthorns.

Another distinctive feature of the U.S. feed grain farming region is its combination of ideal soils and growing conditions. Whereas in many midlatitude mixed-farming agricultural areas, including the eastern United States, western Europe, and the Soviet Union, *podzolic* soils predominate, *chernozem* soils occur in the corn belt. Podzolic soils are moderately fertile, typically acidic, heavy clay soils associated with broadleaf forest areas. Although they are better than the highly acidic, gray-brown podzolic soils developed in coniferous forest areas, they are not as fertile as the chernozems. The latter developed in grassland prairie rather than in areas with a heavy tree cover. Chernozems develop a deep friable black profile that is due to the presence of rich organic matter. Soils in eastern Europe and the Ukraine region of the Soviet Union are also chernozems, but generally they are not as intensively cultivated as in the United States because the moisture supply is not as plentiful. In Europe, plentiful rainfall occurs in areas with podzolic soils rather than in the chernozem area. Therefore, in Europe, poorer soils are utilized more intensively than the best ones, unlike the situation in the United States.

On many farmsteads in the U.S. feed grain belt, the feedlot has become a recent distinctive feature. In fact, just over half the cattle now produced in the United States come from just over 400 large-scale feedlots averaging over 30,000 animals each. Beef cattle and hogs are kept in pens housing a few up to a thousand animals, as we saw earlier. Supported by automated feeding and watering equipment, animals are fattened with high-protein rations in an intensively programmed management environment in these feedlots. Frequently, farmers buy yearling animals from western ranchers rather than breed their own replacements. In this way they specialize in "finishing" the animal for market.

Dramatic variations in scales of operation have emerged in the feed grain area in recent years as corporate farms have supplanted the traditional family farm. Over 60 percent of U.S. corn belt beef production is now finished in about 2000 feedlots, while the remaining 40 percent occurs on no fewer than 130,000 smaller operations. Some large operations now process 100,000 cattle or more at a time. Another emerging trend is for the feedlot itself to migrate westward to Colorado and elsewhere, although it depends on feed shipped from Iowa and Illinois.

This transition of the feed grain belt into an integrated agricultural industrial system has dramatically increased the investment in farms and been associated with increased indebtedness. Because of high interest rates and sluggish market demand, bankruptcies increased in the early 1980s. This situation further solidified the position of the largest operations, which may lead to a new wave of automation in the future. Improved biological knowledge and improved veterinary medicinal products are also on the horizon. Genetic engineering in particular offers great potential for the future by providing animals that may gain weight faster and have a higher proportion of lean tissue. Recombinant-DNA techniques (gene splicing) offers the most promise in this area.

Despite the productivity advantage of the middle western agricultural region, severe problems overtook that farming area in the 1980s as American agricultural exports dropped due to the strong dollar and as high interest rates, related to the rising national debt, severely squeezed the heavily indebted farmer. Farm foreclosures increased and farmland values plummeted in this period. In fact, the average value of cropland in the United States fell 13 percent in 1984–1985. The declines were the steepest in the heart of the feed grain belt, falling 29 percent in Iowa, 28 percent in Nebraska, and 27 percent in Illinois. Nevertheless, farmland values, at $1300 per acre in Illinois, remained double the U.S. average.

Soviet Mixed Farming. Soviet agriculture employs 20 percent of that country's labor force, whereas only 4 percent are similarly employed in the United States. Much more land is under cultivation in the Soviet Union than in the United States—one and one-half times as much—producing less product, about four-fifths of that in the United States. Nevertheless, the Soviet Union leads the world in the production of several cereal grain crops (rye, barley, wheat), cotton, potatoes, butter, milk, and others. Wheat and other small cereal grains dominate Soviet agriculture, whereas feed grains (especially corn) are dominant in the United States. The total harvested grain crop for the United States in 1982 was about 415 million metric tons, of which just over half was corn. At the same time, Soviet output of grain was 180 million metric tons.

Mixed farming in the Soviet Union typically involves wheat and other small-grain production, potatoes, sugar beets, and cattle. As mentioned earlier, corn is conspicuously absent due to the low rainfall and short growing season. Wheat accounts for about half of the grain production in the Soviet Union. Whereas the United States has shifted to soybean production to increase its output of protein, the Soviets have increasingly turned to the production of sunflowers, which, unlike soybeans, can tolerate a cooler climate. Soybean production continues to soar in the United States, while sunflower output has stabilized in the Soviet Union. Sunflowers are generally grown in hilly upland areas in the Ukraine, and winter wheat occupies the flat lowlands.

Rather than discussing wheat output in the Soviet Union in this chapter, livestock meat production will be covered together with a brief mention of potato, rye, and barley output. The Soviet wheat-producing area compares favorably with the North American commercial wheat grain-producing area and will therefore be discussed more appropriately in the next chapter.

Efforts in recent decades to accelerate meat production in the Soviet Union have been only marginally successful (Table 5-5). In this effort a tripling of feed grain consumption occurred but meat production less than doubled. This lack of efficiency in the conversion of grain to meat occurs because of the greater reliance on lower-quality protein sources (potatoes, wheat, rye, etc.) than in the United States, where corn and soybean grains predominate. Lackluster breeding improvements have also restricted Soviet output. Recall the statement made earlier that the same breeds of animals are used for meat and milk. A comparison of the per capita meat production rates shown in Table 5-6 indicates that the levels of the So-

Table 5-6

U.S. and Soviet Meat Production Per Capita, 1980 (kilograms)

	United States	Soviet Union
Beef and veal	44	26
Pork	36	19
Poultry	40	8
Total	120	53

Source: Lester R. Brown, *U.S. and Soviet Agriculture: The Shifting Balance of Power*, Worldwatch Paper 51, (Washington, D.C.: Worldwatch Institute, 1982). Data based on USDA Foreign Agricultural Service Report.

viet Union are roughly half those of the United States for beef, veal, and pork, and one-fifth for poultry.

While Soviet potato, grain, milk, and meat output in 1980 was higher than it was in 1960, production figures in the 1980s are typically equal to or below those of 1970. The tendency in the past decade has been for agricultural output to decline, unlike the situation in the United States (Table 5-6). The reasons for this are complex but generally not limited to environmental factors alone. The popular notion that Soviet agricultural output variability is due to alternating good and poor weather conditions is no longer generally supported. The Soviets now appear to be continuously in need of growing quantities of grain imports to feed the country. Public expectations regarding meat consumption have increased in the meantime, creating additional pressure. Some inefficiencies of the Soviet system will be discussed in a succeeding section, following a discussion of various types of collective agriculture producing units.

Collective Farming. Collectivization of agriculture began earnestly in the Soviet Union in 1928 under Stalin with the First Five-Year Plan. Two major types of organizational units created by collectivization now dominate Soviet agriculture in terms of land management and control: (1) the collective farm, or *kolkhoz,* and (2) the state farm, or *sovkhoz.* Each is quite distinctive in terms of function and location. The kolkhoz exists primarily in traditional agricultural areas, whereas the sovkhoz is more likely to be located in land only recently placed under cultivation in western Siberia and Kazakhstan. The kolkhoz units, although smaller, are more numerous than the sovkhoz units.

A kolkhoz collective unit comprises many farm families having perpetual rights to rent-free state land. Typically, there are 400 to 500 households per kolkhoz. The cultural tie with the past is maintained by the preservation of the individual farm homestead as the basic housing and operational unit. State farms are run more like rural farm factories, complete with dormitory-like apartment-style housing units. These units are state enterprises, and the employees earn

Table 5-5

Soviet Production of Basic Agricultural Commodities, 1960–1982* (millions of metric tons)

Year	Grains	Potatoes	Milk	Meat
1960	123	84.4	61.7	8.7
1965	114	88.7	72.6	10.0
1970	179	96.8	83.0	12.3
1971	174	92.7	83.2	13.3
1972	161	78.3	83.2	13.6
1973	214	<u>108.2</u>	88.3	13.5
1974	186	81.0	91.8	14.6
1975	134	88.7	90.8	15.0
1976	214	85.1	89.7	13.6
1977	188	83.6	<u>94.9</u>	14.7
1978	<u>229</u>	86.1	94.7	<u>15.5</u>
1979	174	91.0	93.3	15.3
1980	182	67.0	90.6	15.0
1981†	165	72.0	88.5	15.2
1982‡	170	82.0	83.3	14.6

Source: Lester R. Brown, *U.S. and Soviet Agriculture: The Shifting Balance of Power*, Worldwatch Paper 51 (Washington, D.C.: Worldwatch Institute, 1982). Data based on USDA Economic Research Service Report.
*Peak production year for each commodity is underlined.
†Preliminary.
‡Worldwatch estimate based on preliminary data.

wages. In the early 1970s, there were just over 32,000 kolkhozes and about 16,000 sovkhozes, but the latter averaged twice the size of the former and therefore the total land under cultivation by each was approximately the same.

The distinctions between the sovkhoz and the kolkhoz have diminished over the years. The kolkhoz worker did not formerly receive wages but now does. Sovkhozes are generally better equipped but less intensively cultivated, owing to their more marginal location in terms of rainfall and length of growing season. In the mid-1970s the average size of the sovkhoz was 40,000 acres, with about one-third of each unit under cropping. About 600 persons worked on each unit. In recent years many weaker kolkhozes, which averaged 13,000 acres in size in the 1970s, have been combined and converted to larger sovkhozes.

A third type of landholding, the *private plot*, although accounting for only a few percentage points of all land, produces a large quantity of dairy, meat, and poultry products. This right to private plots evolved in the past 20 years and became a part of the constitution in 1977. This development followed disappointing results from state farm reforms instituted in the 1950s. Kolkhoz residents produce the bulk of private-plot farm output, which accounts for about one-third of all agricultural production. Those plots function like gardens in terms of size and locational proximity to farm dwellings. Strict limits determine the number of animals a farmer can keep, usually a single cow and pig and several sheep or goats. Restrictions also limit the availability of feed and fertilizer, but the prospect of cash income from product sales provides the needed incentive for farmers to utilize the plots to the best advantage.

Three types of work units exist on both the kolkhoz and the sovkhoz: (1) the *brigade* or *section*, (2) the *ferma*, and (3) the *link* or *zveno*. The brigade forms the nucleus of the permanent work force assigned to a specific territory. A brigade is typically assigned permanent personnel and managers. A comprehensive brigade produces both crops and livestock. A ferma production unit specializes in meat, poultry, or dairy product output. The link refers to a small team assigned to work on a particular crop. Link workers are paid on the basis of crop yields.

Soviet Agricultural Inefficiencies. Soviet livestock numbers (cattle and hogs), although comparable to those in the United States, yield only two-thirds the meat output of the United States. Inefficiencies in the production process and shortages of grain for feed explain the discrepancy. Increasing grain output, a priority in Soviet agriculture for over 30 years, remains an elusive goal. Imports of grain, especially wheat, accelerated greatly in the 1970s to counteract the shortfall. The environmental constraints faced by Soviet agriculturalists are greater than those in the United States, which explains part of the production deficiency, but there are other compelling factors as

well. It has not been for a lack of investment that this weakness exists. Heavy capital investment programs have been instituted, but they have failed to increase efficiency.

Farm areas in the Soviet Union are generally short of labor and housing is inadequate: "What is bad in the city, is worse in the country ... unsurprisingly, drunkenness is even more of a headache in the rural areas than it is in the cities."[2] Promotion by the Kremlin of the output of private plots has increased efficiency somewhat, but this practice is resented by local councils. Critical animal food supply deliveries to private holdings are sometimes delayed and access by farmers to markets to sell goods can be thwarted. Supplies of better quality seeds and opportunities for transportation services are not always available. But there are also limits as to what can be accomplished by the private holdings because of their small size. Rearing livestock is particularly limited. The absence of small machines and tractors, which are not produced in large quantities, also complicates the situation.

The processing and distribution of agricultural products also create headaches in the Soviet Union. Because of long distances, transportation is a problem in all sectors of the economy. But in agriculture, this spatial disparity limits marketing opportunities for fresh products. Only half the roads are paved, and muddy rural roads are often impassible during the harvest season. More than distance alone, the problem of disorganization exacerbates the problem. Grain crops frequently rot in the field or at railroad sidings because of logistical inefficiencies in the transportation sector.

Even the machinery is inefficient. Although the Soviet Union is the world's leading tractor manufacturer, the machines are of poor quality and not as reliable as their U.S. counterparts. The average number of running hours between major engine overhauls in the Soviet Union is half that for American tractors. The Soviets must therefore produce many more parts for repairs than in the United States, and they are generally in short supply. Cannibalizing new machines is often the only way that local operators can get spare parts. "The shortage is so great that new machines are sometimes stripped at farms to supply spare parts. In 1979, the farms in the Altai region received 236 cars, 232 tractors and 142 combines that had been stripped during shipment."[3] Breakdowns during the planting and harvest seasons occur regularly, further decreasing production efficiencies.

Western European Trends

The farming sector of most western European countries comprises a small share of the total gross national product (7 percent in France, 3 percent in Great

[2]"The Good Earth Stubbornly Refuses to Deliver the Goods," *The Economist*, November 15, 1980.
[3]Ibid.

Britain) and total labor force, but it has modernized and become increasingly efficient in the post–World War II era. In the 1960s, the European Economic Community (EEC) developed a common agricultural policy (CAP) for the six original countries, later expanded to nine and then 12. To date, agriculture remains the only agreed-upon common policy. That policy called for a *single market* for farm goods so that they could be imported and exported freely among member nations. A *common tariff* against exports, and a *joint financial responsibility* commitment, also characterize the policy. The latter refers to the way in which costs are shared. Three-fourths of the EEC budget goes to pay for the CAP operation. Surpluses are bought by the organization, funds are available for modernization improvements, and prices are subsidized when market prices drop below specified thresholds.

France is the leading EEC farming nation (Table 5-7). In no small part the CAP program grew out of the French insistence that this program be started first. France also produces the largest agricultural surpluses of member nations and stood to gain the most. Other leading western European agricultural nations are West Germany, Italy, and Great Britain. Farms are much smaller in Europe than in the United States, averaging 25 hectares in most countries. In Italy, the country most dependent on agriculture when the EEC began, with 25 percent of the labor force so employed, farms remain very small, averaging 10 hectares in size. At the other extreme, farms in Great Britain average over 60 hectares.

The biggest surplus in EEC agriculture occurs in the milk and dairy product sector, which consumes over 40 percent of the CAP farm fund price support expenses, mainly to buy surpluses. The leading dairy countries producing this surplus are France, West Germany, Great Britain, and the Netherlands. Denmark and Ireland also produce surpluses but have smaller economies. Agricultural exports, primarily milk, butter, pork, and beef, account for over 40 percent of the exports from the latter two countries.

Although the Soviet Union and the United States dominate the world's dairy cow milk production output, with 35 percent of the total, western European nations as a group contribute over 20 percent of the

total, led by France, West Germany, Great Britain, and the Netherlands (Table 5-8). The only other leading milk-producing nations on a world scale are Poland and India. Given this large milk output, one would also expect to find a large dairy products industry in these areas, which is borne out in Table 5-9, which shows world cheese and butter output.

The United States and the Soviet Union dominate world cheese production, with one-third of the total, while France and West Germany contribute another 17 percent of the total. The Soviet Union, followed by India, leads in butter production. France, West Germany, and the United States round out the list of leading producers.

At the world scale, Italy, the Netherlands, and West Germany are the largest milk importers. The latter two

Table 5-8

Leading Dairy Cow Milk Production Nations, 1982
(thousands of metric tons)

Rank	Nation	Volume	Percent of World Total
1	Soviet Union	89,600	21
2	United States	61,552	14
3	France	34,500	8
4	West Germany	25,550	6
5	Poland	16,720	4
6	Great Britain	15,600	4
7	India	13,800	3
8	Netherlands	12,750	3
	Subtotal	270,072	63
	World total	437,909	100

Source: United Nations, *1982 Statistical Yearbook,* 1985.

Table 5-9

World Cheese and Butter Production, 1984
(thousands of metric tons)

Rank	Nation	Total	Percent of World Total
	World Cheese Production		
1	United States	2,402	20
2	Soviet Union	1,659	13
3	France	1,250	10
4	West Germany	872	7
	Subtotal	6,183	50
	World total	12,337	100
	World Butter and Ghee Production		
1	Soviet Union	1,610	21
2	India	740	10
3	France	605	8
4	West Germany	572	8
5	United States	508	7
	Subtotal	4,035	54
	World total	7,660	100

Source: United Nations, *1984 FAO Production Yearbook,* 1985.

Table 5-7

EEC Member Country (nine nations)
Farm Production, 1978

Rank	Nation	Percent of EEC Farm Production
1	France	28
2	West Germany	22
3	Italy	20
4	Great Britain	11
	Total	81

Source: "Europe's Green and Expensive Land," *The Economist,* November 1980, p. 51.

Table 5-10

Leading World Market Economy Dairy Product Trading Nations, 1983
(millions of U.S. dollars)

	Imports				Exports		
Rank	Nation	Value	Percent of World Total	Rank	Nation	Value	Percent of World Total
			Milk and Cream				
1	Italy	649	14	1	Netherlands	996	22
2	Netherlands	525	11	2	West Germany	923	21
3	West Germany	424	9	3	France	654	15
	Subtotal	1,598	34		Subtotal	2,573	58
	World total	4,619	100		World total	4,456	100
			Butter				
1	Great Britain	517	20	1	Netherlands	686	25
2	Netherlands	380	14	2	New Zealand	364	13
3	Belgium/Luxembourg	303	12	3	France	306	11
4	France	174	7	4	West Germany	275	10
	Subtotal	1,374	52		Subtotal	1,631	59
	World total	2,624	100		World total	2,778	100
			Cheese and Curd				
1	Italy	783	20	1	France	770	20
2	West Germany	762	20	2	Netherlands	738	19
3	United States	383	10	3	West Germany	643	17
4	Great Britain	376	10	4	Denmark	389	10
	Subtotal	2,304	60		Subtotal	2,540	67
	World total	3,863	100		World total	3,797	100

Source: United Nations, *1983 Yearbook of International Trade Statistics*, 1985.

countries, along with France, lead as exporting nations (Table 5-10). They provide fluid supplies to some nations, while importing from other countries to support manufacturing needs. European countries dominate the world's butter and cheese import–export trade. These countries produce a wide variety of cheeses, some of which cater to specialized regional tastes. Only New Zealand, an important butter exporter, prevents western European countries from having a clean sweep in the leading exporters list given in Table 5-10. Denmark adjusted its product mix in recent years, for example, to take advantage of the growing feta cheese market in the Middle East. Great Britain has traditionally been a dairy products importer, but the flow, formerly supplied by Commonwealth countries, particularly New Zealand and Australia, now comes from within the EEC.

SUMMARY AND CONCLUSION

Commercial agriculture in the midlatitudes in North America, western Europe, and the Soviet Union generally conforms to the pattern suggested by the von Thünen model, wherein more intense production occurs near the market and more extensive activities at greater distances. Typically, this involves dairy production near the market, followed by grain production, general farming, and grazing activity in more remote locations. Agriculture in the Soviet Union faces a more restrictive environment than that in North America and western Europe, owing to a shorter growing season, cooler temperatures, and less rainfall.

In western Europe and the United States, specialized animals produce milk and meat in distinctive producing regions, a situation largely absent in the Soviet Union. The dairy industry is highly subsidized in both the United States and western Europe and considerable surpluses are produced. Important advances in mechanization in recent years permitted significantly fewer but larger farms to produce more product than a few decades ago, exacerbating the problem.

A feed grain and livestock belt borders the dairy region to the west in the United States. This area is the premier agricultural region in the world. No comparable producing area exists in western Europe or the Soviet Union. Corn and soybeans serve as feed for the beef cattle and hog production in this area. Hogs are grown in western Europe and the Soviet Union but are fed potatoes and other lower protein feeds than in the United States. Another feature of the feed grain belt in the United States is the feedlot, where very large operations "finish" cattle for the market on high-protein rations.

Agricultural output has decreased in the Soviet Union in recent years due to poor weather conditions and management problems. Labor and housing problems, poor machinery, and the lack of incentives contribute to the inefficiencies of Soviet agriculture. Shortages have been overcome by increasing imports.

SUGGESTIONS FOR FURTHER READING

AMBLER, JOHN, ET AL., eds., *Soviet and East European Transport Problems.* New York: St. Martin's Press, 1985.

ANDREAE, BERND, *Farming Development and Space, A World Agricultural Geography.* New York: Walter de Gruyter, 1981; translated from German by Howard F. Gregor.

BATTLE MEMORIAL INSTITUTES, *Agriculture 2000: A Look at the Future.* Boulder, Colo.: Westview Press, 1983.

BOWLER, IAN R., *Agriculture under the Common Agricultural Policy: A Geography.* Manchester, England: Manchester University Press, 1985.

CENTRAL INTELLIGENCE AGENCY, *U.S.S.R. Agricultural Atlas.* Washington, D.C.: CIA, 1974.

CLOUT, HUGH, *A Rural Policy for the EEC?* London: Methuen, 1984.

DUCHENE, FRANCOIS, ET AL., *New Limits on European Agriculture: Politics and the Common Agricultural Policy.* Totowa, N.J.: Roman and Allanheld, 1985.

DUNN, EDGAR, *The Location of Agricultural Production.* Gainesville, Fla.: University of Florida Press, 1967.

EBELING, WALTER, *The Fruited Plain, the Story of American Agriculture.* Berkeley, Calif.: University of California Press, 1979.

FURUSETH, OWEN J., and JOHN T. PIERCE, *Agricultural Land in an Urban Society.* Washington, D.C.: Association of American Geographers, 1982.

GOLDMAN, MARSHALL I., *U.S.S.R. in Crisis: The Failure of An Economic System.* New York: W. W. Norton & Co., 1983.

KURDIE, ROBERT T., *Agricultural Tractors: A World Industry Study.* Cambridge, Mass.: Ballinger, 1975.

MARSH, J. S., and P. SWANNEY, *Agriculture and the European Community.* London: Allen & Unwin, 1980.

RASMUSSEN, WAYNE D., *Agriculture in the History of the United States: A Documentary History.* New York: Random House, 1975.

SYMONS, LESLIE, *Russian Agriculture: A Geographic Survey.* New York: Wiley, 1972.

von Thünen, J. H. *Der Isolierte Staat in Beziehung auf Landwirtschaft und Nationalalobonomie,* 3rd ed., Berlin, 1875. Part I, which contains the original isolated-state statement, was originally published in 1826. For an English translation, see P. Hall, ed. *von Thünen's Isolated State: An English Version of "Der Isolierte Staat,"* Translated by C. M. Wartenberg, Pergamon Press, New York, 1966.

TRACY, MICHAEL, *Agriculture in Western Europe: Challenge and Response 1880–1980.* St. Albans, Herts, England: Granada, 1982.

WONG, LUNG-FAI, *Agricultural Productivity in the Socialist Countries.* Boulder, Colo.: Westview Press, 1986.

6 Grain Farming and Livestock Grazing

Commercial grain farming is most often associated with the wheat belts of the world and is treated as such here. Based on our earlier discussion of the von Thünen model, livestock ranching should occur in regions less desirable economically for cropping, at greater distances from urban markets. It is therefore not surprising to find grain farming flourishing in interior continental locations and livestock ranching in an adjacent zone farther from the market (recall the patterns on the von Thünen–based map in Figure 5–3).

Both grain farming and livestock grazing as practiced in North America are typically large-scale operations not possible until the interior of the continent was opened up by the railroad in the latter part of the nineteenth century. The low-cost land transportation offered by the railroad allowed products to be moved to growing urban markets cheaply and in vast quantities. Later, in the twentieth century, growing export markets provided additional production incentives.

In the United States, the wheat belt occupies the western portion of the prairie, where rainfall is 20 inches or less a year (Figure 5-1). Short grass and a semiarid climate characterize the western prairie. This area contrasts sharply with the tall-grass prairie humid area found farther east in the higher-rainfall regions in Iowa and Illinois. To the west of the prairie is the steppe, an area of 10 to 20 inches of annual rainfall, extending from northern New Mexico to southern Alberta in North America. Large stands of trees are not common in either the western prairie or the steppe environment because of the lack of rainfall needed to sustain them. The Great Plains area has not traditionally sustained crop agriculture due to the lack of precipitation, but as we will observe, the expansion of irrigation in the area and the introduction of new meat processing techniques have dramatically altered the landscape in recent years.

COMMERCIAL GRAIN FARMING

Commercial grain farmers typically specialize in producing a single crop—wheat—but often rotate production with other grains or forage crops. Such an arrangement occurs on the North American short-grass or dryland prairie in portions of the Soviet Union, southern Asia, western Europe, and South America (see Figure 6-1). The proportion of farmland devoted to cash cropping attains unusually high percentages in commercial grain regions. Cropland, for example, accounts for 80 percent of the total farming area in the valley of the Red River in both North Dakota and Manitoba. In some regions mixed farming also characterizes wheat-producing areas because animal production is emphasized, but the core producing realm remains specialized.

The concept of *comparative advantage* helps explain this specialization in one crop. This locational concept originated in the field of international trade as researchers observed that nations specialized in

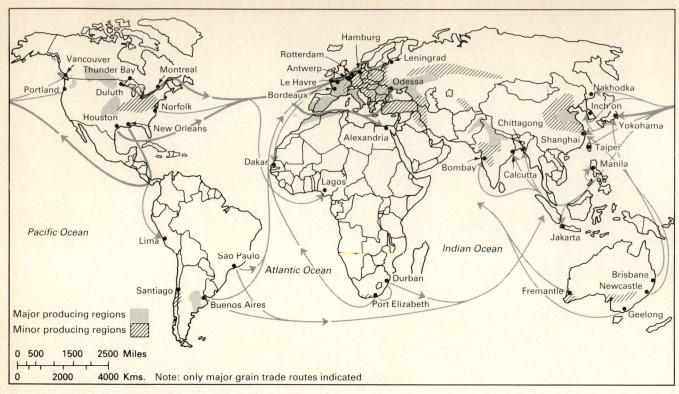

Figure 6-1 Major World Grain Production Areas and Trade Routes. *The widely dispersed grain production regions at the world scale emphasize the dominance of the Northern hemisphere. The flows of grain from North America, Argentina, South Africa, and Australia to Japan, the Soviet Union, Egypt, and Western Europe can be ascertained by the directional arrows which graphically demonstrate the global interdependence of major world regions. (Redrafted from Morgan,* Merchants of Grain, *Viking, 1979)*

special products for trade. This concept works particularly well in the United States in explaining why wheat farmers and corn farmers each tend to specialize in a particular activity. Basically, each agriculture area produces only what it can do best in comparison with competing activities. As an example, consider the farm situations in Iowa and Kansas. Corn yields are higher in Iowa than they are in Kansas, and because Iowa farmers can produce more product by concentrating on corn, they do not emphasize wheat production. Similarly, Kansas farmers have less of a comparative disadvantage by concentrating on wheat rather than producing corn. The following table illustrates the point. It indicates hypothetical productivity levels in terms of yield per acre of wheat and corn in Iowa and Kansas:[1]

Yield per Acre (bushels)

	Iowa	Kansas
Corn	130	100
Wheat	50	45

[1]This example is modified after Harold H. McCarty and James B. Lindberg, *Economic Geography* (Englewood Cliffs, N.J.: Prentice-Hall, 1966), pp. 216–217.

Notice that corn in Iowa yields 130 bushels of output per acre, whereas in Kansas it produces only 100, giving Iowa an advantage of 30 units per acre. In terms of wheat output per acre, the disparity between the two states is far smaller. In fact, the Kansas yield is only 5 units less than that in Iowa. Therefore, Iowa has a comparative advantage for specialization in corn, while Kansas competes better by producing wheat. Nevertheless, corn and other crops, including soybeans, are grown in Kansas as elsewhere, indicating that individual farmers often diversify their production, taking into account more factors than simply average yield.

Physical Factors

Commercial grain farming competes best in areas with at least 12 to 20 inches of rainfall, preferably occurring evenly during the winter, spring, and early summer months. Wheat can tolerate hot, dry summers and in fact prefers a less humid growing season than does corn. Excess rainfall in the harvest season encourages mold and fungus growth.

Two types of wheat grow in distinctive subregions of wheat-producing areas: winter wheat and spring wheat. Distinctions are also made between hard wheat and soft wheat. Hard wheat has a higher gluten content and makes an excellent bread flour. It does not spoil as easily and withstands transportation more readily. Soft wheat has a low gluten content and is less suitable as a bread flour, as it does not rise as well. Soft wheat is better for spaghetti and other pasta products, cakes, crackers, and pastry. In the United States, hard wheat grows better in the more arid Great Plains, while soft wheat grows better in the more humid eastern and middle western climates. Hard wheat also has a higher protein content than that of soft wheat.

Temperature and precipitation factors in combination help explain the distinctive locations of winter and spring wheat-growing areas. Both types prefer a cool, moist season initially, followed by a warm, sunny, dry period during maturation. Farmers prefer to grow winter wheat where they can because yields are generally higher and problems with potentially dry summers can be avoided. Winter wheat production occurs in areas with mild winters and a snow cover to insulate the ground from severe freezing. This need arises because the winter wheat crop is planted in the fall and grows to a grass cover a few inches in height before the growing season ends with the first severe frost in the fall. In spring, growth resumes, taking advantage of early spring rainfall. Harvest occurs during midsummer drought periods.

Spring wheat predominates in areas where winters are too severe for wheat seedlings to survive over winter. It is planted in the spring and harvested in later summer. Spring wheat is typically hard wheat, as is the winter wheat grown in drier regions. Both types of wheat have shorter growing seasons than corn, further extending their growing range, which occurs as far north in Europe as Finland, and into northern Alberta in North America.

In North America there are two major wheat belts: the spring wheat belt and the winter wheat belt (Figure 6-2). The spring wheat belt extends from the Dakotas and Montana northward into southern Manitoba, Saskatchewan, and Alberta. Cold, relatively dry winters cause the ground to freeze in this region, preempting the possibility of growing winter wheat. Although winters are severe, the growing season for small grain is adequate. The rich prairie soil and favorable topography also favor large-field cropping. Although wheat production predominates, oats, barley, and oilseed output are also large. The latter, typically canola (rapeseed), is particularly prominent on the Canadian prairie. Some prairie farmers also specialize in growing certified grass or grain seeds for the marketplace.

Farm units average 1000 acres in the Dakotas, while their Canadian counterparts across the border in Alberta and Saskatchewan average just over 800 acres. The labor force is small on grain farms because planting and harvesting are both highly mechanized and little field work is required during the growing season. Once fields are planted, no labor needs arise other than possible spraying for weed control or fertilization. Frequently, outside contractors with fleets of combines arrive in harvest season to thresh the grain. In fact, the seasonal migration of harvesting crews from south to north has become an institution in the wheat belts. Beginning in June, teams begin the harvest in Oklahoma, moving northward as the season progresses, harvesting the crop in Kansas in mid-season, and going on to the Dakotas and Canadian prairies in August.

Wheat farmsteads are small with few building structures, unlike the situation in mixed-farming and

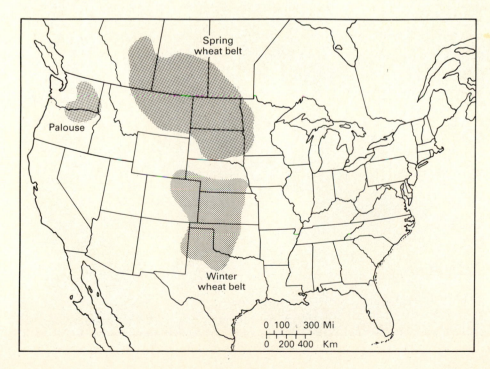

Figure 6-2 North American Wheat Belts. *Two wheat belts, the winter and spring belts, split apart in Nebraska, and a major outlier to the west, the Palouse, produce most of the wheat in North America. Note the location of this area to the west of the corn and soybean regions shown in Figure 5-9. (Source: U.S. Census of Agriculture)*

dairy areas. The lack of animals eliminates the need for barns, sheds, silos, and feedlots for their care. In addition to the family residence, farmers need only machinery sheds and grain storage bins.

Two terms often used to refer to wheat farmers are *sidewalk farmers* and *suitcase farmers,* because of the frequency of absentee owners. Unlike tenant farms, these units are run by their owners, but they do not always live on the land. Sidewalk farmers live in town rather than on the farm itself. Suitcase farmers live more than 30 miles from the county line in which their farm is located, typically in larger cities. Many farmers now commute by airplane to their holdings as necessary.

The most impressive structures on the prairie landscape are the grain elevator storage facilities in towns along the rail lines that serve as the transportation arteries to move grain to the market or export terminal (Figure 6-3). Often, the grain elevator is visible for miles and is the only imposing structure in the towns, which serve mainly as market centers for the farm community. The grain elevator not only stores grain but also serves as the place where the wheat is cleaned, dried, graded, and tested. Many elevators are farmer owned; others are operated by cooperatives, governmental agencies, or the railroads. In sparsely settled portions of the Dakotas, Saskatchewan, and Alberta, towns may be spaced 40 miles apart, but in higher-density areas such as in the winter wheat belts of Kansas, they are typically 7 to 10 miles apart.

Minneapolis and Winnipeg were traditionally the major milling centers in the spring wheat belt. The Minneapolis mills are now closed in favor of those farther south, but the city remains a major control and

storage center for the industry. The names Cargill, Gold Medal, General Mills, and Pillsbury therefore remain prominent in the city. Kansas City, Topeka, Wichita, and Oklahoma City perform milling and storage functions in the winter wheat belt. A list of the leading storage facility cities for grain appears in Table 6-1. Greater shares of wheat are exported today than years ago further weakening traditional processing centers. Spring wheat moves eastward via Duluth and the Great Lakes system, while both winter and spring wheat move southward via the Missouri and Mississippi River systems. New Orleans and Houston on the Gulf coast are both major export centers.

The winter wheat belt extends from Kansas and eastern Colorado southward to Oklahoma. Nebraska serves as a dividing line between the winter and spring producing areas. The landscape in the two areas is similar in many ways, with the grain elevator dominating. Farms in the winter wheat belt tend to be smaller than those in the spring wheat area. Those in Oklahoma and Kansas, for example, average 400 to 600 acres in size, but land values per acre are higher, so overall farm values, from $200,000 to $300,000, are comparable to those in the spring wheat belt.

Livestock production is much more important in the U.S. winter wheat belt than in the spring wheat area. Oklahoma and Kansas, for example, each had about 5 million cattle and calves in 1978, while the Dakotas each had less than half that amount.

Generally speaking, bulk wheat hauls by truck are more expensive than those by rail, which in turn are far more expensive than those by water over long distances. For this reason shippers use water transportation when given a choice for the longer hauls. Of course, truck and rail movements are used at the local

Figure 6-3 Grain Elevator Complex in Kansas. *The most impressive structures on the prairie landscape are the grain elevator storage facilities. The grain elevator is typically the tallest building in both large and small communities along the rail line. (Courtesy USDA)*

Table 6-1

Leading U.S. Grain Storage Centers (millions of bushels)

Rank	City	Capacity
1	Minneapolis	102
2	Enid, Oklahoma	76
3	Kansas City	75
4	Duluth	74
5	Topeka	55
6	Chicago	54
7	Salina, Kansas	50
8	Wichita	46

Source: *The World Almanac and Book of Facts, 1987* (New York: Newspaper Enterprise Association, Inc., 1986).

level to assemble the product at elevators. The attraction of cheaper water transportation has led to the development of major elevator storage complexes and milling operations at *break-of-bulk* locations, where rails intersect major waterways. "Break-of-bulk" refers to a change in mode of shipment at a port or terminal. Traditionally, processing occurred at these locations, which minimized the need for excessive loading and unloading of the product. For this reason elevators and milling occurred at major ocean ports and large inland urban centers such as Minneapolis, Kansas City, and Buffalo. Prior to the opening of the St. Lawrence Seaway in 1959, Buffalo served as the eastern terminus of Great Lakes navigation and therefore became a major grain refining center. In winter months, ice-locked barges in the Buffalo port provided additional wheat storage capacity for the mills.

In the U.S. spring wheat belt a greater portion of the product now also moves westward. Montana wheat, for example, now moves westward to Seattle or Portland for export, first by truck over the Bitteroot Mountains, then by barge from Lewiston, Idaho, on the Snake and Columbia rivers.

Railroads abandoned considerable branch-line mileage in the 1960s and 1970s and in the wheat belt whole systems faced bankruptcy, forcing a decline in the number of small grain elevators located in rural service centers along the lines. Longer, truck-based hauls of grain to larger subterminals built by the railroads replaced the local rail-siding collection system. Unit trains then carried the product to river terminals for transshipment by barge.

The most significant wheat-producing outlier region in the United States is the Palouse area of eastern Washington (see Figure 6-2). Wheat is also produced as a cattle feed on many farms in the dairy and mixed feed grain belts. Only a portion of this grain is sold as a cash crop, as most is consumed locally.

Canadian Wheat Belt

The three Canadian prairie provinces, Manitoba, Saskatchewan, and Alberta, account for three-fourths of Canadian farmland and for most of wheat production (Figure 6-2). This producing area represents an extension of the U.S. spring wheat belt just discussed. Rainfall is higher to the east in Manitoba, where mixed farming is more important, including dairy farms around Winnipeg and hog and beef cattle in the southwest. Saskatchewan is by far Canada's breadbasket, accounting for two-thirds of the total wheat output (Figure 6-4); Alberta is second. In both provinces beef cattle production is growing. Irrigation has increased in recent years, expanding the acreage of vegetables, sugar beets, and forage crops. The Peace River valley region of northern Alberta is a major outlier of the prairie grain and livestock producing region.

Canadian wheat production occurs in a more isolated setting than that of the United States or any other major exporting nation, such as France, Australia, or Argentina. U.S. producers can take advantage of the Missouri–Mississippi River system as well as the Great Lakes waterway, but only the latter option is available to Canadians. Some of the Canadian product moves east to Churchill on Hudson Bay or Thunder Bay on Lake Superior, but water transportation is not used for most of it (Figure 6-5). Much of the Canadian product moves westward by rail through the Rocky Mountains to Vancouver or Prince Rupert for export. Newer, brightly colored yellow and orange hopper cars now carry the load 1000 miles from the interior to ocean terminals on the Pacific coast. These cars hold 100 tons of grain, 30 percent more than the older-style boxcars they are displacing. In addition to the railways, cars are also owned by the Canadian Wheat Board and by provincial and federal government agencies. Indirect subsidies in the form of favorable freight rates, public ownership of the rolling stock that moves the product, and considerable federal and provincial support of research and development benefit wheat producers.

Movement of grain through the mountains and through the Fraser River canyon creates a bottleneck, but engineering innovations and interline cooperation between Canadian National and Canadian Pacific railways have improved the flow in recent years (Figure 6-6). A new bulk commodity port terminal is now under construction on Ridley Island, near Prince Rupert on the northwestern coast of British Columbia. Even without this new terminal, Vancouver and Prince Rupert have storage facilities for over 1 million tons of grain, reflecting the strong dependence of the industry on foreign markets.

Canada, in fact, is the second-leading wheat-exporting nation today, after the United States. Together these two countries account for about two-thirds of world wheat exports (see Table 6-2). Most of the Canadian exports from the west coast go to Japan or China (see Table 6-3). Among market economy nations, Brazil, Egypt, and Italy are also leading wheat-importing nations.

The Canadian Grain Commission, created by the 1971 Canada Grain Act, oversees the inspection, weighing, and storage of grain in Canada. It licenses

Figure 6-4 Wheat Harvest in the Canadian Spring Wheat Belt. *Saskatchewan is by far Canada's breadbasket, accounting for two-thirds of the country's wheat output. Expansive, relatively flat, fields aid in the use of large equipment such as the self-propelled combines shown here. (Courtesy Canadian Consulate General)*

Figure 6-5 Toronto Railyards. *Canadian grain shipped eastward by rail is transferred to ships at Thunder Bay, Churchill, Toronto, and other ports. The railyards in Toronto separate the downtown area, shown here in the background, from its Lake Ontario frontage. (Courtesy of Richard Pillsbury)*

Table 6-2

Leading Market Economy Wheat-Importing and Wheat-Exporting Nations, 1983 (millions of U.S. dollars)

Rank	Nation	Value	Percent of World Total
	Imports		
1	Japan	1,126	11
2	Brazil	697	7
3	Egypt	660	7
4	Italy	549	6
	Subtotal	3,032	31
	World total	9,874	100
	Exports		
1	United States	6,239	42
2	Canada	3,771	25
3	France	2,053	14
4	Australia	1,068	7
	Subtotal	13,131	88
	World total	14,904	100

Source: United Nations, *1983 Yearbook of International Trade Statistics,* 1985.

Table 6-3

Selected Canadian Grain and Oilseed Exports to Asian Nations, 1978–1979 (tonnage shipped from Vancouver and Prince Rupert)

	Japan	China	Other Asian
Wheat	1,351,358	3,419,079	234,151
Barley	962,633		152,560
Canola (rapeseed)	1,120,662		211,754
Total	3,434,653	3,419,079	598,465

Source: Adapted from Anthony Burgess, "Moving Grain from Out West to the Far East," *Canadian Geographic,* 100 (June–July 1980), 23.

grain elevator operators and supervises grain futures trading. With headquarters in Winnipeg, the agency has major offices in Vancouver, Thunder Bay, and Montreal. The Canadian Wheat Board, created under the Canadian Wheat Board Act of 1935, is the sole purchaser and seller of feed grains for export. The board coordinates the movement of wheat to major terminals. Quotas set up by the agency fix the delivery schedules for wheat from farms to local elevators prior to the shipment of the product to the export terminals. In this way, price fluctuations and distribution inefficiencies are kept to a minimum. The United States has no similar unified export system that handles the many services provided by the Canadian Wheat Board.

SOVIET PRODUCTION

The Soviet Union, the world's second-leading wheat-producing nation after China, grew about 15 percent of total world production in 1984 (Table 6-4). Even at this level of production the country has a growing wheat deficit that requires greater quantities of imports. The Soviet wheat-producing region occupies the southern portion of the fertile triangle, with the bulk of the production centered in the Ukraine (Figure 6-7). This wheat-growing area tapers off sharply to the east in Siberia. Wheat accounts for about half of the total grain production in the Soviet Union and about

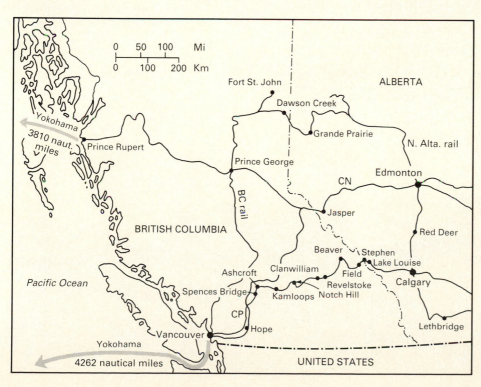

Figure 6-6 Westward Movement Corridors for Canadian Grain. *Movement of Canadian grain westward through the Rocky Mountains and the Fraser River Canyon creates a bottleneck which has been overcome with a complex tunnel switchback network in the Field and Revelstoke areas and interline cooperation between the Canadian National and Canadian Pacific Railways. (Redrafted from* Canadian Geographic *[June/July 1980], p. 22)*

Table 6-4

Major Wheat-Producing Nations, 1984
(millions of metric tons)

Rank	Nation	Production	Percent of World Total
1	China	88	17
2	Soviet Union	76	15
3	United States	71	14
4	India	45	9
5	France	33	6
6	Canada	21	4
	Subtotal	234	65
	World total	522	100

Source: United Nations, *1984 FAO Production Yearbook*, 1985.

90 percent of food grain production. Winter wheat predominates in the Ukraine, but farther east spring wheat must be grown because of the severe winters. Nearly all of the Soviet crop is hard wheat.

The *virgin lands* agricultural program of the 1950s brought vast quantities of additional wheat acreage into production (Figure 6-8). That area produces spring wheat. The potential for drought is greater in this area, causing wide variations in the harvest from year to year. From 1953 to 1958, the Soviets added 29 million hectares of this marginal land in southern Siberia and northern Kazakhstan to the crop inventory. Shelterbelts of trees serve as windbreaks, and strip-cropping innovation techniques serve to counteract drought hazards, but severe dust storms and wind-induced soil erosion remain big problems in the area.

Unlike the situation in the United States, where the reverse is true, over one-half of Soviet wheat is spring wheat. In the period 1950 to 1970, wheat output in the United States and the Soviet Union rose in a roughly parallel pattern. From the mid-1970s to the present a growing disparity in production in the two countries emerged. In the United States during the earlier portion of this period, production controls discouraged output, while the Soviets gave increased production the highest priority. Beginning in 1972, greater levels of imports into the Soviet Union from the United States largely erased the surplus production in the United States and production quotas vanished. Thereafter, production zoomed upward in the United States and downward in the Soviet Union, especially after 1978. "For U.S. farmers, 1981 marked not only a record harvest, but also the first time they doubled the output of their Soviet counterparts—331 million tons of grain to 165 million tons."[2]

As mentioned in Chapter 5, corn accounts for the largest part of the U.S. grain output. Its higher productivity (yield) than that of wheat accounts for some of the disparity in output of the two nations. But there are many others, including less productive land and inefficient use of fertilizer, pesticides, and machinery in the Soviet Union, as discussed in Chapter 5. The Soviets, for example, produce "roughly six tons of grain per ton of fertilizer compared with 16 in the U.S."[3] The increasing disparity in agricultural productivity between the United States and the Soviet Union, 1960–1984 is graphically portrayed in Figure 6-9 on page 74.

Decline in Soviet grain output constitutes more than an academic problem. It has created a food shortage, especially a meat shortfall. To overcome this, the massive wheat import program begun in the 1970s signaled a dramatic break with the earlier part of the century, when the country exported grain. The Sovi-

[2]Lester R. Brown, *U.S. and Soviet Agriculture: The Shifting Balance of Power,* Worldwatch Paper 51 (Washington, D.C.: Worldwatch Institute, 1982).

[3]Ibid.

Figure 6-7 Soviet Wheat Production Belt. *These large harvesting machines are cutting and thrashing wheat for bulk loading onto a truck on a collective farm near Odessa in the Ukraine. (Source: TASS from SOVFOTO/photo by I. Pavlenko)*

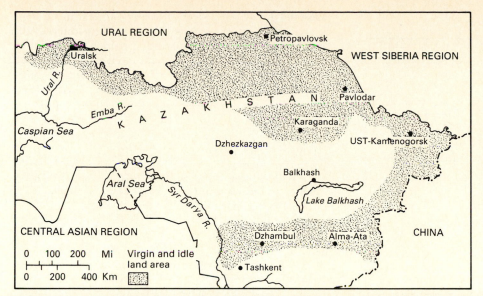

Figure 6-8a Virgin and Idle Lands Region of Kazakhstan. *The virgin lands agricultural programs of the 1950s brought vast quantities of additional wheat acreage into production. Shaded area on map shows the extent of this area which forms a horseshoe pattern around the Caspian and Aral seas and Lake Balkhash.*

Figure 6-8b *Leninsky State Farm in Kazakhstan in virgin lands region. Shelterbelts of trees serve as windbreaks to counteract wind-induced soil erosion in this drought prone area that has brought thousands of additional acres of wheat acreage into production. Source: SOVIET LIFE from SOVFOTO.*

ets are now the world's leading importers of grain, buying twice the quantity purchased by Japan, the second-ranking importer, from overseas markets. The Soviets now also lead all countries in meat imports.

The United States dominates the world grain export flow, accounting for over half of the world trade (Figure 6-1). About one-third of Soviet grain imports came from the United States in 1982, whereas about half came from the United States prior to the wheat embargo imposed by the Carter administration in 1980. The U.S. embargo not only altered the flow to the Soviet Union but also that from other countries, because the Soviets turned increasingly to them for

imports, particularly Canada and Argentina. In the mid-1980s, the embargo was voided and greater wheat flows resumed to the Soviet Union from the United States.

WORLD WHEAT PRODUCTION

On the world scale, other major wheat-producing areas outside the Soviet Union and North America in the northern hemisphere include the Mediterranean area of Europe, North Africa, and the Middle East; France, Romania, and Bulgaria; and Pakistan, north-

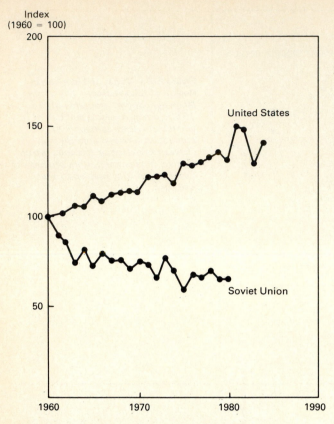

Index
(1960 = 100)

United States

Soviet Union

1960　　　1970　　　1980　　　1990

Figure 6-9 Growing Agricultural Productivity Disparity between the United States and the Soviet Union, 1960–1984. *Using 1960 as a starting point with both nations possessing a productivity level index value of 100, the Soviet Union declined to a level of about 65 in 1980, while the United States' index approached 150. (Source: Chandler,* Worldwatch Paper 72, *Worldwatch Institute, p. 14, 1986)*

Table 6-5

Production of Selected Small Grains, 1984
(millions of metric tons)

Rank	Nation	Production	Percent of World Total
	Barley		
1	Soviet Union	42	44
2	United States	13	8
3	France	12	7
4	Canada	10	6
5	Great Britain	11	6
	Subtotal	75	71
	World total	100	100
	Oats		
1	Soviet Union	15	40
2	United States	7	16
3	Canada	3	7
4	Poland	3	7
	Subtotal	35	70
	World total	43	100
	Rye		
1	Soviet Union	11	36
2	Poland	10	32
3	East Germany	2	7
	Subtotal	23	75
	World total	31	100

Source: United Nations, *1984 FAO Production Yearbook,* 1985.

In Europe and the Soviet Union, these small grains are more heavily used in the human diet, being especially important as a bread flour. They compete well on the poorer soils and cooler climates of Europe and the Soviet Union.

LIVESTOCK GRAZING

Livestock grazing, true to the von Thünen model, typically finds its home most distant from the market, being an extensive form of commercial agriculture (Figure 6-10). In much of the world today, cattle rearing accounts for the majority of this activity, but in western Europe, the Mediterranean region, Australia, and New Zealand, sheep grazing predominates (Figure 6-11 on page 76). Cultural ties to an earlier nomadic tradition and dietary preferences partially account for these differences. The ability of sheep to thrive on rugged terrain and in semiarid climates, or graze efficiently on poor-quality grass and in scrub brush areas, also assists in the explanation. In western Europe, India, and China considerable sheep production occurs in areas outside traditional ranching areas. High population densities and land shortages in these areas require more intensive sheep-farming practices rather than extensive ranch operations.

ern India, and northeastern China (Figure 6-1). The latter three areas were discussed in Chapter 4. In the southern hemisphere significant producing areas are limited to parts of Australia and Argentina. In both countries production is near the coast. Wheat output in Australia extends 1000 miles inland from Adelaide in the general vicinity of the Murray and Darling River basins and in the extreme southwestern reaches of the continent in the interior to the east of Perth. In Argentina a crescent of wheat production lies inland from Buenos Aires on the *pampas,* a vast, flat grassland.

Other Small Grains

Production of other small grains, especially barley, rye, and oats, occurs predominantly in the Soviet Union and Europe (see Table 6-5). Growing conditions and production regions are similar to those for wheat. Of the three grains, only oats and barley are grown heavily in the United States and Canada, where they are an important cattle feed. Production in the upper middle west (North and South Dakota) predominates.

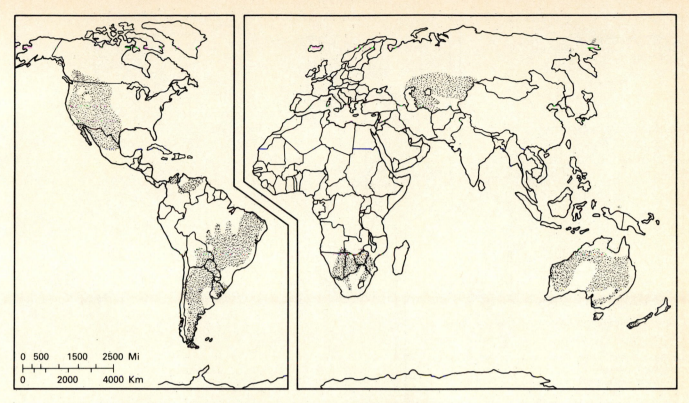

| 0 | 500 | 1500 | 2500 Mi |
| 0 | | 2000 | 4000 Km |

Figure 6-10 World Distribution of Livestock Ranching. *Livestock ranching typically occurs the greatest distance from the market of any form of commercial agriculture, as in the United States. This form of activity is largely absent in Eurasia except in the Soviet Union to the east of the Caspian Sea. Note extensive areas in South America, Africa, and Australia.*

Livestock Production Regions

On a world scale just five regions delimit the majority of livestock ranching. In North America, a broad region extends from western Canada southward through the western United States to central Mexico. In South America the major livestock ranching area extends from the southern tip of Argentina northward through Brazil, encompassing virtually all of the southeastern third of the continent. Venezuela also has a small producing area at the northern end of the continent. The third major producing area lies in southern Africa, and a fourth in Australia and New Zealand, where the highest percentage of this activity occurs, in relation to total land, of any area in the world. East and north of the Caspian Sea in southern Russia lies the fifth and last major zone. In the latter area, livestock ranching has only recently replaced nomadic herding.

Characteristics

In contrast to nomadic herding, livestock ranching restricts animals to parcels of land owned or leased by herdsmen or ranching organizations. The operating units consist not only of the natural pastureland but also the ranch house and acreage devoted to supplementary feed crops. Livestock ranching involves the largest operating units of any type of bioculture. In the United States, ranches frequently exceed 1000 acres and in Wyoming, Nevada, Arizona, New Mexico, and Texas they often exceed 2500 acres in size. One ranch in southern Texas encompasses 865,000 acres—over 1000 square miles. But the world's largest ranches are found in Australia, where several spread over 5000 square miles; one Australian ranch covers an astonishing 12,000 square miles.

A close balance between man and nature occurs in livestock ranching areas, as overgrazing can destroy the food supply. The concept of *carrying capacity* is often used to indicate this relationship. Carrying capacity refers to the number of animals a given amount of land can support with its natural vegetation. The variation is great. Over the dry near-desert area of the southwestern United States, 100 acres or more may be required to supply forage necessary for just one steer. In steppes and mountain meadows the carrying capacity generally varies from 25 to 75 acres per steer. On the eastern margins of the Great Plains the capacity improves to 3 to 5 acres per steer. These figures can also be phrased in terms of "animal units," whereby to measure carrying capacity, one steer is equated with one horse or five sheep. If an area has a carrying capacity of 10, a rancher can count on successfully raising 10 steer or 10 horses, or 50 sheep or any combination thereof, as long as the 10-unit figure is not exceeded.

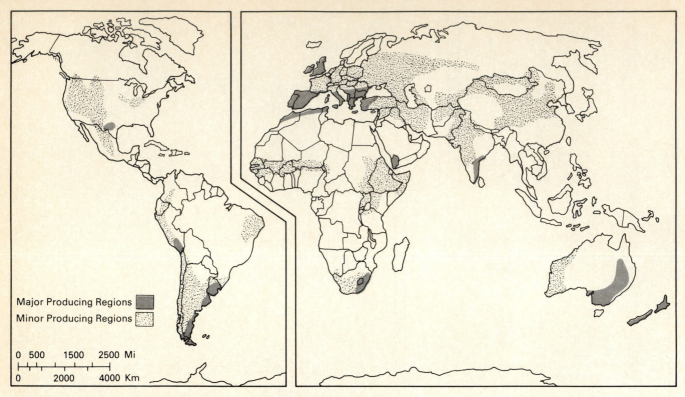

Figure 6-11 Sheep Grazing in Major World Regions. *Sheep are less demanding than cattle in their forage requirements and can compete better than beef cattle in arid regions. Secondly, sheep are often reared in the same areas as cattle and in mixed farming areas, creating a wider production area than that for cattle. Note the extensive production region in the Soviet Union, China, Australia, the Mediterranean, and the United States.*

Low population densities characterize livestock ranching regions. Density ranges of between 2 and 25 persons per square mile are normal. Settlements serve primarily as service centers for the rural farm population. They are typically small in size and few in number. In the American West these settlements are very widely separated. A spacing of 100 miles between communities often occurs, and in some instances this distance can increase to 300 miles, as it does in Nevada.

Environmental Setting

Most commercial grazing occurs in drylands. Dry climates prevail in the five largest ranching areas described earlier. Precipitation is significant in its effect on both natural vegetation and alternative possibilities for bioculture. Generally, it is held that 10 to 20 inches of annual rainfall marks the limit for unirrigated farming in the midlatitudes. Beyond that line, the drier regions are too risky for cropping and generally are devoted to livestock ranching. Most of the plants that can survive in drylands are not valuable enough to cultivate. But animals can live off such scattered vegetation. This is the basic reason that animal husbandry rather than cropping is the prevailing form of bioculture on dryland that is not irrigated.

Livestock ranching occurs in regions of grassland and desert shrub (Figure 6-12). But the correlation between ranching and these vegetation types is not complete since nomadic herders also occupy vast stretches of such environments and some livestock ranching occurs in forest zones. The type of grass cover in ranch areas varies greatly throughout the world. In the low latitudes are the tropical grasslands, generally termed savannas—such as the *llanos* of Venezuela, the *campos* of Brazil, and the *gran chaco* of Argentina–Paraguay–Bolivia. Generally, these tropical grasses are tall, coarse, and fibrous, so are not very edible, and their carrying capacity is low. Yet some tropical grasses, which are softer and more palatable, can support animals fairly well and would provide rather good range land if their spreading could be encouraged. In a few places, strains of lusher African grasses have been sown in forest clearings in the tropics and in some instances, seem to be thriving well enough to provide pasturage.

Sheep and cattle are only slightly deterred by rough land, and goats are famous for their ability to scale steep slopes. There is little direct correlation, then, between ranching regions and landforms; ranching occurs on plains, hills, plateaus, and mountains, in lowlands, and in highlands. If mountains are high enough, they pass the tree line, above which grasses are unrivaled. If the country at these altitudes

Figure 6-12 Cattle Ranching in the High Plains of the Western United States. *Cowboys have always been associated with cattle grazing on the range. This cattle drive scene is from the drylands of eastern Colorado. (Courtesy Denver Public Library, Western History Department)*

is not too rough, excellent mountain pastures provide plentiful food supplies for grazing animals.

In the midlatitudes, variations in temperature are associated with a distinctive practice, called *transhumance,* in which herders move their animals to different altitudes with a change in season. This system prevails in regions where the climate has one severe season, say, a very cold winter. The usual practice is to graze the herd on lowland forage in the winter when the mountain pastures are inaccessible and then in summer to move the herd to the slopes to permit replenishment of lowland grasses. This technique occurs in the sheep-grazing areas of southern Wyoming and western Colorado and in mountainous areas of Europe (see Figure 6-13). Transhumance is not customary in mild climates where the highland meadows can feed a herd year around. In such places the lowlands can be given over to more remunerative forms of bioculture, and ranching is largely confined to the highlands. This is the case in southern California and Nevada.

In the tropics, the combination of heat and humidity handicaps commercial grazing (particularly in Venezuela, Brazil, Paraguay, Bolivia, and northern Argentina); furthermore, animal diseases flourish in that climate, as do flies, ticks, and other pests. Native animals that have developed a resistance to their environment do not produce the high-quality wool and meat demanded by the market.

Origins and Relationships with Farmers

Livestock ranching is a newcomer on the world map. Whereas nomadic herding has been practiced for millennia, commercial grazing is little more than 200 years old. Not until the period 1750 to 1800 did world markets generate enough demand for wool, hides, and meat to support a commercial animal economy based on natural vegetation in the drylands. The advent of refrigeration in the 1880s gave the industry another boost, especially by opening up the Argentinian product to world market.

In two quite different ways, livestock ranchers find themselves involved with farmers. In some areas they ship their cattle to farmers in humid regions for fattening before final sale. This relationship holds particularly in North America, where the ranchers of various western states collaborate with corn belt farmers, as mentioned in Chapter 5, but increasingly the feedlot is moving to the prairie as well. On the other hand, as population has increased, the demand for food has risen, and new farms have been staked out, often on the humid margins of range country. In this instance, the tillers of the soil have steadily invaded grazing

Figure 6-13 Transhumance in Switzerland. *In autumn, flocks of sheep are led down from alpine pastures by way of rocky trails along streams to Belknap for shearing. (Courtesy Swiss National Tourist Office)*

land, tempered primarily by the availability of irrigation water.

Annual rainfall in the borderlands between humid climates and dry climates is a rather unreliable affair. The critical boundary sweeps back and forth across a climatic transition zone several hundred miles wide over a period of years. For example, during one 5-year period on the U.S. Great Plains in the 1920s, the boundary between the dry and the humid climates oscillated from western Iowa to central Montana. In Canada's prairie provinces about 80 percent of the variation in annual grain output is associated with weather conditions. Tilling the soil is clearly a risky business in this transitional area. In U.S. economic history repeated waves of farmers have invaded this zone—their hopes aroused by a series of a few unusually wet years—only to be driven to bankruptcy during subsequent drought years. Left behind have been two monuments to human ignorance of nature's ground rules: abandoned farmsteads and ruined land.

Without irrigation, indeed, cultivating these drylands is extremely risky. Once the sod is broken by the plow or exposed by overgrazing, the soil is no longer anchored by grass roots, and upon drying out, the soil particles become vulnerable to wind erosion. Dust bowl areas, where wind erosion strews dust around similar to that in sand dunes areas, are the inevitable result unless there is enough rainfall to enable a grass cover to take root immediately.

Farmers in Canada and the United States increasingly practice *low-till* or *no-till* agriculture in much of

this transition area and the technique may spread throughout the prairie. This approach saves energy costs associated with plowing and disking fields, reduces the risk of erosion as stubble from earlier plantings is not removed and stabilizes the soil, and saves moisture. On the negative side no-till agriculture requires the application of additional chemical herbicides to control weeds, and specialized planting equipment.

EVOLUTION OF U.S. RANGELAND

A distinct break between two periods in the evolution of the commercial grazing economy in the western United States occurred in 1880. Prior to that time the *open-range* predominated and afterward, the *organized-ranch* period. During the open-range period, which prevailed from the middle of the eighteenth century, commercial grazing on communal land characterized the western half of the country. Hides, wood, and tallow were far more important commodities than meat, which was either marketed in dried form or left to rot on the carcass after the hide was removed. The animals on which this economy was based were introduced from abroad. Many of the cattle had been brought by the Spaniards by way of Mexico.

During the open-range period, there were no operating units such as those that now exist. Animals were semiwild and were left to graze on the open range, little of which was owned by the people who owned the animals. At that time the federal government still held most of the land in the public domain. Harvest season was round-up time, and the herdsmen and their cowboys roamed hundreds of miles corralling the wandering animals. Branding was an essential practice in these days, for only by the brand symbol burned into the hide of an animal could a grazer claim possession of it.

Railroads had made only a few penetrations of range country prior to 1870 (Figure 6-14). Most of the animals were driven long distances to the nearest railway station. In the earliest days cattle drives from Texas, Oklahoma, Colorado, and other grazing states to the northeastern or middle western market states often occurred. Those were also the days of unbridled independence, of rudimentary law enforcement. Each cowman and sheepman was a law unto himself. Western novels, movies, and television shows still embody the spirit of that age, or at least a reasonable facsimile thereof. No limits to the number of herdsmen or to the number of animals, or their wanderings, existed. Such uncontrolled grazing could exceed the carrying capacity of the land, but less overgrazing occurred in the days of the open range than developed later with the fenced ranch.

The organized-ranch period that began around 1880 promoted the expansion of a well-planned commercial grazing economy. Several characteristics differentiate this phase from the preceding one. Meat became the main cash item, wool and hides second-

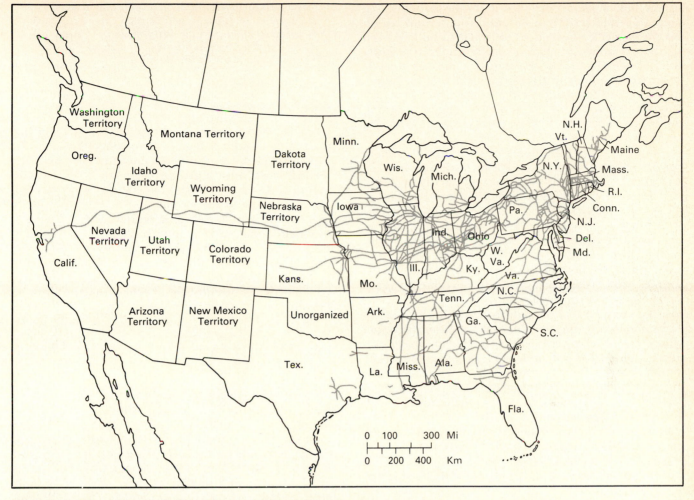

Figure 6-14 Railway Map of the United States, 1870. *Railroads barely penetrated range country prior to 1870, necessitating long overland cattle drives from various regions such as Texas, Okalhoma, and Colorado to the nearest railroad station. Note that a single trans-continental railroad traversed the country at that time, severely limiting service west of Iowa. (Courtesy of the Association of American Railroads.)*

ary. Emphasis in the marketplace turned to beef, and cattle became the most numerous animals, but sheep and goats remained important on the western margins of the region.

The chief new feature was the organized *ranch* itself. Private operators purchased and fenched large tracts and built ranchsteads. The fence played a key role in this transition to the organized ranch. Essentially, a fence protects a piece of land by preventing interlopers from bringing in unwanted animals. Further, the fence prevented the herds from wandering and interbreeding with poorer strains of animals; it also protected against marauders. Ranchers put in windmills, which became conspicuous features of the landscape, to power pumps in the wells drilled to augment the natural drinking water supply. Similarly, they supplemented the natural forage supply by growing crops that tided the animals through the winter in regions with dormant grasses in the colder months.

As the railway network expanded, threading its lines into the grazing regions, new markets opened up. These new markets increased the demand for meat products, making possible larger-scale grazing operations which placed more pressure on the land. Overloaded pastures in dry years led to overgrazing and greater wind erosion of the soil. At the same time, the government restricted grazing on public lands—forests as well as grasslands—renting out the privilege each year to ranchers in accordance with the capacity of the range. In 1934, the government's policy was formalized in the Taylor Grazing Act, which divided the public domain into numerous grazing districts in which privileges could be leased and rules established for grazing.

Another development in the organized-ranch period was a closer collaboration between ranchers in the west and farmers in the midwest and east, as discussed earlier. Since the short grass of ranch country is not as effective in fattening cattle as the corn belt, ranchers began selling yearlings to midwestern farmers for fattening. In the days when the main cash product was hides, this movement created no handicap. When meat became the premium product, the animals either had to be shipped to a humid region for fattening, or fattening feeds had to be imported to the drylands—or grown locally under irrigation. Since the final markets were mostly in the east, it made sense to ship the lean steers eastward, first for fattening, and later for slaughtering. At the turn of the century, farmers as far east as New York state were getting animals through the Buffalo stockyards for fattening on their farms.

Long-distance shipments of live cattle by rail to major terminal markets and meat-packing plants in Cleveland, Chicago, and Omaha gradually replaced the overland drives. The feedlot system described in Chapter 5 maintained the same directional flow as well, but gradually, refrigerated cars began picking up processed carcasses farther west in the eastern plains, nearer the growing areas. Feedlots moved westward as well. Those feedlots in Texas and Oklahoma, for example, now feed cattle locally produced grain sorghum and cottonseed cake from irrigated fields—not the corn associated with the finishing process in Iowa. Gradually, the terminal market concept began disappearing as buyers came to feedlots to make purchases. Videotapes of cattle sent to buyers' offices or remote video or tele-auctions held at the feedlot now frequently replace the intermediate shipment stage to a stockyard. A recent innovation in the industry, which further reduces distribution costs, is to break down the slaughtered animal nearer the feeder site by removing bone and fat and ship the product in boxes rather than as a carcass. In this way, meat packing now occurs on the high plains in former dust bowl areas.

> By discarding nearly one-quarter of the weight which is fat and bone and shipping the meat cut-up in boxes, Iowa Beef [Processors, Inc.] claims to save two million gallons of fuel oil every year on a volume of 3.9 million cattle that it processes into boxed beef . . .
>
> The company owns beef slaughter and processing complexes at Dakota City, Nebraska; Emporia, Kansas; and Amarillo, Texas and it operates additional slaughterhouses at Luverne, Minnesota; Ft. Dodge and Dennison, Iowa; and at West Point, Nebraska. These locations, in the heart of the grain-fed cattle market, typify the modern location pattern of the industry. The High Plains region currently produces more cattle than it slaughters and Iowa Beef is building a new slaughter/processing facility near Garden City, Kansas, that will surpass all other plants in size. In addition to those sites, the company operates two more plants in the northwest, at Pasco, Washington, and at Boise, Idaho. The Pasco plant also manufactures an edible tallow product that is consumed by the Idaho–Washington potato processing industry as an ingredient of precooked french fries.
>
> The spectacular growth of Iowa Beef in the last decade has come largely from acceptance of the boxed beef idea. The company now produces about fifteen percent of the total output of the U.S. beef industry, consuming approximately 100,000 cattle every week. When the new western Kansas complex opens it alone will require 4,000 head each day.

The movement of large-scale meat packing to such places as Amarillo and Garden City would have been thought impossible forty years ago. What will likely become the T-bone steak capital of the world was once the heart of the Dust Bowl.[4]

SHEEP PRODUCTION

As mentioned previously, sheep are less demanding than cattle in their forage requirements (Figure 6-13). Therefore, they can compete better than can beef animals in areas with a meager grass cover on the arid margins. Over 1 billion sheep are now tended throughout the world. Production is very widespread in developed and developing regions. Sheep have traditionally dominated animal husbandry in Mediterranean areas, extending from Spain to Italy, Yugoslavia, Turkey, Iran, and northern Africa. About one-fifth of the world's sheep are located in this region. England and Wales account for another concentration in Europe.

The world's leading concentration of sheep today occurs in the Soviet Union. Production areas are very extensive, but concentrations occur in the mountainous trans-Caucasus areas and in central Asia to the east of Tashkent. The second leading producing nation today is Australia, with China ranking a close third, followed by New Zealand. Australia, in fact, now dominates the world lamb and wool market. In the United States, a minor producer by world standards, sheep husbandry centers on the Edwards plateau region in Texas. Other concentrations occur in northern California and in upland areas of the mountain states of Colorado, Wyoming, and Utah. South Dakota completes the list of major producing states.

IRRIGATION

Dramatic changes in agricultural output occur in semiarid grazing areas with the introduction of irrigation. The more intensive and profitable farming operation this innovation affords typically involves a switch from animal husbandry to cropping or a change to a more profitable crop based on optimal dollar returns per acre. Introduced crops include citrus, cotton, sugar beets, and grapes. Several of these specialty crops are discussed in Chapter 7. Some irrigated land also produces grain (corn, wheat), cattle fodder (sorghum), hay (alfalfa), and various vegetables for human consumption (Table 6-6).

Kansas and Nebraska farmers with irrigated land, for example, typically convert wheat fields to corn acreages. About two-thirds of the corn harvested in these two states is irrigated. Alfalfa hay is another favored crop for irrigated land in the west. It serves as

[4]John C. Hudson, "Great Plains, U.S.A.—The 1980's," in John Rogge, ed., *The Prairies and the Plains: Prospects for the 80's*, Manitoba Geographical Studies 7, Department of Geography, University of Manitoba, Manitoba, Canada, 1981, pp. 7–8.

Table 6-6

Irrigated Corn and Wheat Acreages in Selected States, 1982 (thousands of acres)

Rank	State	Total Irrigated Acreage	Total Acreage	Percent Irrigated
		Irrigated Corn		
1	Nebraska	4,335	6,519	66
2	Kansas	775	1,162	67
3	Colorado	739	760	97
4	Texas	629	1,097	57
		Irrigated Wheat		
1	Texas	897	5,087	18
2	California	717	929	77
3	Kansas	715	11,664	6

Source: U.S. Bureau of the Census, *1982 Census of Agriculture.*

a winter feed supplement for beef cattle and as a staple for the dairy industry. Farmers continued with cattle rearing in many instances following the introduction of irrigation, especially in Texas, Oklahoma, and Florida. Feedlot operations, as discussed in the preceding section, benefit substantially from the introduction of irrigation. As Figure 6-15 indicates, much of the West has continued to function as pasture and range land following irrigation.

The majority of the production of several food and industrial crops now occurs on irrigated land in the United States. About 60 percent of potato and sugar beet production occurs on irrigated acreages. Over half of the commercial vegetable production in the United States, much of it from California, also comes from irrigated land.

The largest concentration of irrigated land on a world scale occurs in the intensive rice-growing economies of east and south Asia. Therefore, it is not surprising that three of the top five countries in total ir-

Figure 6-15 Irrigated Pastures and Farmland in the United States. *Dramatic changes in agricultural output occur in semiarid grazing areas with the introduction of grazing. As indicated in (a), much of the West has continued to function as pasture as this map shows. (Source: U.S. Census of Agriculture)*

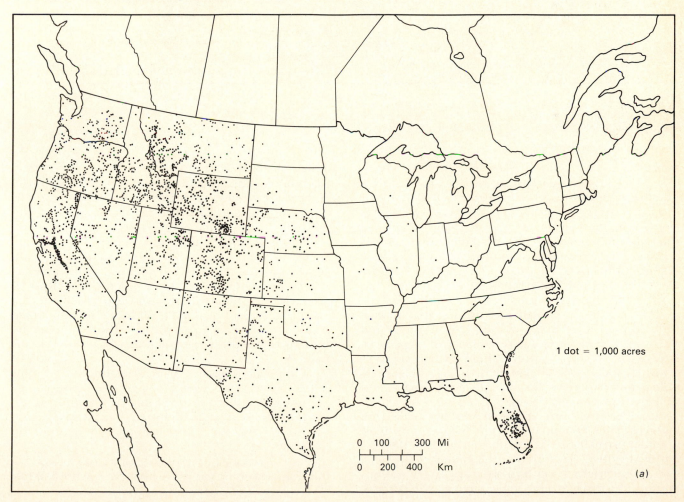

1 dot = 1,000 acres

0   100   300 Mi

0   200 400   Km

(a)

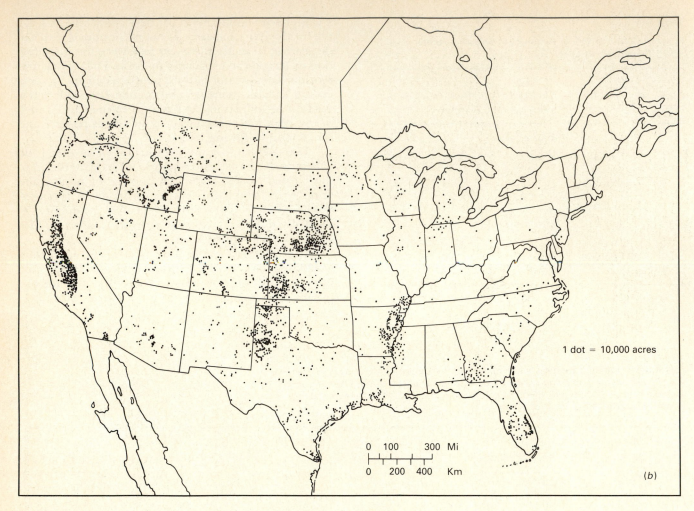

Figure 6-15 (cont.) *Note irrigated farmland concentrations in map (b) above in the Mississippi Valley, Nebraska, western Kansas, the panhandle of Texas, Southern Idaho, and the Central Valley of California. (Source: U.S. Census of Agriculture)*

rigated land are China, India, and Pakistan (Table 6-7). The United States and the Soviet Union are the other leading nations, each with about 19 million hectares (2½ acres) of land under cultivation. In the Soviet Union, the majority of irrigated land lies in central Asia, but it is also important in the trans-Caucasus and Ukraine. Furrow irrigation has traditionally dominated in the Soviet Union, but a transition to sprinkler-type irrigation, preferred in the United States, has occurred in the past decade.

The number of irrigated farms increased dramatically in the United States in the 1970s; in the period 1974–1978 alone there was an increase of 28 percent. Improvements in irrigation technology and access to water partially explain this increase. Sprinkler technology improvements, for example, have made it possible to be more selective in applying water over short periods and on short notice in western states with highly variable precipitation regimes. One of the most popular sprinkler irrigation technologies is the center-pivot system. Anchored at the center of the field, the self-propelled center-pivot sprinkler sweeps around in a circular arc. Since it misses the corners,

square fields become circular producing units, creating a fascinating landscape change (see Figure 6-16). In fact, the circles created by center-pivot irrigation may supplant the grain elevator as the most distinguishing feature of the plains landscape.[5] The first installations occurred in central Nebraska in the late 1950s. This innovation then spread in a classical diffusion process through the plains regions and beyond. Center-pivot systems now operate in at least 39 states. In 1970, there were over 7000 systems in the Great Plains, and nearly 30,000 by 1976, an increase of over 300 percent in six years. Nebraska alone, the leading center-pivot state, had nearly 12,000 systems in 1976.

California leads the list in acres under irrigation in the United States, even though center-pivot systems are not widely used there (Table 6-8). California agriculture is discussed in Chapter 7. Texas ranks a close second in irrigation. Other leading states, also in the

[5]Tom L. McKnight, "Great Circles on the Great Plains: The Changing Geometry of American Agriculture," *Erdkunde*, 33 (1979), 70–79.

Table 6-7

Irrigated Farmland, by Nation, 1984
(thousands of hectares)

Rank	Nation	Quantity	Percent of World Total
1	China	45,144	21
2	India	39,500	19
3	United States	19,831	9
4	Soviet Union	19,146	9
5	Pakistan	14,720	7
	Subtotal	138,341	65
	World total	213,376	100

Source: United Nations, *1984 FAO Production Yearbook*, 1985.

plains or mountains regions, include Idaho, Colorado, Kansas, and Montana. Note that all these states lie to the west of the humid/dry climate border discussed earlier.

INTERNATIONAL TRADE

Livestock ranchers in the major world regions depend for their livelihood on two markets: western Europe and Anglo-America. In the early days of commercial grazing the demand was for hides, wool, tallow, and only slightly for meat—and even that in dried form. Before the days of refrigeration, fresh meat simply could not stand the long trip from the western United States to the Atlantic seaboard, from Buenos Aires to Berlin, or from Capetown to London. The population explosion and the increased purchasing power attributable to the industrial revolution in these two market areas triggered the development of commercial animal economies.

Commercial grazing areas are so far from their markets that cheap transportation is essential to their

Table 6-8

Leading Farm Irrigation States, 1982 (millions of acres)

Rank	State	Acreage Irrigated	Percent of U.S. Total
1	California	8	17
2	Nebraska	6	12
3	Texas	6	11
4	Idaho	4	7
5	Colorado	3	7
	Subtotal	27	54
	U.S. total	49	100

Source: U.S. Bureau of the Census, *1982 Census of Agriculture.*

Figure 6-16 Center Pivot Irrigation in Minnesota. *The circles created by center-pivot irrigation may supplant the grain elevator as the most distinctive feature of the western landscape. Two quarter-sections of corn are being irrigated with this self-propelled system. One 110-foot well serves both fields on this Minnesota farm. (USDA—SCS photo by Russell V. Jongewaard)*

existence. Fortunately, for the most distant ones (Australia, southern Africa, and South America), cheap ocean transportation is possible. Railroads converge from the grazing country on a few port cities such as Buenos Aires, Montevideo, Salvador (Brazil), Lobito (Angola), Capetown (South Africa), and Freemantle (Australia). In the early days of commercial grazing in the United States, southern Californians shipped hides and tallow all the way around Cape Horn to New York and the east coast. Galveston and New Orleans dockworkers still load ships with wool and hides, but by and large the American grazing region is dependent on truck movements rather than rail and waterways.

As late as 1900 the United States was the world's leading exporter of fresh beef and a major exporter of wool. Europe was the principal customer. Europe is still an important customer, but the United States has reversed position, to become an importer of wool, hides, skins, and meat. This reversal resulted from several forces. The American market has expanded so much that there is no longer a surplus for export, other forms of agriculture have invaded ranch territory, and vigorous competition from more recently developed ranches in the southern hemisphere captured the European market.

Most surplus wool, skins, and meat must be transported across international boundaries to reach their markets. In some countries these products account for very high percentages of total exports:

Uruguay	81%
New Zealand	52%
Australia	37%
Argentina	33%
Paraguay	26%
South Africa	11%

The six countries listed above all have midlatitude grasslands except for Paraguay. Of the three items—wool, skins, and meat—meat provides over one-half of the export income in four of these countries. In Australia and South Africa, wool provides over one-half of the total export income from these commodities.

Japan and Italy dominate the wool import markets, accounting for about one-third of total world imports (Table 6-9). Australia and New Zealand, on the other hand, export nearly two-thirds of the world dollar value of wool.

There is a paradox involving regions of nomadic herding and those of livestock ranching. Since the foremost market for wool and hides is western Europe, it would seem that the nearest dry regions of steppe vegetation would serve that market. Northern Africa and the Middle East satisfy these environmental requirements and lie from 500 to 3000 miles from Europe. Moreover, the Mediterranean Sea links them to Europe. But environmental factors are not enough.

Table 6-9

Leading World Market Economy Wool-Importing and Wool-Exporting Nations, 1983 (millions of U.S. dollars)

Rank	Nation	Value	Percent of World Total
	Imports		
1	Japan	631	19
2	Italy	413	13
3	France	363	11
4	Great Britian	343	11
5	West Germany	300	9
	Subtotal	2,050	63
	World total	3,256	100
	Exports		
1	Australia	1,637	43
2	New Zealand	722	19
3	South Africa	279	7
	Subtotal	2,638	69
	World total	3,781	100

Source: United Nations, *1983 Yearbook of International Trade Statistics,* 1985.

Here is a vivid instance of the effect of cultural forces. In northern Africa, although it is much nearer Great Britain, people continue practicing nomadic herding. Australians, on the other hand, have bridged the 10,000 miles from their continent to Great Britain with wool and hide shipments. The tie to Europe on the part of Australia is due partially to its traditional status as a member of the Commonwealth and the close sociocultural connection with Great Britain.

SUMMARY AND CONCLUSION

Grain farming and livestock grazing remain major enterprises in midlatitude farms. These regions produce tremendous quantities of foodstuffs for a rapidly growing world population. Exports of these products to needy third-world and centrally planned countries provide a lifeline of support. The large-scale expansion of these activities awaited the growth of urban markets in the late nineteenth century, associated with manufacturing expansion and the deployment of comprehensive railroad networks. The notion of comparative advantage in the context of the von Thünen agricultural model assists in understanding the relative positioning of these activities, together with environmental constraints. In recent years the rapid expansion of irrigation has both assisted these industries, located as they are in semiarid environments, and promoted diversification of production through the introduction of specialty crops such as citrus and cotton as well as corn, wheat, and alfalfa. In addition to cattle rearing, sheep husbandry remains a major grazing activity on a world scale.

The major market for cattle and sheep products is the United States and western Europe. Whereas the United States was formerly a major meat exporter, imports are now more important. The United States and Canada are the biggest wheat exporters. Europe remains a net importer of grain. The Soviet Union, although the foremost wheat producer, must supplement that production with massive imports to meet the dietary needs of the country. For the future, the prospects are good for a strong grain and livestock industry worldwide because these two industries serve as major food suppliers.

SUGGESTIONS FOR FURTHER READING

BLOUET, BRIAN W., and FREDERICK C. LUEBKE, eds., *The Great Plains: Environment and Culture.* Lincoln, Neb.: University of Nebraska Press, 1979.

BLOUET, BRIAN W., and MERLIN P. LAWSON, *Images of the Plains: The Role of Human Nature in Settlement.* Lincoln, Neb.: University of Nebraska Press, 1975.

COWARD, E. WALTER, JR., *Irrigation and Agricultural Development in Asia.* Ithaca, N.Y.: Cornell University Press, 1980.

ERICKSON, JOHN R., *Panhandle Cowboy.* Lincoln, Neb.: University of Nebraska Press, 1980.

HEWES, LESLIE, *The Suitcase Farming Frontier: A Study in the Historical Geography of the Central Great Plains.* Lincoln, Neb.: University of Nebraska Press, 1973.

HUDSON, JOHN C., *Plains Country Towns.* Minneapolis: University of Minnesota Press, 1985.

MEALOR, W. THEODORE, JR., and MERLE C. PRUNTY, "Open Ranching in Southern Florida," *Annals,* Association of American Geographers, 66 (1976), 360–376.

ROGGE, JOHN, ed., *The Prairies and the Plains: Prospects for the 80's,* Manitoba Geographical Studies 7, Department of Geography, University of Manitoba, Manitoba, Canada, 1981.

SPRAGUE, HOWARD B., ed., *Grasslands of the United States: Their Economic and Ecologic Importance;* A Symposium of the American Forage and Grassland Council, Ames, Iowa: Iowa State University Press, 1974.

WEBB, WALTER PRESCOTT, *The Great Plains.* New York: Gunn, 1931.

WEBB, WALTER PRESCOTT, *The Great Frontier.* Austin, Tex.: University of Texas Press, 1952.

WORTHINGTON, E. BARTON, ed., *Arid Land Irrigation in Developing Countries.* Oxford: Pergamon Press, 1977.

7 Specialty Crops and Poultry

The major midlatitude commercial crops under cultivation on a world scale, in addition to grains, include cotton, tobacco, citrus, and various other horticulture and vegetable products. The discussion here focuses on the aforementioned products as well as poultry. Accounting for the location of these activities using the von Thünen formulation generally does not work due to the specialized growing requirements for most of the crops individually and their widespread distribution as a group. Fresh vegetables are widely grown near all major markets because of perishability considerations and their ability to compete favorably on relatively more expensive land near the market. This aspect of the industry supports the von Thünen conceptualization, but vegetable production is not limited to the immediate market area of the city. Using greenhouses, transplanting techniques, hydroponics, and irrigation, producing areas are greatly expanded in many parts of the world.

In this chapter we initially explore the California specialty agriculture region, not only the most important such area in the United States, but also in the world. This discussion will be succeeded by a more-in-depth examination of four specialty crops—cotton, tobacco, grapes, and citrus—and a section on the poultry industry. In each instance emphasis is placed on U.S. production patterns, especially the situation in California and the southeast. The world production for each of these four major crops is also covered. Most of the output of these crops, with the notable exception of cotton and tobacco, occurs in areas with Mediterranean-type climates—hence the label "Mediterranean agriculture" (Figure 7-1).

The term "Mediterranean agriculture" derives from an association of the largest area with this type of farming—the region that encircles the Mediterranean Sea. Other areas characterized by similar agricultural environments include the Central Valley of California, central Chile, and the extreme southern tip of Africa.

Rainfall totals are low in Mediterranean agricultural areas, typically ranging from 10 to 40 inches annually. Unlike much of the rest of the world, rainfall totals are higher in winter than in summer months in Mediterranean areas. Summers, in fact, are typically very dry. Temperatures are mild and the growing season almost continuous in this region. Nearness to water bodies is one reason for the mild climate. These areas often border steppe and desert zones. The natural vegetation reflects the unfavorable growing conditions—trees often have thick, oily leaves which help retain moisture, typified by the olive and fig tree.

Given the moisture problem in Mediterranean areas, agriculturalists have developed *dry farming* techniques to compensate for drought periods. Fields are cultivated in the fall to loosen soil and maximize the absorption capacity of winter rains. This aeration process is frequently repeated during the winter and spring. During the summer, shallow harrowing is practiced to retard moisture loss.

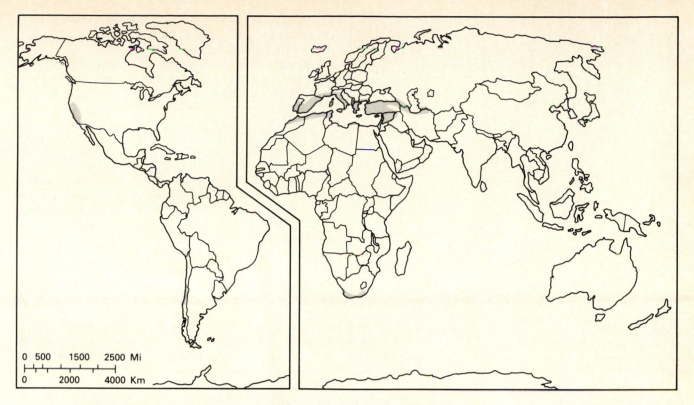

Figure 7-1 Mediterranean Agriculture Regions. *Named after the area most closely identified with this type of agriculture, the Mediterranean Basin, one also finds similar activity in extreme South Africa, Chile, and California. Hot dry summers and cool winters occur in Mediterranean areas. Grapes, citrus crops, and livestock grazing are some of the important activities in these areas.*

SPECIALTY AGRICULTURE IN THE UNITED STATES

In the United States, specialty agriculture practically encompasses the entire coastal margin of the country. Recall, for example, the U.S. map in Figure 5-3. Truck farming (fresh vegetables), tobacco, and citrus form a peripheral band of agriculture along the east coast, while a broad mix of vegetable, fruit, and nut agriculture provides a similar array on the west coast in California. On the Gulf coast, particularly Texas and Louisiana, rice and sugarcane production occurs.

California Corporate Farming

California leads the country in vegetable production. An impressive array of crops are produced in the Central Valley, the primary agricultural region of the state. California is the leading U.S. grower of such products as asparagus, cauliflower, spinach, tomatoes, and strawberries. Over half of the U.S. lettuce, broccoli, and cauliflower output comes from California. The same dominance occurs with many fruit products—melons, peaches, plums, apricots, lemons, avocadoes, grapes, figs, and dates—and with nuts—almonds, pistachios, and walnuts. About 40 percent of U.S. fruit and nut output comes from California, as do over half of the canned and frozen fruits and vegetables.

The reason that California competes so strongly for specialty agriculture relates to its favorable year-round Mediterranean-type growing season; good soil characteristics; a favorable irrigation system; efficient, well-managed, highly mechanized coporate farms; and a good marketing and distribution system for the products.

Agriculture, in fact, is California's largest business. The term *agribusiness* best describes this booming enterprise, which in many ways operates on the scale of the tropical plantations discussed in Chapter 8. On the other hand, the variety of crops produced (over 200) and the lack of dominance of any single activity makes the area unique in this age of specialization and large-scale enterprise. "Only four crops—cattle, dairy products, cotton and grapes—each account for more than 5 percent of gross farm income."[1]

The California Central Valley producing region, sometimes referred to as the Great Valley, stretches nearly 500 miles southward from Redding (see Figure 7-2). The southern half of the area, the San Joaquin

[1]"California's Agribusiness," *The Economist*, December 13, 1980, p. 68.

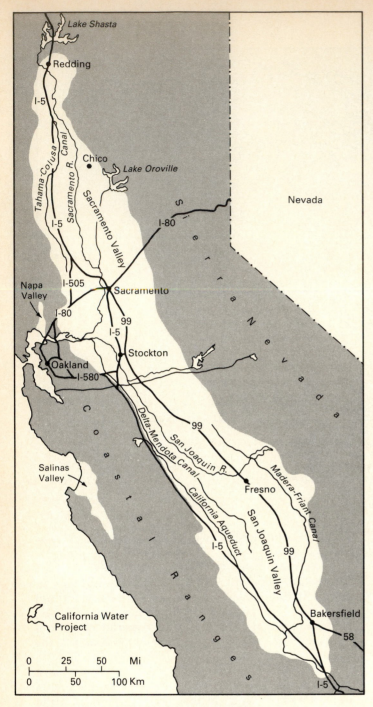

Figure 7-2 Central Valley of California. *Sometimes referred to as the Great Valley, this region stretches nearly 500 miles from Redding in the north to Bakersfield in the south. The Sacramento River Valley anchors the northern half of the Central Valley and the heavily irrigated San Joaquin River Valley anchors the southern portion. An impressive array of crops are produced in the valley, the primary agricultural region of the state.*

Valley, is heavily irrigated. The long growing season, favorable soils, and the availability of water create an unparalleled crop-producing environment. The northern portion of the Central Valley, the Sacramento Valley, does not have as much irrigation, but its use is increasing as farmers switch from wheat and hay crops to more intensive agriculture, including rice

and tomatoes. Other major California farming regions include the Imperial Valley in the south and various coastal valley inlets.

Some of the most productive farmland in years past, such as the Santa Clara Valley south of San Francisco, is succumbing to urbanization. The encroachment of high-technology office and research activity into this Silicon Valley region has displaced many farmers (see Chapter 16). Those not driven out of business have converted their land to fruit and nut production, which yields a higher return on investment and can compete more effectively with urbanization. Land devoted to walnut production commanded values of $5000 to $6000 per acre in 1980, whereas that for vegetables in the same area averaged less than $3000 per acre.

In addition to the urbanization threat to agricultural land, the biggest long-term problem confronting the California agribusiness manager is water. Elaborate public works canals move water from the surplus region in the north to the deficit area in the south. Associated problems include a water table that is dropping, growing competition from urban and industrial users who want a greater share of the water, and growing quantities of salt deposits in the soil due to irrigation and poor drainage, which threatens long-term fertility.

Heavy dependence on migrant labor presents another problem. The care of vegetable crops in California and elsewhere requires a large number of laborers, especially during the harvest season. On most vegetable plantations, the migrant labor force can outnumber the permanent force six- to eightfold. For example, a 2000-acre operation with a permanent labor force of 25 (tractor drivers and irrigators) may have 200 seasonal migratory workers, mostly Mexican nationals, who weed and harvest the crop. Some of these workers are illegal aliens, referred to locally as "undocumenteds." Wage and benefit programs for them often meet only minimal standards, and housing accommodations generally are quite meager. Attempts to unionize the labor force have been only partially successful.

Central Valley Irrigation Project

The history of water projects in California reflects the history of agriculture because the driving force has been irrigation in the San Joaquin Valley. Early agriculturalists practiced cattle ranching and wheat farming in this area in the latter half of the nineteenth century, but the industry declined by the turn of the present century. Farmers gradually turned to more intensive agriculture as they discovered efficient ways to irrigate fields. By 1906, there were over 500 wells in the valley used for irrigation purposes. The numbers grew dramatically by 1920, when there were 11,000 wells, and went up again to nearly 24,000 in 1930. Over 1.5 million acres were irrigated by groundwater in that region by 1940.

It became clear in the early 1900s that groundwater alone could not provide enough water for irrigation needs. Engineers proposed supplementing groundwater with surface water from new reservoirs and an elaborate canal network. Local canals and distribution systems were built by local authorities and lobbying began for two major statewide programs: (1) the Central Valley Project and (2) the State Water System.

The Central Valley Project was first proposed in 1920 but not approved by the legislature until 1933. The idea was to sell bonds to build dams and canals together with pumping stations and hydroelectric plants. Cash generated by the sale of electricity would pay for the program as then envisioned. But the Great Depression intervened, forestalling the possibility of state financing, so the program gradually became a New Deal effort under the auspices of the Federal Bureau of Reclamation. World War II delayed the program further, but in 1951, the Shasta Dam on the Sacramento River and the Contra Costa Canal were completed. Eventually, an elaborate irrigation network unfolded as a part of this predominantly federal program.

The other major effort, the State Water System, aimed to serve both the needs of the San Joaquin agriculture region and the southern California urban water districts in the greater Los Angeles basin. Although this program was originally proposed in the early 1950s, it took until the early 1970s for the system to come into existence, following the completion of the Oroville Dam on the Feather River and the San Luis Dam on a creek of the same name.

The largest user of the water created by this program was southern California, which received its supply via the California aqueduct. The San Joaquin Valley became the second biggest user. Unlike the earlier Central Valley Project, state funds primarily funded this system and user fees paid most of the cost. The system has not been completed; additional canals and dams are contemplated as water needs continue to multiply.

COTTON

Cotton production occurs widely throughout the world in areas with warm subtropical growing environments, typically on the equatorial side of the midlatitudes. Most of the production occurs in the northern hemisphere, with the notable exception of the output in Brazil and Argentina (Figure 7-3). The leading producing countries today include China, the United States, and the Soviet Union (Table 7-1). The length of the cotton staple produced in various areas differs, with each staple length yielding threads with different qualities and tensile strength which has a market in a particular fabric or textile item. This situation leads to considerable movement of cotton in international trade among major producing areas regardless of the level of production or the size of the domestic market.

Figure 7-3 World Cotton Production Regions. *Cotton production occurs in areas with warm subtropical growing environments, typically on the equatorial side of the midlatitudes. Most of the production occurs in the northern hemisphere, paced by the output in the United States, the Soviet Union, India, and China.*

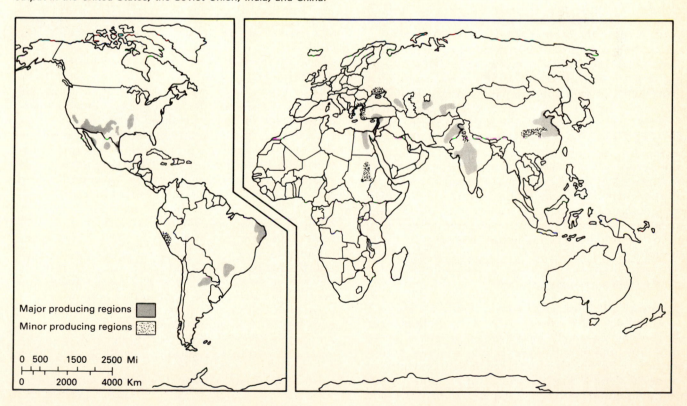

Major producing regions
Minor producing regions

0 500 1500 2500 Mi

0 2000 4000 Km

Table 7-1

Leading Cotton-Producing Nations, 1984
(thousands of metric tons)

Rank	Nation	Production	Percent of World Total
1	China	6,077	34
2	United States	2,894	16
3	Soviet Union	2,400	14
4	India	1,250	7
5	Pakistan	990	6
6	Brazil	618	4
7	Turkey	586	3
8	Egypt	390	2
	Subtotal	15,205	86
	World total	17,794	100

Source: United Nations, *1984 FAO Production Yearbook*, 1985.

Four main groups of cotton fibers can be distinguished based on the length and quality of the staple:

1. Very long and fine staple ($1\frac{1}{4}$ inches), grown widely in Egypt
2. Long staple (1 inch), grown in the United States
3. Medium staple ($\frac{7}{8}$ inch), grown in the United States and the Soviet Union
4. Short staple (under $\frac{7}{8}$ inch), Asiatic variety with no irrigation

Cotton production increased dramatically following the invention of the cotton gin in 1793 by Eli Whitney and with the coming of the industrial revolution of the early nineteenth century. World production levels continued rising in the early twentieth century, following mechanization and irrigation advances. The boll weevil, however, badly decimated production in the second quarter of this century.

In the period 1950–1980, cotton production doubled worldwide, owing primarily to increased yields per acre. In the United States, there have been no improvements in yields in the past 15 years. Problems of weather, insects, and weeds have kept the yield at a stable level. As production costs have increased in the past decade due to higher energy, fuel, fertilizer, herbicide, and pesticide prices, competition with other crops has increased. Soybeans, for example, are now generally more competitive than cotton in the southeastern United States. In the mid-south (Mississippi delta), cotton maintains a competitive edge over soybeans. In the southwest (Arizona), competition with grain sorghum is keen, and in the far west, grain crops compete strongly.

Environmental Considerations

Originally a tropical crop, cotton requires a growing season of about 200 frost-free days. Usually, the longer the growing season, the longer the staple length. There must also be 5 months of high temperatures, preferably above 70°F. Growth stops with temperatures below 60°F. Plant activity is typically three times as fast with temperatures of 90°F compared to 68°F. Cloudy weather, though, prevents the cotton boll from ripening. Unless irrigated, at least 20 inches of rainfall are needed annually, with 8 inches of this distributed through the growing season. Excess moisture in the growing season, however, promotes attacks by fungi and bollworms. Cotton grows on a large variety of soils; it does particularly well on less rich ones that limit excess vegetative growth at the expense of boll growth. The deep taproots need oxygen during the growing season, so long periods of heavy rains and crusty soil at the surface can be detrimental.

Demise of the Cotton Belt

The importance of cotton, which had dominated much of the Plainsland South between the Civil War and the First World War, began to decline dramatically during the 1920's. Before 1930, for example, more than 40 percent of the cropland harvested in the six states of Arkansas, Louisiana, Mississippi, Alabama, Georgia, and South Carolina had produced cotton but, by 1970, the figure had dropped to 16 percent and the actual acreage had shrunk from more than 15 million acres to less than 4 million. In 1900, these six states produced more than 60 percent of the nation's cotton but, by 1970, they were producing only 40 percent.[2]

The primary factors associated with the demise of cotton included the depletion of the soil, which meant lower yields, disease, primarily the boll weevil, and a gradual retreat of production to the best land and to areas where mechanization worked best, typically on upland terraces in major river valleys.

The limits of the old cotton belt and the location of major specialty crops now grown in the region are shown in Figure 7-4. The predominance of the *plantation system* in the American cotton belt before the Civil War is well known. There seems to be a widespread impression, however, that the plantation system disintegrated with the abolition of slavery. To be sure, the nature of the plantation labor force did change radically, but the landholdings remained generally intact and have persisted to the present time. Indeed, the term "plantation" is still common coin in the south to apply to an agricultural enterprise that meets the following criteria: (1) A landholding considerably larger (averaging between 700 and 800 acres) than a "family farm" (a farm operated by a single family). (2) A division of labor and management duties. (3) Specialization in commercial production. Rarely, however, is the specialization in just a single item; rather, groups of two or three specialties are the norm today, principally some combination chosen from cotton, rice, sugar, soybean, pecans, peanuts, tobacco, and hogs. (4) Location somewhere in the south, where the plantation tradition is part of the cultural heritage, namely, the Mississippi valley and a broad belt from the Brazos River in Texas to the James River in Virginia. (5) Centralized location of dwellings (in a "plan-

[2]John Fraser Hart, *The South* (New York: Van Nostrand, 1976), p. 33.

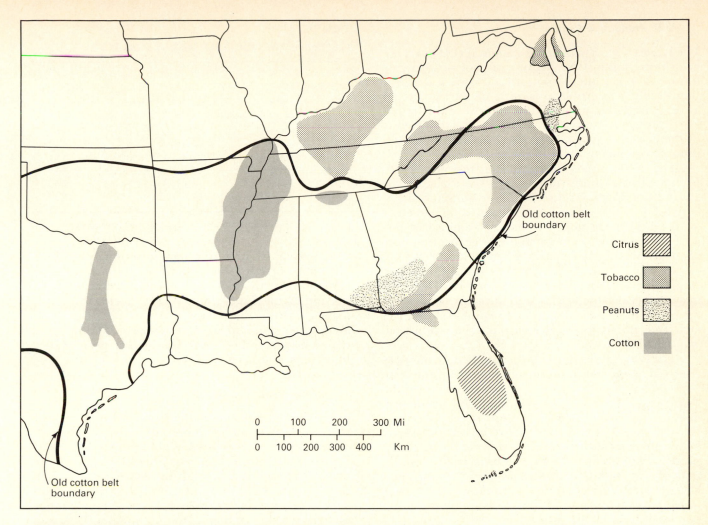

Figure 7-4 Specialty Agriculture in the Southeastern United States. *In place of the old cotton belt in the South, one finds large areas of general farming (shown here as blank areas on this map) along with several specialty production regions. Two major areas of tobacco production occur in the Carolinas and Kentucky; citrus occupies an area of Central Florida; peanuts identify with South Georgia along with tobacco, and cotton is raised on both the floodplains of the Mississippi Valley and terraces of the Trinity and Brazos river valleys.*

tation village") and of farm operations. (6) A large input of cultivation power per acre.

The term *neoplantation* applies to a type of plantation operation that has evolved since World War II. What mainly distinguishes this type from traditional plantations is a high degree of mechanization. This mechanization process was also associated with the migration of rural labor to cities in response to the lure of higher wages in industrial plants and an increase in labor and other production costs relative to market prices. To resolve these two problems, farm operators substituted machines for labor. On one plantation, for example, three tractors replaced 14 teams of mules and the labor force decreased from 16 to 7

The functional focus of the neoplantation is the tractor station, which also shelters harvesting, cultivating and accessory machinery, spare parts, repair tools, and fuel. Neoplantation occupance eliminates most share-paid labor. Cash-wage labor, which works machines under central direction, and which is applied as a unit to the whole acreage of the holding in a manner approximating the employment of gangs of slaves on large fields under ante-bellum occupance, predominates on these holdings today. The tractor station is always located close to the manager's or owner's residence, a location factor suggesting the intimate interrelationship of centralized management and cultivating power. Thus its location approximates closely that of the mule under ante-bellum occupance.

The neoplantation differs from the ante-bellum unit in a few significant spatial aspects. The proportion of woodland is less, generally, than on the ante-bellum farm.... and beginnings of sensible legume-livestock rotations are implied by the crops now grown. There are fewer houses than on the ante-bellum unit of the same acreage, and their arrangement customarily is more linear (along roads) than rectangular and compact.[3]

[3]Merle Prunty, Jr., "The Renaissance of the Southern Plantation," *Geographical Review*, 1955, pp. 459–491.

U.S. Production Regions

The largest cotton-producing state in recent years has been Texas, but in 1982, because of poor weather conditions in Texas, the output of California led the nation (Table 7-2). About one-third of the acreage in Texas is irrigated, whereas 100 percent of the output is irrigated in Arizona and California. Practically none of the cotton is irrigated in Mississippi or other traditional producing areas in the southeast.

The mid-south cotton-growing region extends from southern Illinois to northern Louisiana on the alluvial floodplain (Figure 7-5). This bottomland soil is rich but needs protection from flooding as well as artificial drainage improvements. Memphis, the major city in the region, became the largest cotton market. Memphis today remains the headquarters of the Cotton Foundation and National Cotton Council of America, even though the role of the city as a cotton market has declined. Rice and soybean production have now increased in the delta region. Arkansas, in fact, leads the nation in rice output, followed by Louisiana and California.

The Blackland Prairie of eastern Texas comprises a second major cotton-producing region. Cotton competes most favorably on the lower river terraces of the Trinity and Brazos river valleys. A third major producing region is centered in west Texas, extending into southwestern Oklahoma. This region, in the Lubbock sphere of influence, leads the nation in cotton productivity. Water from deep wells irrigates this portion of the High Plains, using center-pivot systems. The water table is dropping steadily due to this pumping, increasing the cost of irrigation and making production of cotton less profitable.

Table 7-2

U.S. Cotton Production, 1982 (millions of bales)

Rank	State	Production	Percent of Acreage Irrigated	Percent of U.S. Total Production
1	California	2.9	100	25
2	Texas	2.7	27	24
3	Mississippi	1.6	6	14
4	Arizona	1.2	100	11
	Subtotal	8.4	—	74
	U.S. total	11.4	35	100

Source: U.S. Bureau of the Census, *1982 Census of Agriculture.*

A fourth major cotton region occurs in the Central Valley of California. Cotton cropping leads the agricultural activity in five counties of the southern portion of the San Joaquin Valley. Yields are much higher in that area than in the traditional production areas in the south. A major outlier cotton-producing region occurs in the heavily irrigated Imperial Valley in extreme southern California.

World Production Regions

Cotton, the leading fiber crop produced in the Soviet Union, is also the most important irrigated crop. All Soviet cotton is irrigated. Beginning in the late 1960s Soviet cotton production has exceeded that of the United States. It is now a leading Soviet export product. About 90 percent of the crop comes from central Asia, predominately to the southeast of the Aral Sea on irrigated alluvial floodplains of the Amu Darya and

Figure 7-5 Major Irrigated Cotton-Production Regions in the United States. *In addition to the irrigated cotton in the Mississippi River Valley, one finds major concentrations on the high plains of west Texas, southern Arizona, and the Central and Imperial Valleys of California.*

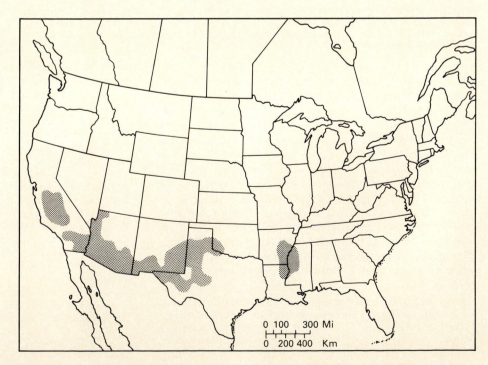

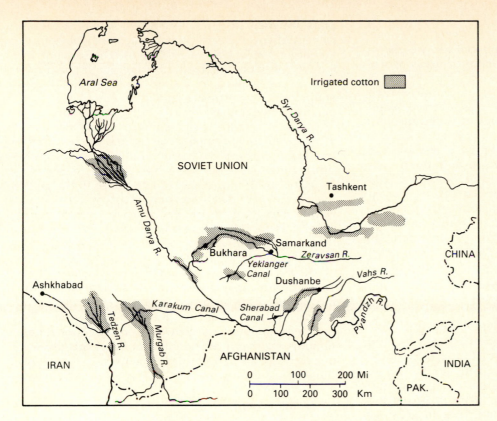

Figure 7-6 Irrigated River Valleys in Soviet Central Asia. *Since the late 1960s Soviet cotton output has exceeded that of the United States, and is now a leading export. Virtually all the crop comes from irrigated areas southeast of the Aral Sea and alluvial floodplains fed by canals and rivers such as the Amu Darya and Syr Darya. (Redrafted from Central Intelligence Agency. U.S.S.R. Agricultural Atlas, 1974)*

Syr Darya rivers (Figure 7-6). A former lake basin, now covered with rich alluvium to the southwest of Tashkent leads in production. A secondary producing area lies to the west of the Caspian Sea in Azerbaijan. China also produces large quantities of cotton, primarily on floodplains of major rivers in the northeastern section of the country. Other major producing areas include western India, Pakistan, and the Nile valley of Egypt.

International Trade

The largest cotton-exporting nation among the market economy countries is the United States, which dominated the market with 38 percent of the export trade in 1983 (Table 7-3). Egypt, Pakistan, and Turkey are other major exporters, each contributing 4 to 7 percent of total world trade. Significantly, only the Soviet Union, the United States, and Egypt are major exporters among the top seven producing nations. In China, India, and Pakistan, cotton is primarily consumed domestically. The major importing nations lie in east Asia (Japan, South Korea, and Hong Kong) or western Europe (Table 7-3). Japan and South Korea together import just over one-third of the product, while Italy, West Germany, and France import one-fourth.

Cotton's share of the world fiber market declined slightly in the 1970s as competition with synthetics increased, but still accounts for over half of the market. Cotton consumption increased worldwide in this same period due to population growth. The former dispar-

ity in price between cotton and cheaper synthetics largely disappeared following energy price increases in the late 1970s that inflated the cost of petroleum-based materials. The textile and apparel industries associated with cotton and synthetic fibers are examined in detail in Chapter 14.

Table 7-3

Leading World Market Economy Cotton-Trading Nations, 1983 (thousands of U.S. dollars)

Rank	Nation	Value	Percent of World Total
	Imports		
1	Japan	1,145	23
2	South Korea	529	11
3	Italy	481	10
4	West Germany	403	8
5	France	302	6
6	Hong Kong	234	5
	Subtotal	3,094	61
	World total	4,924	100
	Exports		
1	United States	1,860	38
2	Egypt	321	7
3	Pakistan	334	7
4	Turkey	213	4
	Subtotal	2,728	56
	World total	4,879	100

Source: United Nations, *1983 International Trade Statistics*, 1985.

TOBACCO

Unlike many specialty crops, tobacco can be grown widely in tropical as well as midlatitude areas. World trade is smaller than that for several other crops because many areas are self-sufficient. Imports often supplement homegrown crops because of the demand for tobacco products using blends of different types of leaves.

Tobacco plants are very responsive to climatic and soil conditions. Leaves on plants grown in the tropics are dark and strong in flavor, while Mediterranean products such as those grown in Turkey are known for their aromatic qualities. Tobacco grown on the coastal plain of eastern North Carolina produces bright yellow or yellow-brown leaves when cured. Tobacco produced on the interior Piedmont (Kentucky) has a much darker leaf. Binder tobacco, which has a firm leaf used to hold together the tobacco in a cigar, is usually grown in cooler northern climates, such as the Connecticut valley or south central Wisconsin.

Various techniques for drying tobacco also affect its flavor. Tobacco-producing regions, except tropical areas, have traditionally had distinctive wooden barns on the landscape where this drying occurs. Some barns are airtight, whereas others use ventilation processes. Airtight barns use flue-curing or fire-curing techniques to dry leaves that are hung in $1\frac{1}{2}$-story barns. Flue curing involved heat moving through flues. Fire curing traditionally involved burning wood on the earth floor of the barn to produce heat. Some modern curing now occurs in airtight prefabricated aluminum-clad units using natural gas forced-air heat (Figure 7-7).

Tobacco fields are typically small (about 100 acres) and do not occupy large acreages of the farms of which they are a part. Usually, very small garden-sized plots are used to sprout seedlings, which are fumigated and covered with plastic, then transplanted to the larger fields. Growing tobacco remains labor intensive, but more mechanization in planting and harvesting has evolved in recent years. In the United States, tobacco output is guided by complex federal policies, involving production allotments calculated in terms of pounds, price supports, and marketing controls. These allotments can be sold by the farmer to other producers. Such federal involvement in subsidizing the tobacco industry has come under increasing criticism in recent years in response to the growing evidence of health-related problems associated with smoking cigarettes. Nevertheless, a vocal lobby of agricultural interests in tobacco-growing states continues to argue strongly for government protection of the industry.

In the future, an economical use of tobacco as a protein source may provide an alternative market for the product, but consumers may reject this use due to its past negative image as an addictive substance. Experiments demonstrate that tobacco provides an outstanding source of protein, but perhaps this use will be practical only in underdeveloped areas, where food supplies are most critical.

U.S. Production Regions

Tradition plays a strong role in explaining the location of tobacco production regions. Market factors have been important considerations, including access to buyers and auction houses, as is the federal government production allotment system. Tobacco production had its origins in the United States in tidewater Virginia on colonial plantations. From there production spread to North Carolina and inland to Kentucky, exhibiting a classic diffusion pattern. To-

Figure 7-7 Flue-Cured Tobacco Barns. *Increasing labor costs have led to newer replacement curing barns for the traditional one and one-half story flue-cured tobacco barns shown on the right. A newer, more efficient bulk curing facility is shown on the left. Note the use of propane gas heat in both types of barns on this farm near Warrenton, North Carolina. (Courtesy of Richard Pillsbury)*

bacco competed well on smaller plots and in rougher upland terrain in the latter region. Minimal competition occurred in these areas with other crops.

Today there remain two core regions of tobacco production, one focused on the eastern Carolina coastal plain, and the second in Kentucky, west of the Appalachians (Figure 7-4). The only significant outliers of these core areas are much smaller producing areas in south Georgia, southeastern Pennsylvania, and Maryland. The Carolina producing region represents the premier tobacco cultivation area of the world. The optimal soil conditions found there are not duplicated anywhere else. As a result, tobacco from the Carolinas is in strong demand both domestically and overseas for cigarette production. The stability of the producing area is also influenced by the government subsidy and allocation system, which strictly regulates output.

World Production Regions

The largest tobacco-producing countries in the world are China and the United States (Table 7-4). China produces 25 percent of the world total, while the United States contributes 13 percent. India, Brazil, the Soviet Union, and Turkey also produce heavily. Tobacco in China is produced on small parcels at many locations along the southeast coast and in scattered inland locations. Production in India occurs predominantly in coastal locations. An extensive tobacco-producing region occurs in the Soviet Union on the northern margins of the fertile triangle to the east of Moscow. In Turkey, production occurs on coastal margins of the Aegean and Black seas.

International Trade

European countries, the United States, and Japan import by far the most tobacco (Table 7-5). The United States and Great Britain are also major exporters, accounting for nearly one-half of world exports. A major adjustment to the trade pattern has occurred in the pattern in recent years. In 1979, Japan supplanted

Table 7-4

Leading Tobacco-Producing Nations, 1984
(thousands of metric tons)

Rank	Nation	Production	Percent of World Total
1	China	1,526	25
2	United States	791	13
3	India	497	8
4	Brazil	415	7
5	Soviet Union	350	6
6	Turkey	210	3
	Subtotal	3,789	62
	World total	6,205	100

Source: United Nations, *1984 FAO Production Yearbook*, 1985.

Table 7-5

Leading World Market Economy Tobacco-Importing and Tobacco-Exporting Nations, 1983
(millions of U.S. dollars)

Rank	Nation	Value	Percent of World Total
	Imports		
1	West Germany	548	15
2	United States	473	13
3	Japan	438	11
4	Great Britain	435	11
	Subtotal	1,894	50
	World total	3,864	100
	Exports		
1	United States	1472	39
2	Brazil	466	12
3	Great Britain	238	6
	Subtotal	2,176	57
	World total	3,775	100

Source: United Nations, *1983 International Trade Statistics*, 1985.

Great Britain as the leading single market for U.S. exports. Other significant tobacco-exporting countries are Brazil and Great Britain.

GRAPES

Of the three main uses for grapes, wine is the most important. Their use as fresh table fruit and as dried raisins remains secondary. More closely than any other crop, grape cultivation is associated with *mediterranean agriculture*. Grapes adapt well to dry farming techniques. Grape vines are typically planted several feet apart. Deep taproots facilitate maximum use of available moisture. The vines themselves also shade the ground in summer, keeping the soil relatively cool, retarding moisture loss. The hot summer sun and dry air, together with the droughty growing environment, interact to stimulate a favorable sugar/acid balance in the grapes (Figure 7-8).

Optimal growing environments for grapes are extremely important for vintage wines. Although conditions in production areas vary widely in temperature regimes, particular strains of grapes respond best to precise climate, soil, and slope combinations. Many vineyards produce their best product on slopes that permit proper air drainage and optimal sun angles. Frost during budding, flowering, or harvest can be disastrous. Warm, dry days are required during harvest. With so many environmental factors that influence the crop, it is no wonder that the quality of the product varies in the same vineyard from year to year.

With the exception of Greece and Turkey, where raisin production is also important, most countries surrounding the Mediterranean produce great quantities of wine from their grapes. Notice the close

Figure 7-8 Rhine River Valley Wine Region. *In this scenic river valley world famous grapes are produced on relatively steep slopes where microclimates and soil conditions are favorable for producing vintage quality products (T.A.H.)*

correspondence between Tables 7-6 and 7-7, which indicate that Italy, France, and Spain are leading grape- and wine-producing nations. Italy and France produce about 40 percent of the world's wine. These countries also lead in per capita wine consumption, with rates registering 10 times as high as those in the United States.

Soviet production occurs in the Ukraine, Moldavia (Romanian border area), and in a belt from the Crimean peninsula to the Caucasus. About 90 percent of U.S. grape/wine production occurs in California. The Napa and Sonoma valleys in the San Francisco

area are the most famous producing areas. Grape production in Argentina, the only significant grape and wine producer in the southern hemisphere, occurs in an irrigated desert area on the eastern foothills of the Andes Mountains.

International Trade

Specialization in wine production of one type—white, red, dry, sweet, and so on—is the norm. Therefore, it is not surprising to find that the largest production areas are also the largest importers (Table 7-8). The

Table 7-6

Leading Grape-Producing Nations, 1984
(thousands of hectares)

Rank	Nation	Production	Percent of World Total
1	Spain	1,710	18
2	Soviet Union	1,400	15
3	France	1,292	13
4	Italy	1,125	12
5	Turkey	640	7
6	Argentina	317	3
7	United States	313	3
	Subtotal	6,797	71
	World total	9,611	100

Source: United Nations, *1984 FAO Production Yearbook*, 1985.

Table 7-7

Leading Wine-Producing Nations, 1984
(thousands of metric tons)

Rank	Nation	Production	Percent of World Total
1	Italy	7,000	21
2	France	6,447	20
3	Soviet Union	3,800	12
4	Spain	3,554	11
5	Argentina	2,000	6
6	United States	1,620	5
	Subtotal	24,421	75
	World total	32,759	100

Source: United Nations, *1984 FAO Production Yearbook*, 1985.

Table 7-8

Leading World Market Economy Wine-Importing and
Wine-Exporting Nations, 1983 (millions of U.S. dollars)

Rank	Nation	Value	Percent of World Total
	Imports		
1	United States	955	25
2	West Germany	618	16
3	Great Britain	595	16
	Subtotal	2,168	57
	World total	3,771	100
	Exports		
1	France	1,552	44
2	Italy	785	22
3	West Germany	343	10
4	Spain	303	9
	Subtotal	3,159	85
	World total	3,518	100

Source: United Nations, *1983 International Trade Statistics Yearbook*, 1985.

demands for imports range from cheap table wines to "prestige" products for the connoisseur.

West Germany, normally one of the world's leading importers, provides a good example of large-scale demand for wines not locally produced. Germany exports its white wines in volume, ranking among the ten leading exporters, but it produces very little red wine. Its large imports represent acquisition of red table wines from Yugoslavia, Spain, Hungary, Morocco, and Tunisia as well as smaller quantities from France and Italy. Again, French imports partially reflect wines not produced in volume at home: low-priced dessert wines from Spain, Portugal, and Greece. The Soviet Union's substantial imports of wine, mainly from Algeria, also supplement inadequate local production. During the 1970's, the U.S.S.R. alternated with France as the world's major wine importing country.[4]

CITRUS

Citrus crops, primarily orange, grapefruit, and lemon, represent the most commercialized form of Mediterranean agriculture. The producing areas for citrus are much more restricted than are those for grapes. A smaller share of total production occurs in the Mediterranean region because of water shortages. Irrigation must be available to sustain productive citrus yields in the Mediterranean. Italy is the only significant orange and lemon producer in the Mediterranean area (Table 7-9). Greece is the sole major lemon producer in that region (Table 7-10).

[4]Harm de Blij, *Geography of Viticulture*, Department of Geography, University of Miami, Miami Geographical Society, Coral Gables, Fla., 1980, p. 109.

Table 7-9

Leading Orange-Producing Nations, 1984
(thousands of metric tons)

Rank	Nation	Production	Percent of World Total
1	Brazil	13,372	34
2	United States	6,566	17
3	Italy	1,700	7
4	Mexico	1,600	4
5	China	1,495	4
	Subtotal	26,043	69
	World total	39,679	100

Source: United Nations, *1984 FAO Production Yearbook*, 1985.

The United States has traditionally served as the world's leading citrus-producing country, but a series of freezes in the winters of 1984 and 1985 severely curtailed production in Florida. As a result, Brazil now leads in world production (Table 7-10). From a peak of about 10 million metric tons a year, production in the United States plummeted to just under 7 million metric tons in 1984. The United States continues to lead world production in lemon and lime output, but the market share in Italy, Mexico, and India is very competitive (Table 7-10). The United States holds a commanding lead in grapefruit output, with about one-half the world production. Second-ranking Israel grows only 11 percent of the total.

Four major producing regions dominate U.S. citrus activity. The most prominent and best known region lies in the rolling uplands of central Florida. That area does not have a Mediterranean-type climate because it receives greater amounts of precipitation during the summer months. Nonetheless, some irrigation is used in Florida citrus production. In recent years, production has moved southward, away from the rapidly urbanizing Orlando area, and to avoid colder weather. Newer producing areas in the Lake Okechobee area are not ideal because of drainage problems. Citrus sales account for nearly one-third of Florida's

Table 7-10

Leading Lemon- and Lime-producing Nations, 1984
(thousands of metric tons)

Rank	Nation	Production	Percent of World Total
1	United States	787	16
2	Italy	690	14
3	Mexico	600	12
4	India	500	10
5	Argentina	320	7
6	Turkey	300	6
	Subtotal	3,197	65
	World total	4,907	100

Source: United Nations, *1984 FAO Production Yearbook*, 1985.

Table 7-11

Leading U.S. Citrus-Producing States, 1982
(thousands of acres)

Rank	State	Production	Percent of U.S. Total
	Oranges		
1	Florida	706	77
2	California	172	19
3	Texas	26	3
4	Arizona	14	2
	Subtotal	918	100
	U.S. total	918	100
	Grapefruit		
1	Florida	166	69
2	Texas	48	20
3	California	19	8
	Subtotal	234	97
	U.S. total	241	100
	Lemons		
1	California	51	67
2	Arizona	20	27
	Subtotal	71	93
	U.S. total	77	100

Source: U.S. Bureau of the Census, *1982 Census of Agriculture.*

Table 7-12

Leading World Market Economy Orange-Importing and
Orange-Exporting Nations, 1983 (millions of U.S. dollars)

Rank	Nation	Volume	Percent of World Total
	Imports		
1	France	202	17
2	West Germany	168	14
3	Great Britain	114	9
4	Canada	97	8
5	Netherlands	94	8
	Subtotal	675	56
	World total	1,214	100
	Exports		
1	United States	221	20
2	Spain	207	18
3	Cuba	141	13
4	Morocco	135	12
5	Israel	107	10
	Subtotal	811	73
	World total	1,126	100

Source: United Nations, *1983 International Trade Statistics*, 1985.

agricultural production. In good years Florida produces over three-fourths of the nation's oranges and two-thirds of the grapefruit (Table 7-11). Of the major citrus products, only lemons are not heavily produced in Florida. California produces the majority of that fruit. Florida oranges are increasingly used for processing into frozen juice concentrate, whereas the California product makes a better fresh fruit. By contrast, grapefruit is sold mainly as a fresh fruit.

Variations in grapefruit harvest times occur in major producing areas, extending the fresh fruit season. Whereas fall output is high in Florida, winter harvests occur in Texas and Mexico. In the summer months Arizona and California output dominates. Southern hemisphere production in the summer also balances out the annual demand for citrus in the United States. The output of Brazil is particularly important in this regard. Brazilian production occurs primarily along the northeastern coast of the country, north of Rio de Janeiro.

International Trade

West European countries and Canada lead in orange imports (Table 7-12). In addition to the United States, Mediterranean countries are the major exporters, led by Spain, Morocco, and Israel. Cuba is also a significant exporter. Lemon and grapefruit trade patterns are similar to the orange movement, with two exceptions. These products primarily flow from the Mediterranean area to western Europe (Table 7-13), but Ja-

pan is the leading importer, registering 21 percent of the total. Japan has traditionally had stiff import restrictions on oranges, which largely keeps that portion of the citrus crop out of the country. Second, Turkey and Italy replace Morocco and Israel as grapefruit-exporting nations. Canada, Japan, and western Europe are major importers.

POULTRY

Poultry, predominantly chickens, can be found in virtually all farm areas in the world. All types of subsistence agriculturalists, commercial farmers, and plantation systems, from tropical to arctic climates, have chickens. Many chickens are raised as sidelines on farms specializing in other activities. Some are kept as scavengers and are not cared for in a formal manner. The concern here is not with this small-scale activity but with the farms that specialize in poultry for sale. Even with this part of the industry it is hard to explain why activity is located as it is. Access to the urban market, proximity to feed sources, and the absence of competition with other farm activities partially explain the situation. The poultry industry can compete in remote areas more effectively than other animal farming activities because less feed is needed to produce a pound of meat. It takes about 2 pounds of feed to yield a pound of meat with chickens, 4 pounds of feed to yield a pound of pork, and 8 to 10 pounds of feed per pound of beef. Less feed has to move to poultry areas per pound of meat yield. Consequently, it could be argued that poultry farming can compete on lower-quality land.

Table 7-13

Leading World Market Economy Lemon/Grapefruit-Importing and Lemon/Grapefruit Exporting Nations, 1983 (millions of U.S. dollars)

Rank	Nation	Value	Percent of World Total
	Imports		
1	Japan	188	27
2	France	120	17
3	West Germany	93	13
4	Great Britain	55	8
	Subtotal	456	65
	World total	711	100
	Exports (1979)		
1	United States	216	36
2	Spain	88	15
3	Israel	61	10
4	Turkey	45	7
5	Italy	44	7
	Subtotal	454	75
	World total	607	100

Source: United Nations, *1983 International Trade Statistics*, 1985.

On a world scale, the largest number of chickens occurs in the Soviet Union, China, and Brazil. In these areas poultry is an extremely important source of protein in the diet. The United States ranks fourth in the number of chickens and Japan fifth. A similar ranking of countries occurs for hen egg production. The leading nations include China, the United States, the Soviet Union, and Japan.

Since there is no critical growing season for chickens and because commercial producers buy most of their feed, physical factors are not important in explaining distributions. In fact, some might argue that poultry farmers take advantage of cheaper, rougher, more isolated land, leaving the flatter, more expensive land to crop farmers. Poultry production in the Shenandoah Valley of western Virginia, northern Georgia, northern Alabama, Arkansas, southern Mississippi, and to some extent the Delmarva peninsula of Maryland illustrate these relationships (Figure 7-9). In the Shenandoah Valley and in Arkansas, a large number of farmers concentrate on turkey production.

The correlation between urban markets and chickens becomes stronger when one looks only at egg production rather than the industry as a whole, which includes meat-type chickens, broilers, and turkeys. The northeastern quadrant of the United States has a significant share of the nation's population and produces most of its eggs. Southern New Jersey, for example, traditionally served as a major production region for the New York and Pennsylvania market. California is also a major egg-producing state. Several large metropolitan markets, however, are not served by nearby chicken farms, including Chicago, St. Louis, and Minneapolis.

Much of the explanation for the pattern of poultry production at the local level lies with *contract farming*. Contract farming is a relatively new type of agriculture conceived by managers of commercial feed corporations. In the 1930s, these businesses, sensing the increasing demand for broiler chickens, began to line up farmers who would agree to buy the corporation's feed, concentrate on raising broiler chickens,

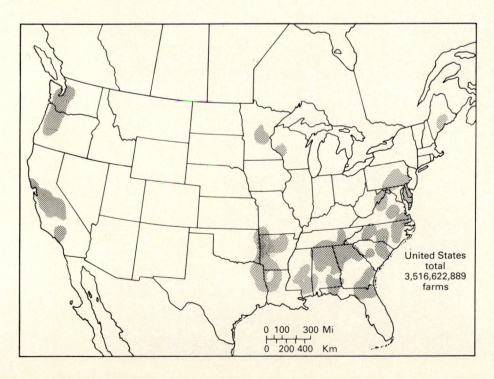

Figure 7-9 Major Poultry Broiler Production Areas in the United States. *Poultry farmers take advantage of cheaper, more isolated land to produce chickens. Southern producing areas have the advantage of being accessible to the middle west feed source and yet are still easily accessible to the northern market. Note concentrations in the Carolinas, northern Georgia and Alabama, Mississippi, Louisiana, and Arkansas.*

United States total 3,516,622,889 farms

and market all their birds through distributors with whom the feed corporations also had contracts. In a sense, then, the contract-broiler farmer is no longer an independent producer whose profit depends on the difference between expenses for feed (and other production costs) and income from an unpredictable market. Rather, the farmer is like an employee in a factory, a hired technician who is paid a stipulated fee, regardless of whether prices rise or fall, by a company to which a certain output is guaranteed.

To save transportation and distribution costs and to expedite the shipment of birds to market, commercial feed manufacturers contract with many farmers in a few areas rather than with a few farmers in many areas. Poultry feed is shipped in carload lots to the farmers in these specialty areas; the largest farms even have their own railroad sidings, and several smaller farms are often serviced by one siding. In selecting the areas that would respond most readily to broiler specialization, the commercial feed manufacturers tended to choose regions of "problem farms."

Understandably enough, prosperous farmers in the corn belt and dairy belt were not enthusiastic about the idea of contract poultry farming. Most of the farmers who responded to this scheme were located in areas characterized by rather backward farm methods, small acreages, rough terrain, or poor soil. This new scheme seemed an ideal solution for these farmers. Equipment would be provided and methods specified under the terms of the contract, small acreages would suffice, rough land would serve well enough for broiler houses, and the quality of the soil would be immaterial. Almost any part of the south, except the more lucrative cotton, tobacco, vegetable, and fruit areas, was perfectly suitable. Even an area located rather far from a sizable urban market was satisfactory, for the higher transport costs were offset by lower production costs.

For the nation as a whole, 90 percent of the poultry is now produced by only 23,000 farms, a few of which are huge and many of which operate under contracts held by commercial feed corporations. Much California poultry production now comes from contract farms, which are known locally as poultry ranches.

The success of contract farming in the poultry business has led agriculturalists to scrutinize its possibilities for other types of production. But critics have also noted its negative side. It is argued that the system exploits laborers. Often, corporations own not only the animals, the feed, the land, and the buildings and machinery, but also control the marketing and price paid for farm output, which virtually enslaves the farmer to the corporate giant (Figure 7-10). Wages paid are modest at best considering the hours worked.

Most poultry feed comes from crops grown in the middle west feedgrain belt. Feed is expensive. In chicken-producing areas the expenditure for feed exceeds 50 percent of the value of the animal products sold. Southern producing areas have the advantage of

Figure 7-10 Egg and Poultry House. *This chicken house in north Georgia is typical of the decentralized contract poultry operations found in the southern Appalachia and the Ozarks.*

Table 7-14

Inventory of Hens and Pullets of Laying Age in the United States, 1982 (billions)

Rank	State	Number	Percent of U.S. Total
1	California	39	13
2	Georgia	24	8
3	Pennsylvania	21	7
4	Arkansas	20	6
5	Indiana	18	6
6	North Carolina	15	5
7	Alabama	15	5
8	Florida	14	4
	Subtotal	166	53
	U.S. total	311	100

Source: U.S. Bureau of the Census, *1982 Census of Agriculture.*

being easily accessible to the middle west feed source and yet still easily accessible to the northern market, with their location at or near the southern end of the Appalachians. Fully 44 percent of the hen and pullet inventories in the United States are in the south. Five of the eight leading states are southern. Leading areas outside this area include California, Pennsylvania, and Indiana (Table 7-14). It is significant to note, however, that the eight leading states account for only half of the total production, the remainder of which is widely distributed throughout the country.

SUGGESTIONS FOR FURTHER READING

DE BLIJ, HARM, *Geography of Viticulture,* Department of Geography, University of Miami, Miami Geographical Society, Coral Gables, Fla., 1980.

DE BLIJ, HARM, *Wine—A Geographic Appreciation.* Totowa, N.J.: Rowman & Allanheld, 1983.

DE BLIJ, HARM, *Wine Regions of the Southern Hemisphere.* Totowa, N.J.: Rowman & Allanheld, 1985.

GOODALL, MERRELL, ET AL., *California Water: A Political Economy,* Montclair, N.J.: Allanheld, 1978.

HADWIGER, DON, and ROSS B. TALBOT, eds., *Food Policy and Farm Programs,* Proceedings of the Academy of Political Science, 34, No. 3 (1982).

HART, JOHN FRASER, *The South,* New York: Van Nostrand, 1976.

PRUNTY, MERLE C., and CHARLES S. AIKEN, "The Demise of the Piedmont Cotton Region," *Annals,* Association of American Geographers, 62 (1972), 283–306.

SUMMARY AND CONCLUSION

Specialty agriculture, despite the wide variety of crops falling under the umbrella—ranging from vegetable, fruit and nut, grape, and citrus food crops to an industrial crop—cotton—and a crop fitting neither of these groups—tobacco—all grow primarily in semitropical dry areas, and many are irrigated. Mediterranean-type dry farming techniques characterize areas not irrigated. Most of the production occurs distant from markets, although some specialized output of vegetables and fruits occurs near the market, supporting the von Thünen model. The contribution of poultry to the specialty agriculture picture remains rather unique. Although poultry production also occurs widely, it appears to be concentrated in poorer agriculture areas with easy access to both feed supplies and markets, such as the situation that occurs in the southern Appalachians in the United States.

Much of the output in specialty farms occurs on very large scale units, such as those found in California. On the other hand, production can also occur on small-scale units, such as the highly regulated tobacco farms of the Carolinas. The outlook for these activities remains very favorable, with the biggest uncertainty created by the growing evidence that plentiful supplies of cheap water for irrigated units will decrease. In many areas, especially in California and the U.S. southwest, water will become much more expensive in the future as government subsidies on rates dwindle and the competition for urban and industrial uses increases.

8 Commercial Agriculture in the Tropics

In previous chapters relating to commercial agriculture, attention focused on activities in North America, western Europe, and the Soviet Union, where the majority of the markets are domestic. Commercial agriculture is also practiced in the tropics, but the mix of crops is quite different, the scale of activity considerably smaller, and until very recently, almost all of the produce was sold to foreign markets.

Until this century, a large share of commercial crops grown in the tropics were produced on widely dispersed foreign-controlled plantations. Today, those estates are not the only producers of cash crops, nor in a number of cases are they the major producers. In many parts of the world, commercial crops are cultivated by indigenous peasant farmers or smallholders (introduced in Chapter 3), who collectively provide large quantities of produce for the world market. In this chapter we necessarily discuss both plantations and smallholders, because they grow the same crops for essentially the same market.

The tropical plantation is one of the world's oldest systems of commercial agriculture. At the same time, it occupies less space today than any other form of commercial cropping. Precise figures do not exist, but it is estimated that cash crops produced in the tropics for export, both on plantations and by smallholders, are grown on approximately 228 million

acres. This amounts to about 10 percent of the earth's cropland.

Since A.D. 1500, the products from over a dozen tropical crops have been in constant demand by people in the temperate middle latitudes. Without a morning cup of coffee or tea, for example, it is unlikely that half the adult population of the developed world could function during the workday. The total amount of money earned in the sale of tropical cash crops is enormous. According to the Food and Agricultural Organization, in 1983 seven crops alone (coffee, sugar, cocoa, rubber, tea, palm oil, and bananas) generated more than $30 billion in export sales (see Table 8-1). Despite the fact that these particular cash crops are cultivated in a relatively small area, their annual value of exports in the late 1970s was $3 billion higher than the amount earned in foreign trade by all cereal grain crops. However, it should be kept in mind that cereal grain crops are primarily consumed domestically and do not enter the export market.

Most tropical cash crops are produced in third-world countries. In 1980, these nations earned over $65 billion from all agricultural exports. About 48 percent of these exports emanated from Latin America, including the Caribbean, while 35 percent came from Asia, with the remainder from Africa and Oceania. With only a few exceptions, the cash crops entering world trade are destined for the U.S. and European markets.

Note: Chapter authored by Sanford Bederman, Professor of Geography, Georgia State University, Atlanta, Ga.

Table 8-1

Leading Tropical Cash Crops in World Export Trade, 1983

Cash Crop	Total World Export (thousands of metric tons)	Dollars Earned (millions)
Sugar	29,592	10,636
Bananas	6,227	1,324
Coffee	4,051	9,636
Palm Oil	3,938	1,743
Natural Rubber	3,433	3,321
Cocoa	1,261	2,051
Tea	933	1,844

Source: United Nations, *1983 FAO Trade Yearbook*, 1984.

PLANTATION TYPES AND DEFINITIONS

Social scientists of almost every persuasion have studied plantations, and the literature reveals over 20 different definitions. Plantations have been identified by the climatic regions they are in, their size, the crops they grow, and their orientation for crop export. Some definitions are set strictly in an economic framework and suggest simply that plantations are a technique for organizing land and labor in the tropics and subtropics to supply midlatitude markets with certain products.

Some scholars regard the plantation as more of a political and social institution than a strictly economic one. This adds a sociocultural component to the definition of the plantation and encompasses the idea that it is an intrusion of one society into the land and life of another. This is a clear reference to the colonial heritage of plantation economies. Large-scale agriculture most often was forcibly introduced into areas under the domination of European colonial powers. For this reason plantations can be regarded as a "frontier institution," thrust into a less developed society for the purpose of economic and social exploitation.

Rather than attempt to define what a plantation is, a working description of the plantation system is provided. This description stresses location, scale of operation, management, and the organization of production. Traditionally, commercial plantations occupy relatively large units of land, usually found in sparsely populated areas of the tropics. They employ large numbers of imported, unskilled, and low-paid laborers who, through careful supervision, concentrate on producing one or more crops for export. In regard to supervision, E. P. Thompson once called the system *military agriculture*. The capital required to finance these large-scale agribusinesses comes from European and U.S. corporations. Resident managers provide the operation's business and technical skills and, until recent years, they have been foreigners. Although crops are produced mainly by intensive hand

labor, they must be harvested in a careful and organized way, and they have to be processed in some manner before leaving the estate. Factories are therefore common entities on the plantation landscape.

The world's plantations are located along tropical coasts, inland near navigable rivers, or along railroads extending into the interiors (see Figure 8-1). Inasmuch as the products are exported, these locations help account for relatively efficient and inexpensive transportation from factory to port. Geographical location and environmental factors also help contribute to the fact that plantations often specialize in a single crop. This is known as *monoculture,* and it is a critical and distinctive feature of what was the old plantation system.

Although plantation estates around the world incorporate relatively large holdings, they vary considerably in size. In Sri Lanka, for example, an estate must contain at least 10 acres to qualify legally as a plantation. In Sarawak, Indonesia, the figure is 1000 acres. Many of the rubber plantations in Malaysia comprise only about 100 acres. At the other extreme, the Firestone Company rubber plantation at Harbel, Liberia, includes almost 136,000 acres.

Plantations have evolved consistently over the past centuries, but in the last 30 years or so they have changed in surprising ways. The usual classification of plantation farming as strictly a tropical institution can no longer hold in light of developments elsewhere. Those characteristics considered essential to the plantation—crop specialization, advanced cultivation and harvesting techniques, large operating units, centralized management, labor specialization, massive production, and heavy capital investment—have become increasingly associated with farms in the middle latitudes. By the same token, the character of the tropical plantation has begun to change. Monoculture has declined in some locales, and crops usually associated with cooler climates have been modified and are being cultivated in the tropics. While the plantation economy was originally geared exclusively to exporting goods to foreign markets, domestic markets in the tropics have recently increased in importance. Finally, third-world governments commonly promote them as an important element of their overall rural development strategies. Notwithstanding these important modern developments, the plantation is undeniably the product of the colonial past. In much of the third world, it remains as a symbol of oppression, economic exploitation, and cultural domination by Europeans.

HISTORICAL ANTECEDENTS

Plantations have existed for over 500 years. They were created in response to a demand in Europe for foods, spices, fibers, and beverages that, because of climatic constraints, could be produced only in the tropics or subtropics. Over the centuries the demand for most

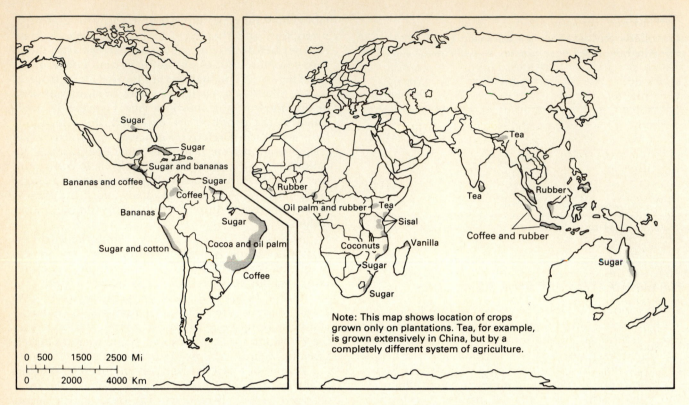

Figure 8-1 Major Plantation Areas of the World. *The majority of the world's plantation activity occurs in the tropics and subtropics. Note the coastal location for most of the products identified on the map. This proximity to water and port facilities minimizes transportation costs.*

of these items has increased with the growth of world population and with the insatiable needs of modern western society. During that time, North America, especially the United States, has equaled or even surpassed Europe as a market for tropical agricultural products.

Latin America

Plantations exist on every continent possessing a tropical climate. The plantation system, however, is considerably older in tropical America than in Asia, Oceania, and Africa. Actually, the first plantations date to the fifteenth century establishments located on several islands off the Guinea coast of Africa. From there, the Portugese introduced the system to northeast Brazil to produce sugarcane. This system of sugar production diffused northward to the Caribbean islands. When plantations were introduced into the subtropics of North America in the seventeenth century, indigo, tobacco, and cotton were the major cash crops.

European colonizers and settlers directly operated the plantation system in Latin America. Imported slaves and indentured laborers did the work, as discussed in Chapter 2. These plantations produced what were then considered luxuries for an elite European market. Sugar and indigo are examples of

these luxury items. This situation prevailed from approximately 1500 to about 1850. In the middle of the nineteenth century, three separate developments disrupted the *latifundium* (broad estate) system in the Americas. The abolition of slavery led to a labor crisis that took years to resolve in many parts of Latin America. The political order of the region was also shaken during the independence movement of the Spanish colonies after 1820. Finally, this period also involved increased competition from expanding agricultural economies in other parts of the world producing products for the same markets.

Asia

Commercial crops from Asia and Africa came to Europe in the nineteenth century, but they were grown primarily by indigenous peasant farmers. Commercial agricultural development of the Asian tropics did not begin in earnest until the Suez Canal was opened in 1869. This, together with steam replacing sails, made Asia closer to Europe in travel time. The increase in Asian produce also coincided with a new mass demand in Europe, owing to the effects of the industrial revolution. This demand was particularly acute for certain products grown in Asia, especially rubber and vegetable oils.

Whereas American plantations were commonly

family owned, the new estates in Asia were financed, controlled, and staffed by commercial companies based in Europe. These estates were created specifically to compete with already established smallholders. As noted above, slavery did not exist at the end of the nineteenth century; therefore, estate managers often solved their labor needs with what they considered to be the next best thing—*indentured laborers.* An indentured laborer, through contract, binds himself or herself to work for another at low or no wages for a specified period of time. This agreement often entails work in return for payment in kind, such as the provision of food and lodging by the plantation owner.

Africa

Africa was the last region of the tropics to experience plantation development. As in Asia, the demise of slavery and the success of the industrial revolution led European entrepreneurs to search for additional products to satisfy rising demand. At the same time, they sought alternative places to grow their agricultural crops. When European colonial powers subdivided Africa politically a hundred years ago, the resulting colonies not only provided agricultural products for the home country, but also became closed markets for finished industrial goods from the European motherland (discussed in Chapter 2).

The plantation system never caught on in Africa. In fact, only in the German colonies of Togo and Cameroon were the economies totally dominated by plantations. It should be noted, however, that some commercial agriculture on a large scale did exist in east Africa during the colonial period. It was more often the case that cash crops in most of colonial Africa were produced by indigenous smallholders. The bulk of Nigeria's and Ghana's cocoa, for example, were grown by these small-scale peasant farmers and not on plantations. This was also true for coffee produced in Ivory Coast, Tanzania, and Kenya. The situation in postcolonial Africa in the late twentieth century is essentially the same. In only a few African countries today will one find a thriving plantation economy.

The history of plantations can thus be divided into three distinct phases. The old plantation system flourished in the Americas before the nineteenth century. These privately owned estates were only barely integrated into the world economy, for they specialized in producing luxury products for sale to the European elite. The new plantation system developed in Asia to meet the demands of a modern industrial society in Europe, one that required raw goods for the factory just as much as food for consumption. In this way, plantations in faraway lands became an integral part of a global economy largely directed from Europe. Finally, the third phase of plantation development is underway presently and features attempts by third world nations to gain control over the production and marketing of the agricultural goods grown on their land.

TROPICAL AND SUBTROPICAL COMMERCIAL CROPS

A vast number of commercial cash crops are grown in tropical and subtropical countries. Only a few of them, however, are described in the case studies in this section (Table 8-2). Some crops, such as the oil palm, cocoa, bananas, coffee, and rubber, are produced solely in the tropics, whereas the cultivation of others, such as cotton, tobacco, sugarcane, groundnuts (peanuts), and tea, extends to areas that lie in more temperate zones.

These crops can be divided into three types, which are distinguished by relative maturation time and longevity of production. Many cash crops are perennials and have life cycles of more than 2 years. Natural rubber, coconuts, oil palm, tea, cocoa, and coffee are all *tree crops* and take years to mature, but afterward they are productive for long periods. The second category (also perennials) are more rapidly maturing *field crops,* thus do not require the extended level of continued maintenance needed by tree crops. Sugarcane, henequen and sisal, and bananas are examples of this group. Finally, some cash crops are *annuals;* that is, they require a single planting and are harvested within a year. Included here are cotton, jute, tobacco, and groundnuts.

These botanical distinctions are quite important. Tree products are particularly well suited to large-scale plantation production, since so much capital investment is required and because profits will not be seen for many years, until the trees mature. A small-scale planter simply cannot afford to keep a tract of land unproductive for much more than a year. By the same token, annual crops are better suited for the smallholder, since they allow for greater flexibility in planting followed by a harvest the same year. It is for

Table 8-2

Leading Producing Nations for Selected Tropical Cash Crops, 1984 (thousands of metric tons)

Cash Crop	Total World Production	Leading Producing Nation	Production of Leading Nation
Sugarcane	935,769	Brazil	241,518
Bananas	41,113	Brazil	6,968
Coffee (green or roasted)	5,210	Brazil	1,353
Palm oil	6,810	Malaysia	3,717
Tea	2,218	India	645
Cocoa	1,660	Ivory Coast	411

Source: United Nations, *1984 FAO Production Yearbook,* 1985.

this reason that the plantation system almost completely retreated from the subtropics during the last century, leaving the production of annuals grown there (tobacco and cotton, for example) mostly in the hands of smallholders. This was particularly true of the American south.

The life span of tree plants varies considerably. For coconut trees it is 100 years, whereas healthy rubber trees can be tapped for more than 30 years. Many trees, if not properly tended, rapidly decline in yield after a few years. If the plantation manager makes a miscalculation, it could mean either having no output at a time of high prices, or just as bad, having excess production in periods of depressed prices.

Another way to categorize tropical cash crops is by their ultimate use. Many are consumed as food and beverages. These include bananas, coffee, tea, sugarcane, cocoa, vanilla, pineapples, and pepper and assorted spices. Some crops are used in the manufacturing of fibers and cloth. Cotton, jute, henequen and sisal, and manila hemp fit this category. Finally, some cash crops provide important industrial materials. Included in this group are rubber and pyrethrum.

The leading plantation crop in terms of tonnage exported is sugar, followed by bananas, coffee, palm oil, and natural rubber (see Table 8-1). These products also generally lead in the value of exports, with the exception of bananas, which are superseded by cocoa and tea in dollar volume. Because space limitations preclude a discussion of every important tropical cash crop, in this section we present very brief discussions of only five, including bananas, sugarcane, tea, rubber, and oil palm products.

Bananas

Together with oranges and apples, the sweet banana is one of the western world's most popular hand fruits. Although there are hundreds of varieties of bananas, the most important in world trade are from the species *Musa sapientum*. The plantain, from the species *Musa paradisaica,* is an extremely popular staple cooking banana throughout the tropics, but very little of its production is sold to foreign markets and it is generally not considered a commercial cash crop.

In the early sixteenth century the banana, native to southern Asia, was brought to the western hemisphere, where it now serves as both a cash crop and a subsistence food. By the late nineteenth century it had become an important commercial crop in the lowlands encircling the Caribbean Sea. Since the mature fruit ripens very rapidly, it was only with the advent of refrigerated boats that it has been possible to ship bananas from the tropics to Europe and North America all year around. Because such transportation is costly, those producing regions nearest the major markets have trade advantages.

The banana plant, actually an oversized herb related to the lily family, matures in about a year after planting. It will grow successfully in regions that ex-

perience heavy rainfall (65 to 200 inches) and high temperatures (75 to 85°F). No dry season is necessary. Soils must be well drained, however, to keep the roots from becoming waterlogged. Because it depletes soil nutrients so rapidly, the banana has been called a soil killer.

Several natural hazards plague the growing of bananas. Because it does not possess any woody tissue, the plant is easily uprooted by heavy winds, a common occurrence in the tropics. It is also very susceptible to a number of diseases, which in the past have caused heavy crop losses. The most serious of these are Panama disease and leaf spot disease, also known as sigatoka. Both are transmitted by fungoids, but Panama disease not only kills existing plants, it also infects the soil, making it impossible to grow bananas again for a number of years. As yet, no satisfactory method of controlling Panama disease has been developed. Leaf spot disease can be controlled by spraying plants with fungicides. The prevalence of Panama disease required large plantations in Central America to relocate early in the present century. In the 1920s the crop growth shifted from the Caribbean side to the Pacific lowlands, where a drier climate helped to inhibit the disease. In some parts of the world, varieties resistant to Panama disease have been cultivated successfully.

In 1984, the total world production of bananas was slightly more than 41 million metric tons, of which over 6 million metric tons found their way into world trade. Many of the largest banana-producing nations, including the first- and second-ranking countries, Brazil and India, consume most of their fruit domestically. While just under half the world's bananas are grown in Asia, only a small share are sold abroad. Honduras, for example, exports almost as many bananas as all Asian countries combined. The bulk of the world's bananas entering international trade emanate from Latin America, with Costa Rica being the largest exporter (Table 8-3), followed closely by Ecuador and Colombia.

Table 8-3

Leading Exporting Nations for Selected Tropical Cash Crops, 1983

Cash Crop	Leading Exporting Nation	Metric Tons Exported (thousands)	Dollars Earned (millions)
Sugar	Cuba	6,746	4,717
Coffee (green or roasted)	Brazil	931	2,078
Rubber (natural)	Malaysia	1,560	1,579
Palm oil	Malaysia	2,906	1,285
Cocoa	Ivory Coast	360	530
Tea	India	209	439
Bananas	Costa Rica	1,009	239

Source: United Nations, *1983 FAO Trade Yearbook,* 1985.

Sugarcane

Although dieticians and dentists decry the consumption of sugar, it remains a major element in the diet of populations around the world, both rich and poor. Sugar is described as a luxury energy food, and its per capita annual consumption ranges between 5 and 100 pounds, depending on one's personal predilection, wealth, and cultural background. The per capita consumption of sugar in the United States in 1982 was 74 pounds, with the food and soft-drink industries the largest users. Ice cream manufacturers, food canners, bakers, and confectioners rely heavily on this product. Sugar is also an important ingredient in the production of liquor, but approximately one-third of sugar consumption in the United States is purchased for home use.

Sugar is obtained from several different plants, but the most important sources are sugarcane and sugar beets. The bulk of the sugar that enters world trade, however, is refined from sugarcane. Sugarcane is a perennial plant reproduced from stem cuttings, and requires a total of 50 to 65 inches of rainfall a year, with a dry period to allow for final ripening. Continuous rainfall encourages stalk growth but limits the formation of sucrose. Growth usually ceases when temperatures drop below 65°F and is best above 80°F. The first crop, called plant cane, requires 15 to 24 months to mature; thereafter, the harvests are annual. Succeeding harvests are called ratoon crops, but their yields decline in 2 to 3 years. Once yields are no longer satisfactory, growers burn the stubble and begin the planting cycle again.

When cane is cultivated on large plantations, the crop is processed by modern methods in factories. The chemical process of refining sugar is extremely complex and requires no less than five steps: (1) extracting the sugar juice from the cane, (2) clarifying the juice, (3) evaporating it, (4) crystalizing the sugar, and (5) purifying the sugar. It usually takes 10 tons of cane to produce 1 ton of raw sugar.

The total world production of sugarcane approached 1 billion metric tons in 1984. Only 4 percent of refined sugar entered the world trade market, nearly one-half of it from the Americas. The major producer of sugarcane is Brazil, followed by India and Cuba. Despite being third in overall production, Cuba leads all countries in the export of sugar, most of it going to the Soviet Union (Table 8-3).

Whereas many tropical nations produce sugar, the economies of the islands of the Caribbean Sea are almost completely dominated by the crop. Sugar is the mainstay of the economy of Mauritius and Reunion (in the Indian Ocean), as well as a number of islands in the Pacific. Cane is also cultivated in large quantities in Peru, Argentina, the Philippines, Australia, China, and Hawaii (Figure 8-2).

Tea

Tea is the world's most popular beverage, being favored by at least half the world's population. It is the national drink of China, Japan, India, Sri Lanka, Great Britain, and the Soviet Union. Many African and Near Eastern countries also prefer tea to coffee, its nearest rival. However, in many third-world countries of Asia and Africa, coffee commands a higher price and subsequently becomes a status drink. In these countries it is consumed mostly by the upper class. Because unbrewed tea leaves contain over twice the amount of caffeine found in an equal weight of coffee beans, they are a major source of the base product for medicinal caffeine.

Tea is native to the subtropics of Asia. It is now grown widely in both the subtropics and high-elevation locations within the tropics (Figure 8-3). Tea is

Figure 8-2 Sugarcane Field in Hawaii. *Sugarcane and pineapple production formed the backbone of traditional commercial agriculture in Hawaii. This view of sugarcane fields near the historic city of Lahaina on the island of Maui. Steep slopes of the foothills of the volcanic West Maui Mountains in background. (T.A.H.)*

an evergreen plant (actually, a tree crop pruned to form bushes) that grows best where temperatures range between 55 and 90°F. Few plants require as much moisture as tea. Over 100 inches of precipitation well distributed throughout the year is desirable. Humidity must also be high to assure abundant leaf formation. Also, tea cultivation is extremely sensitive to soil quality. Some specialists claim that characteristic tea flavors, like those of wine, are the result of differences in types of soil.

Young tea plants are started in nurseries under protective conditions and are transplanted in the fields when they are a year old. Plucking leaves begins usually when plants are three years old. The highest yields are achieved at six years. Tea bushes are constantly pruned to generate tender new leaf growth, which is the only part of the plant that is harvested. Once tea leaves are plucked, they must be processed quickly at a nearby factory (see Figure 8-4). Depending on how complete the process is, two types of "made" tea are manufactured. The leaves that are completely processed (i.e., withered, fermented, and fired) make black tea, the major product in world trade. Green tea is produced when withered leaves

undergo a steaming or scalding to stop fermentation from occurring. Green tea is extremely popular in Japan and China, where it is almost completely produced by small-scale farmers.

Several of the world's largest tea-growing nations are located largely, if not entirely, outside the tropics. Most tea cultivation occurs in Asia, with India, China, and Sri Lanka being the most important producers. The total world production of "made" tea in 1984 amounted to 2.21 million metric tons. India, China, and Sri Lanka are the largest growing nations, but tea has also become a major cash crop in Kenya, and small quantities are now cultivated in west Africa. India ranks as the major exporter of this beverage, shipping over 209,000 metric tons to foreign markets in 1983 (Table 8-3), followed by China and Sri Lanka.

Rubber

Processed latex from wild South American rubber trees first came into use in the nineteenth century as a waterproofing substance and pencil eraser product. But rubber did not attain major importance until the early twentieth century when a huge demand was cre-

Figure 8-3 Tole Tea Estate on Mount Cameroon. *Because tea is obtained from new tender growth, the plant must be pruned constantly. It is much easier to pluck the leaves when the plant is bush-like and waist high. Note Mount Cameroon in the background. (Courtesy Sanford H. Bederman)*

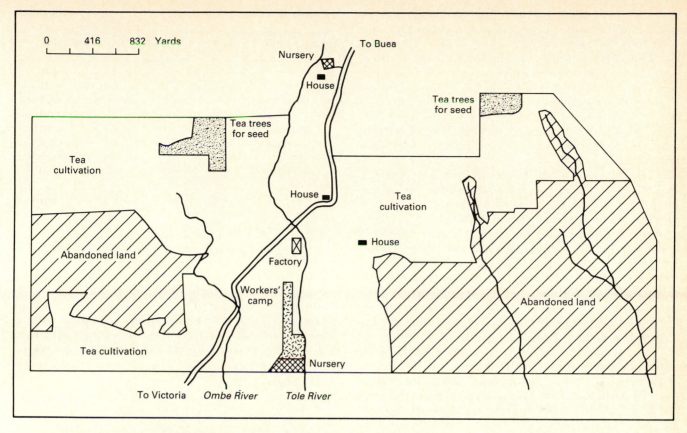

Figure 8-4 Map of Tole Tea Estate on Mount Cameroon. *On plantation estates one typically finds a centrally located factory and space devoted to workers' dwellings. (Courtesy Sanford H. Bederman)*

ated by the automobile and bicycle tire industries. Consumption increased rapidly, but natural rubber production could not satisfy expanding industrial demand. A synthetic rubber industry developed in Europe and North America during World War II to fill that need. Today, more rubber is produced chemically than by agricultural methods.

The primary rubber plant, *Hevea braziliensis*, requires a hot, humid environment where rainfall is no less than 70 inches and preferably at least 100 inches, and where average temperatures stay above 75°F. Extremely high temperatures, those above 95°F, usually desiccate the plant and discourage the flow of *latex*, the white liquid sap from which commercial rubber is made. The soil must be well drained, for rubber trees will not grow with wet feet.

The rubber tree is indigenous to Brazil. In the nineteenth century, the Brazilian government attempted to preserve its monopoly over the industry and banned the export of the plant. Nevertheless, in 1876, an Englishman by the name of Henry Wickham successfully smuggled several thousand rubber seeds out of the country. He propagated the seeds first at Kew Gardens in London, and when he later planted them in Ceylon, now called Sri Lanka, he successfully launched the rubber industry in Asia. Within a short time, especially after N. H. Ridly (a botanist working in Singapore) discovered that prolonged tapping

would not kill the plant, rubber was introduced into Malaya and Indonesia.

In 1905, the year the first shipment of rubber was exported from Malaya, now called Malaysia, Brazil was producing 99 percent of the world supply. Today, Brazil accounts for less than 1 percent. Malaya and Indonesia possessed a much larger and more efficient labor pool, easier access to ocean shipping, and an environment for rubber growing that was equal to Brazil's.

It was only a matter of time before southeast Asian countries came to dominate the industry. The world exports of rubber in 1983 totaled 3.43 million metric tons, of which 46 percent was processed in Malaysia (Table 8-3). Most of the world's natural rubber cultivation is undertaken in southeast Asia, but substantial production also occurs in Liberia, Zaire, and Nigeria.

Oil Palm

Vegetable oils are extracted from many different crops. Some are annuals, whereas others are tree crops. Everyone is familiar with corn oil, soybean oil, and olive oil; however, few people realize that on a per acreage basis the most productive of all oil-yielding plants is the African oil palm, *Elaeis guineensis*. Not only are palm oil products a major ingredient in cooking oils and margarine, but like coconut oil products, they are also used in the manufacture of soap. Kernel cake, another by-product, produces a high-protein animal feed.

The oil palm is native to the forests of west Africa, but has been introduced quite successfully into other parts of the tropical world. The most important oil palm growing areas lie within 10 degrees of the equator. The plant grows best where temperatures are over 80°F throughout the year and where rainfall is abundant and well distributed. The oil palm tree requires well-drained soils.

The oil palm usually begins to bear fruit within 3 to 5 years after planting and reaches full bearing age after 8 years. The life span of the plant is about 50 years. Immediately after the oil palm fruit (consisting of dozens of palm nuts; Figure 8-5) is harvested, it is taken to a factory, where oil is immediately extracted from the fleshy pericarp of the palm nut. Every palm nut contains a kernel from which a high-quality "hard" oil is obtained. This is a prized ingredient by soap manufacturers. As part of the factory process, kernels are extracted from their shells, bagged, and sold for later processing in developed countries.

Until very recently, the entire production of oil palm came from tropical Africa, but now the plant is grown on a large scale in Malaysia, Indonesia, and in parts of Latin America. The total world production of palm oil in 1984 was 7 million metric tons. Of that amount, Malaysia produced just over half. Malaysia is also the leading exporter of palm oil, selling nearly 3 million tons in 1983 (Table 8-3). India imports more than any other nation, exceeding a half million tons a year. Malaysia is also the largest producer and exporter of palm kernels.

In review, cash crops produced in third-world countries find their primary market in developed countries, mainly the United States, Europe, and Japan. The leading importers of these cash crops are shown in Table 8-4. Among these countries, India is the only one that is part of the developing world. Note that sugar accounts for the largest tonnage shipped, followed by bananas and coffee. In terms of dollar values, sugar ranks first, followed by coffee.

Table 8-4

Leading Importing Nations for Selected Tropical Cash Crops, 1983

Cash Crop	Leading Importing Nation	Metric Tons Imported (thousands)	Dollars Spent (millions)
Coffee (green or roasted)	United States	998	2624
Sugar	Soviet Union	6023	4085
Rubber (natural)	United States	684	655
Cocoa	United States	217	349
Tea	Great Britain	184	352
Bananas	United States	2458	592
Palm oil	India	647	324

Source: United Nations, *1983 FAO Trade Yearbook,* 1985.

THE PLANTATION'S ECONOMIC AND SOCIAL ENVIRONMENT

The plantation possesses certain common characteristics that qualify it as both an economic and a social institution of continued relevance in the modern world. It is a continuing source of livelihood for many third-world peoples, and this status should be examined in light of both economic and social factors.

Figure 8-5 Oil Palm Fruit. *The oil palm fruit consists of dozens of nuts. The harvest shown in this photograph has just been sawed from trees and is on its way to the factory where edible oil is produced from the fleshy pericarp. (Courtesy Sanford H. Bederman)*

The plantation, unlike the manor of the Middle Ages, is an outward-looking economic unit designed to carry on production at a high-profit level. It is also fundamentally an economic institution where the goal of securing profits often comes at the expense of those who work the land. Indeed, economist George Beckford has charged that the plantation system has all but ignored the social aspirations, family relationships, and even the religious beliefs of plantation workers. What he claims is partially reflected in the spatial design of most plantation estates. For example, the typical estate includes residential units for laborers (see Figure 8-2) and, according to critics, this acts against family integration and creates a situation of dependence on management by laborers.

The need for a large, disciplined, and unskilled labor force is constant for the plantation, and this sometimes causes special problems. When plantations are established in sparsely inhabited areas, labor must be imported from other regions. Before the nineteenth century, this was accomplished through slavery or indentured servitude. Plantations now attract migrant laborers, who unable to find jobs at home, search elsewhere for steady work. If the colonial plantation exploited workers, then today's commercial estates in many developing areas, regardless of who owns and manages them, do the same thing.

There is no question that the worker's lot has improved somewhat by the creation of trade unions and through the enactment of local social legislation. Nevertheless, unemployment (and underemployment) rates are extremely high in third-world nations and for the itinerant worker, having a low-paying job is better than having no job. Those countries that are well on their way to development and producing crops with high profits are better able to provide for their plantation laborers. People involved in plantation work in Malaysia, for example, earn more than those in other sectors of the economy. Even when workers are earning relatively good wages, there is the concern of steadily accelerating rates of migration from one region within a country to another. This movement, mostly of able-bodied men between the ages of 16 and 35, disrupts many economic and social activities in the migration source areas.

The plantation, whether it is owned by a foreign corporation or by a local government, must ensure a steady supply of workers. A labor stoppage or shortage at a critical moment in the tending, harvesting, or processing of a crop can be economically disastrous. Wages for labor, although low for individuals, constitute anywhere from 50 to 80 percent of the total plantation expenses because of the requirement of a large number of workers.

In spite of large annual earnings, the plantation mode of agriculture suffers from a variety of chronic problems. Some of the troubles are attributed to the botanical characteristics of certain crops and their susceptibility to diseases and other natural hazards. Also, world price levels of tropical cash crops are highly volatile, and when prices are depressed, managers are compelled to lay off laborers or lower their wages. On many occasions, plantations simply have failed. The main reason plantations are vulnerable is that they do not have the flexibility to shift production easily or quickly from a crop with a low world market value to one that brings higher prices.

The modern plantation employs advanced agricultural systems and often engages in vigorous research and development programs, but this was not always the case. The colonial plantation system practiced highly destructive agricultural methods such as overcultivation, poor drainage systems, and incorrect selection of crops that did not suit the environment. Consequently, the soil often was ruined, leaving a long-term damaging impact on land fertility and productivity. Compelled to relocate, these businesses left behind almost permanent desolation. Fortunately, plantations today are maintained with much less likelihood of destroying fragile tropical soils, but reckoning with this problem creates great expense.

All of these concerns—labor needs, price fluctuations, and environmental constraints—have convinced many governments in developing countries to promote policies that would reduce their dependence on large estates and to increase the production of cash crops by peasant farmers.

CASH CROP PRODUCTION BY SMALLHOLDERS

In many parts of the tropical world, cash cropping has always been performed by a large number of indigenous peasant farmers (Figure 8-6). Smallholding, as this economic activity is called, has had a particularly long history in west Africa and southeast Asia. During the early part of this century, there were developments that caused the smallholding system to challenge plantations in areas where the latter had been dominant for centuries. For example, smallholding became an attractive alternative to plantations in parts of Latin America in the 1920s and 1930s when the prices of many agricultural goods were severely depressed.

Peasant farmers, unlike the plantation, have few fixed costs. In areas where it is impossible to secure the services of a large labor pool, small-scale agriculture is the only alternative. The smallholder also has the option in bad times of returning to subsistence farming, and switching back to cash crops again during periods of high prices.

So, just as the plantation system has certain advantages attributable to economies of scale, the smallholder can often apply a smaller level of inputs and still generate a profit. It is important to note that most peasant farmers, especially those in Africa, engage in the production of cash crops primarily as a supplement to subsistence farming activities. Approximately 60 percent of the labor force and 70 percent of the cultivated land in Africa are still used for subsistence production. It depends on the region whether the activity of cash cropping assumes a greater importance than that of subsistence farming.

Smallholders have begun to specialize in the cultivation of annual crops. It is also the case that in some regions plantations are opting not to grow certain crops, and where this happens, smallholders pick up the slack. It is claimed that ecological, political, and social factors in west Africa have combined to result in some smallholders producing their crops as efficiently as plantations or state farms. Nonetheless, modern agricultural techniques are very rarely applied by individual peasant farmers, and this results in both lower yields and poorer-quality products.

Unfortunately, peasant farmers in the postcolonial period have experienced serious problems created by their own governments. Until about 25 years ago, in order to sell their produce, most smallholders had to work through foreign export–import firms. These agencies acted as middlemen between the indigenous farmer and the world market. In recent years, however, government-sanctioned agricultural marketing boards have been created to supervise export trade in agricultural commodities. The original task of mar-

keting boards was to help farmers by stabilizing their income. The record of marketing boards is rarely good. They tend to act against farmers by favoring middlemen wholesalers, and they rarely assist farmers in improving their productivity. Furthermore, marketing boards have become refuges for inept political appointees who know very little about agriculture and less about international economics. It is understandable why they are so unpopular with peasant farmers.

Real tensions exist not only between peasant farmers and their governments, but also between peasant farmers and plantation owners. In some areas there is extensive competition for land and labor, for access to processing factories controlled by plantations, and for government support. The disposition and terms of land usage is a lingering major issue in Sri Lanka, Fiji, Indonesia, Mauritius, and in some of the islands in the West Indies.

The plantation and smallholding systems can be (and often are) reconcilable. Indeed, both systems of cultivation can gain mutual benefit through cooperation. The exchange of labor for access to processing and marketing facilities is often a common practice

Figure 8-6 Smallholders Preparing Bananas for Marketing in Western Cameroon.
These peasant farmers, because of their proximity to nearby commercial banana estates, can sell their small output for export to the same buyers at the same time that the plantation markets its product. (Courtesy Sanford H. Bederman)

between smallholder and plantation manager. In some regions the development of several smallholder units around a nucleus estate of at least a minimum economic size has been encouraged. In cases such as this, smallholders are even provided plant material by the plantation. This scheme tends to boost output for all parties, decrease overhead costs for the plantation, and eliminate needless duplication of effort on the part of the smallholders surrounding the estate. Although it would seem to be a natural grass-roots economic option, all too often it is incumbent on third-world governments to initiate cooperative arrangements between smallholders and plantations.

TROPICAL AGRICULTURE AND ECONOMIC DEVELOPMENT

Cast in a historical perspective, it seems clear that the third-world countries that specialize in agriculture can be divided into two broad types, distinguished by their prevailing land tenure. There are those economies where the peasant farmer had the opportunity to develop commercially feasible smallholdings, as in tropical Africa and southeast Asia. On the other hand, there are those tropical lands (Central America, for example) that have been dominated by a plantation economy characterized by severe inequalities in social well-being between planter and worker.

It would be an easy matter to distill all the dilemmas of developing countries into the choices between promoting plantations or smallholders. It seems that many developing nations have rejected the idea of further expanding their traditional agricultural sector in favor of enlarging the industry, manufacturing, and service sectors, which are known as "modern" activities. Unfortunately, only a few developing nations have been successful with this strategy, and many have failed dismally in their attempt to nurture modern sectors. While this was happening, very important agricultural activities have been neglected.

It is not surprising that many third-world countries have become frustrated with the process of development and feel that the present international economic order is stacked against them. On the one hand, they can never really compete with western industrial nations in manufactured products, yet, on the other hand, to be dependent on the export of cash crops is a constant gamble. There is the continual concern that price levels for most tropical cash crops will never stabilize. Furthermore, as modern technology advances, cheaper substitutes may be found for agricultural products, just as rubber, sugar, and many fiber crops have been partially supplanted by synthetics. Finally, there is the overwhelming sentiment among governments of developing nations to avoid becoming dependent on any one crop or groups of crops.

The plantation is thus an ambivalent symbol. Some countries that previously had promoted smallholdings are now shifting their policy to increasing output by developing plantations. In the process,

leaders hope that the additional income will "trickle down" to all levels of society. Nigeria is one of these states. Officials there have asserted that the evils associated with plantation agriculture are not inherent in the system. Some of these same officials have regretted that their colonial rulers, the British, had not established plantations and thus did not leave them with a legacy of large-scale agriculture.

When some countries disbanded existing plantations and sought to create smallholdings, it was quickly discovered that total agricultural production plummeted. Yet there are continued criticisms that the plantation system encourages income inequalities since it fundamentally denies a large part of the rural population from having a real stake in the land.

The paradox is quite evident to developing countries today. The economics of agriculture may suggest the advantage of large-scale farming, whereas social factors such as the question of opportunities and equalities, and in some cases even economic factors such as land productivity, may dictate smallholding as an approach to agricultural development. The major problem confronting most developing nations is whether to spur development, production, and growth, or to ensure a fairer and more equitable distribution of wealth in society. Although they would like to achieve both at the same time, it might be difficult to do so in terms of planning and policy execution. Whatever reforms the third-world nations decide to undertake, they will have to begin in the countryside and with agriculture since the vast majority of the population resides in villages and relies on farming for a livelihood.

SUMMARY AND CONCLUSION

Agricultural products used for food and industrial purposes have been grown in the tropics for close to five centuries. Demand by people in Europe and North America for such crops as sugar, rubber, coffee, tea, cocoa, and a variety of tropical fruits was met during this time primarily by large foreign-owned and managed plantations which utilized vast numbers of unskilled, poorly paid, and mostly transient laborers. In parts of southeast Asia, Africa, and Latin America small peasant farmers also provided cash crops for foreign export.

When the colonial era came to an end after World War II, most of the newly created nations were located in the tropics. These nations continued to satisfy the demands of people in the temperate midlatitudes by supplying agricultural products. This situation created problems because leaders in many third-world countries had previously condemned the plantation as an exploitative economic mechanism, yet to earn foreign exchange, many of these leaders have now been forced to endorse large-scale agricultural production. This is an indication of how government policies with regard to agricultural development have changed in the past two decades, but the system of

commercial agricultural production has not seen much change. Despite many economic, political, and social problems associated with it, commercial agriculture in the tropics will continue to be practiced both on large-scale plantation and small-scale peasant-owned farms. One question that remains is how important smallholding will become in the years ahead.

SUGGESTIONS FOR FURTHER READING

Beckford, George I., *Persistent Poverty: Underdevelopment in Plantation Economies of the Third World.* New York: Oxford University Press, 1972.

Bederman, Sanford H., and Mark DeLancey, "The Cameroon Development Corporation, 1947–1977: Cameroonization and Growth," in N. Kofele-Kale, ed., *An African Experiment in Nation Building: The Bilingual Cameroon Republic since Reunification.* Boulder, Colo.: Westview Press, 1980, pp. 251–278.

Courtenay, P. P., *Plantation Agriculture,* 2nd ed. London: Bell and Hyman, 1980.

Galloway, J. H., "Tradition and Innovation in the American Sugar Industry, c. 1500–1800: An Explanation," *Annals, Association of American Geographers,* 75 (1985), 334–351.

Graham, Edgar, and Ingrid Floering, eds., *The Modern Plantations in the Third World.* New York: St. Martin's, 1984.

Grigg, David, *The Agricultural Systems of the World: An Evolutionary Approach.* Cambridge: Cambridge University Press, 1974, pp. 210–240.

Manshard, Walther, *Tropical Agriculture: A Geographical Introduction and Appraisal.* London: Longman, 1974.

Morgan, W. B., *Agriculture in the Third World: A Spatial Analysis.* Boulder, Colo.: Westview Press, 1978.

Myint, H., *The Economics of the Developing Countries,* 5th ed. London: Hutchinson, 1980.

9 Fishing and Forestry

In this chapter we examine two primary activities: fishing and forestry. Both involve gathering or harvesting a raw material from nature and each has declined in relative importance in the world economy in the twentieth century. Nevertheless, they remain extremely important as sources of employment for a substantial number of persons, a major item in world trade, and either a major source of food (fish) or an important building material (forests).

FISH PRODUCTION

Fish provide an important source of protein in the diet for much of the world's population. For many parts of the third world, fish are the predominant source of high-quality animal protein, the sole source in some instances. Dependence on fish is heaviest in China, Japan, southeast Asia, subsaharan Africa, and parts of Latin America. Fish also serve a growing role in the diet in many developed countries, including the United States, even though western-style diets place more emphasis on beef, pork, and poultry. The developed world also has a high demand for fish meal as an animal food and for fertilizer.

Worldwide demand for fish remains high, and considerable competition exists to gain access to productive ocean waters which produce approximately 90 percent of the world's total supply. Only 10 percent of the fish supply comes from inland waters. Annual fish production rose about 5 percent per year in the post-World War II era, from a level of about 19 million tons in 1947 until the early 1970s, when output plummeted, largely due to the failure of the anchovy harvest off the coast of Peru (Figure 9-1). Since that time fish catches have grown slowly, rising at an average rate of only 1 percent per year.

Mounting evidence gathers that the worldwide sea harvest may be reaching peak limits, in the absence of new techniques and more selective harvesting. Aquaculture and ocean farming offer one opportunity to overcome these production limits. Technological improvements that can potentially raise yields include the use of acoustic resonators to track fish, stronger nylon nets, and on-board freezing and canning, which reduces waste. International cooperation will become more critical in the future to avoid overfishing and assure equitable catch allocations. The United Nations Law of the Sea program is deeply involved in discussions to allocate this ocean supply.

Maximum yearly harvests from the ocean, based on current conditions, approximate 100 to 150 million metric tons. In 1980 the total world catch was just over 72 million metric tons. In 1982 the total rose to 76 million metric tons. The most desirable fish for human consumption are already being harvested at optimum sustainable levels. Only less desirable edible fish and "trash fish" can be harvested in significantly higher quantities. Trash fish are those harvested solely for processing into poultry, cattle, or pet food or fertilizer.

Greatly increased oil prices in the late 1970s ham-

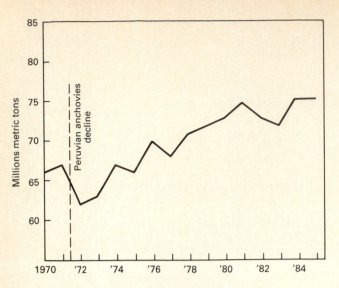

Figure 9-1 World Fishing Harvest, 1970–1985.
Since the dramatic fall in fish production in 1972 due to the failure of the anchovy harvest off the coast of Peru, catches have gradually increased, leveling off at about 75 million metric tons annually by 1985. The most desirable fish for human consumption are already being harvested at optimum sustainable levels, making significantly larger catches unlikely in the future. (Source: United Nations)

pered fishing fleets, which are very energy intensive enterprises. One rule of thumb in the industry suggests that it takes 1 ton of fuel (oil) to catch 1 ton of fish. For this reason, larger, long-distance trawlers are not as competitive today as more locally oriented smaller rigs, particularly those less than 80 feet in length (Figure 9-2).

Locational Considerations

Commercial fishing activity occurs predominantly in northern hemisphere ocean waters. Coastal margins of the middle latitudes are the biggest producing areas. In these waters one finds the largest concentrations or schools of marketable species. Tropical waters produce many fish as well, but they are less desirable for eating because of their higher oil content. Schools of single species rarely occur in tropical waters. More varieties are intermingled in tropical waters, making commercial gathering less efficient.

Fish prosper in shallow coastal waters. Access to a plentiful food supply partially explains this association. Mineral and organic matter eroded from continents accumulates most profusely in coastal marine waters. Sunlight penetration, critical to the photosynthesis process required by plants, assures continuity in the food chain in coastal areas. Plant nutrients thus collect on the continental shelf, with few such organisms living in depths beyond 200 feet.

Water circulation patterns promoted by upwelling currents and convectional mixing provide plenty of oxygen to sustain life in coastal zones as well as assuring a favorable dispersal of nutrition. Together these processes create an ideal living environment for fish. Upwelling refers to the drifting of surface currents away from land masses, leaving a "vacuum" which is replaced by nutrient-rich water surging in from greater depths.

The convergence of cold-water and warm-water currents occurs predominantly in the middle latitudes along the coastlines. The most significant interactions are those between the Labrador current and the Gulf Stream in the northwest Atlantic, and between the Kamchatka Current and the Japanese Current in the northwest Pacific (Figure 9-3). Convectional mixing also promotes the blending of water beneficial to fish culture through a natural process of cooler water displacing warmer water. This vertical movement of water occurs most dramatically during winter months, when surface cooling increases the density of water. The cooler, denser water then sinks, churning up warmer water below.

The largest producing areas for commercial fish

Figure 9-2 Tuna Boat. *Due to higher energy costs, larger long-distance trawlers are not as competitive today as smaller rigs, such as the one shown here. This tuna vessel stationed in San Diego has an observation tower and radar for locating schools of tuna. The small craft on the rear deck pulls the net away from the boat and encircles the fish which are then hauled on board and stored in large holds. (Courtesy Clair A. Shenk)*

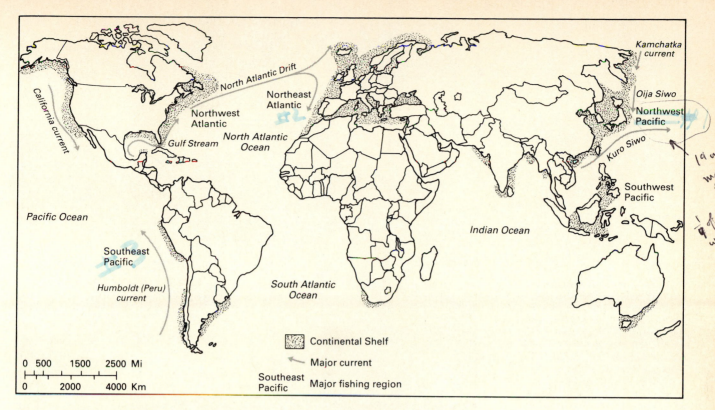

Figure 9-3 **Major World Commercial Fishing Regions.** *The largest producing areas for commercial fish are shallow waters on the continental shelf. The leading fishing region in the world occurs in the Northwest Pacific, followed by the Northeast Atlantic, and Southeast Pacific. Note the convergence of ocean currents in these areas.*

are shallow waters on the *continental shelf* (Figure 9-3). The term "continental shelf" refers to the gently sloping section of the ocean bottom adjacent to the land. It can vary from a few miles to over 150 miles in width. The depth of water in this area varies widely but rarely exceeds 500 to 1000 feet. At the outer extremity of the shelf, a sharp dropoff called the continental slope occurs, where depths of 12,000 feet are common.

Elevated portions of the continental shelf, banks, typically located several miles offshore, provide ideal environments for fish activity. The Dogger Bank in the North Sea encompasses an area of over 20,000 square miles. The Grand Bank and Georges Bank are two major fishing areas off the North Atlantic coast (Figure 9-4 on page 118).

A complex food chain provides fish with their sustenance. At one end of the chain microscopic plant organisms such as diatoms drift in the water. Diatoms absorb nutrients from the water, and in the presence of sunlight, convert them into substances that can be assimilated by animals such as small crabs and barnacles. These plants and associated microscopic animals are collectively known as *plankton*. Some commercially harvested fish such as herring and mackeral live directly on plankton, while others, such as tuna and shark, prey on the plankton eaters. Most fish live near the surface, but about one-third of the harvested take are bottom feeders and dwellers. Cod, snapper, flounder, lobster, crab, and squid all fit the latter description.

Leading Fishing Areas

In countries with a scarcity of arable land for agriculture and a coastal setting, a strong incentive has existed for the labor force to turn to the sea for a livelihood. Nordic countries (Norway, Denmark, Iceland, and Greenland) and Japan provide classic examples. In many cases fishing offers part-time employment in conjunction with farming, owing to the seasonal nature of both activities in northern latitudes. For others, fishing means year-round work. Some large ocean vessels serve as canneries as well as fishing rigs and stay at sea a year or more at a time.

The top ten fishing nations in 1983, which accounted for about half of the world catch, form the basis of Table 9-1. Japan tops the list, with a total production of over 11 million metric tons. Nine other nations also caught over 2 million metric tons of fish in 1983. In this group Norway lost ranking during the 1970s, falling from fourth to eighth by 1980, but rebounding to sixth position in 1983. Denmark and Peru dropped off the top 10 list, being replaced by Thailand and Indonesia.

In 1980, the traditional regional leader in fish production, the northwest Pacific, retained its rank-

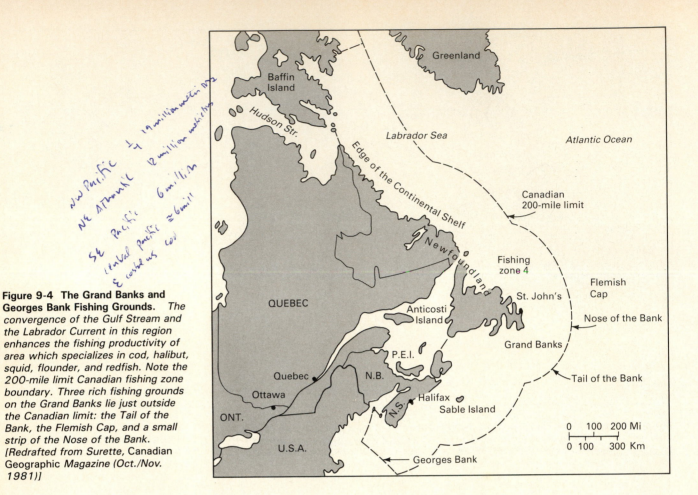

NW Pacific ± 19 million metric in 1972
NE Atlantic 12 million metric tons
SE Pacific 6 million
Central Pacific ± 6 mill
& costal us cod

Figure 9-4 The Grand Banks and Georges Bank Fishing Grounds. *The convergence of the Gulf Stream and the Labrador Current in this region enhances the fishing productivity of area which specializes in cod, halibut, squid, flounder, and redfish. Note the 200-mile limit Canadian fishing zone boundary. Three rich fishing grounds on the Grand Banks lie just outside the Canadian limit: the Tail of the Bank, the Flemish Cap, and a small strip of the Nose of the Bank.* [Redrafted from Surette, Canadian Geographic *Magazine (Oct./Nov. 1981)*]

ing. Production in the area approached 19 million metric tons, just over one-fourth of the total world take. That area extends southward from the outer Aleutian Islands in the north to the central Pacific north of the Philippine Islands. Americans are most familiar with the salmon and king crab produced off the Alaska coast. Herring, halibut, Alaskan pollock, and cod are also found in these northern waters. In the warmer water closer to China and Japan, tuna, mackerel, sardines, and shellfish predominate.

The northeast Atlantic Ocean and adjacent waters of the Arctic, extending from Norway to Great Britain and Iceland, comprise a second major fishing region. That area produced about 12 million metric tons of fish in 1980. The world's best cod-fishing grounds occur in these waters. Herring and haddock are also major contributors to the catch (Figure 9-5).

A third major producing area, the southeast Pacific, lies off the west coast of South America. Production there exceeded 6 million metric tons in 1980. The southwest Pacific area is best known for the anchovy harvest off the coast of Peru. The northward-flowing Peru current provides an ideal environment for the anchovy culture because it is associated with a coastal upwelling of nutrient-rich colder water laden with plankton on which the anchovy feeds. The success of the anchovy harvest lifted Peru to the status of the world's top fishing nation in the 1960s, with much of the catch exported as fish meal.

By the early 1970s overfishing depleted the stock, and when coupled with periodic setbacks from climatic changes, overfishing dramatically reduced the catch (Figure 9-1). An extremely complex natural phenomenon associated with a warming current, the El Niño, periodically limits the upwelling of colder food-rich water and reduces the anchovy supply. The Peruvian government instituted management con-

Table 9-1

Top 10 Fish-Catch Nations, 1983 (millions of metric tons)

Rank	Nation	Quantity	Percent of World Total
1	Japan	11	14
2	Soviet Union	10	13
3	China	5	7
4	United States	4	5
5	Chile	4	5
6	Norway	3	4
7	India	3	4
8	South Korea	2	3
9	Thailand	2	3
10	Indonesia	2	3
	Subtotal	46	61
	World total	76	100

Source: United Nations, *1983 Yearbook of Fishery Statistics,* 1984.

Figure 9-5 German Fish Trawler. *The Northeast Atlantic offers the world's best cod-fishing grounds, along with plentiful supplies of herring and haddock. In this photograph, a bed trawler net is being emptied on the deck of the German fishing vessel. (Courtesy German Information Center)*

trols on the harvest and nationalized the industry in the early 1970s following a disastrous El Niño. As normal conditions returned, the harvest increased later in the decade, but another El Niño setback occurred in 1982–1983, further destabilizing the industry.

The west central Pacific places a close fourth as a fish-producing area. The catch approached 6 million metric tons in that region in 1980. This fish-producing area in southeast Asia extends in the north from the Philippines and Indonesia southward to the Australian coast. A wide variety of fish flourish in this extensive region, including redfish, bass, mullet, mackerel, oyster, crabs, and tuna. This area, together with the Indian Ocean, comprises a major marine environment not being fished at maximum level. Statistics on catch levels are unreliable in these areas due to the large number of active independent fishermen in the fragmented island realm.

The fifth major producing area includes the Grand Banks and Georges Bank area of the northwest Atlantic east of Maine, Massachusetts, Nova Scotia, and Newfoundland (Figure 9-4). The convergence of the Gulf Stream and the Labrador Current in that region enhances productivity. Cod is the major fish, but halibut, squid, flounder, and redfish are also important. In 1977 the Canadian and American governments extended their fishery limits to 200 miles, which then placed most of the water in these banks off limits to foreign fishermen without permission. In Canadian waters, foreigners caught half the fish before this change, whereas their share dropped to one-fourth more recently. Canadians allow foreigners to fish for surplus fish within the 200-mile limit, but primarily for exotic fish not in demand domestically.

The leading fishing states in the United States in terms of value of production include Alaska, California, Louisiana, Texas, and Florida. Gulf of Mexico fishing, which accounts for the majority of production associated with the latter three states, consists mainly of shrimp and menhadden. Menhadden primarily provides a source for oil, fertilizer, and animal feed. Sponges contribute to the Florida catch. Louisiana leads U.S. fish output in terms of tonnage due to its large menhadden take. In fact, menhadden is the most important species in the United States in terms of quantity taken.

International Trade

Imports and exports of fish represented a $14 billion business in 1980. Most of this activity, by far, occurs among the developed nations, which are net importers of fish, even though they are also the major producers. Developing countries serve as exporters. The leading importer of fresh, chilled, and frozen fish is the United States, which accounts for 22 percent of the trade. Japan is a close second, importing 15 percent of the total. Other major importers include several developed countries of western Europe. Canada, Denmark, South Korea, the United States, Iceland, and Norway dominate as major exporters of fresh fish. The movement of salted, dried, and smoked fish is about one-fifth the level of fresh fish activity but is important to the economies of the major exporters, which include Norway and Iceland. Japan is the leading importer of this processed fish.

Law of the Sea

The emergence of nationalized fishing grounds dramatically shifted the share of fishing catches in favor of developing countries in recent years. Many nations now subscribe to the treaty drafted by the Third United Nations Law of the Sea Conference concluded in 1982, even though it does not officially take effect until ratified by 60 nations. The United States has not signed this treaty, due to the fact that this third world–dominated group seeks more concessions than the United States will agree to. Since the late 1970s, many countries have adopted a territorial sea limit extending 12 miles from shore, a 24-mile contiguous zone, and a 200-mile exclusive economic zone (EEZ) as called for in the Law of the Sea treaty.

The United States claims a territorial sea of only 3 miles but does recognize 200-mile fishery zones, as do Canada, Mexico, and the Soviet Union, among others. One hundred third-world countries now control over two-thirds of the world's fish reserves. Among the countries benefiting from new EEZs, one can count Fiji, whose catch tripled from 1977 to 1983, and Mexico, where the catch doubled. Overall, developing countries now account for over half the world catch, whereas they accounted for only 45 percent of the total in 1977. The small, inexperienced fishing fleets operating from these areas may not be able to sustain recent gains in catches in the future. The best fishing areas belong to the United States and Canada, and Canada is now the leading fish exporter.

Within the 200-mile zone, the Law of the Sea convention states that coastal states can manage fisheries to assure optimum yields. Domestic fishermen have first claim on the fish. Surpluses, if any, can be made available on a fee basis to foreigners. The U.S. Department of State allocates foreign fish taken in U.S. waters with shares assigned on the basis of the sizes of past harvest. Japan, for example, received 79 percent of the 1979 allocation and the Soviet Union, 17 percent. The actual foreign fish take in the U.S. Fish-

Table 9-2

Foreign Fish Take in U.S. Fishery Conservation Zone, 1979 (metric tons)

Rank	Nation	Total (Metric Tons)	Percent of World Total
1	Japan	1,109,115	68
2	Soviet Union	282,491	17
3	South Korea	127,357	8
4	Poland	56,812	4
5	Canada	26,260	2
	Subtotal	1,602,035	98
	World total	1,640,997	100

Source: Robert L. Stokes, "The New Approach to Foreign Fisheries Allocation: An Economic Appraisal," *Land Economics,* 57 (1981).

ery Conservation Zone in 1979 appears in Table 9-2. Most of the fish came from Alaskan waters (note the dominance of the Japanese take). In 1980 the United States cut off the Soviet allocation almost entirely in retaliation for its invasion of Afghanistan.

In addition to nationalized fishing grounds, provisions in the Law of the Sea treaty address military and commercial navigation, overflights, communications, continental shelf gas and oil, the prevention of pollution, seabed mining, marine scientific research, and the settlement of disputes. The United States' dissatisfaction with the agreement hinges on resistance to the transfer of technology to other nations and problems with the way the deep-seabed mining authority is constituted.

Aquaculture

Aquaculture, the commercial raising of plants and animals in water, promises to become big business in the fishing industry. Today, aquaculture accounts for less than 1 percent of world fish production, but it is important in coastal realms of China and inland waters of Japan. Fish are also raised as a sideline in rice paddies in these areas. In recent years Japanese farmers have permanently converted some rice paddies into huge dug-out tanks for fish production. They have increased the output of ponds and lakes by using floating cages and building underwater pens ringed by nets. The Japanese fish farmer is supported by cooperatives that purchase supplies, feed and breeding stock, and assist in marketing the product. In the United States, commercial production of catfish has become big business in many southern states (Figure 9-6).

Since fish are cold blooded and adapt to water temperatures rather than wasting energy to keep warm, they are 30 to 50 percent more efficient than cattle and hogs as sources of protein. For this reason and the fact that they can be grown at various depths in the same body of water, fish produce very large yields in a small space. In coastal Singapore, for example, mussels yield 45 times as much protein per acre as soybeans. Evidence also mounts that fish grow faster in warmer water, which bodes well for aquaculture in tropical environments of the third world.

Lingering Issues

Pollution of coastal marine waters remains a threat to world fisheries. Heavy metals such as cadmium, lead, and mercury reach coastal waters via river discharges in industrial areas, and oil spills occur in shipping lanes and coastal waters. In the Mediterranean, for example, high mercury levels found in tuna and swordfish are 10 times those of the Atlantic, owing to the greater residency time of pollutants in this enclosed marine basin.

Potential solutions to the marine pollution problem have received much attention, such as curtailing industrial emissions, prohibitions on the dumping of untreated effluent into streams, and stricter controls over seafaring vessels, particularly supertankers. Disaster preparedness crews provided by private indus-

try and governments have been assigned to cleanup duties following oil spills in marine waters. Techniques for use in cleanup operations include the application of chemicals that promote absorption and emulsification and spreading fertilizers that promote the growth of oil-consuming bacteria. On the other hand, there is mounting evidence that it may be better to leave oil spills alone, encouraging natural dissipation, rather than to undertake elaborate cleanup operations.

Overlapping claims of control of rights over marine waters continue to be a source of contention between nations. While guidelines based on the Law of the Sea framework have alleviated much tension, problems continue. One trouble spot occurs in the South China Sea. Historically, China claimed these southeast Asian waters, which encroach on claims by Vietnam, Indonesia, and other countries in the region. Japan and the Soviet Union continue to have differences over the waters between the Island of Hokkaido and the Kurile Islands. Japanese fishermen

Figure 9-6 Commercial Catfish Farm. *The commercial production of catfish has become big business in many southern states. This aerial perspective of the Hildinger Hodge catfish farm in Bonham, Texas shows a large scale operation involving over 20 ponds for the growth of fish. (USDA-SCS photo by John Doak)*

have been accused of currying favor with the Soviets by providing them with consumer goods in return for access to Soviet waters in the area.

Consumption Economics

Fish will continue to benefit from favorable economic conditions in relation to other animal protein foods. In world diets at present, fish are twice as important as poultry as a source of protein,

and more than half as important as the meat of cattle, pigs, sheep, horses and goats put together. Of the 76m[illion] tonnes [metric tons] annual catch, 54m (71%) is for human consumption (split about evenly between rich and poor countries). This compares with 105m tonnes of red meat and 25m tonnes of fowl. On top of that, 22m tonnes of fish is fed to livestock as fishmeal.

Fish has an intrinsic cost advantage over other kinds of food. The raw material is free. The running costs of a small (30–80 foot) trawler are around $200,000 a year. During the year, such a vessel will land, perhaps, 600 tonnes of cod or 1,500 tonnes of sardines. So, in theory, the average cost of a tonne of fish varies between $130 and $330. Compare that with other kinds of meat: in Argentina, the wholesale price of beef is above $500 a tonne; in America, a tonne of pork costs more than $1,000. And it takes only one-tenth as much energy to fuel a trawler to catch and refrigerate a pound of cod as it does to produce the soybean meal that cattle need to put on a pound in weight.

This sort of calculation leads people to argue that fish is the ideal source of protein for the world's poor. In poor countries, it is often the main source of animal protein to eke out the staple diet of rice or maize. Thirteen African countries (including some land-locked ones) get more than half their protein from fish. If countries are ranked by reliance on animal protein derived from fish, 39 of the first 40 places go to developing countries.[1]

Future Prospects

In addition to the potential offered by aquaculture as a source of increased production, better management of existing catches, and changing dietary habits in favor of more plentiful species, will enhance the role of fish as a source of protein in the future. Improved management can promote better use of the by-catch, the fish that are unintentionally captured or those thrown back as more profitable species are caught. Reducing waste and spoilage will also enhance yields.

Two fish species gaining favor by experts as good prospects for greater human consumption in the future are krill and squid. Krill are small shrimplike creatures found in Antarctic waters that must be processed prior to consumption in order to make them competitive, because of the large size of the shell in relation to the meat. Krill now serve a role in the food chain only as a food for whales, seals, and squid. Owing to their enormous numbers, up to 200 million tons of krill could be harvested annually. A declining whale population has also led to greater surpluses of krill. Squid are particularly plentiful in the northwest Pacific. They are already heavily consumed in Asia, but western diets have traditionally shunned squid.

The future of fish in the human diet remains bright. As that demand increases there may be less use made of fish for animal food and more used for direct human consumption. Most certainly, third-world nations will begin to produce more fish for their domestic markets rather than for export (Figure 9-7).

[1] "World Fishing Flounder," *The Economist*, June 23, 1984, p. 71.

Figure 9-7 Catamaran Fishing Vessel. *Production of fish in the third world is expected to grow in the future. The Law of the Sea guidelines have increased the territorial control rights for these areas and technology advances will further enhance their harvest capabilities. This rather crude catamaran being launched at Mutham, India has been mechanized with overseas and government counterpart funds, aimed at improving commercial fishing ventures. (FAO photo by D. Mason)*

This will help make them more self-sufficient and their countries less like that of a plantation economy dependent on the uncertain and fluctuating markets in the developed world, as discussed in Chapter 8.

FORESTRY

In this section we discuss the world timber industry. Timbering is the most widespread and dominant type of forestry activity—the broader form of which also includes the gathering of tree products such as sap, nuts, and bark, among others. The latter were discussed in Chapter 3. Eight nations produced 60 percent of the total world roundwood (logs) in 1983 (Table 9-3). These countries produced nearly 2 billion cubic meters of products in 1983. The United States and the Soviet Union lead in production, but several less developed nations, such as India, China, and Brazil, follow closely behind.

On a world scale half the wood consumption occurs as fuel, including cooking and heating uses, and half for industrial purposes (boards, pulp, veneer), but extreme variations occur by region. In the tropics fuel use predominates, whereas industrial consumption prevails in the middle latitudes. Fuel uses account for over 80 percent of lumber consumption in Brazil, Indonesia, India, and many other Asian and African countries. These countries lead the world in fuelwood production (Table 9-4). Similar large shares of wood find industrial uses in western Europe, Canada, and the United States. The United States, the Soviet Union, and Canada lead in industrial wood production (Table 9-4).

Global Belts

Commercial forests occur in two huge global belts, as shown on Figure 9-8 on page 124. The first virtually encircles the world in the higher latitudes of the northern hemisphere. The second forest-gathering

Table 9-3

Leading Total Roundwood (Log)-Producing Nations, 1983 (millions of cubic meters)

Rank	Nation	Production	Percent of World Total
1	United States	438	14
2	Soviet Union	356	12
3	India	233	8
4	China	232	8
5	Brazil	220	7
6	Canada	142	5
7	Indonesia	122	4
8	Nigeria	86	3
	Subtotal	1,829	60
	World total	3,042	100

Source: United Nations, *1983 Yearbook of Forest Products*, 1985.

Table 9-4

Leading Fuelwood- and Industrial-Roundwood-Producing Nations, 1983 (millions of cubic meters)

Rank	Nation	Production	Percent of World Total
Fuelwood			
1	India	213	13
2	Brazil	162	10
3	China	155	9
4	Indonesia	114	7
	Subtotal	644	39
	World total	1,633	100
Industrial Roundwood			
1	United States	336	24
2	Soviet Union	273	19
3	Canada	136	10
	Subtotal	745	53
	World total	1,409	100

Source: United Nations, *1983 Yearbook of Forest Products*, 1985.

realm lies in the tropical equatorial zone, including a large part of South America and central Africa. Apart from these two general regions, commercial forest gathering is of significance only in small zones in Japan, southeastern United States, Chile, southeastern Australia, New Zealand, and in some of the countries flanking the Mediterranean Sea. At one time, trees blanketed 25 to 30 percent of the earth's land, but today the closed tree cover, defined as tree crowns covering 20 percent or more of the ground when viewed from above, encompasses only 15 to 20 percent of the landscape. Moreover, closed forests are apparently disappearing at a removal rate of 25 million acres per year.

The forest resource of the midlatitudes is distributed according to the regional patterns of hardwoods and softwoods. A hardwood deciduous tree belt occurs on the side toward the equator of the midlatitude forest region. This belt includes oak, chestnut, hickory, maple, birch, and beech trees (Figure 9-9 on page 124). Only a few softwoods occur in the region. Hardwoods serve as raw materials for railroad ties, furniture, and specialty items (tools, wagons). Originally, hardwoods covered the eastern United States and Europe from the Baltic to the Mediterranean, with the belt narrowing as it extended eastward into western Siberia. Most of China and Korea, too, were once forested with hardwoods, just as parts of Japan and Manchuria remain. In the southern hemisphere this hardwood belt has a limited counterpart: the narrow coastal fringes of Chile, southern and eastern Australia, New Zealand, and the southern tip of Africa.

A softwood coniferous belt occupies the poleward flanks of the midlatitude forests. In Eurasia it forms the world's single largest forest zone, stretching 7000 miles from Scandinavia to eastern Siberia, where it fans out broadly to nearly 2000 miles. A similar vast region of softwoods cover North America from Alaska

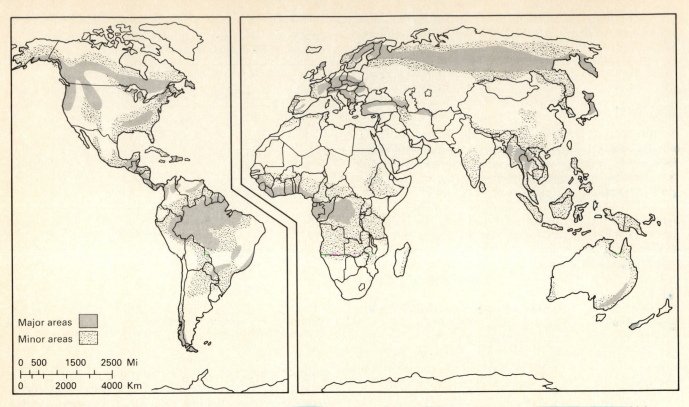

Figure 9-8 Major World Forest Regions. *Commercial forests occur in two huge global belts, virtually encircling the world at higher latitudes in the northern hemisphere and again in the tropical equatorial zone as shown here. At one time nearly one-third of the earth's surface was tree-covered, but a long-standing removal process continues to withdraw 25 million acres per year from the inventory worldwide.*

Figure 9-9 Midlatitude Deciduous Forest. *This upland oak forest is in High Rock Park, Staten Island, New York. Double trunks as shown at right often occur with oaks following disturbance (e.g., fire or cutting). (Courtesy Mary M. Thacher)*

to Labrador; on the west it reaches southward along the Pacific coast, extending even into Mexico (Figure 9-10). On the east there is a detached segment in the southeastern United States. The most common trees in the softwood belt include pine, spruce, fir, and larch. They produce paper pulp and lumber framing materials for buildings. Mixed forests, consisting of both hardwoods and softwoods, typically occur as a transition strip between the hardwood and softwood belts.

Mature trees in the midlatitudes are generally smaller than those in the tropics. Yet there is one notable exception to this general rule. The earth's tallest trees grow in the fir and pine forests of northern California, where giant redwood sequoias attain heights of 300 feet and diameters of 20 feet. Douglas fir trees in this region grow 250 feet tall and have trunks that measure 6 feet across.

Forest cover in the United States and Europe remains essentially stable, if much reduced from preindustrial days, due to active conservation and reforestation programs. Forest management in Europe allows for considerable cutting even with a relatively small

tree inventory by world standards. By contrast, most of Asia faces a timber scarcity. Large parts of the Middle East, north Africa, China, and India lost their trees centuries ago. The biggest net losses today occur in Latin America and Africa.

Although accurate data do not exist to indicate the volume of timber standing in the world's forests, the U.S. Forestry Service compiles estimates for the United States. These data indicate that four states dominate the U.S. saw-timber volume: Oregon, Washington, California, and Alaska, which contain respectively 19, 13, 11, and 7 percent of the nation's total forest resource—50 percent altogether.

Industrial Timbering

Lumbering in the midlatitude region provides roughly 80 percent of the world's industrial wood. The leading producers include the United States, the Soviet Union, and Canada (Table 9-4). The low level of demand for industrial wood in the tropics partially explains this situation. In much of the tropics, settlement is sparse and economies remain nonindustrialized. Tropical

Figure 9-10 Seedling Regeneration in Softwood Forest. *A vast softwood coniferous belt covers North America from Alaska to Labrador, extending southward into Mexico on the west coast. In this photograph seedlings grow in an old clearcut harvest area in Glacier Peak National Forest, Washington. (Courtesy Weyerhaeuser Company)*

forests occur in remote locations in relation to potential markets. Since transport costs for lumber are high per unit of weight, the demand for tropical industrial wood is low except for specialty types such as mahogany, which is highly desired by industrial nations.

The forest resource of the tropics contrasts sharply with that of the midlatitudes. In the heart of the tropics lies the *rain forest,* the most luxuriant and variegated community of plants on the face of the earth. The unique features of the rain forest most directly related to forest gathering include the profusion of species, "confusion of stands," large sizes of trees, and hardness of woods.

The remarkable variety of species in the rain forest yields a broad range of products. The warm, humid tropical environment, which encourages rapid maturation and frequent mutations, resulting in numerous different species, may explain the abundance of plant types. The "confusion of stands" in the rain forest refers to the heterogeneous supply of timber in a given forest, which will frequently contain 15 to 30 different species. Why the stands of timber are mixed rather than uniform is obscure, but the economic consequence is clear: confusion of stands increases the cost of gathering the forest products.

Figure 9-11 Tropical Rain Forest. *The most notable features include rich tree diversity, numerous lianas (vines) and epiphytes (e.g., orchids, ferns); and the tall column-like structure of the tree trunks which soar to great heights. (Courtesy of Thom McHugh)*

The large size of tropical trees offers an advantage that partially counteracts the absence of pure unmixed stands. Many trees soar 150 to 180 feet into the air. As a consequence, the volume of timber that can be harvested from an acre is great. The large size of its trees is one reason Brazil ranks so high among nations in quantity of wood logged.

The largest continuous rain forest region in the world occurs in the Amazon basin of Brazil (Figure 9-11). One-half of the Amazon basin remains forested but considerable encroachment into the area has occurred from the east and south in recent decades as roads penetrate previously inaccessible areas. Logging interests and smallholding settlers practicing slash and burn agriculture have cut back the rain forest cover considerably, but the removal is not as great as often reported. By comparison, in Indonesia, which claims the largest rain forest acreage in Asia, much of the virgin resource has been harvested. Logging increased by six times in that area in the 1960s and 1970s. Much of this timber flows to Japan. Lumber interests then plant fast-growing pine species, useful for pulpwood production and manufacture into paper and cardboard boxes.

A principal tropical timber in strong demand, mahogany, comes primarily from Latin America, including Haiti and the Dominican Republic, although some comes from the Guinea coast of Africa. Other important woods include cedar, teak, ebony, and balsa. Cedar, a soft fragrant wood, originates mainly in Mexico, Central America, and Brazil. Teak, a strong hard wood, contains an unusually high amount of oil, which acts as a natural preservative in both fresh and salt water. Accordingly, teak is highly valued in the construction of wooden boats and as pilings for wharves and piers. Teak trees rarely occur outside southeast Asia. One of the earliest woods gathered commercially in the tropics was ebony, a hard, jet-black, fine-grained wood from central Africa, Madagascar, and southeast Asia. Ebony carvings of chest figurines and handles make popular craft exports.

Most tropical woods are hard and very heavy—indeed, some even too heavy to float. In fact, since almost all of these woods are too hard to work easily, many tropical countries with a surplus of trees import wood from the midlatitudes for building materials and other purposes. One exception is balsa, a tropical wood that is very light; many natives of the tropics have used balsa logs for centuries to construct rafts and canoes.

Midlatitude Forests

The forest resource of the midlatitudes reflects fewer species, purer stands, and a greater mixture of softwoods and hardwoods than its tropical counterpart. The purity of stands has important economic consequences, permitting foresters to concentrate on gathering one kind of wood. Being able to harvest whole groves, referred to as clear cuttings, rather than isolated trees encourages mass-production techniques, with all the resulting savings of time and expense.

Water Corridors

Waterways provide essential transportation corridors for commercial forestry activity. In both the tropics and midlatitudes, streams and rivers supply the main arteries for the movement of lumber products. Logs float downriver to ports and sawmills year-round in the tropics, but this movement is only a seasonal pursuit in the higher latitudes. The orientation of rivers in major timber regions dramatically affects their potential for commercial exploitation. In many countries streams flow in the general direction of the final market (southward), as in Sweden, Finland, and southern Canada. Such favorable accessibility conditions keep production and marketing costs low. In the Soviet Union and on the west coast of the United States, rivers flow in the opposite direction of the market, and higher transportation costs decrease the attractiveness of the forests.

In the Soviet Union several major rivers flow northward, including the Ob, Yenisey, and Lena. Frozen river conditions in winter and flood-stage problems in springtime, as thaws come first downstream while northerly portions remain ice blocked, hamper the forestry industry in the Soviet Union. For these reasons vast resources of trees in Siberia remain untouched due to inaccessibility.

In developed forestry producing areas, railroads and trucks also move timber. Forest products comprise a major commodity shipped on the trans-Siberian railway, as is the case in Alaska, northern Canada, and the U.S. northwest and southeast. Irregular terrain, although favorable to the growth of timber due to less competition with other types of land uses, handicaps harvest. Steep slopes hinder access by workers and tree removal itself. In some inaccessible areas helicopters provide the only means of transportation—for both the laborers and timber products.

Fuelwood Scarcity

The declining forest cover in many third-world countries creates a severe rural wood famine. In many areas of south Asia and Africa, peasant scavenging for wood poses a major effort. Where wood cannot be found, dried cattle dung serves as a substitute fuel. This practice, in turn, indirectly decreases the productivity of the soil by preempting its use as a fertilizer.

Inefficient wood stoves also contribute to the excessive use of wood in many third-world nations. Other cultural practices also divert precious fuel wood away from household use. In India, for example, more efficient cremation techniques for dead persons would save considerable wood and be less expensive.

In addition to the immediate human suffering from a lack of wood, the depletion of forest resources is associated with increased erosion, siltation, and flooding. The long-term productivity of land can be threatened by encroaching *desertification,* which many experts link directly to the declining forest cover (Figure 9-12).

Figure 9-12 Desertification.
Depletion of forest cover has been associated with increased desertification in arid and semi-arid areas. In this scene a farmer is watering a tree in a drought-affected area of Senegal. (Courtesy United Nations/Carl Purcell)

Perhaps the biggest threat to forested lands in the third world is agricultural encroachment. In some areas forest clearing occurs for cropland expansion, while in others it is associated with cattle raising, shifting cultivation, or simply rural squatter migrant encroachment. Multinational-controlled logging firms and plantation investors often underwrite this expansion, while in other cases it is promoted by government programs attempting to settle frontier areas. In the latter case it is often easier for the government to distribute land by encouraging the settlement of marginal tree-covered areas rather than to implement land reform by reallocating property already under cultivation. This policy not only jeopardizes peasant families economically but also depletes the forest cover.

Some third-world areas now promote agroforestry systems that have considerable merit. Shifting cultivation can be made compatible with increasing production by adapting the crop-fallow system described in Chapter 3. In this instance purposeful tree cultivation becomes a part of the fallow land cycle designed to assist fertility regeneration.

Finding a solution to the wood shortage in the third world will not be easy. Continuing population growth in the region adds to the problem, as does price inflation due to the shortages. In many countries more attention must be placed on protecting and enhancing the output of forests that serve domestic needs and less on export-oriented holdings largely managed by multinational corporations. A few coun-

Table 9-5

International Trade in Wood Products, 1983 (millions of U.S. dollars)

	Imports				Exports		
Rank	Nation	Value	Percent of World Total	Rank	Nation	Value	Percent of World Total
			Roughwood (Unmilled)				
1	Japan	3,005	59	1	Malaysia	1,541	34
2	South Korea	428	8	2	United States	1,195	26
3	Italy	288	6	3	Indonesia	286	6
4	West Germany	244	5	4	Ivory Coast	196	4
5	France	194	4	5	Gabon	148	3
	Subtotal	4,159	82		Subtotal	3,366	73
	World total	5,062	100		World total	4,562	100
			Shaped Wood (Milled)				
1	United States	2,808	26	1	Canada	3,338	36
2	Great Britain	1,380	13	2	Sweden	1,210	13
3	Italy	912	8	3	United States	919	10
4	Japan	842	8	4	Finland	704	8
5	West Germany	779	7	5	Austria	501	5
	Subtotal	6,721	62		Subtotal	6,672	71
	World total	10,924	100		World total	9,396	100
			Pulp and Waste Paper				
1	United States	1,500	19	1	Canada	2,506	32
2	West Germany	1,191	15	2	United States	1,755	22
3	Japan	930	12	3	Sweden	1,101	14
4	Italy	664	8	4	Finland	586	7
5	France	649	8	5	Brazil	311	4
6	Great Britain	629	8		Subtotal	6,259	70
	Subtotal	5,563	69		World total	7,918	100
	World total	8,055	100				
			Fuelwood (Including Charcoal)				
1	Italy	21	21	1	Spain	15	22
2	West Germany	12	12	2	France	4	6
3	France	10	10	3	United States	4	6
4	Japan	6	6	4	Sri Lanka	3	4
	Subtotal	49	49	5	Philippines	3	4
	World total	99	100		Subtotal	29	43
					World total	67	100

Source: United Nations, *1983 International Trade Statistics Yearbook,* 1985.

tries have made this successful transition such that equilibrium returns to domestic wood consumption and forest inventories.

The key to enhancing production in these countries may lie in the area of more effective community-based tree programs. In both South Korea and China good results have occurred with such programs at the grass-roots level. The South Korean program began in the early 1970s with the *saemaul* or new community movement.[2] Cooperatives mobilized villages to plant public and private holdings. A profit-sharing plan encouraged widespread participation of private landholders in the program. Village forestry associations connected with a national federation formed a communication linkage between the government and the villagers. The federation provided policy guidelines, technical assistance, and funding for the program. As

[2]Erik Eckholm, *Planting for the Future: Forestry for Human Needs*, Worldwatch Paper 26 (Washington, D.C.: Worldwatch Institute, 1979).

a result of this effort, by 1977 over a half million hectares of forests had been planted for fuelwood in South Korea.

International Trade

The geographic separation of forest product demand and supply regions makes international trade an important factor in the forestry industry. Most of the trade occurs among highly developed countries (Table 9-5). Unlike underdeveloped areas, they can afford to buy products in short supply.

Five nations account for nearly all (82 percent) of the importing of rough (unmilled) wood (Table 9-5). These countries include (1) Japan, which imports over half the world total, and (2) South Korea and several western European countries, which each import 4 to 8 percent of the total rough wood (Figure 9-13). Malaysia, in turn, is the leading exporter of rough wood (34 percent of the total), followed by the United States and three Asian and African countries.

Figure 9-13 **Log Export Ship.** *Japan imports more than one-half of the world's rough (unmilled) wood, while Malayasia and the U.S. are the leading exporters. These logs are being shipped from Longview, Washington to Japan. (Courtesy Weyerhaeuser Company.)*

Shaped wood (milled) imports again flow predominantly to western European countries and Japan, but the United States is the single leading importing country, accounting for 26 percent of the total in 1983 (Table 9-5). Canada leads the shaped wood export business, claiming over one-third of the total, followed by Sweden, the United States, Finland, and Austria.

Pulpwood is the third most important wood commodity involved in world trade (Table 9-5). This $8 billion business in 1983 involved large imports to the United States, West Germany, Japan, Italy, France, and the United Kingdom. Canada exported about one-third of the wood pulp. The United States and Scandinavian countries also contributed significant export quantities. In addition, Japan also imported over $1 billion of pulpwood logs and wood chips not accounted for here. These came largely from the United States and Australia.

Western Europe and Japan remain deficit fuelwood regions and import heavily (Table 9-5). Notice that there are no third-world countries among the leading fuelwood importers. They generally lack the resources to finance such trade even though they have major deficits. Leading fuelwood exporters include Spain, France, the United States, Sri Lanka, and the Philippines.

In summary, Japan, with about $5 billion in trade in 1983, leads the world in wood imports, while the United States and Canada, each with a volume of about $5 billion in 1980, lead the export side. Roughwood is by far the largest item involved in trade, nearly a $10 billion business in 1983, counting both imports and exports, while fuelwood was the least significant, involving less than $200 million in trade in 1983.

Prospects

The notion and practice of reforestation expanded rapidly in the post–World War II era in the northern United States and northern Europe on land once cleared for farming. In addition, abandoned farmland reverted naturally to tree cover at a growing rate. In Japan, where 75 percent of the country is mountainous and forested, strict laws prohibit tree cutting. Instead of depleting that resource, it is national policy to import industrial timber. Similar positive developments are now occurring in the third world.

Sustained-high-yield forests require that the rate of cut not exceed the rate of growth. This balance has already been achieved in Japan, West Germany, and Sweden. The United States has also achieved this objective—not in every part of the nation yet, but for the country as a whole. In the western United States, where the best quality wood occurs, cutting greatly exceeds growth. The reverse is true in the south and especially in the northeast, where growth is now twice the rate of cut. This net gain in the east, however, is mostly from seedling and pulpwood trees, while the net loss in the west occurs with saw timber.

Increasing levels of air pollution may decrease forest yields in many industrial areas in the future. *Acid rain* poses the greatest danger. The problem is most acute in the northeastern United States, eastern Canada, and in Central Europe, particularly in West Germany, Poland, and Czechoslovakia (Figure 9-14). Sulfur and nitrogen oxide pollutants emitted from industrial and coal-fired electrical power-generating plant smokestacks cause this pollution problem, which is growing in intensity. Although increased levels of these sulfur and nitrogen oxide pollutants in the soil can be beneficial to tree growth initially, they eventually become toxic to plants and attack the root system. Heavy metals and gaseous pollutants in the air also attack leaves, stunting tree growth.

For the world as a whole, the bulk of future forestry supply increases must come from the tropics because land and climatic conditions there will permit greater output than elsewhere. Intensive forestry plantations can provide some of this supply, but village woodlots around cities and towns must also be promoted. The overall forestry problem will not be

Figure 9-14 Acid Rain-Induced Forest Depletion. *Acid rain has produced yellow needles on Norway spruce in the Black Forest of West Germany. Note crosses placed by environmentalists to call attention to the problem. (Courtesy of Tom McHugh)*

solved unless a comprehensive approach is taken to the fuelwood supply issue, which includes the cessation of agricultural land encroachment and coordinated land settlement programs. International trade of industrial timber will remain very strong, but substitutions of other materials for wood will increase as prices rise and demand intensifies, especially in developing countries.

SUMMARY AND CONCLUSION

The world fishing and forestry industries experienced considerable adjustments following the energy and inflation crises of the 1970s. The outlook for both industries is far brighter today than the situation was a decade ago. The United Nations Law of the Sea conventions greatly assisted the fishing industry, much as conservation and reforestation projects have brought stability to the forestry realm. In both cases the third world promises to become a more active provider in the future, with more emphasis placed on domestic rather than export needs. Trade will continue to flow predominantly toward the developed countries in both industries, even though they are also major producers. The United States, Japan, and Canada play pivotal roles in both industries.

SUGGESTIONS FOR FURTHER READING

BENE, J. G., H. W. BEALL, and A. COTE, *Trees, Food and People: Land Management in the Tropics.* Ottawa: International Development Research Centre, 1977.

EARNEY, FILLMORE C. F., *Petroleum and Hard Minerals from the Sea.* London: Edward Arnold, 1980.

ECKHOLM, ERIK, *Planting for the Future: Forestry for Human Needs,* Worldwatch Paper 26. Washington, D.C.: Worldwatch Institute, 1979.

EDWARDS, R. L., and J. B. SUOMALA, JR., "World Fisheries in the Twenty-First Century," *Technology Review,* 83 (February–March 1981).

HAYTER, ROGER, "Corporate Strategies and Industrial Change in the Canadian Forest Product Industries," *Geographical Review,* 1976, pp. 209–228.

POSTEL, SANDRA, *Air Pollution, Acid Rain, and the Future of Forests,* Worldwatch Paper 58. Washington, D.C.: Worldwatch Institute, 1984.

SHOLTO, DOUGLAS J., and ROBERT A. DE J. HART, *Forest Farming.* Boulder, Colo.: Westview Press, 1985.

SMITH, NIGEL, *Wood: An Ancient Fuel with a New Future,* Worldwatch Paper 42. Washington, D.C.: Worldwatch Institute, 1981.

SMITH, ROBERT W., "The Maritime Boundaries of the United States," *Geographical Review,* 71 (1981), 395–410.

STOKES, ROBERT L., "The New Approach to Foreign Fisheries Allocation: An Economic Appraisal," *Land Economics,* 57 (1981), 568–582.

WHITE, PETER T., "Tropical Rain Forests: Nature's Dwindling Treasures," *National Geographic,* 163 (January 1983), 2–48.

WISE, MARK, *The Common Fisheries Policy of the European Community.* New York: Methuen, 1984.

Coal, Electricity, Water, and Nuclear Energy Resources

10

Energy, the stored ability to perform work or produce heat, comes in many forms—electrical, mechanical, or nuclear, among others. The major renewable sources of energy include wood, water, and solar, while fossil fuels and hydrocarbons (petroleum and natural gas) comprise a nonrenewable grouping. See the accompanying box for energy definitions.

ENERGY TRANSITIONS

Beginning with the industrial revolution, a trend emerged toward the greater consumption of higher-quality fossil fuels and a relatively reduced use of renewable forms, such as wood and water, as an energy source. These nonrenewable energy forms produced more energy per unit of weight, needed less upgrading, and burned cleaner. The progression from the use of wood to coal, then oil and natural gas to uranium, illustrates this transition. In recent years we have entered yet another era, the beginning of a conversion back to renewable forms. This transition that has just begun may take a hundred years to complete. During this period many short-term aberrations in the use of various energy forms will occur, depending on prices, government policy, international trade relations, domestic supplies, and shortfalls, to name a few.

Whereas some areas, including the Soviet Union, are largely self-sufficient in energy resources, many areas are notoriously deficient, including Japan and western Europe. This situation has created consider-able tension in recent years as energy consumption levels zoomed upward, prices increased, and nations began rethinking their energy priorities. Among the changes occurring has been the "second coming" of coal, in recognition of its relative abundance and flexible use both for generating electricity and as a source of steam power (Figure 10-1). In this chapter we discuss the changing role of coal, electricity, nuclear, and water energy sources. The related fortunes of petroleum and natural gas are discussed in Chapter 11.

The present worldwide dependence on oil may or may not have reached a maximum, but it certainly will crest by the end of the century. The uncertainty that marked the energy picture in the later 1970s and early 1980s made it difficult to know whether the peaking of petroleum production that occurred in 1979 reflected a watershed year or a temporary pause. In the early 1980s, for example, world petroleum demand decreased as consumption declined in response to earlier price jumps. Prices then fell through the mid-decade period, but the long-term trend in consumption was not clear.

What is known is that during the long-term transition to renewable resources dependence on coal, the most abundant fossil fuel, will increase. The future of nuclear energy is also unclear. Pressure from environmentalists and safety issues cloud its future following a period of growing use in the 1960s and 1970s. One author has suggested that the importance of the energy transition we are now entering is as significant as the Neolithic period "transition of the food system

132

Energy is typically measured in terms of quantities of fuel burned. Power is the rate at which work is performed, usually defined as work divided by time. The most universal measure of power is the metric watt unit. A watt is the rate of doing mechanical work in time. Other terms that measure the rate of doing work and their equivalents include horsepower (745.7 watts), joule (1 watt-second), and kilowatt (1000 watts). A "megawatt" is 1000 kilowatts or 1 million watts. This term is often used to indicate the electrical generating capacity of power stations, or the capacity of an electrical system. The term "horsepower" reflects the power exerted by a horse in pulling. It refers to the force required to raise 33,000 pounds at the rate of 1 foot per minute.

The terminology found in energy studies can be confusing. Wherever possible we will use coal equivalent comparisons. Typically, these will be expressed as metric ton equivalents. A metric ton is 1000 kilograms of 2204.6 pounds. Each energy field has developed its own measure to assess energy use or production. For example, oil companies use the term "barrels of oil equivalent" (BOE). Electric utilities typically use kilowatt hours as a measure. Many government studies attempt to standardize comparisons by using the "British thermal unit" or Btu. The British thermal unit refers to the quantity of heat required to raise the temperature of 1 pound of water 1 degree Fahrenheit.

from hunting and gathering to animal husbandry and farming."[1]

ENERGY INTERDEPENDENCE

The interrelatedness of the present world system is well illustrated in the case of energy. Dependency relationships cut across planned, developed, and developing nations. Disparities in energy consumption have already been mentioned. Because developing countries depend more than the world as a whole on

[1]Wolfgang Sassin "Energy," *Scientific American*, September 1980, 132.

high-quality energy (petroleum), and because most do not have their own supplies or the means to pay for them, they are very vulnerable to world energy supply and price changes. The growing debt burden facing third-world countries in the early 1980s occurred largely because of energy price increases and associated problems related to higher interest rates. Third-world countries originally shifted from wood to oil and gas consumption rather than evolving through an intermediate stage of dependence on coal because petroleum was relatively inexpensive, transportation was easier, and environmental problems could be avoided. But this shift became enormously expensive following the price jumps of the late 1970s, and foreign aid assistance could no longer be guaranteed to fill the gap.

Figure 10-1 Martiki Strip Coal Mine in Martin County, Kentucky. *Kentucky produces more strip-mined coal than any other state. At this mine, Appalachia mountain tops are scalped and moved to adjacent valleys exposing seams of coal. This mine now yields two million tons of coal all year at the site of the former town of Threefolds, Kentucky which was moved. The owners, Mapco Coals, Inc., reclaimed the land in compliance with the standards of the Service Mining Control and Reclamation Act of 1977, once coal has been mined. (New York Times Photos)*

Developed countries reevaluated their own priorities in this period in the face of rapidly increasing energy costs and often cut back on their assistance to third-world nations.

Developing countries actually suffer from energy malnutrition in two forms.[2] Not only are high-quality energy sources in short supply, but so are traditional forms. Scarcities of traditional fuels pose as great a problem as access to higher-quality forms. In some countries wood fuel for home cooking is itself as expensive as the food for which it will be used as heat for cooking. Recall also from Chapter 2 the problems caused by diverting for use as a fuel animal dung that could be available as a fertilizer. This situation restricts agricultural productivity, not to mention the greater erosion potential it creates. In essence, the more that price increases for higher-quality fuels force greater dependence on traditional fuels, the more trapped consumers become by shortages.

Conservation offers one way to escape the trap, but it is a more appropriate option for highly developed areas than for developing regions. Developed countries can finance conservation measures and make adjustments more easily. Cooking stoves are often very inefficient in the third world, and the appropriateness of western technologies to solve these and other inefficiencies themselves often cause bottlenecks. Fortunately, oil-exporting nations have helped cushion oil price increase shocks for the third world by increasing economic aid and offering barter arrangements involving commodity swap options for payments rather than requiring hard cash.

ENERGY PRODUCTION AND CONSUMPTION LEVELS

World energy production nearly quadrupled in the 30 years between 1950 and 1980. Roughly equal shares of production now occur in the developed, developing, and centrally planned economies as groups, whereas two-thirds was claimed by the developed world alone in 1950. Per capita consumption is another matter. Although consumption doubled in this time period on the world scale, developing countries started with such a low level in 1950 that they still lag behind the developed world dramatically. Developed areas consumed 15 times as much energy on a per capita basis in 1980 than did the third world. Planned economies operate at a consumption level one-third that of the developed world.

In the relatively affluent nations of Europe, Australia, New Zealand, and Japan, each person consumes the equivalent of 5 tons or more of coal a year. Americans and Canadians consume more than twice this amount, about 11 tons of coal equivalent annually. In most developing countries, however, consumption of conventional commercial fuels, such as oil, natural gas, coal, and electricity, is barely a quarter

ton of coal equivalent annually with, very roughly, an equal amount of energy obtained from fuelwood and other traditional, noncommercial energy sources. With three-fourths of the world's people, third-world countries consume just one-fourth of the world's energy production.[3]

Electricity production on the world scale clearly reveals the continuing disparity between developed and developing nations. Electricity production increases of fourfold occurred on the world scale during the period 1950–1980, paced by the growth in the developed world (see Table 10-1). Tenfold increases occurred in the developing and planned economies, but their sparse output level in 1950 was so low that the gap actually widened between them and the developed world by 1979. In fact, total electricity production in both the developing and planned areas in 1980 was only slightly higher than that of the developed areas in 1950. The share of electricity produced by developed areas decreased from 90 percent to 68 percent, in the period 1940–1980, whereas that in developing and planned economies combined jumped from 10 percent to 31 percent. A problem with electricity as a source of power is that considerable energy is lost in the conversion process. Generating facilities and transmission lines, for example, yield losses averaging 18 percent in the United States.

As discussed earlier, third-world energy production is complicated by the overwhelming emphasis on liquid (primarily petroleum) fuel (Table 10-2). In 1980, 86 percent of energy production in developing nations occurred in this category. The problem with this one-sided output is that it is very expensive fuel and not evenly distributed. Most is found in the Middle East and it is not readily available to other third-world countries, owing to financial constraints. Petroleum produced in the developing countries is primarily exported to the developed world. By comparison, the developed world consumes far more energy and has a more balanced production spread among solid, liquid, and gas energy sources, with each type contributing nearly equal shares. Moreover, the developed world has excellent reserves of solid fuels (particularly coal), which will become increasingly important in the near term. The planned economies are also better positioned for the future than are the developing countries. Hydroelectric resources have been developed in third-world areas, as we will discuss later, but most of this power is used for mineral resource extraction in remote areas.

COAL INDUSTRY

Coal, the black gold backbone of the industrial revolution, experienced a renaissance in the 1970s after 30 years of relative neglect at the hands of petroleum and natural gas. Actually, world production levels of

[2]Ibid. p. 110ff.

[3]Will Knowland and Vaclav Smil, "Developing Energy Dilemmas," *Focus*, 32, No. 1 (September–October 1981), 4–5.

Table 10-1

Electricity Production by World Regions, 1950–1980
(millions of metric tons of coal equivalent)

Year	Developed Economies	Percent Share	Developing Economies	Percent Share	Planned Economies	Percent Share	World
1950	38	90	2	5	2	5	42
1960	68	80	7	8	10	12	85
1970	113	73	20	13	22	14	155
1980	207	68	48	16	45	15	300

Source: United Nations, *1980 Yearbook of World Energy Statistics,* 1981.

coal have grown consistently since the early nineteenth century, but its relative share of the energy pie declined steadily after World War II as conversions to petroleum accelerated. Coal use has experienced three major overlapping phases in the past 200 years. In phase 1 coal provided steam power for industry and railroads. This market largely disappeared after World War II. In phase 2 heavy demand came from the steel industry. (See the discussion of the iron and steel industry in Chapter 15.) After World War II, as phase 3 began, the electrical power industry began dominating, especially in highly developed areas. This electricity market remains strong worldwide if somewhat diminished in highly developed areas.

The energy crisis of 1973 started the turnaround to coal and it is now expected to contribute a greater share of world energy forms. Following the oil price increases in the 1970s, for example, the demand for "steam coal" to power thermal electrical generating plants increased dramatically. If not burned directly for heat, coal can be used as a raw material for syn-

thetic fuels, which are also expected to gain prominence by the turn of the century. In this category coal gasification appears to have the most potential.

In the early 1980s, the international economic downturn lessened the demand for coal as industrial output declined and electrical energy consumption tapered off. This short-term situation passed, but uncertainties remained for the industry. Petroleum prices increased, in the late 1980s and coal demand gained strength. The petroleum share of the energy picture will inevitably decline from its present 40 percent share worldwide to around 20 percent in the next 50 years.

Presently, the largest share of coal consumption goes to electrical power generation (Figure 10-2 on page 136). Worldwide, about two-thirds of electricity came from coal thermal generating plants in 1980. At the same time, industry users consumed about one-fourth of coal output, with the iron and steel industry consuming the largest share. The weaker state of the steel industry in the 1980s and technological advances that require less fuel mitigate against greater industrial use for coal. Nevertheless, many fuel conversions from oil to coal are occurring in steel and other manufacturing areas. The cement industry, an intensive energy user, has almost completely switched from oil to coal as a power source.

Growth in the demand for coal by the residential and commercial markets will be marginal in the future. One bright spot in this area relates to the trend in some parts of Europe, most notably in Denmark, West Germany, Sweden, and Poland, to *district heating* (DH) arrangements whereby a central coal-fired plant provides area-wide heat. Perhaps more popular will be a trend toward the use of electric heat provided by heat pumps for which the electricity will be provided by coal. Solar applications for home heating, particularly passive solar water heating, will also increase.

Table 10-2

World Energy Production, 1950 and 1980
(millions of metric tons of coal equivalent)

Production	Year	Percent of Total (by sector)			
		Solid	Liquid	Gas	Electricity
World total					
9,338	1980	28	48	20	3
2,560	1950	58	31	10	2
Developed world					
3,419	1980	33	30	31	6
1,741	1950	60	25	13	2
Developing world					
2,821	1980	5	86	8	2
340	1950	12	87	1	1
Planned economies					
3,099	1980	45	34	20	2
479	1950	84	13	3	0

Source: United Nations, *1980 Yearbook of World Energy Statistics,* 1982.

Types and Places

Coal comes in a variety of forms depending on the degrees of compression and or the depth of burial. Stage 1 in the production of coal occurs with the decay of plants in bogs. This process yields *peat.* Compressed peat yields *lignite,* a low-grade coal often called brown coal in Europe. When coal is buried deeply,

Figure 10-2 Thermal Electric Power Station. *Massive coal-fired thermal electric generation stations in the western Pennsylvania coal fields provide power for northeastern United States. Note the coal mine mouth which supplies fuel to this station in the foregound, the twin cooling towers, and smokestacks. (Courtesy Richard Pillsbury)*

moisture expelled, and the material subjected to increased temperatures, a firmer *bituminous* coal results. It is the most popular coal in commercial use, often referred to as soft coal. Further compression yields *anthracite,* the highest-quality hard coal. The high-grade coal used in the steel industry is frequently called *coking coal.* It results from the heating of coal in the absence of oxygen, which burns off volatile gases. *Steam coal* is the name given to coal used to produce heat in thermal electrical power generating plants. Steam coal is not as high in quality as coking coal.

The various coal forms occur predominantly in middle latitudes of the northern hemisphere. The United States, eastern Europe, the Soviet Union, and China have the greatest quantity of estimated reserves. South America and Africa, except for South Africa, are notably deficient regions. Lignite is mined chiefly in Europe, especially Germany, and in the western half of the Soviet Union, especially the Tula field south of Moscow.

The United States, China, and the Soviet Union lead the world in coal production, accounting for two-thirds of the world output in 1983 (Table 10-3). Great Britain, the early leader in world coal production and the major producer in western Europe, now ranks seventh in world rankings, after South Africa and India. The British mines, slow to modernize, now have competitive technology and will probably remain a major producer, but labor problems and strikes continued in the 1980s. In Poland, production dropped from a peak of over 200 million metric tons annually to 163 million in 1981 due to labor unrest and the lack of mechanization, but output recovery returned by 1984.

U.S. Production

The United States, the world production leader, has also traditionally led in exports, but in the mid-1980s Australia surpassed the United States as the leading coal-exporting nation. Production increases have occurred steadily in the United States since 1960, reversing a trend of continuous annual production declines in the years following World War II. Coal became more important in the 1960s due to mine mech-

Table 10-3

Leading Hard Coal–Producing Nations, 1983
(millions of metric tons)

Rank	Nation	Production	Percent of World Total
1	United States	710	25
2	China	688	24
3	Soviet Union	487	17
4	Poland	191	7
5	South Africa	140	5
6	India	133	5
7	Great Britain	119	4
	Subtotal	2,468	87
	World total	2,829	100

Source: United Nations, *1983 Yearbook of World Energy Statistics*, 1985.

anization improvements, the greater competitiveness of strip mines, and transportation efficiencies created by the introduction of unit trains. Uncertainties clouded the industry during this period as well. Nuclear energy advances, artificially low government-regulated prices on natural gas and gasoline, petroleum import regulations, and environmental laws affecting both the production (mining) and consumption (burning) of coal all contributed to the unsettled situation facing the industry.

Petroleum price increases following the 1973 Arab oil embargo generally helped the coal industry. Environmental restraints on production have eased in recent years and the industry itself has responded to environmental concerns, further enhancing its competitive position. Regulations on coal combustion do remain problematical for the industry because coal

Figure 10-3 Major Coal Fields and Flows in the United States. *The largest share of U.S. coal is mined in the eastern half of the country. The dominance of the Appalachian field extending from Pennsylvania to Alabama is quite apparent from the map. Note the importance of the Ohio-Mississippi River systems for coal movement and the rail flows that tie the production area to markets. (Modified after Cuff and Young,* The United States Energy Atlas, *2nd ed., New York: Macmillan, 1986, p.56.)*

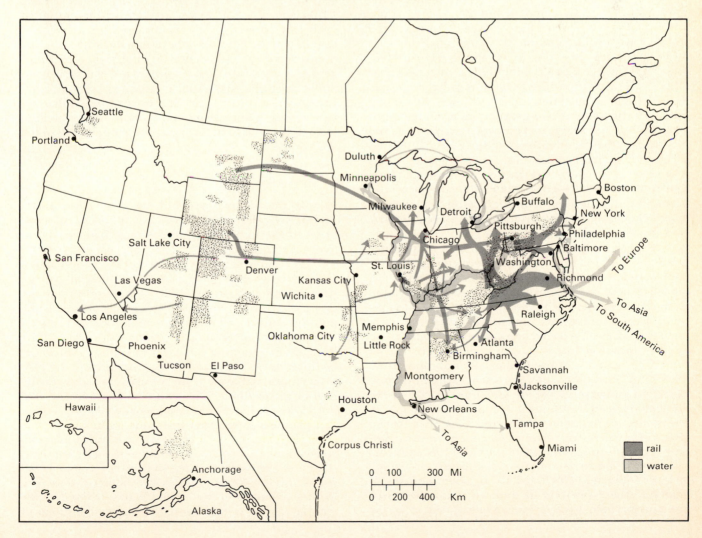

products, regardless of their sulfur content, face the same pollution control restrictions. Acid rain problems affecting the eastern United States, Canada, and the Black Forest region of southern West Germany, among other industrial areas, have been traced to coal-fired smokestacks, further exacerbating the problem.

Future prospects for the coal industry in the United States nevertheless remain bright. Coal prices continue lower than comparable oil prices. Deregulation of natural gas prices has helped the industry, as it accelerated conversions to coal for electricity generation and industrial power. The coal industry also aims to increase exports to 13 percent of total production in the near future, up from the 7 percent share of production in the late 1970s.

The largest share of U.S. coal is mined in the eastern half of the country. The Appalachian field, which extends from northwestern Pennsylvania to Alabama, is one of the most productive coal-mining regions on earth; it alone accounts for just over half of U.S. production (Figure 10-3 on page 137). Kentucky and West Virginia lead in production, but Wyoming now outproduces Pennsylvania (Table 10-4).

The midwest produces about 20 percent of U.S. coal. Considerably more (25 percent) originates in the Rocky Mountains field, mostly in Utah, Montana, Wyoming, and Colorado. In fact, the largest mines in the country are now predominantly in the west, which has nine of the top 10—four in Wyoming, two in Montana, and one each in Washington, Arizona, and New Mexico. Near many of these large western mines energy firms have built huge thermal power generating plants. In the mid-1970s a great boom in mine output occurred in that region in response to environmental restrictions that favored lower-sulfur western coal and large unit trains began hauling coal to power plants throughout the country (Figure 10-4). Since that time the retrofitting of smokestacks with scrubbers and changing pollution guidelines allowing the competitiveness of eastern coal to return have placed the west at a disadvantage.

Western coal is generally at a transportation disadvantage in relation to that produced in eastern mines and is of lower quality in terms of ability to produce heat, notwithstanding its lower sulfur content. As prices increase in the future, western coal may again gain a competitive advantage. By 1990 it is anticipated that western coal movements will equal those of the Appalachians. States west of the Mississippi contain just over half the U.S. reserves and nearly 85 percent of low-sulfur coal. To reach parity in output, a sixfold increase in production must occur in the west, as Appalachian output is expected to double in the same period. Several east and west coast ocean port terminals were planned or under construction by private railroad interests in the early 1980s to assist the export business, but the stabilizing of energy prices in the mid-1980s led to an abandonment of many of these projects. If and when upward price pressures return, greater movement of coal slurry by pipeline will also assist in gaining a larger market share for the industry.

Over half the coal mined in the United States today comes from surface mines, with about 40 percent obtained from underground sources. The share of open strip mines in the middle west and west is higher than in the east. Electric utilities consumed 16 percent of U.S. coal in 1947, about 74 percent in the late 1970s, and an estimated 80 percent of the output in 1980. Government policy encourages increased use of coal in electrical power generation and for industrial use to decrease dependence on oil. Utility companies generally make this conversion more willingly than industrial users. Industrial consumers, as individual firms, typically consume less energy per dollar invested than utility plants and often find switching from petroleum to natural gas less expensive than the shift to coal.

Soviet Production

As with the rest of the world, the Soviet Union has traditionally depended on coal as a fuel source. But between the mid-1950s and the mid-1970s, coal and hydrocarbons reversed roles as leading energy sources. Coal provided two-thirds of the supply in the mid-1950s whereas petroleum and natural gas assumed this stature by the mid-1970s. Beginning in the mid-1970s, Soviet policy became pro-coal in orientation but consumption changes came slowly. The goal is to use coal for about one-third of energy needs. The difficulty in attaining this level of production occurs because reserves are located in Asia, far from the core urban-industrial area to the west of the Urals in the European portion of the country. High transportation costs for moving Siberian coal westward led Soviet planners to electrify railroads, reducing the necessity of hauling fuel to power the trains. Second, conversions from oil pose problems, as elsewhere.

Pressures remain strong to produce as much coal from traditional coal-producing areas as possible. Unlike the situation in the United States, where the residential market is the largest, the biggest consumer of

Table 10-4

U.S. Coal Production, by State, 1984
(millions of short tons)

Rank	State	Production	Percent of U.S. Total
1	Kentucky	165	19
2	West Virginia	131	15
3	Wyoming	131	15
4	Pennsylvania	72	8
5	Illinois	64	7
	Subtotal	728	64
	U.S. total	890	100

Source: U.S. Bureau of the Census, *Statistical Abstract of United States*, 1986.

Figure 10-4 Coal Unit Train. *The largest coal mines in the U.S. are now in the west. A great boom in production occurred in the region in the 1970s and large unit trains began hauling coal to power plants throughout the country. This unit train hauls 11,000 tons of Wyoming coal per trainload to a San Antonio, Texas, public utility company. (Courtesy Southern Pacific)*

coal in the Soviet Union is heavy industry, which is concentrated near these traditional producing areas. But the quality of that coal in terms of seam thickness and volume of recoverable reserves in these areas does not meet the need in quality or quantity. Soviet coal has always had a reputation for inferior quality, owing to greater dependence on lower-quality bituminous, lignite, and peat sources than in the west. The Soviets also use significant quantities of oil shale, particularly for electrical power generation. The reason for this dependence is not only the lack of higher-quality reserves but also poor accessibility to them. Highlighting this locational problem, the average haul length for coal increased from 695 kilometers in 1975 to 819 kilometers in 1980, an 18 percent increase in shipment distance.

Major Producing Areas. The Donets basin producing region retained its lead as the major Soviet coal producing area in the early 1980s (Table 10-5). But production apparently peaked in that area in the mid-1970s, and by the 1990s it will probably be replaced by the Kuznetsk region as the leading coal-mining area. The Donets basin (Donbass) is the oldest producing area in the country and has a favorable location in relation to the market (Figure 10-4). That area has produced the highest-quality coal mined in the USSR, but the deeper discontinuous seams that must now be harvested greatly increase costs and decrease the competitiveness of the area.

Mining in the Kuznetsk (Kuzbass) area dates to the late nineteenth century (Figure 10-5). It offers thick, continuous seams of coal, most of which can be sur-

Table 10-5

Leading Coal-Producing Regions in the Soviet Union, 1982 (millions of metric tons)

Region	Production	Percent of Soviet Total
Donets basin (Donbass)	200	28
Kuznetsk basin (Kuzbass)	147	20
Ekibastuz	70	10
Karaganda	48	7
Kansk-Achinsk	37	5
Pechora	28	4
Subtotal	530	74
Soviet total	718	100

Source: Robert G. Jensen et al., eds., *Soviet Natural Resources in the World Economy* (Chicago: University of Chicago Press, 1983).

face mined. Dramatic production increases occurred in this region in the 1970s despite its poor location. Coal from this region must be back-hauled up to 2000 miles westward to the Urals and other manufacturing centers.

A third producing area, Karaganda, plays a much smaller role in coal production. Its product mainly serves a local iron and steel industry. Deep shaft mines produce the coal, which is not of good quality because of its high ash content. It is likely that not much future growth in production will occur in this area, which is also relatively isolated from major market centers. Seven percent of Soviet coal came from the Karaganda region in 1982 and it had been eclipsed by the newer Ekibastuz producing area (see the discussion below).

The Pechora field suffers from an isolated location within the Arctic Circle. Permafrost constantly causes handicaps. Its production was 4 percent of the market in 1982. The South Yakutan basin is also handicapped by its far eastern location. In 1974, the Japanese signed an agreement to buy imported coal from this site, but even exports to the east are handicapped by long rail hauls. A new town, Heryungri, is under construction, serving the mining areas, as well as a new port east of Vladivostok at Nakhodka-Vostochny. This port, which also exports saw timber, will become the leading bulk product seaport in the country. The competitiveness of this coal mining area is enhanced by low cost strip mining opportunities.

Figure 10-5 Major Coal Fields in the Soviet Union. *Most Soviet coal reserves are located in Asia, far from the core urban-industrial area of the country west of the Urals. Note the prevailing westward flows of coal transported mainly on electrified rail lines. The leading coal-producing areas are the Donets Basin (Donbass) and Kuznetsk Basin (Kuzbass). (Redrafted from Central Intelligence Agency,* U.S.S.R. Energy Atlas, *1985)*

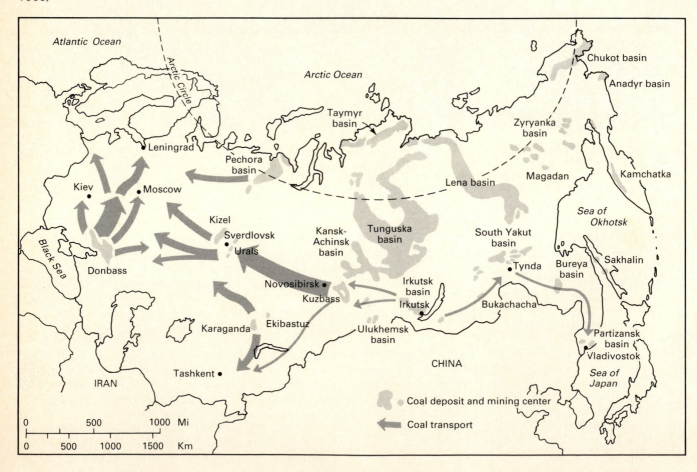

Two rapidly expanding coal-producing areas also deserve attention because they represent bold new initiatives in tapping reserves in Siberia. Both are designed to serve primarily as sources of coal for mine-mouth electrical generating stations, overcoming the need to move coal over long distances. The Kansk-Achinsk basin in southern Siberia now produces about 5 percent of the national output. A power line has been constructed to link the area to the Kuznetsk area power grid. The Ekibastuz basin, which provided the story for the book by Solzhenitsyn on the Gulag archipelago forced-labor camp, is of growing significance as well. Located in northeastern Kazakhstan, it lies closer to the market and now has become the third leading producing region in the country.

Technological improvements in Soviet coal mines have not kept pace by world standards in recent years. Obligations to supply eastern Europe with coal resources accelerated in 1980 with the declining output experienced in Poland. But the Soviets already exported large quantities of coal to eastern Europe. Fully 100 percent of the Bulgarian supply emanates from the Soviet Union, 60 percent of the East German consumption, 50 percent of the Czechlosovakian coal, and 20 percent of the Hungarian need.

Electric Power Consumption. Electrical power generation has received high priority in the Soviet Union as an instrument for economic development. A unified power grid throughout the country was envisioned from the 1920s onward. This approach increased efficiency and reliability but also created a dual economy because rural agricultural areas received little priority access. This neglect encouraged low-priority agricultural sectors to develop their own small-scale generating facilities. The current Soviet agriculture intensification program emphasizes productivity increases, and electrification has received a higher priority as a result.

The major electricity consumer in the Soviet Union is heavy industry, whereas residential/commercial consumption is more important in the United States. As a result, the Soviets have more regular demand patterns without the peaking that occurs in the United States. Many plants in the Soviet Union produce both heat and power, unlike their U.S. counterparts. Nevertheless, gross power consumption in the Soviet Union remains under half that in the United States.

The Soviets place less reliance on coal in electrical power generation than does the United States. About one-fourth of the electrical power fuel comes from natural gas in the Soviet Union. Petroleum and gas pipeline distributions systems remain largely undeveloped in the Soviet Union, which makes gas and oil relatively more attractive for conversion to electrical power generation because the electrical transmission distribution system is more comprehensive than pipeline service.

The deficit in energy resources west of the Urals encourages the transmission of electricity to the area rather than the movement of mineral resources. Environmental concerns also mitigate against greater use of coal. By contrast, solid fuel, mainly coal, does provide the dominant fuel source in Siberia, the Far East, and in Kazakhstan, where it is found in abundance.

Chinese Production

China, the second-leading world producer (Table 10-3), has modernized coal mines in recent years with western equipment and is now poised for greater output. As elsewhere, electrical power generation needs require this added production. China also anticipates exporting more coal in the future, especially to Japan. But the most significant feature of the China coal production today is that it harkens back to patterns evolving in Britain in the late nineteenth century, wherein coal production and urban-industrial growth occurred hand in glove.

In China a far higher share of total energy needs come is from coal than in other major production/consumption areas. Over two-thirds of the country's energy derives from this source. As recently as 1957 over 90 percent of energy came from coal, but competition from petroleum has increased. Most of the coal reserves in China occur in the north (Figure 10-6 on page 142). Shanxi has about one-third and Inner Mongolia one-fourth of the total reserves.

The distinctive feature of recent coal policy in China, reminiscent of nineteenth-century Britain, involves the emergence of major industrial centers near the mines. Two major reasons have been suggested for this situation, which contrasts dramatically with the rather rural and isolated settings for most of the newer coal mine areas in the United States and the Soviet Union.[4] First, the lower productivity per worker in the Chinese mines requires more labor. Second, the rail system remains very inadequate in China. Moving massive quantities of resources to industrial areas by rail as occurs in other countries is simply not possible in China. Although one-third of the railroad rolling stock moves coal, the distances covered are short.

The policy to consume the product where it is mined fits in well with the urban policy aimed to develop the interior of the country. Moreover, the problem of an inadequate interregional electrical power transmission line network can be circumvented with the spreading out of urban-industrial centers near the coal resource base whereby power can be generated locally. In fact, a policy goal has emerged to create eight major coal bases in eight different provinces: Datong (Ta-tung) in Shansiui, Huolinhe (Hu-lin-ho) in Inner Mongolia, Huainan-Huaibei (Huaipei) in Anhui, Yanzhou (Yen-chou) in Shandong (Shantung), Xuzhou (Su-chou) in Jiangsu (Klangsu), Liupanshui in

[4]Sen-Dou Chang, "Modernization and China's Urban Development," *Annals*, Association of American Geographers, 71 (1981), 204, 206.

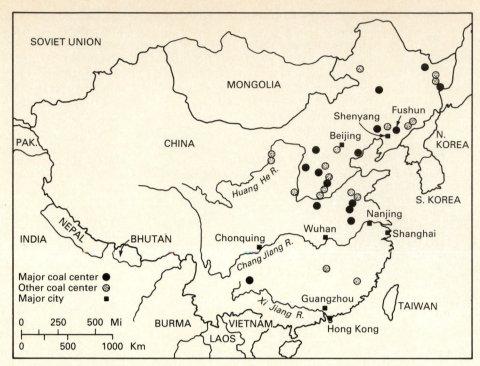

Figure 10-6 Major Coal Mining Areas in China. *Most of the coastal reserves in China occur in the north. Moving massive quantities of coal to industrial areas by rail is not possible in China due to the sparsely developed railroad network. A decentralized development policy has resulted, whereby the product is consumed at urban industrial centers near the mine, many of which are shown on the map. (Redrafted from Sen-Dou Chang,* Annals, Association of American Geographers, *71 (1981), p. 207)*

Guizhou (Kwechou), Pingdingshan in Henan, and Kailuan in Hegei. All these locations represent new mines since 1949 except Kailuan and Datong.[5]

Coal Trade

While Japan as a nation leads in coal imports, western Europe is also an important importing region (Table 10-6). Over one-third of world coal imports flow to western Europe. While France and Italy are the leading European importers, several other countries, such as the Netherlands and Denmark, import 100 percent of their coal needs. A portion of the flow occurs among western European countries themselves now that former trade barriers have been broken down by the Common Market. Western Europe, Japan, and Canada together import over half the coal involved in world trade.

As mentioned earlier, Australia now leads in coal exports, with one-fourth of the world flow. Australia tripled exports of coal in the 1973–1984 period. Australian coal moves predominantly to Japan (71 percent in 1979). Australia anticipates further increases in exports by the turn of the century. By 1990 it is anticipated that Australian exports will even exceed domestic consumption.

The United States is also a major exporter of coal to Japan, but roughly equal shares also go to Canada and western Europe. Poland and South Africa rank third and fourth, respectively, as exporting nations, followed by the Soviet Union. Much of the South African output (60 percent) goes to Europe, with another 20 percent destined for Japan. Most of the Polish and Soviet coal stays within the eastern bloc. Sixth-

ranking Canada has the potential of expanding production and becoming a larger exporting nation in the future. The Canadian reserves are located in the interior, in Alberta and British Columbia, creating problems with access to both foreign and domestic markets. Greater levels of Canadian exports will require better rail transportation through the mountains and improved port terminal facilities, advances which are now being realized.

Table 10-6

Leading Coal-Importing and Coal-Exporting Nations, 1984 (millions of metric tons)

Rank	Nation	Quantity	Percent of World Total
Importing			
1	Japan	86	28
2	France	22	7
3	Italy	21	7
4	Canada	20	7
	Subtotal	149	49
	World total	305	100
Exporting			
1	Australia	77	25
2	United States	74	24
3	Poland	43	14
4	South Africa	38	13
5	Soviet Union	26	9
6	Canada	25	8
	Subtotal	283	93
	World total	305	100

Source: International Energy Agency, *Coal Information 1985,* Paris, 1985.

[5]Ibid. pp. 204–206.

WATER RESOURCES

Water as an economic resource? We do not often think of water in this context. Only when it is in short supply or of poor quality do we place water in its proper perspective. Perhaps its plentiful quantity at the world scale (80 percent of the earth's surface is water) and its relatively cheap cost contribute to its casual treatment. Notwithstanding this image problem, water has historically played an important role in economic development and continues to be a vital factor in agriculture, industry, and urbanization. Perhaps no mineral is so versatile. Water serves us for irrigation, as a power source, as a mode of transportation, as an industrial product, and as a source of nutrition.

To an economic geographer the influential role of water as a power source cannot be overstated. The mineral-generating capacity of water is also significant. Earlier we covered irrigation and agriculture, and the significance of the Law of the Sea in allocating mineral resources in ocean waters. Water power and mineral yield will be emphasized here.

Water Power

Unlike other minerals such as coal, which is burned to produce energy, water has no capacity to produce energy within itself. Water creates power by its motion—usually by falling vertically. The greater the distance of the fall, the greater the power potential. Typically, this potential occurs in upland areas or in mountainous environments with human-made dams.

Power to run machines at the time of the industrial revolution in North America and western Europe came primarily from water. In the late eighteenth century factories sprung up at waterfalls or dam sites suitable for turning waterwheels. The water-

wheels powered shafts, which in turn powered lathes, looms, and millstones. Diversion canals running parallel to rivers, called *millraces,* provided additional waterwheel power sites (Figure 10-7). In New England alone it has been estimated that there were 10,000 waterwheel-power sites in the early nineteenth century.

Water power use declined rapidly in the mid-nineteenth century as steam power replaced the waterwheel. Steam power offered more locational flexibility for manufacturing. Coal power could not only be provided at existing manufacturing centers, but at new locations as well. Many former water-tied manufacturing sites simply converted to steam power. But the railroads also provided opportunities for manufacturing to move away from water sites to more centrally located transportation junctions. The expansion of manufacturing in Philadelphia in the steam age provides a classic example of this association.

Water reemerged as a major power source as electric motors replaced the steam engine. The first water-power-generated electricity dates to 1882. By 1925, 40 percent of the electricity produced in the United States came from hydropower plants located at dam sites. As with steam power before, electricity loosened the tie between the power source and final use. Electricity produced at one location could be transmitted to cities miles away. Water power has since declined relatively as a source of electricity generation worldwide, accounting for only 15 percent of the total electricity produced.

Dependence on water power as a source of electricity varies widely. It provides the most important source of electricity in third-world countries, even though it is not uniformly available. Access to electricity eludes the majority of the population in third-world countries, as only 12 percent of the population currently has access to electrical power. Fortunately,

Figure 10-7 Tanner's Mill Complex in Hall County, Georgia. *Located on Walnut Creek, a fork of the middle Oconee River, the mill complex once consisted of a grist mill and flour mill, a cotton gin, textile mill, a dam with watergate and raceway, a millpond, an iron truss bridge, an owner's house with a barn and several outbuildings, a miller's house, and a storage building. The complex was developed between the mid- and late-nineteenth century. The three-story grist and flour mill shown here dates from ca. 1886. A wooden raceway carried water from the pond shown in distance to turn the breast-shot waterwheel which is 20 feet in diameter and has five-foot wide metal floats. (Photograph by James R. Lockhart for Georgia Department of Natural Resources, Historic Preservation Section)*

there is great potential for expansion of hydroprojects in these areas.

Several industrial countries are also very dependent on hydropower, including Norway (99 percent), New Zealand (75 percent), Switzerland (74 percent), and Canada (57 percent). In South America nearly three-fourths of the electricity comes from hydropower. Of the leading electrical-power-consuming nations, Canada alone provides an example of a highly developed country producing over one-half of its electrical power from hydroelectric sources, and is the only major consuming nation that does not produce at least two-thirds of its electrical power needs from thermal sources (Table 10-7). Most developed countries now obtain about 10 percent of their electrical power from nuclear sources. Canada and the Soviet Union fall below this level, while Japan and West Germany, owing to a paucity of domestic energy resources, exceed the average (Table 10-7).

Europe can boast that it has harnessed a greater share of its potential for producing hydroelectricity than any other continent (see Table 10-8). Japan, the United States, and the Soviet Union have also been world leaders in hydropower development. Asia, Africa, and South America have the lowest share of their water power potential developed. Some countries, particularly Switzerland, now export major quantities of electricity to other nations. The potential for such a situation developing in other areas such as Nepal is also great. In both of these cases the mountainous setting creates advantageous locations for hydroprojects.

Worldwide, the best sites for hydropower development have already been tapped. In some areas, such as the Soviet Union, the United States, and Canada, emphasis has been placed on large projects, while in other areas, such as China, small dams and generating facilities are preferred. Big dams have been built only since 1930. The largest such impoundment in the world in terms of water capacity is the Aswan Dam on

Table 10-7

Electrical Power Generating Capacity by Nation, 1983
(millions of kilowatts of installed capacity)

| Nation | Percent of Total Capacity by Source | | | Total Capacity |
	Thermal	Hydro	Nuclear	
United States	78	12	10	675
Soviet Union	74	19	7	294
Japan	67	21	11	159
West Germany	79	8	13	87
Canada	33	57	10	90
Total world capacity	1,534	523	191	2,251
Share of total world capacity (percent)	68	23	9	100

Source: United Nations, *1983 Energy Statistics Yearbook*, 1985.

Table 10-8

Hydropower Development, by Major World Regions, 1980

Region	Developed Capacity	Share of Potential Capacity Developed
Asia	53,079	9
South America	34,049	8
Africa	17,184	5
North America	128,872	36
Soviet Union	30,250	12
Europe	96,007	59
Oceania	6,795	15
World total	363,000	17

Source: Daniel Dendney, *Rivers of Energy: The Hydropower Potential*, Worldwatch Paper 44 (Washington, D.C.: Worldwatch Institute, 1981), p. 9.

the Nile River in Egypt. The most powerful project to date in terms of generating capacity is the Itiapu between Brazil and Paraguay on the Parana River, which opened in the early 1980s. Four of the top ten hydrogenerating plants in the world today are in South America, three in the Soviet Union, and two in Canada (see Table 10-9).

The Soviet Union has harnessed both natural and human-made sites, including the Dnepropetrovsk Dam on the Dnepr River in the rather flat Ukraine, and two installations at Kuybyshev and Volgograd on the Volga River. But the largest Soviet projects are on the Yenisey River (Table 10-9). Japan, South Korea, and Manchuria dominate hydroelectric power production in the Far East. Indeed, Japan's developed capacity almost equals its potential waterpower. Short river lengths in that island complex inhibit water power development. A mountainous spine that rapidly gives way to plains near the coast restricts suitable sites.

Prominent among the world's water power installations are some very tall dams, several of which exceed 500 feet in height—one in the Soviet Union reaches over 1000 feet in height. Rivers flowing over these dams are often relatively small compared with those at other installations, which have greater capacity (see Table 10-9). Western Europe, with almost one-third of the world's hydroelectric capacity, has none of the world's sixty-four largest hydroelectric projects. The United States, Canada, the Soviet Union, and Latin America complete a list of the top five areas in terms of operating capacity.

The small-scale dam projects preferred in China are multipurpose and well suited to the electrical supply needs of workshop-based commune production facilities. These small reservoirs provide for water storage, flood control, irrigation, and aquaculture. Since 1968, over 90,000 small-scale hydrounits have been built in China, mainly in the southern half of the country. These facilities average only 72 watts of generating capacity and yet account for nearly one-half of the total electrical power generation in the country.

Table 10-9

World's Largest Hydroelectric Generating Plants, 1986

Rank	Plant	River Basin	Ultimate Megawatts	Year of Initial Operation
1	Itaipu, Brazil/Paraguay	Parana	12,600	1983
2	Guri, Venezuela*	Orinoco	10,060	1968
3	Tucurui, Brazil*	Tocantins	8,000	1983
4	Grand Coulee, United States	Columbia	6,494	1942
5	Sayano-Shushensk, Soviet Union	Yenisey	6,400	1980
6	Corpus Posadas, Argentina/Paraguay*	Parana	6,000	1990
7	Krasnoyarsk, Soviet Union	Yenisey	6,000	1968
8	La Grande 2, Canada	La Grande	5,328	1982
9	Churchill Falls, Canada	Churchill	5,225	1971
10	Bratsk, Soviet Union	Yenisey	4,600	1964

Source: *The World Almanac & Book of Facts, 1987* (New York: Newspaper Enterprise Association, Inc., 1986).

*Planned or under construction.

In recent years the emphasis on large-scale dam power production facilities shifted to third-world countries. A secondary shift has been to develop sites with enormous potential even though they are remote from population centers. Mineral extraction and smelting provides much of the incentive for many of these projects. Mining-related projects explain hydroelectric initiatives in the Amazon region, New Guinea, Siberia, and Quebec. Technology improvements now also permit power to be transported over greater distances. Transmission lines can extend from 500 to 1000 miles from the generating facilities to consumption sites. Maximum transmission distances of less than half this amount were the norm 30 years ago. As an illustration of greater flexibility in electricity flows, consider that New York City in 1984 received about 12 percent of its power from Quebec, an unprecedented development.

Aluminum production requires considerable power resources. The trend in that industry is for production to shift from the developed world to developing nations with growing electrical power facilities to take advantage of cheaper electrical power. The Japanese, for example, are closing down smelters and shifting production to Australia, Indonesia, and Papua–New Guinea. New plants in Brazil, Egypt,

Figure 10-8 Norris Dam Hydropower Project. *Norris Dam, started in 1933 and completed in 1936, was the first dam built by the Tennessee Valley Authority. Located on the Clinch River in east Tennessee, it is 265 feet high, and 1,860 feet long. The storage reservoir has a capacity of two and a half million acre-feet of water. The power station includes two 50,000 kw units. (Courtesy of Tennessee Valley Authority)*

Ghana, Tanzania, and Sumatra illustrate this point. A fuller discussion of the aluminum industry occurs in Chapter 12.

International financial institutions favor large dam facilities even though they may not be best for the developing country targeted by the project. Such programs mean more business for contractors and heavy electrical machinery manufacturers (turbines, etc.) headquartered in developed countries. In turn, this bias reinforces the dependence developing nations have on developed countries for loans and ongoing supervisory and maintenance support.

Mineral Resources in Water

For every square mile of land, there are almost 3 square miles of ocean on the earth's surface. Most of the water surface is concentrated in the Pacific Ocean—approximately 65,000,000 square miles. If we could shift all the landmasses in the world into the Pacific, we would still have 9000 square miles of water left—the area of North America. The other oceans, in order of size, are the Atlantic, 32,000,000 square miles; the Indian, 28,000,000; and the Arctic, 6,000,000. In addition, there are numerous smaller bodies, such as the Mediterranean, Caribbean, China, Red, Yellow, Okhotsk, Bering, Baltic, and Japan seas, as well as Hudson Bay and the Gulf of Mexico.

Oceans are deep as well as vast. The greatest known depth, 36,291 feet, is south of the Mariana Islands in the western Pacific. The water world is indeed gigantic and so are its resources. Recall the discussion of mining minerals in Chapter 9 in connection with Law of the Sea negotiations. One single cubic mile of seawater contains, on the average, 166 million tons of dissolved minerals. Despite this large figure, the sea averages 96.5 percent water and only 3.6 percent dissolved minerals. The numbers vary from region to region. In the rainy tropics, for instance, the percentage of minerals is slightly lower, as rainfall dilutes the ocean. By comparison, the ocean in the dry subtropics has a higher mineral content due to reduced rainfall and higher evaporation rates.

The most abundant elements dissolved in seawater are sodium, chlorine, magnesium, calcium, potassium, and sulfur; the most prevalent minerals are:

Sodium chloride (common salt)	78.0%
Magnesium chloride	11.0%
Magnesium sulfate	5.0%
Calcium sulfate	3.0%
Potassium sulfate	2.5%
All others	0.5%
Total dissolved minerals	100.0%

Extracting minerals from seawater is an expensive business, with few exceptions. Chemists have identified some 50 elements in seawater, but technologists have succeeded in extracting only a few at costs competitive with the mining of these minerals from rocks. Seawater is evaporated in huge reservoirs in the open sunshine to produce residues of sodium chloride (common salt) on the shores of the Persian Gulf, China, Japan, and the Philippines. Bromine is now produced commercially by means of the same general techniques. In 1941 scientists discovered methods for commercially extracting the light metal magnesium from seawater. During World War II much of the magnesium used in airplanes, star shells, incendiary bombs, boats, and even lightweight buildings was mined from the sea, a cubic mile of which contains 4 million tons of the metal.

NUCLEAR ENERGY

Nuclear energy as a means to generate electricity represents a recent technological breakthrough, as the first commercial nuclear electrical generating plant opened in the mid-1950s. Located in the United States, this plant was very small (60 megawatts) by current standards. Interest in nuclear power grew rapidly as second-generation reactors (300- to 500-megawatt units) gained prominence in the 1960s. By the 1970s, third-generation plants (over 800 megawatts) appeared.

Over 50 nuclear reactors existed in the United States in the mid-1980s, located in 28 states, largely clustered in three areas: (1) the northeastern megalopolis extending from Massachusetts to northern Virginia, accounting for one-fourth of U.S. total capacity; (2) the Chicago area; and (3) the upland south, in a belt extending from the North Carolina–South Carolina border westward across the northern Georgia–Alabama–Tennessee state borders. The Tennessee Valley Authority and Duke Power Company account for most of these facilities.[6] Each of the plants in the United States has a location in close proximity to a plentiful supply of water, owing to the need for vast quantities to serve as a coolant for the reactor (Figure 10-9).

Although the United States retains a lead in the worldwide production of nuclear power, France, the Soviet Union, and Japan are also large producers (Figure 10-10). Many different design technologies have been used throughout the world, but the light-water reactor perfected in the United States dominated new construction in the 1970s. As mentioned in the preceding section and noted in Table 10-7, nuclear power rarely averages above 10 percent of a nation's electrical generating capacity.

The outlook for nuclear power is now much more subdued than it was two decades ago. The high cost of construction due to inflation and the need for redundant safety systems is part of the reason for the growing pessimism about the future role of nuclear energy. Another part of the concern is focused on en-

[6]John M. Ball, *An Atlas of Nuclear Energy*, Department of Geography Research Series, No. 6, Georgia State University, Atlanta, 1984.

Figure 10-9 Big Rock Point Nuclear Electric Station, Michigan. *This plant, located on the shores of Lake Michigan, includes a high-power density, boiling-water reactor dating to 1962. Owned by Consumers Power Company, the facility provides electric service to nearly 1 million customers. (Courtesy Consumers Power Company)*

vironmental and safety issues. The fear of meltdown, the potential for radiation leaks and for human or mechanical error, and problems associated with the disposal of radioactive waste have all contributed to growing public criticism of nuclear energy power reactors.

In the 1980s, in fact, the cancellation of orders for nuclear power reactors became very widespread in the United States. The last order for a new reactor in the United States was filed in 1978 and total cancellations exceeded 100 by 1984. An accident at the Three Mile Island reactor in Pennsylvania in 1979 reinforced negative public opinions about nuclear reactors as power sources, as did the Chernobyl nuclear accident near Kiev in the Soviet Union in 1986.

In West Germany, too, a vocal antinuclear movement has forced the cancellation of projects. Concern about the feasibility of nuclear energy, however, is not universally shared. In France and Japan, for example, a much more positive atmosphere exists, partly because of the lack of feasible alternative energy sources.

SUMMARY AND CONCLUSION

Shifts in sources, a demand for higher-quality forms, and rapidly rising prices have created a dynamic energy environment in recent decades. A turnaround in the importance of coal as an energy source and the tremendous growth in the demand for electrical energy are two significant developments in the past decade. A decline in the relative role of natural gas and petroleum resources is now on the horizon and the situation will become more pronounced in the future. The developed world with its wealth and greater access to technology has and will adapt more easily to energy changes. At the same time the third world is hampered by a lack of physical resources, suitable technology, and financial resources to adjust to energy shifts.

Transportation accessibility remains a problem for many countries, inhibiting effective utilization of resources for domestic use and foreign trade. The Soviet Union, Australia, Canada, and to some extent

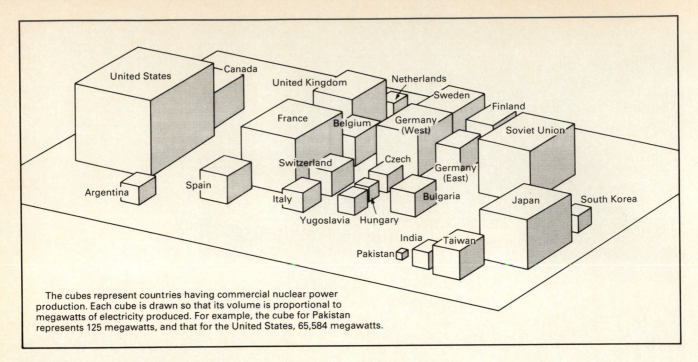

The cubes represent countries having commercial nuclear power production. Each cube is drawn so that its volume is proportional to megawatts of electricity produced. For example, the cube for Pakistan represents 125 megawatts, and that for the United States, 65,584 megawatts.

Figure 10-10 World Nuclear Power Production, by Nation, 1984. *The United States retains a lead in worldwide production of nuclear power, while France, the Soviet Union, and Japan are also large producers. Nuclear power production is very strong in Western Europe due to the shortage of other energy resources. (Courtesy Borden D. Dent)*

China all experience this problem. China has opted to concentrate urban-industrial development near coal resources to overcome this problem. The Soviet Union has opted to convert the maximum quantity of energy to electricity to combat accessibility problems. The United States alone appears to be in an excellent position in terms of both quality coal reserves and their accessibility to both the domestic and foreign market.

In terms of water resources the third world may be better off overall. Many third-world countries depend on hydroelectric projects to supply electricity. China has been a pioneer in emphasizing diversified small-scale hydroelectric projects that supply irrigation, aquaculture, and electricity needs.

Eventually, the world will shift back to a renewa-
ble energy resource base, but not until well into the next century. Solar energy may well become more important in the future as this shift is made. Water power will continue to provide a secondary energy role except in special circumstances where plentiful potential exists, as in Switzerland. Water is one of the least understood mineral resources. Hydrodams provide important electrical power generating capabilities in many areas, especially in the third world. Japan, China, and the Soviet Union have also made good use of hydroelectric resources. Finally, nuclear energy, once thought to be a breakthrough for inexpensive energy, has not lived up to earlier expectations. High construction costs and environmental and health concerns have led to lesser confidence in its potential.

SUGGESTIONS FOR FURTHER READING

CALZONETTI, FRANK J. and BARRY D. SOLOMON, eds. *Geographical Dimensions of Energy.* Boston: Reidel Publishing Company, 1985.

CONGRESSIONAL RESEARCH SERVICE, *The Energy Fact Book,* Committee Print 96-IFC-60. Washington, D.C.: U.S. Government Printing Office, 1980.

COOK, EARL, *Energy: The Ultimate Resource?* Resource Paper 77-4. Washington, D.C.: Association of American Geographers, 1977.

COUNCIL ON ENVIRONMENTAL QUALITY, *The Global 2000 Report to the President.* Washington, D.C.: U.S. Government Printing Office, 1980.

CRABBE, DAVID, and RICHARD MCBRIDE, *The World Energy Book.* Cambridge, Mass.: MIT Press, 1979.

CUFF, DAVID, and WILLIAM YOUNG, *The United States Energy Atlas.* 2nd ed. New York: Macmillan Publishing Company, 1986.

DEUDNEY, DANIEL, *Rivers of Energy: The Hydropower Potential,* Worldwatch Paper 44. Washington, D.C.: Worldwatch Institute, 1981.

DIENES, LESLIE, and THEODORE SHABAD, *The Soviet Energy System.* New York: Wiley, 1979.

"Energy: Facing Up to the Problem, Getting Down to Solutions," *National Geographic,* Special Report, February 1981.

EZRA, DEREK, *Coal and Energy.* New York: Wiley, 1978.

FORD, DANIEL F., *Three Mile Island: Thirty Minutes to Meltdown.* New York: Penguin Books, 1983.

GIBSON, DUNCAN L., *Energy Graphics.* Englewood Cliffs, N.J.: Prentice-Hall, Inc., 1983.

GORDON, RICHARD L., *Coal in the U.S. Energy Market.* Lexington, Mass.: D. C. Heath, 1978.

HARTWICK, JOHN M., and NANCY D. OLEWILER, *The Economics of Natural Resource Use.* New York: Harper & Row, 1985.

HOFFMAN, GEORGE W., *The European Energy Challenge: East and West.* Durham, North Carolina: Duke University Press, 1985.

KNOWLAND, WILL, and VACLAV SMIL, "Developing Energy Dilemmas," *Focus,* 31, No. 1 (September–October 1981).

LOFTNESS, ROBERT, *Energy Handbook.* New York: Van Nostrand, 1978.

OPENSHAW, STAN, *Nuclear Power: Siting & Safety.* London: Routledge Kegan Paul, 1986.

PITT, G. J., and G. R. MILLWARD, *Coal and Modern Coal Processing: An Introduction.* New York: Academic Press, 1979.

PRYDE, PHILIP, *Nonconventional Energy Resources.* New York: Wiley-Interscience, 1983.

WAGSTAFF, H. REID, *A Geography of Energy.* Dubuque, Iowa: Wm. C. Brown, 1974.

WHITE, GILBERT F., ed., *Environmental Effects of Complex River Development.* Boulder, Colo.: Westview Press, 1977.

11 Petroleum and Natural Gas

No part of the world economy identifies more closely with supply and demand uncertainties and political overtones than do the petroleum and natural gas industries. These products also have the distinction of being the most important commodities in international trade today. Major petroleum companies are the largest multinational firms and money transfers arising from oil payments represent the largest block of transactions in the financial world. The industry has also acquired an oligopolistic structure due to the dominance of a few firms in the industry. Petroleum also provides one of the best examples of a truly global industry, wherein prices are reported almost exclusively in international terms.

Petroleum and natural gas supplies are very unevenly distributed, yet in very strong demand to sustain high levels of economic activity in developed nations. Developing nations also seek greater access to petroleum, even though their fragile economies can scarcely afford the price when domestic sources are not available. Given the limited sources of supply and the widespread demand, it is not surprising to note that political considerations have gained prominence in the allocation process.

The present importance of oil is a relatively recent phenomenon, dating back only to the early twentieth century. The automobile industry created this increase in demand, which in turn, changed the world economy; and world economies continue to change because of oil. Price increases in the 1970s left many third-world nations reeling. Their foreign debt

zoomed upward to unprecedented levels as a consequence of these higher prices. Even many major oil-producing nations such as Mexico and Nigeria suffered dramatically from price and demand uncertainties in the industry. Additional problems arose as demand stabilized and prices began dropping again in the mid-1980s, reducing revenues major producers had come to depend on to finance development and repay large foreign debts.

Among the major consuming nations, only the Soviet Union can claim self-sufficiency in oil. The Soviet Union and the United States have always been major producers, and both remain leaders today despite the fact that the largest quantity of reserves lie in the Middle East (Table 11-1). Reserves in the United States account for only 4 percent of the world total, while the Soviet Union has an 11 percent share, both dwarfed by the Middle East, which has one-half of world reserves, paced by Saudi Arabia, which alone has one-fourth of the reserve capacity. The U.S. economy is slightly more dependent on petroleum than the world as a whole and at the same time is the single leading consuming nation, using about one-fourth of the world supply (Table 11-2). The United States now imports a large share of this need.

CRUDE OIL PRODUCTION

The ebb and flow that characterizes production levels of petroleum in the major producing nations is no

Table 11-1

Crude Oil Reserves, 1985
(millions of barrels)

Area		Percent of World Total
North America	13	
Mexico		8
United States		4
Canada		1
South America	6	
Venezuela		4
Western Europe	3	
Norway		2
Great Britain		1
Eastern Europe	12	
Soviet Union		11
Africa	8	
Libya		3
Nigeria		2
Algeria		1
Middle East	50	
Saudi Arabia		24
Kuwait		10
Iraq		6
Iran		5
Abu Dhabi		5
Far East	5	
China		3
Indonesia		1
Subtotal	97	
World total	100	

Source: World Oil, August 1986.

Table 11-2

Leading Oil-Consuming Nations, 1984 (million tonnes)

Rank	Nation	Consumption	Percent of World Total
1	United States	724	26
2	Soviet Union	448	16
3	Japan	215	8
4	West Germany	111	4
5	Great Britain	89	3
6	France	86	3
7	China	86	3
8	Italy	85	3
	Subtotal	1,844	60
	World total	2,845	100

Source: The British Petroleum Company, BP Statistical Review of World Energy. London, 1985.

better portrayed than through a historical perspective. Growing political governmental involvement in determining output, prices, consumption, and trade patterns accelerated after 1970 as producing nations in the developing world sought to retain for their own benefit a greater share of the value of the product, which had heretofore been largely controlled by foreign multinationals.

The Early Years

For thousands of years people knew about petroleum. But not until 1857 was oil produced commercially. Historical records disagree about where and when the first production occurred but the trade journal World Oil indicates that Romania in 1857 warrants this recognition. In that first year wells in the vicinity of Ploesti, 40 miles north of Bucharest, yielded 2000 barrels of petroleum—the entire world output. Home and street lighting provided the primary market for oil at that time.

For years, native Indian Americans and European settlers living in Pennsylvania and elsewhere in the northeastern United States also knew about oil from seepage they observed on the surface. In 1859 an experimental well struck oil near Titusville, Pennsylvania, north of Pittsburgh. By the next year Pennsylvania wells yielded 500,000 barrels. The densely settled Atlantic seaboard provided a ready market for this production for home lighting purposes. Thus both the United States and Romania became the early producing leaders (Table 11-3).

World production increased to 30 million barrels by 1880, paced by the U.S. output, with Pennsylvania contributing 80 percent of the world's total. Russia also surpassed Romania at that time, following the discovery of oil in the Caucasus, east of the Black Sea in 1863. In 1871 the discovery of another field in Baku near the Caspian Sea began an era of dominance for that area lasting for the next half-century.

Twentieth-Century Expansion

Paced by growing automobile sales, especially in the United States, the demand for gasoline for transportation generated a larger demand for oil in the early twentieth century. Europe continued to be a major deficit region and the United States and Venezuela major exporters. Fortunately, many new source areas emerged, both in traditional producing countries and in other countries, as market demand continued to accelerate.

In the United States, Ohio displaced Pennsylvania as the leading producer at the turn of the century. Both states pumped from the same Appalachian field. More important for the future than the shifts in eastern production were the new fields that appeared in the mid-continent region (Texas, Oklahoma, and Kansas), and in California. By 1920, Oklahoma led the United States in production. The Gulf coast field in Texas and Louisiana also became a major producer in 1920 (Figure 11-1).

Table 11-3

Crude Oil Production by Leading Nations, 1880–1985
(percent of world production)

Region	1880	1900	1920	1940	1950	1960	1970	1980	1985
Europe									
(including Soviet Union)	10	54	6	13	9	17	17	25	31
Great Britain								3	5
Romania		1	1	2	1	1	1		1
Soviet Union	10	51	4	10	7	14	15	20	22
North America	90	44	87	65	55	37	25	20	22
Canada		1			1	3	3	2	3
United States	90	43	64	63	52	34	21	14	17
Mexico			23	2	2	1	1	3	5
South America			1	13	17	17	10	6	7
Venezuela				9	14	14	8	4	3
Africa						1	17	10	9
Libya							7	3	2
Egypt							7	1	2
Algeria						1	2	2	1
Nigeria							2	3	3
Middle East			2	4	17	25	31	31	20
Iran			2	3	6	5	8	3	4
Iraq			1	1	1	5	3	4	3
Saudi Arabia					5	6	8	16	6
Kuwait					3	8	6	2	2
Far East		3	4	4	2	3	2	7	9
Indonesia			3	3	1	2	2	3	2
China								4	5
World production (millions of barrels annually)	30	149	689	2,150	3,803	7,663	16,677	21,891	19,710

Source: *World Oil,* August 15 issues for 1957, 1961, 1971, and 1982; February 15, 1983; February 1986.

Russia emerged as the world leader in oil production in the early twentieth century, pumping more crude oil than the rest of the world combined. Most of the output moved to western European markets. This early Russian dominance in the industry was short-lived because domestic political and economic turmoil curtailed production in pre-revolutionary years.

Mexico, Venezuela, and the Middle East production centers also became major suppliers of oil in the early 1900s (Table 11-3). Mexico contributed over 20 percent of the world production in 1910. The largest volume came from fields in the vicinity of Tampico on the eastern coast.

Mexico's presence as an early producing leader was also short-lived. Not until 1970 did post-1920 production return to World War I levels in Mexico. In 1921 an event occurred with far-reaching consequences—the *Queretaro Convention.* A long-standing Mexican policy traditionally vested subsoil rights to the landowners. Oil prospectors, including representatives of companies from California, Great Britain, and the Netherlands, then purchased these subsoil rights from the individual owners. The Queretaro Convention reversed this policy and transferred all petroleum rights to national ownership. Lengthy litigation between the government and the oil companies ensued, reducing operations in Mexico. In 1938 the Mexican government expropriated all property and equipment of foreign oil companies, terminating their activity. The 1940 production data show that Mexico, which 20 years earlier pumped 157 million barrels—23 percent of the world's share—was down to 44 million barrels, only 2 percent of the world's total. It took many decades for output to recover from this setback.

Venezuela appeared for the first time as a major producer on the world oil map in 1920. Production focused on the northwestern part of the country on the eastern shores of Lake Maracaibao. While the Venezuelan government both owned and controlled the production of all subsoil minerals in the nation, it rec-

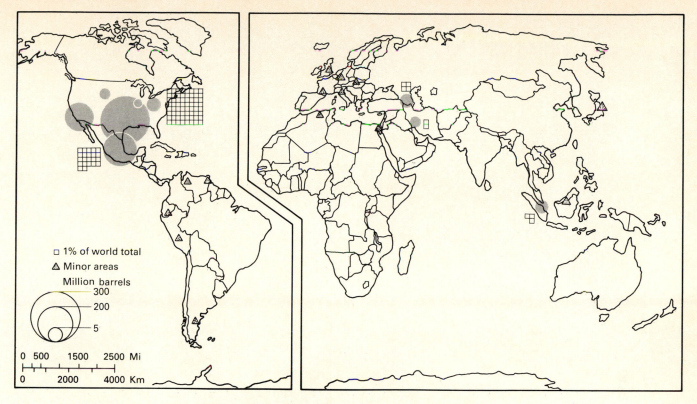

Figure 11-1 World Oil Production in 1920. *The United States led all producers in 1920, paced by the output of the mid-continent region in Texas, Oklahoma, and Kansas, while Mexico contributed nearly one quarter of world output.*

ognized that a strong mining economy required both natural resources and the skills and equipment to develop them. In the early days of petroleum production, western European companies provided the latter assistance. But unlike the Mexican situation, the government did not obstruct this partnership.

In the Middle East, as in Venezuela, the early petroleum economy blended resources from two regions: (1) the Middle East contributed the raw material, and (2) the western world the knowledge to exploit it. The very company names—Anglo-Iranian and Aramco (Arabian-American Company)—demonstrate the joint-venture arrangements that occurred. U.S., Dutch, and British multinational firms controlled exploration, drilling, and production. About 40 percent of the crude oil from the Middle East flowed westward by pipeline to ports on the Mediterranean Sea for shipment and refining in Europe; a slight amount also moved by tanker from Persian Gulf ports to the same destination. But over half the oil was refined in large factories in new Gulf settlements and then moved by tanker, mostly through the Suez Canal to Europe.

In 1940 the United States remained the dominant producer, accounting for over 60 percent of world production (Figure 11-2 on page 154). The Soviet Union and Venezuela each contributed about 10 percent of total output. Soviet exploration in the 1930s uncovered another field west of the Urals, in the heart of the Volga drainage basin, that rapidly gained prom-

inence. The Middle East, led by Iranian production, remained a minor contributor.

Post–World War II Surge

Worldwide, petroleum pumping rose to unprecedented levels by 1960 (Table 11-3). The tremendous growth in automobile sales that spread from the United States to western Europe and Japan promoted this expansion. So did industrial fuel conversions from coal to oil. Nevertheless, the world production pattern for 1960 remained basically the same as that for 1950. Old established areas continued to dominate. But output had nearly doubled over the previous decade due to growing demand. The contrasts in production levels are dramatically portrayed by the range in sizes of the proportional circles on the maps for 1940 and 1960, as shown in Figures 11-2 and 11-3 on page 154. Only by massive increases in export-import flows could the growing demand in several deficit regions be met. This situation set the stage for increasing tensions in the world market by the end of the decade. Gradually, political and military factors began over-shadowing resource availability in accounting for oil production and distribution.

The U.S. share of world production fell relatively in 1960 as output grew elsewhere, particularly in the Middle East. Nevertheless, the United States remained the world production leader, accounting for over one-third of total output. Texas alone produced almost 1

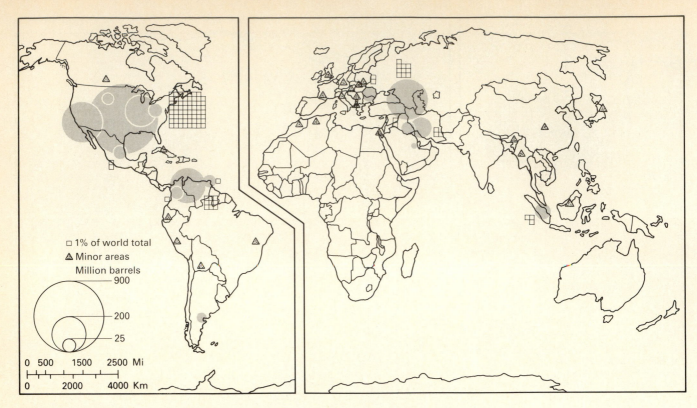

Figure 11-2 World Oil Production in 1940. *Even with vastly larger output, the United States' share of output remained the same in 1940 as in 1920, despite the collapse of Mexican production. Soviet and Venezuelan expansion in this period made up for the Mexican loss.*

Figure 11-3 World Oil Production in 1960. *World output more than tripled in the 1940–1960 period, paced by tremendous increases in the United States and Middle East, but the United States' share dropped to about one-third of the world total. The Soviet Union and Venezuela also remained major producers.*

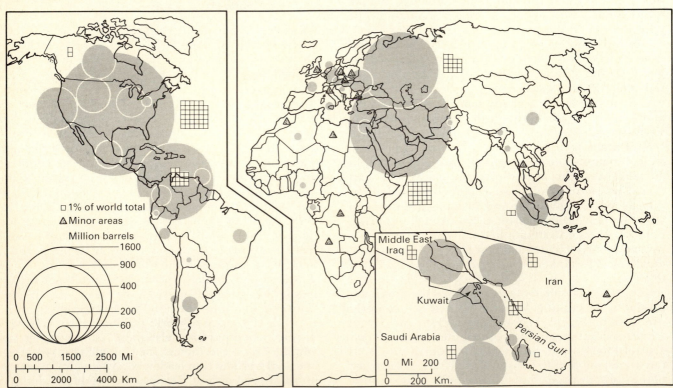

billion barrels in 1960. Other leading states included Louisiana, California, Oklahoma, and Wyoming. Even that level of output did not satisfy market demand. For the first time the United States became a major oil importer in the early 1960s. Most of the imported oil came from Venezuela.

The Middle East became the second-ranking producing region in 1960, moving ahead of Venezuela and the Soviet Union. Pumping in the traditional producing areas of Iran, Iraq, and Saudi Arabia, supplemented by extremely productive wells in Kuwait, provided a broader base of operations in that region (Figure 11-4). In fact, well productivity in the Middle East far exceeded the yield elsewhere. Average wells in Kuwait produced 3700 barrels a day, compared with 200 in Venezuela and 11 in the United States. Kuwait soon outpaced the output in Iran, Iraq, and Saudi Arabia because of its rich resources.

The Volga-Ural field in the Soviet Union, the "second Baku," experienced an oil boom by 1950. Production centered in the vicinity of Tartar, Bashir (Bashkir), and Kuibyshev (Keybyshev), with lesser wells near Saratov and Stalingrad (Volgograd) to the south.

This area rapidly accounted for 30 percent of the Soviet production, and by 1954 eclipsed Baku output. The strategic location of this field in the heart of the Soviet Union between two great industrial regions, the core industrial center and the Urals, provided an ideal setting. Moreover, the main railway and waterway systems served the area providing low distribution costs.

Indonesia's oil wells recovered from the World War II devastation by 1950. Crude oil flowed from Indonesian wells at the rate of 60 million barrels a year at the beginning of the war. With Japanese evacuation in 1945 came demolition of facilities such that the 1946 yield trickled off to only 2 million barrels. A production rebound followed rapidly.

Canada enjoyed an oil boom in the western prairie provinces in the 1960s, particularly Alberta (Figure 11-5 on page 156). Petroleum was first discovered in Leduc, south of Edmonton, in 1947. Canada produced 3 percent of the world's output in 1960. Unfortunately, Canada's petroleum is considerably farther from the center of the country's population than is the case in the United States, requiring greater transportation investments.

Figure 11-4 Oil Drilling Camp in Saudi Arabia. *Well productivity in the Middle East by 1960 far exceeded the yield elsewhere. Average wells in Kuwait produced 3,700 barrels a day, compared with 200 in Venezuela and 11 in the United States. (Courtesy EXXON Corporation)*

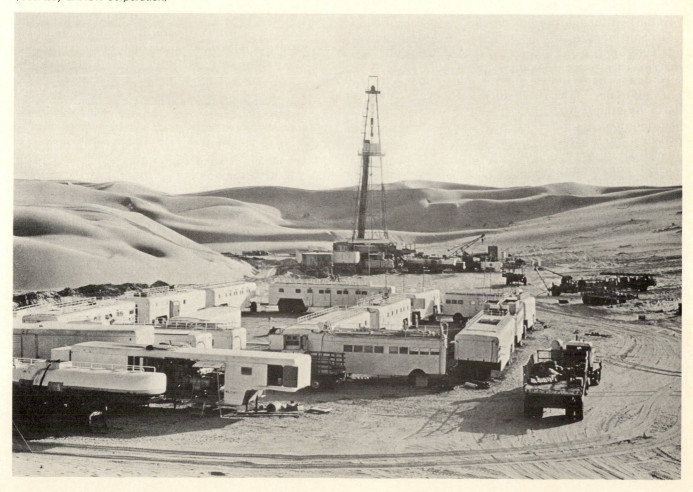

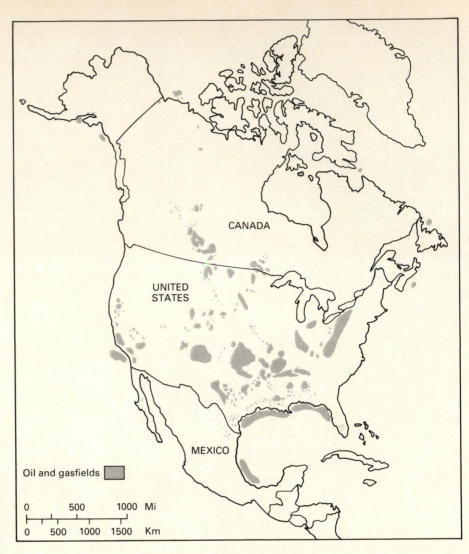

Figure 11-5 Oil and Gasfields in North America. *Oil fields are widely dispersed in the United States, including the Gulf Coast, West Coast, Mid-Continent, and Alaskan fields. Note the Canadian concentration in the western prairie and Mexican fields in the Gulf of Mexico. (Source: U.S. Geological Survey)*

CANADA

UNITED STATES

MEXICO

Oil and gasfields

| 0 | 500 | | 1000 | Mi |
| 0 | 500 | 1000 | 1500 | Km |

In summary, world oil movement patterns indicated only two major regions of surplus oil in 1960, the Caribbean and the Middle East. A profound change that became obvious at the time was the disappearance of the United States from the list of exporters, ushering in an era of growing imports, at the same time as domestic output continued to expand.

Growing Political Involvement

An event that would have far-reaching international implications, yet was largely ignored at the time, occurred in the Middle East in 1960. At a conference in Baghdad, the Organization of Petroleum Exporting Countries (OPEC) was founded. Originally comprised of five nations (Iran, Iraq, Kuwait, Saudi Arabia, and Venezuela), OPEC sought to provide a mechanism to deal with the multinational companies that controlled oil production and distribution in their countries.

Multinational petroleum firms successfully resisted reforms desired by host countries until 1960 by threatening to redirect oil purchases to other areas, leaving host countries largely impotent to deal with their problems. These multinational companies, often referred to as the *Seven Sisters,* managed 100 percent of the oil production and owned 80 percent of the output in the Middle East in 1960. The Seven Sisters included five firms headquartered in the United States: Exxon, Standard Oil of California (SOCAL), and Mobil (all three successors to Rockefeller's Standard Oil); Gulf; and Texaco.[1] Only two of the Seven Sisters—BP (British Petroleum), formerly Anglo-Iranian, headquartered in Great Britain, and Shell from the Netherlands—were not U.S. based. The term "Seven Sisters" was popularized in the 1950s by Enrico Mattei, Chairman of ENI, the Italian State Oil Company, when he was trying to establish a strong presence in the market for his firm. The Seven Sisters are also called the "majors."

Over the years the multinationals established interlocking ownership and concession arrangements

---

[1]Anthony Sampson, *The Seven Sisters: The Great Oil Companies and the World They Made.* (New York: Viking Press, 1975).

with the host countries, which meant that they continued to control prices. They also established vertical integration in the industry, gaining complete control of all activity from wellhead to refinery and final consumption, including pipeline, barge, and ocean tanker transportation, providing a classic example of an oligopolistic system. The majors in effect controlled the resource, its processing, and its distribution.

During this era production flourished as never before, allowing the multinationals to develop very complacent attitudes. But events changed rapidly. New discoveries in the Soviet Union increased the Soviet potential as an exporter. Elsewhere, independent firms entered the market, increasing competition with the majors. These and other factors led to a decline in oil prices that cut back royalty income to host countries, further alienating them and increasing the pressure on the majors to give up a greater share of oil revenues.

Political and military tension in the Middle East, centered on the Arab–Israeli conflict, also created an increasingly difficult working environment in the region for multinationals in the 1950s and 1960s. The Suez Canal was nationalized and closed in 1956 by President Nasser of Egypt, interrupting oil flows via the Canal to western Europe. This situation created uncertainties in oil flows and encouraged the deployment of supertankers to carry oil around Africa, but it also encouraged the development of western Mediterranean supplies. Newly discovered fields in Libya and Algeria provided an alternative source for Europe that helped overcome the problem of supply logistics created by the Suez Canal shutdown. By 1968 Libya outproduced Kuwait and became the fourth largest exporter after Venezuela, Iran, and Saudi Arabia.

OPEC countries obtained more and more concessions from the multinational producers in this period. Initially, OPEC functioned primarily as an information clearinghouse. A 50:50 sharing formula formerly used to distribute revenues between the host countries and the multinationals was challenged more and more as oil prices fell. Prices reached a bottom in 1970 of $1.56 a barrel. Prices are discussed more completely later.

Multinationals eventually agreed to financial concessions that did not cost them directly as pressures from host countries increased. They typically passed along increased costs created by larger royalty and fee payments as tax credit deductions. Increasing production also helped them increase profits even though their margins declined. But OPEC's appetite for more concessions continued to grow. Its membership also grew. Libya and Indonesia joined the Organization in 1962, Abu Dhabi in 1967, Algeria in 1969, and Nigeria in 1971. Eventually, the host countries negotiated three-fourths of the crude price for themselves. The OPEC strategy gradually became "after cash must come control."

Middle Eastern nations began nationalizing oil operations in the 1970s. Nevertheless, the multinationals were still strong. They continued to manage 70 percent of the Middle East oil business in 1970 and owned 40 percent of the crude production at that time. In June 1973, Libya nationalized its oil industry, followed by Iraq in October of the same year.

War again broke out in the Middle East on October 7, 1973. As before, exports from the pipeline terminals in the region stopped. On October 16 the representatives to OPEC of oil-producing "Gulf" states met in Kuwait and began a unilateral action, precipitating the infamous *Arab oil embargo* of 1973. OPEC raised prices 70 percent and ordered a production cutback. A demand that Israel withdrew to her 1967 cease-fire lines was accompanied by a threat to impose an embargo on sales.

The perceived pro-Israeli attitude of the U.S. government made it a primary target for OPEC complaints. By November 1973 an embargo was imposed on sales to the United States, Canada, and the Netherlands, among others. A 25 to 30 percent production rollback further curtailed output. By December 1973 OPEC invoked a staggering 130 percent price hike. Clearly, power had shifted from the multinationals to OPEC, and the era of cheap energy came to an abrupt end.

Multinationals attempted to thwart the impact of the embargo by redirecting flows of oil to those countries most affected with supplies from other areas under their control. This effort only hastened the pressure for the nationalization of production in the Middle East, which accelerated in the mid-1970s. In 1979, the Iranian revolution crisis led to nationalization in Iran. Nationalization also occurred in Nigeria the same year. Among major Middle East producers, Saudi Arabia remained an exception to this nationalization process. In the early 1980s four major U.S. firms—Exxon, Standard Oil of California, Mobil, and Texaco—continued to play an integral role in the Saudi Arabian oil production.

OPEC production peaked in 1973 and again in the late 1970s but declined abruptly thereafter until 1983, when output stabilized (Figure 11-6 on page 158). Due to continuing dependence on the OPEC-controlled supplies by importing countries in the 1970s and early 1980s, Middle East politics continued to influence prices throughout this period. At regular intervals prices continued to increase, especially in 1979, but in the early 1980s the price bubble burst (see the discussion in the following section). U.S. production remained steady in this period.

Energy Crisis Adjustments

The rapid price increases of the 1970s benefited OPEC producers and other major exporters greatly, including the Soviet Union. The surge in prices created an enormous cash flow increase. In major consuming nations, especially major importing areas such as Japan, Western Europe, and the United States, con-

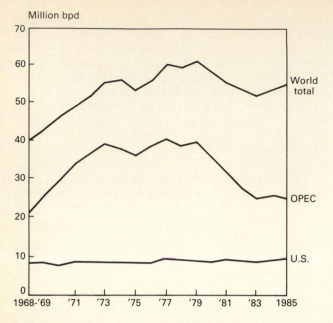

Million bpd

Figure 11-6 *World oil production and OPEC output increased in parallel fashion in the 1970s, peaking at the end of the decade. Since that time the OPEC share of world output has shrunk. Production in the United States during this period has remained relatively constant. (Source:* World Oil, *February 1983 and 1986)*

servation measures and research and development on alternative fuel sources increased. Conversions back to coal as an energy source accelerated and synthetic fuels (synfuels) programs received serious attention. Exploration for new oil sources accelerated worldwide. Offshore production grew rapidly as well.

The United States retained its world production lead in 1970, with its 3.5 billion barrel annual output, but the world share of this production dropped to 20 percent as an even broader base of major producers emerged (Table 11-3). The Soviet Union's share of world production remained relatively stable in that period at 15 percent. Newcomers to the production leader list included Egypt, Algeria, and Nigeria, all in Africa. Offshore production in the North Sea increased, but Great Britain and Norway had not yet experienced the boom to come later in the decade.

Western Siberia emerged as a major producing region in the Soviet Union in the 1970s. The cold swampy region of Tyumen province along the Ob River claimed most of the activity. The Samotlor oil field became the most productive region in the country, culminating in 1978 with its emergence as the single leading source area in the Soviet Union (Figure 11-7).

In 1980, increases in production in western Siberia leveled off, but the Samotlor field retained its lead ranking with an output of 3 million barrels a day. The Fedorovo and Mamontovo fields each produced at about one-fourth of the Samotlor level, but together the three fields accounted for three-fourths of Tyumen production and one-third of total Soviet output. Altogether there were over 60 fields in pro-

duction or in preparation in western Siberia in the early 1980s.

A disadvantage of the western Siberian oil region, in addition to a severe climate, is its physical location far from the European market region. The Ob River basin producing area lies 1500 miles east of the Volga-Urals region which it displaced. The distribution of oil from this isolated location required massive pipeline investments. The Soviets are now pioneering the development of very large 56-inch pipelines linking this area with markets as far away as eastern Europe.

Gas fields in western Siberia are even farther north than the oil areas. Lying in the Arctic, they are particularly isolated. Pipelines serving that area will eventually extend to western Europe, opening the way for greater gas exports.

Before it became evident that western Siberian oil would be so lucrative, the Soviets began opening up gas fields in arid central Asia to the south. They also signed an agreement with Iran to begin importing natural gas via pipeline in 1966.

The Soviets had little choice but to push production in western Siberia once the full extent of the resources became known. European Russia experienced a net energy shortfall in the mid-1970s. While producing only 9 percent of total petroleum in 1970, western Siberia output doubled from 1970 to 1975, accounting for 30 percent of Soviet oil in 1975. Production peaked in the Volga-Urals in the mid-1970s and began a slide by the end of the decade. Paced by the output of western Siberia, the Soviets became the world's leading producer in 1974, a role the country still maintains (Table 11-3). In 1985 the Soviet Union output exceeded that in all Middle Eastern countries combined, and matched that for all North American producers (the United States, Canada, and Mexico) as a group. The major producing areas, paced by the west Siberian region are shown in Table 11-4.

Offshore oil production accounted for nearly one-fourth of world output in the early 1980s. Saudi Arabia led in offshore drilling activity (Figure 11-8 on page 160), accounting for over 20 percent of the total. Abu Dhabi contributed another 10 percent, giving nearly one-third of the output of offshore oil to the Middle East. The second-leading offshore producer is Great Britain, owing to the largesse of the North Sea producing area (Figure 11-9 on page 161). The economic boom of offshore oil on western Europe, which also benefited Norway, Denmark, West Germany, and the Netherlands, is somewhat offset by political bickering. Disputes over taxes, who to sell to, and quotas have an uncanny resemblance to the problems discussed earlier with the Middle East.

Venezuela and the United States each contribute about 8 percent of the world offshore output. In the United States, production centers on the Gulf of Mexico. The trend in that area is for production to shift southwest away from Louisiana, toward Texas. Australia is also an important offshore producer. The Bass Strait/Gippsland Basin south of Melbourne is now the leading offshore source in that area.

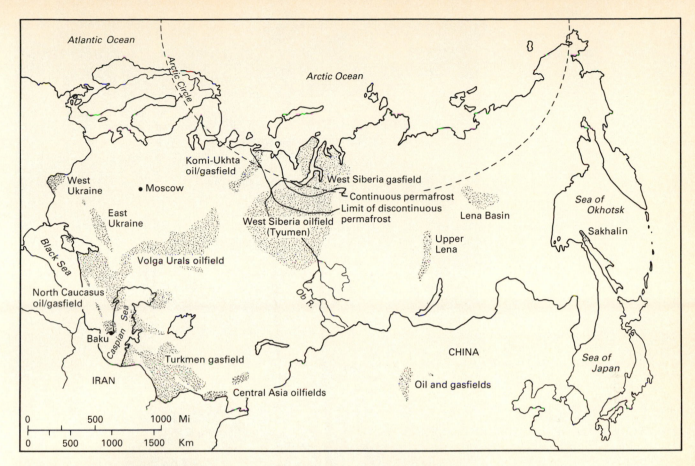

Figure 11–7 Major Soviet Oil Producing Regions. *Western Siberia emerged as a major producing region in the Soviet Union in the 1970s. This inhospitable area lies 1500 miles east of the Volga-Urals region which it displaced. Gasfields in Western Siberia are even farther north than the oil producing areas, lying in an area of continuous permafrost astride the Arctic Circle. Note the original Soviet producing area in the Baku vicinity and the North Caucasus oil and gas field. (Redrafted from Central Intelligence Agency,* U.S.S.R. Energy Atlas, *1985, p. 16)*

In western Europe the switch from coal to oil came much later than the U.S. conversion, owing largely to the lack of indigenous petroleum supplies. As recently as 1962, only Italy and the Netherlands used more oil than coal. By 1972, oil consumption dominated in all western European countries except the Netherlands, where a large-scale conversion to natural gas had occurred. Local gas supplied from the Groningen field, the world's most lucrative area, explains this transition. When coal predominated as an energy source, Europe relied on domestic resources, but with the oil era came greater dependence on both foreign supplies and suppliers, largely U.S.-dominated firms.

European dependence on foreign supplies lessened in the late 1970s as North Sea production accelerated. From 1976 to 1979 annual production in western Europe rose from 29 million tons to 150 million tons. Total output in the future could easily triple. Moreover, new-found offshore reserves also exist in Spain and Greece.

Despite these local reserves, western Europe has increasingly become more dependent on the Soviet Union for imports (Table 11-5 on page 161). This relationship will intensify in the future as natural gas imports increase. Larger Soviet shipments to Austria, Greece, and Portugal occurred in the early 1980s, reflecting this dependency relationship.

Price Shocks and Recession

Until the early 1970s the price of crude oil had remained relatively stable since the 1920s. In 1929, for

Table 11-4

Leading Soviet Oil Producing Areas, 1985
(millions of metric tons)

Rank	Area	Production	Percent
1	West Siberia	372	63
2	Volga	70	12
3	Komi ASSR	18	3
4	North Caucasus	10	2
	Subtotal	470	80
	Total	595	100

Source: Soviet Geography, Review and Translation, April 1986.

Figure 11-8 Oil-drilling Platform in the Persian Gulf. *Offshore production accounted for nearly one-fourth of world output in the early 1980s. Saudi Arabia, with a 20 percent share, leads in offshore drilling activity. (Courtesy EXXON Corporation)*

example, the U.S. price was $1.53 a barrel. The price fell during the depression to $0.99 a barrel, where it remained during the 1930s. Prices did not strengthen again until the 1950s. Price discounting soon returned as production and surpluses mounted. Prices remained stable again from 1960 through 1971. In the early 1970s OPEC nations became the source for over 85 percent of petroleum exports. This situation provided the basis for extracting higher prices, as discussed earlier.

In 1971 OPEC raised the crude price 30 percent from the previous base of about $1.50 (Figure 11-10 on page 162). The prices used here are for so-called Arab light crude, sometimes called "marker" oil because of its use as a benchmark price. The Tehran Agreement which established this raise also provided for future price hikes. Even bolder action followed when OPEC countries raised prices fourfold in the October 1973–

January 1974 period. Later in 1974, OPEC nations raised tax and royalty rates. Other increases followed. A 10 percent increase came in October 1975, and in 1977 OPEC introduced a two-tier pricing system. At that point the price was $11.51 a barrel. Following several more small hikes in 1976 and 1977, a 15 percent increase for 1979 was announced in December 1978. In April 1979, this price jump raised the price to $14.55. By June a barrel of crude sold for $18.00, followed by another jump to $24.00 later in 1979.

The dramatic price increase initiatives for OPEC crude caused retail prices to skyrocket worldwide. The increases continued. By May 1980 the price was increased to $28.00. But breaks in OPEC unity also began to occur at that time. Unified pricing became difficult to enforce and discounting increased as intra-OPEC rivalries mounted. Attempts to overcome this problem included establishing production ceiling

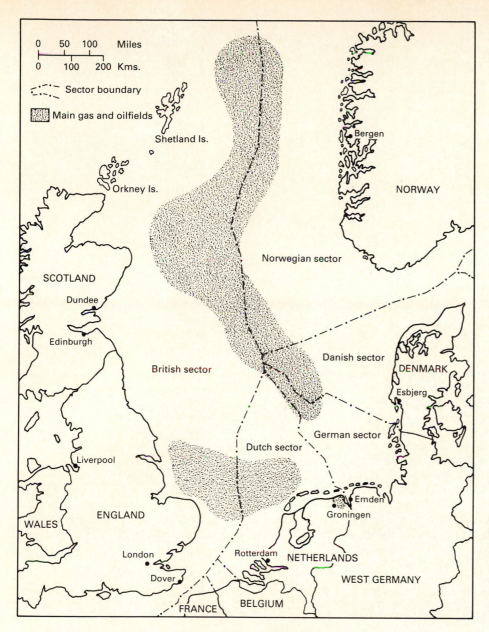

Figure 11-9 North Sea Oil and Gas Producing Areas. *While Great Britain commands second place in offshore output, after Saudi Arabia, the largesse of the North Sea producing area also greatly benefits Norway, Denmark, West Germany, and the Netherlands. (Source:* The Economist, *June 12, 1982)*

Table 11-5

Nations Receiving Soviet Petroleum Exports, 1983*

Western Europe	Eastern Europe	Other
Austria	Bulgaria	Brazil
Belgium	Czechoslovakia	Ethiopia
Denmark	East Germany	India
France	Finland	Japan
Great Britain	Hungary	Mongolia
Greece	Poland	Morocco
Italy	Yugoslavia	Vietnam
Netherlands		
Spain		
Switzerland		
West Germany		

Source: *USSR Energy Atlas* (Washington, D.C.: Central Intelligence Agency, 1985).

* Alphabetical listing by region of those receiving more than 15,000 barrels/day.

quotas among OPEC nations. This last round of increases developed at the time of the Iranian Revolution in 1978–1979, which led to export stoppages by the end of 1978. Those interruptions led to shortages around the world even though shipments began again in the spring of 1979 when a new Iranian government came to power. This situation encouraged OPEC member countries again to accelerate price increases.

Even though the price increases earlier in the decade had significant impacts, the late 1970s raises were the most inflationary. Earlier price hikes, in retrospect, occurred at the same pace as inflation. The late 1970s increases were much higher.

In the fall of 1981 prices of light Arab crude were hiked to $34.00, where they remained through the end of 1982. In the meantime world oil prices began to fall because of growing surpluses. In early 1983 the price of Arab market oil dropped to $29.00, owing to

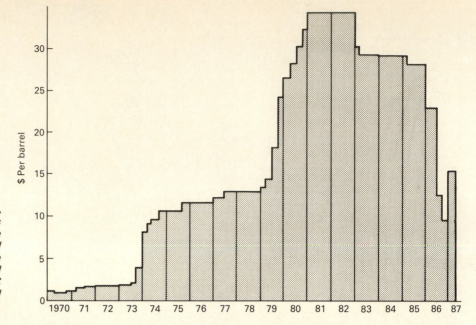

Figure 11-10 Crude Oil Price Changes, 1970- 1987. *Crude oil prices increased dramatically in two time periods in the 1970s, both associated with OPEC price hikes. The price bubble burst by early 1983, followed by further cuts in 1986 when OPEC lost its ability to control prices. Upward adjustments again returned in 1987.*

a continuing glut of product on the market. By implementing this reduction, OPEC countries attempted to prevent further price discounting. New quotas limiting production of member countries prevented further price softening. But the major significance of this cut was that for the first time in a decade OPEC countries lost their ability to force price increases. By 1985 a further adjustment lowered OPEC prices to $26.00, owing to growing worldwide surpluses. A series of cuts in early to mid-1986 cut this price to the $12.00 range, thus returning world prices to their 1977-1978 level. Nevertheless price increases returned by 1987, suggesting that oil would soon again sell for more than $20.00 (Figure 11-10).

Oil prices in many countries were far higher in 1980 than the Middle East prices just discussed. Western Europe had the highest refined oil prices due to heavy government taxation. Prices of $74.00 a barrel occurred in Europe, where a tax rate of up to 39 percent also resulted in the highest retail prices in the world. In Japan, refined crude sold for $55.00 a barrel, also due to heavy taxation. In the United States, controls traditionally kept the price of domestic crude lower than world prices, giving the country lower retail prices.

Retail gasoline price increases generally paralleled OPEC price hikes in the 1970s, as suggested earlier. In 1973, prices in the United States were much lower than those in Europe and a far smaller portion of the price was accounted for by taxes (Table 11-6). In 1980, U.S. prices remained much lower than those in Europe even though they had tripled in the preceding decade. Taxes on petroleum products in the United States increased far less than those in Europe. In Italy gas prices stood at the highest level in Europe in 1980—$3.38 a gallon. Taxes were also the highest in Italy—$2.66 a gallon in 1980. Retail prices for gasoline declined briefly in the mid-80s in both Europe and the United States, but rates in Europe remained higher than those in the United States.

Even though a large worldwide surplus emerged in the early 1980s, energy problems continued because of economic uncertainty. Inflation, partially fueled by oil price hikes, became a big issue at the beginning of the decade. By 1983 this inflation rate subsided in developed areas but remained a problem in developing areas. Due to higher prices, exploration and drilling had also become overextended.

Table 11-6

Retail Prices for Regular Gasoline in Selected Nations, 1973–1983 (U.S. dollars per gallon)

	Nation					
	United States		France		Italy	
Date	Price*	Tax	Price*	Tax	Price*	Tax
10/73	0.40	0.12	1.14	0.77	0.80	0.61
1/76	0.58	0.12	1.60	0.90	1.38	0.91
1/77	0.60	0.12	1.90	1.15	2.21	1.59
1/78	0.62	0.13	1.99	1.21	2.21	1.59
1/79	0.68	0.13	2.31	1.59	2.21	1.59
1/80	1.10	0.14	2.78	1.66	2.92	1.87
7/80	1.22	0.14	2.94	1.68	3.38	2.16
1982	1.30	—	2.56†	—	2.91†	—
1983	1.25	—	2.38†	—	2.89†	—

Source: Congressional Research Service, Library of Congress. *Energy Factbook: Data on Energy Resources, Reserves, Production, Consumption, Prices, Processing, and Industrial Structure.* Washington, D.C.: U.S.G.P.O., 1980; and PennWell Publishing Company, *International Petroleum Encyclopedia.* Tulsa, 1985.
  *Including tax.
  †Premium gasoline.

A major worldwide recession occurred in 1981–1982. By the summer of 1982, this recession hit the U.S. oil industry. Drilling, pipeline, and engineering firms went bankrupt, layoffs increased, and the steel industry experienced a fallout in the demand for the products it supplied the petroleum industry, especially seamless pipe for wells and pipelines. Synfuel programs, largely subsidized by the federal government, were cancelled, including programs dealing with oil shales, coal gasification and liquefication, tar sands, and ethanol production. In the United States, Houston and Denver were both hard hit by this energy recession.

The Canadian oil recession began earlier than that in the United States. A national energy program, introduced in Canada in 1980, called for energy self-sufficiency by the turn of the century. The private sector doubted that the program would work but the "Canadianization" program began anyway. Essentially, the goal was to (1) increase Canadian ownership of oil industry firms to majority status, (2) encourage conservation by imposing large tax increases, (3) develop a synfuels program, and (4) encourage alternative fuel substitutions away from oil.

The pressure for increasing Canadian ownership levels exerted by the government in the late 1970s depressed the stock prices of the many U.S.-based firms operating in Canada, primarily in Alberta, the Arctic, and Newfoundland. In two years, 17 significant acquisitions of foreign firms occurred, greatly increasing Canadian ownership levels. But this activity occurred just before interest rates skyrocketed and many firms assumed too much debt. Dome Petroleum nearly went bankrupt in 1982, for example. The Alando and Cold Lake synfuels programs in Alberta were phased out as the recession intensified.

The biggest impact of Canadianization of the oil industry involved the curtailing of exploration and drilling. Many rigs formerly operating in Canada headed south for the more lucrative laissez-faire environment in the United States. The Canadian government relented with some tax concessions as the recession mounted. The economies of Calgary and Edmonton were particularly adversely affected by the economic downturn. In 1983 it appeared that the worst was over and that more joint ventures and farmouts to American firms would return. The future production potential of east coast offshore areas, the Beaufort Sea, and the Arctic Islands area began to look brighter, particularly for natural gas, as economic prosperity returned.

By 1985 Canada scrapped its energy independence program. A new approach relied on market forces to manage the industry rather than government intervention. The 50 percent Canadian ownership requirement at the production stage remained, but subsidies to Canadian firms were drastically reduced, in favor of tax credits to stimulate exploration.

Mexico also experienced serious setbacks as a result of the worldwide recession in the early 1980s. High interest rates and the largest public and private debt in the world created a serious strain on the economy. High interest rates nearly bankrupted the country. This problem, together with a 60 percent inflation rate and a 40 percent unemployment level, led to a severe crisis. A peso devaluation of 45 percent occurred in February 1982, followed by another, smaller devaluation later that summer.

Much of the Mexican debt resulted from the high cost of importing machines and technology to produce more oil and gas. Despite financial strains, great successes also occurred. In 1982, Mexico became the leading exporter of crude oil to the United States (Table 11-3). Most of this production today occurs in the southern zone in the Campeche Sound region, and in Cheapas-Tabasco. The latter has about one-third of Mexican oil reserves and one-half the gas.

Recent U.S. Trends

In the early 1980s, U.S. imports of crude oil continued falling dramatically, a trend that began in 1978 (Figure 11-9). In 1982 imports declined to their lowest level in a decade. Only the United Kingdom, Canada and Mexico shipped more oil to the United States in 1982 than in the year before. Saudi Arabia, the leading supplier in 1981, fell to second place in 1982. With the import decline also came a lesser dependence on OPEC oil (Table 11-7). In 1982, for the first time since 1973, the majority of U.S. oil imports came from non-OPEC countries, ushering in a new era of trade relationships.

Crude oil production in the United States increased slightly in the mid-1980s (Figure 11-6). The Kuparuk field in Alaska, which came into production in late 1981, largely explains this gain but output also remained strong in the lower 48 states. Texas remained in the lead, if by a narrower margin, due to production declines, while Alaska moved ahead of

Table 11-7

Leading Sources of U.S. Petroleum Imports, 1981 and 1982

Nation	1982		1981	
	Rank	Percent of Total	Rank	Percent of Total
Mexico	1	13	3	9
Saudi Arabia*	2	12	1	19
Nigeria*	3	10	2	10
Canada	4	9	4	8
Great Britain	5	9	6	6
Venezuela*	6	8	5	7
Subtotal		61		59
OPEC contributions to U.S. imports		43		55

Source: World Oil, February 15, 1983, p. 67.
*OPEC nation.

Table 11-8

Leading U.S. Crude Oil–Producing States, 1985
(millions of barrels annually)

Rank	State	Production	Percent
1	Texas	876	27
2	Alaska	665	20
3	Louisiana	520	16
4	California	416	13
5	Oklahoma	164	5
	Subtotal	2641	81
	U.S. total	3255	100

Source: *World Oil*, February 1986.

Louisiana, California, and Oklahoma to control second place (Table 11-8). Only unleaded gas consumption increased in 1982, among all petroleum products, largely due to a 20 percent drop in prices, which peaked in the spring of 1981. The overall decline in petroleum consumption in the early 1980s dampened the pressure for price increases, but by 1986 and 1987 consumption levels and import increases had returned again.

In 1975, recognizing a vulnerability to foreign supply uncertainties, Congress created the Strategic Petroleum Reserve to stockpile petroleum. By 1981 over 100 million barrels of oil had been pumped into five underground salt dome caverns in Louisiana and Texas. The goal is to increase this reserve to three-fourths of a billion barrels to create a hedge against future embargoes.

Mergers and Continued Uncertainty

Petroleum price fluctuations in the 1970s and early 1980s vastly changed the financial operating environment for oil companies. Generally, marketing and refining became less profitable and reserves became a more valuable asset, but stock prices did not always reflect this situation. Companies with large oil reserves were often perceived as more valuable than those devoid of reserves, which in turn fueled merger activity. Due to higher operating costs many experts believed that it would be cheaper to buy reserves on Wall Street through merger activity than to expand exploration activity. It was also observed that a breakup of oil company assets would yield additional benefits by the purchasing group, by selling assets, merging operations, and consolidating labor forces.

Eight major mergers occurred with American firms in the 1979–1984 period (Table 11-9). The two largest deals, the Gulf/Chevron buyout in 1984, and the Getty/Texaco deal in 1984, each exceeded $10 billion each. These mergers led to additional employment contractions in an already depressed oil economy. The Texaco/Getty merger led to a layoff of 14,000 employees, primarily in Texas and California. The Gulf headquarters in Pittsburgh was shut down following its merger with Conoco as operations were consolidated in San Francisco.

A final comment about the U.S. petroleum industry should include a discussion of the changing regulatory climate. In the early years of government regulation in the United States the concern focused on breaking up oligopolistic firms to ensure competition. In 1907 the Interstate Commerce Commission gained jurisdiction over domestic petroleum and gas pipelines. In 1911 the Supreme Court dissolved the Standard Oil Company, but policies soon changed to favor the producer. Special tax treatment for crude oil extraction began in 1913 with the initiation of the oil depletion allowance. This concession gradually became more favorable to the industry, culminating with a boost in the depletion allowance to 27.5 percent in 1926. This subsidy remained in force for over 40 years. Further assistance to domestic producers came from crude oil import restrictions beginning in the 1950s. A voluntary import control program gave way to a mandatory system in that decade as well.

The oil depletion allowance was phased out in the 1970s. Price controls emerged in 1973, remaining in place for about a decade. All this federal intervention, and the heightened public concern over shortages and the uncertainty created by a growing dependence on imports, led to the creation of the Cabinet-level Department of Energy in 1977. Price controls became very complicated, with the designation of "old" and "new" oil categories creating a two-tier pricing system by the end of the 1970s. Old oil referred to base production levels of a well subject to price controls. New oil, or the production in excess of the base output level of a property, could increase in price.

Price controls on refined product prices began in 1976. By 1980 Congress passed a windfall profits tax to return the difference in market prices between former ceiling-price levels and uncontrolled prices to government coffers. The Reagan administration eliminated these controls in 1981, reflecting a changing government priority away from the use of controls in deference to the use of taxes and market prices to regulate the industry.

Table 11-9

Largest Mergers of Petroleum Firms, 1979–1984
(billions of dollars)

Company/Acquirer	Transaction Amount	Year
Gulf/Chevron	13.2	1984
Getty/Texaco	10.1	1984
Conoco/duPont	7.4	1981
Marathon/U.S. Steel	6.5	1981
Mobil/Superior	5.7	1984
Texasgulf/Elf Aquitaine	4.2	1981
Cities Service/Occidental	4.0	1982
Belridge/Shell	3.6	1979

Source: PennWell Publishing Company, *International Petroleum Encyclopedia.* Tulsa, 1985.

REFINERY LOCATIONS

Unlike many manufacturing processes, petroleum refining does not involve weight gain or loss. Virtually all the product can be used. Some volume increase does occur during the process, but this is not an important locational factor. Refineries can locate near the raw material, at the market, or at intermediate break-of-bulk locations such as coastal or port locations, because no clear advantage occurs from a single orientation. Facilities are capital-intensive operations and highly automated. They are also frequently linked with other petrochemical operations to share technology and expertise. For this reason, refineries frequently are located near the resource (Figure 11-11). A potential disadvantage of locating refineries near oil fields is the short life expectancy of the well, but traditionally a large number of wells have been clustered in the most productive well-field regions. Some observers suggest that higher evaporation rates for refined products also encourages refining near the source.

After World War II a trend intensified in the United States to place refineries at market or at break-of-bulk locations. Greater reliance on imports partially explains this trend, as well as the expanding pipeline network to move the product. In some cases break-of-bulk and market locations occur at the same locations, as in the case of refineries situated in Philadelphia and Montreal. In southern California, refineries occur near both the wells and the market. In the Los Angeles/Long Beach areas, refineries can be found in Wilmington, Torrance, El Segundo, and Carson.

Middle East oil fields typically lie near the coast and refineries occur at port locations (Figure 11-12). In Europe, refineries also cluster near port locations, such as those in Rotterdam and the Thames estuary. Pipelines, river barges, and railroads converge on Rotterdam, making the port an ideal focus for production. In Europe, industry, not the automobile, as is the case in the United States, consumes a majority of the product, further reinforcing the advantage provided by a clustering of refineries in major manufacturing centers. As can be seen in Table 11-10 nearly half of the refined product in North America is gasoline, whereas only one-fourth takes this form in Japan and western Europe. In western Europe, middle

Figure 11-11 Strathcona Refinery in Edmonton, Alberta. *This facility, located on the banks of the North Saskatchewan River, produces both oil and gas. It is served by a large rail yard. Note downtown skyline in distance. (Courtesy EXXON Corporation)*

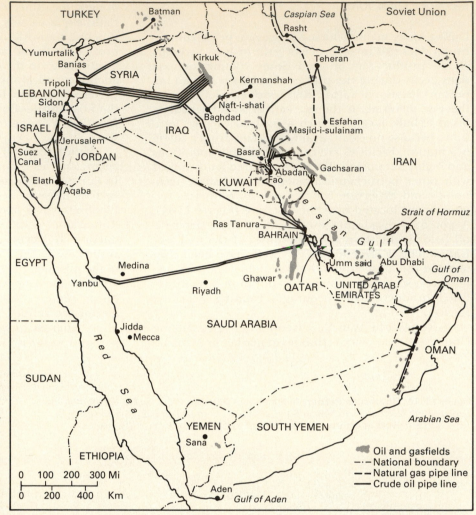

Figure 11-12 Middle Eastern Petroleum and Natural Gas Fields and Facilities. *Oil and gas fields in the Middle East are clustered in the Persian Gulf area and in the interior desert of Iran and Iraq. Refineries cluster at port locations on the Persian Gulf and the Mediterranean Sea. Note the pipeline connections to the Mediterranean coast. (Redrafted from* International Petroleum Encyclopedia, *Tulsa, 1986, p. 169)*

distillates (kerosene, jet fuel, diesel) predominate, while fuel oil has a slight output lead in Japan.

Market locations for petroleum refining have the advantage of allowing the refinery to provide the appropriate mix of products for local consumption and the ability to draw supplies from more than one field. Large refinery capacities in the United States therefore occur in the vicinity of many major metropolitan areas in addition to those already discussed. Pipeline connections make interior cities such as Chicago just as prominent as coastal locations for such activity.

Environmental considerations mitigate against locating new refinery facilities near densely populated areas. Fumes, spills, explosion hazards, and the need for huge storage facilities encourage the use of sites on the outskirts of urban areas, such as Whiting, Wood River, Joliet, Robinson, and Lemont in the Chicago area and Marcus Hook near Philadelphia. In New Jersey, refineries exist at Perth Amboy, Linden, and Paulsboro.

Gulf coast states continue to claim the highest concentration of U.S. refineries, but the industry declined in this area in the 1980s as it did throughout the country (Table 11-11). In greater Houston, refinery locations include Texas City and Baytown (Figure 11-13). Corpus Christi, Port Arthur, and Beaumont also have significant refining capacity. Refineries in Baytown and Port Arthur experienced significant production cutbacks in the mid-1980s, while statewide more than 30 refineries closed in Texas in the first half of the decade. In Louisiana, Lake Charles, Baton Rouge, Belle Chasse, Garyville, Norco, Chalmette, and

Table 11-10

Refinery Yield Comparisons, 1981
(percentages by volume)

Product	North America	Western Europe	Japan
Gasoline	48	25	24
Middle distallates*	28	34	31
Fuel oil	11	26	35
Other (including loss)	13	15	10
Total	100	100	100

Source: The British Petroleum Company, *BP Statistical Review of World Petroleum.* London, 1982.

*Kerosene, jet fuel, diesel fuel.

Table 11-11

Refinery Retrenchment Trends in the United States
in the 1980s

| 1982 Rank | State | Number of Plants | |
		1982	1985
1	Texas	56	33
2	California	40	30
3	Louisiana	30	16
4	Oklahoma	13	5
5	Wyoming	10	6
6	Kansas	9	7
7	Pennsylvania	9	8
	Total	167	105

Source: PennWell Publishing Company, *International Petroleum Encyclopedia.* Tulsa, 1982, 1985.

Convent each have large capacities. Refinery output adjustments reflect a decreasing demand for fuel, reduced imports of crude product, and an increase in refined product imports. Refined materials comprised one-third of total petroleum imports in 1982, up considerably from the one-fourth share in 1981. As the realignment in the industry continued it became obvious that new state-of-the-art refineries abroad, particularly in the Middle East, could produce refined product such as gasoline much more cheaply than could domestic producers. For this reason, domestic consumption of imported gasoline reached a new high of 10 percent of total demand in 1985.

The United States and the Soviet Union continue to dominate the oil refining industry, together accounting for over one-third of the world refining capacity (Table 11-12). Japan also maintains a significant

Figure 11-13 Baytown Refinery, Texas. *Gulf coast states continue to claim the highest concentration of U.S. refineries, but the industry declined in the 1980s as it did throughout the country. (Courtesy of EXXON Corporation)*

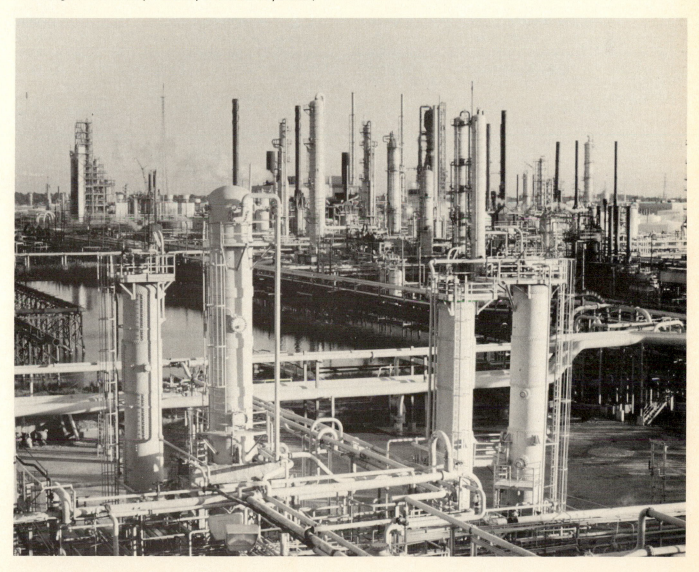

Table 11-12

World Oil-Refining Capacities
by Nation, 1985

Rank	Area	Percent of World Total
1	United States	21
2	Soviet Union	16
3	Japan	6
4	Middle East	5
5	Italy	4
6	France	3
7	West Germany	3
8	China	3
9	Great Britain	3
	Subtotal	64

World refining total	74,753,000 barrels/day

Source: PennWell Publishing Company, *International Petroleum Encyclopedia.* Tulsa, 1985.

refining capacity at widely scattered locations, but with concentrations occurring in coastal industrial regions on the island of Honshu, including Chiba, across Tokyo Bay from Tokyo, and Yokkaichi, near Nagoya. Several western European countries also maintain significant refining capacities, but several are now overshadowed individually by expanded output in the Middle East.

INTERNATIONAL TRADE

International trade flows of petroleum by sea reflect traditional areas of surplus and deficit (Figure 11-14). Most of the flow emanates from the Middle East, moving primarily to western Europe, with Japan also receiving a major share (Figure 11-15). Major anomolies in the pattern of waterborne trade in comparison with total consumption levels occur with the Soviet Union, Canada, and Mexico. Pipelines move much of the product from these areas to contiguous markets. The trade map also fails to capture the enormity of oil use in the United States, which is supplied by a very large domestic output in addition to imports. The most accurate representation of final consumption levels shown by the water-based flow map occurs with western Europe and Japan, which lack domestic sources. A final point is that the Soviet Union remains in the best situation among major consuming nations in terms of petroleum self-sufficiency, using about 15 percent of world oil.

NATURAL GAS

While the gas utility industry dates to the early nineteenth century, mass marketing of its present form—natural gas—only goes back to the post–World War II era. *Manufactured gas* dominated until that time. Man-

ufactured gas is produced by heating coal. Gas works firms producing this product typically were located near the consumer, serving a local market. Natural gas, on the other hand, came of age with the deployment of a comprehensive pipeline network after World War II, first in the United States (Figure 11-16).

Natural gas burns clean, is easy to use, and has traditionally been cheap to buy, but it is not without problems. It can be used for direct heating due to its purity and does not require extensive refining, but storage and distribution present complications. Pipelines overcame these shortcomings, but a continuous feed from the well to the final user must be established. Conversions from coal and petroleum to natural gas accelerated in the 1950s and 1960s in the United States as a result of these desirable qualities. The Soviet Union is now undergoing a similar transition to natural gas that occurred in the United States 30 years ago.

Crude oil generally occurs with natural gas, but gas is also found separately. The geography of mining natural gas is quite different from that of crude oil for the simple reason that whereas oil is easily trapped at the wells and lends itself to shipment via barrels, tank cars, trucks, barges, or ocean tankers, gas readily escapes at the wells, and what is trapped can only be transported by pipelines or converted to a liquid form (LNG) under pressure, a form that is more difficult to transport.

Gas is generally an economic product only in those mining areas close enough to the market to accommodate pipeline connections, but gas does move in liquid form in international trade. Where there is not a local market for the product, gas is often wasted by burning it off (flaring) at the well. Most of the world's natural gas is produced in the United States, the Soviet Union, Canada, and western Europe, where there is a readily accessible market (Table 11-13 on page 171). The tremendous jump in production in the Soviet Union, traditionally the second-leading producing nation, thrust the country into first place in 1984, with a 37 percent share of world output. The United States, Canada, and the Netherlands followed.

As with oil, regulations traditionally controlled natural gas prices in the United States and they remained low well into the 1970s. Supplies appeared plentiful in the United States after World War II and the huge backlog of reserves provided confidence for the future. Drilling and exploration declined by the late 1950s, however, in response to the lack of incentives due to low prices and plentiful supply. The U.S. Supreme Court placed prices of domestic natural gas sold interstate under the regulation of the Federal Power Commission. Gas sold within the producing state (intrastate) did not fall under the control guidelines, and therefore drilling and production continued to expand as prices rose.

Price lids discouraged interstate gas movement, not to mention the exploration cutbacks to serve those markets. Nevertheless, the demand for interstate gas shipments kept increasing as conversions to gas from

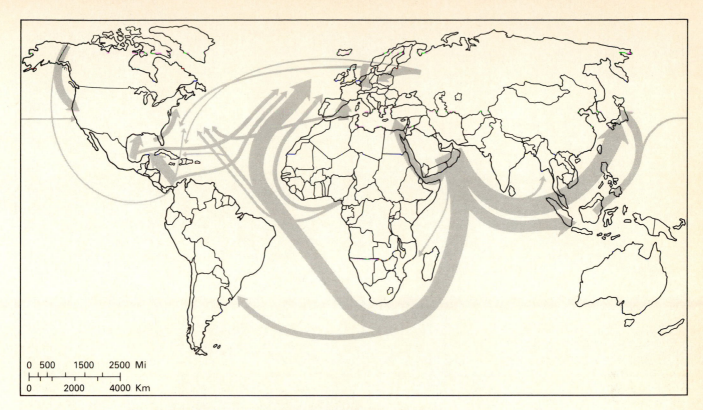

Figure 11–14 Major Ocean-based Oil Flows. *International trade flows of petroleum by water reflect traditional areas of surplus and deficit. Most of the flow emanates from the Middle East, moving primarily to western Europe, with Japan also receiving a major share. This trade map does not account for international pipeline flows, as between the Soviet Union and eastern and western Europe or the United States and Canada. (Adapted from* BP Statistical Review of World Energy, *1981)*

Figure 11–15 Kharg Island Petroleum Terminal. *This island houses tanker docks connected to Iranian oil fields by pipelines. The Iran-Iraq war disrupted operations in the Persian Gulf in the mid-1980s, but the Persian Gulf continued to supply a large share of the world petroleum supply. (Courtesy EXXON Corporation)*

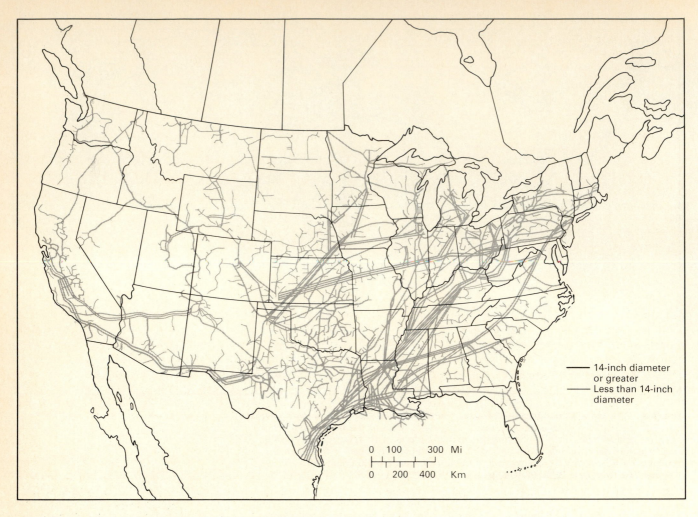

14-inch diameter
or greater
Less than 14-inch
diameter

0   100   300 Mi

0   200   400   Km

Figure 11–16 Natural Gas Pipeline Network in United States. *The elaborate petroleum pipeline network in the United States numbers 225,000 miles, while the natural gas system is four times as large. The flow occurs primarily from Texas and Louisiana coastal areas to the northeast markets. Secondary concentrations link California with the Southwest and the western Kansas and Oklahoma area with the upper middle west and northeast.*

coal and petroleum continued. This trend accelerated after the Mine Health and Safety Act in 1969 led to increased coal mining costs and the Clean Air Act of 1970 required pollution abatement measures in coal-fired industrial plants. Coupled with severe winters in the early 1970s, which also depleted reserves, these factors combined to create interstate gas delivery shortages in the 1970s.

Natural gas shortfalls in the United States became especially severe in the winter of 1977, when many users faced intermittent cutoffs. No intrastate shortages occurred, as higher prices provided incentives to produce for that market. As Figure 11-17 shows, natural gas production peaked in the United States in 1973, a few years after oil output reached its highest level. In response to the pressure from shortages, deregulation in the late 1970s led to higher prices and interstate movements again increased.

The petrochemical industry in the Gulf region has a large share of the gas market. For this reason Texas

and Louisiana consume nearly one-third of the natural gas in the United States. California counts as another big user largely because of strict environmental pollution laws in the state that work against coal. A final large consuming region is the industrial northeast, where both residential heating and industrial use provide a big market.

As mentioned earlier, the Soviet Union leads the world in natural gas output. The western Siberian Tyumen province claims two-thirds of Soviet reserves. The Soviets have recently pressed ahead with a massive pipeline project to connect that region with the lucrative western European import market. The potential for receiving hard currency from the sale of this gas provides the incentive. This source of income is all the more important to the Soviets as oil prices drop and exports decline. Soviet gas export sales reached $1 billion in 1976 for the first time, rising to $2 billion in 1978 and $5 billion in 1980. In 1980 sales to western Europe exceeded those to eastern Europe

Table 11-13

Natural Gas Production Shares, by Leading Nations, 1971–1984
(millions of tons of oil equivalent)

Nation	1971 Rank	1971 Production	1971 Percent of World Total	1975 Rank	1975 Production	1975 Percent of World Total	1980 Rank	1980 Production	1980 Percent of World Total	1984 Rank	1984 Production	1984 Percent of World Total
United States	1	561	56	1	491	43	1	495	37	2	439	30
Soviet Union	2	191	19	2	260	23	2	380	28	1	527	37
Canada	3	58	6	3 (tie)	69	6	3 (tie)	69	5	3	65	5
Netherlands	4	31	3	3 (tie)	69	6	3 (tie)	69	5	4	62	4
Subtotal		841	84		889	78		1,013	75		1,093	76
World total		1,010	100		1,132	100		1,342	100		1,444	100

Source: The British Petroleum Company, *BP Statistical Review of World Energy.* London, 1985.

for the first time. In addition to the large natural gas flow from the Soviet Union to eastern and western Europe, the second largest movement is from the North Sea producing realm to western Europe. The flow from Canada to the United States is the third largest movement.

Liquefied natural gas (LNG) increased in importance in recent years but not at the rate originally foreseen at the outset of the energy crisis in the early 1970s. Higher prices dampened the demand for LNG. Its shipment is also expensive, due to the need for pressurization and special containers for transport. Natural gas is liquefied by cooling it to −258°F and by pressurizing it to a level greater than atmospheric pressure. This process greatly reduces the volume of the product, to 1/600 as much space as that of natural gas itself, per heating unit. Shipments of LNG go to Japan from the Middle East, southeast Asia, and Alaska. In the future Japan plans to import more LNG from Australia and Canada. Some LNG also flows from Algeria to the United States and from Libya to western Europe.

Deregulation of natural gas prices in the United States proceeded at a slower pace than for oil. The Natural Gas Policy Act of 1978 set an overall framework for decontrol so that gas prices would rise to crude oil price equivalents. Section 107 of the 1978 program immediately deregulated gas produced in wells below 15,000 feet, but not for shallower depths. For this reason producers' interest in deeper wells increased immediately. Bigger rigs in Oklahoma and elsewhere began exploiting such reserves as a result.

A source of uncertainty for natural gas in the future relates to the impact of upward price adjustments. Many observers indicate that production can stagnate as producers switch to other fuels due to a less price competitive situation, but the advantages accruing to a cleaner-burning fuel remain and will continue to enhance the demand for natural gas.

Figure 11–17   Gas and Oil Production Trends in the United States, 1965–1985. *Natural gas production peaked in the U.S. in 1973 despite growing demand because price controls reduced incentives on interstate shipments. By 1977 severe shortfalls occurred but deregulation and higher prices led to larger flows by the early 1980s. Production declined again during the national recession of the early 1980s and the energy recession that followed. Oil output remained much more even in this period. (Source:* World Oil, *February 1983 and 1986)*

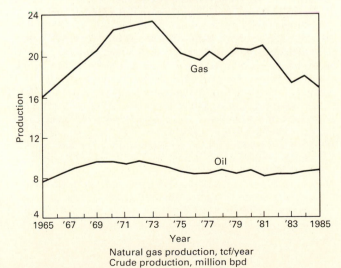

Natural gas production, tcf/year
Crude production, million bpd

SUMMARY AND CONCLUSION

The petroleum and natural gas industries serve as highly visible global activities producing a product in strong demand worldwide but with limited sources of supply. Considerable shifting has occurred among major production regions in the past century and the political factor has become more central in accounting for movement patterns of these products. Nations have also become more interdependent as reliance on imported product has grown.

Production of crude oil in the United States paced world output over the years, but domestic demand grew even faster, necessitating the purchase of imports to supplement domestic production for the past thirty years. During the peak years of this import activity in the late 1960s and early 1970s, the Middle East served as the major exporter of product to the United States. A producer cartel, OPEC, gradually gained a greater share of the control of the industry in that period, wresting control from U.S. and European-dominated multinationals. This greater involvement of OPEC culminated with an oil embargo in 1973, triggering a worldwide energy crisis and significantly higher fuel prices. Following a series of price hikes by OPEC over a period of years, stabilization in demand and price returned in the early 1980s and the influence of OPEC waned. During this period, conservation practices and shifts to coal and other fuels dampened the demand for the product. Many importing nations, including the United States, also shifted away from OPEC sources of supply. Mexico, for example, became the largest petroleum supplier for the United States.

Other major adjustments in world petroleum production also occurred in the 1970s and 1980s. Production zoomed upward in the Soviet Union, following the successful harnessing of the western Siberian field, and the country became a major world exporter, serving many energy-starved eastern and western European countries directly by pipeline, and other nations by tanker. Funds derived from petroleum and natural gas sales became a major source of foreign currency for the Soviet Union. A second major innovation occurred with the tapping of offshore petroleum resources on the Gulf coast of the United States, the North Sea, and the Middle East, among others in this period.

As with petroleum, the United States and the Soviet Union serve as major natural gas producers. Although an attractive fuel source, more limited transport options occur with natural gas than with petroleum. Generally, it is economical only to mine the product in areas near the market or with direct pipeline connections. It is an efficient source of energy, burns clean, and is therefore in strong demand when available.

No gas or oil moves anywhere in the world today without government sanction. Such is the critical nature of these products to national welfare. Government regulation takes many forms around the world. In some areas, such as Mexico, the Middle East, and the Soviet Union, oil is produced and distributed by state-owned organizational units. Elsewhere, private-sector firms provide this function. Even in the United States and Canada, however, government controls exert a major impact on the industry despite recent deregulation initiatives.

Uncertainty in the industry due to recent price fluctuations and a growing surplus, especially in the case of oil, created a depressed market for petroleum in the United States in the mid-1980s. Mergers among major producers added to the unsettled operating environment of oil firms as surviving firms maneuvered to control a greater share of reserves. Many domestic refineries closed in this period as more refining activity shifted to the third world. The Soviet Union also became a major exporter of petroleum and natural gas products in this period, especially via pipeline to western Europe. The Middle East remained a major supplier as well, but no longer played an arbitor's role in pricing or allocation as it did a decade earlier in the heyday of OPEC.

SUGGESTIONS FOR FURTHER READING

AHRARI, MOHAMMED E., *OPEC: The Failing Giant.* Lexington, Ky.: University of Kentucky Press, 1986.

AL-OTAIBA, MANA SAEED, *OPEC and the Petroleum Industry.* New York: Wiley, 1975.

CHAPMAN, KEITH, *North Sea Oil and Gas: A Geographical Perspective.* North Pomfret, Vt.: David & Charles, 1976.

COMMITTEE FOR ECONOMIC DEVELOPMENT AND THE CONSERVATION FOUNDATION, *Energy Prices and Public Policy.* New York: Committee for Economic Development, 1982.

DIENES, LESLIE, and THEODORE SHABAD, *The Soviet Energy System: Resource Use and Policies.* Silver Spring, Md.: V. H. Winston, 1982.

HALLWOOD, PAUL, and STUART SINCLAIR, *Oil, Debt and Development.* London: Allen & Unwin, 1981.

KALT, JOSEPH P., *The Economics and Politics of Oil Price Regulation: Federal Policy in the Post-embargo Era.* Cambridge, Mass.: MIT Press, 1981.

MOSLEY, LEONARD, *Power Play: Oil in the Middle East.* New York: Random House, 1973.

NORENG, OYSTEIN, *Oil Politics in the 1980s: Patterns of International Cooperation.* New York: McGraw-Hill, 1978.

PAINTER, DAVID S., *Oil and the American Century.* Baltimore: Johns Hopkins University Press, 1986.

PAUST, JORDAN J., and ALBERT BLAUSTEIN, *The Arab Oil Weapon.* Dobbs Ferry, N.Y.: Oceana, 1977.

PEEBLES, MALCOLM W. H., *Evolution of the Gas Industry.* New York: New York University Press, 1980.

RUSSELL, JEREMY, *Geopolitics of Natural Gas.* Cambridge, Mass.: Ballinger, 1983.

RUSTOW, DANKWART, *Oil and Turmoil: America Faces OPEC and the Middle East.* New York: W. W. Norton, 1982.

SAMPSON, ANTHONY, *The Seven Sisters: The Great Oil Companies and the World They Made.* New York: Viking Press, 1975.

TETREAULT, MARY ANN, *Revolution in the World Petroleum Market.* Westport, Conn.: Quorum Books, 1985.

TURNER, LOUIS, *Oil Companies in the International System.* London: Allen & Unwin, 1978.

TUSSING, ARLON R., and CONNIE C. BARLOW, *The Natural Gas Industry: Evolution, Structure, and Economics.* Cambridge, Mass.: Ballinger, 1984.

YERGIN, DANIEL, and MARTIN HILLENBRAND, eds., *Global Insecurity: A Strategy for Energy and Economic Renewal.* Boston: Houghton Mifflin, 1982.

12 Ferrous and Nonferrous Mineral Extraction and Processing

Unlike the energy-related minerals of energy, discussed in Chapter 11, the three minerals presented here find use only as raw materials for further processing into manufactured products. This trio of minerals, iron ore and two nonferrous ores—bauxite and copper—comprise the mainstay of metals production in developed countries and the basis for their heavy-manufacturing industry.

As a group, these products illustrate classical locational tendencies in mineral ore recovery. Since high-quality, easily recovered sources of these ores are relatively scarce worldwide, countries with plentiful, accessible reserves have found ready markets for the product. As price increases, technology advances, and depletion of reserves has occurred, extraction has increased in less accessible locations. The biggest recent locational adjustment facing the industry has involved expansion in countries with smaller domestic demand for ore, such as Jamaica, Brazil, Guinea, Zambia, and Zaire, due to the presence of high-quality reserves that could be marketed abroad. Since many of these production centers occur in the southern hemisphere and markets lie in the north, this movement is frequently referred to as a north-south trade relationship.

Unlike the fuel products discussed in Chapter 11, most minerals experience several stages of processing, refining, concentration, and smelting, which adds more complexity to their production pattern. In many cases the initial processing occurs near the raw material source, with successive stages of production positioned closer to the market or near other resources required in the production process, such as close proximity to cheap electrical power. Often, successive stages occur in different countries, as is the case with the aluminum industry. Environmental considerations have also led to locational shifts as the greater awareness of the environmental pollution potential of these processes became an issue and abatement measures took affect, often leading to the closure of older, less efficient plants.

Copper ore and bauxite (the raw material for aluminum), typically experience concentration near the mine, followed by refining and smelting elsewhere. Iron ore traditionally moved more directly from the mine to a manufacturing complex, usually a steel production center. The lower-quality iron ore produced in many areas today, however, does require further processing near the mine before entering commerce.

Increasingly fewer consuming nations possess adequate domestic mineral resources to satisfy their industrial needs. International trade in minerals overcomes these deficits and provides for closer ties among supplier and user nations. The situation differs slightly in each developed nation, but none is totally self-sufficient today. Europe, for example, is essentially devoid of all three mineral resources discussed here, with one exception—iron ore—as both Sweden and France are significant producers. Japan faces a similar weak raw material base. The United States possesses excellent copper reserves but lacks high-quality iron ore and bauxite. The Soviet Union remains relatively self-sufficient as a producer of these minerals, except for bauxite shortages.

In this chapter we explore locational and behavioral characteristics of these three mineral resource industries. The discussion includes an analysis of recent changes in the procurement policies of Japan and the United States, the biggest users. The demise of multinational firms as unilateral actors in the mineral industry and the rise of joint venture programs with host nations are discussed. The growing role of cartels and third-world governments in the production process also merits attention. Following a review of each product separately, in the final section of the chapter we provide an integrating overview of the industry today from the perspective of growing global interdependence.

IRON ORE MINING

The geography of iron ore production traditionally provided a striking resemblance to coal mining patterns. Indeed, in many of the world's most lucrative producing areas, both coal and iron deposits occur in close proximity. Significant modifications to the list of leading iron ore–producing and iron ore–trading countries in recent years has nevertheless changed these relationships. The relative share of world production attributable to western European countries and the United States, for example, has fallen dramatically.

As with coal, there are several grades of iron ore, based on purity and mineral content. *Hematite,* the highest-grade ore, is reddish in color. The interme-

diate grade, *magnetite,* is characteristically black. *Taconite,* the lowest-quality ore, requires crushing, magnetic separation, and heating to convert the ore to pellets that are competitively marketable. In the United States, where the best quality ores have been depleted, less desirable taconite-grade ores now provide the primary source (Figure 12-1).

Worldwide, iron ore production declined slightly in the early 1980s following years of modest growth in the 1970s. Demand for steel, the major market for iron ore, fell as a recession stymied business activity and structural changes altered the economies of traditional producers. Owing to changes in market demand precipitated by less competitive prices for U.S.– and western European–produced steel, manufacturing has gradually shifted to newly industrializing countries (NICs), located primarily in east and southeast Asia, including South Korea, and Malaysia.

Production of iron ore in the Soviet Union, which has led world production for 25 years, remains very strong (Table 12-1). As recently as 1972, the United States ranked second in production, while Australia and Brazil moved up to third and fourth positions, respectively. By 1981, output in Brazil and Australia exceeded that of the United States, and by 1985 China also exceeded the United States in production (see Table 12-1). We explore production trends in major producing areas in succeeding sections.

Soviet Production

Open-pit iron mines predominate in the Soviet Union and ore quality has declined in recent years, especially that produced in the Ukraine (Table 12-2). Pro-

Figure 12-1 Taconite Iron Ore Pellets. *Taconite pellets have been produced in the United States since the mid-1950s. Pelletization makes it feasible to use lower-grade ores which have increasingly accounted for a greater share of the mined product in the Mesabi Range. (Courtesy American Iron and Steel Institute)*

Table 12-1

World Iron Ore Production, (millions of gross tons)

Rank	Nation	Production	Percent of World Total
1	Soviet Union	242	30
2	Brazil	95	12
3	Australia	90	11
4	China	75	9
5	United States	48	6
	Subtotal	550	68
	World total	799	100

Source: Iron Ore, 1985 (Cleveland, Ohio: American Iron Ore Association, 1986).

Table 12-2

Soviet Iron Ore Production, 1982 (millions of metric tons of usable ore)

Rank	Region	Output	Percent of Soviet Total
1	Krivoy Rog	105	43
2	KMA*	40	16
3	Kazakhstan	25	10
4	Urals	24	10
	Subtotal	215	88
	Soviet total	244	100

Source: Shabad, Theodore, *Soviet Geography:* Review and Translation. V. H. Winston and Sons, April 1983, p. 328.
*Kursk Magnetic Anomaly.

duction costs also rose as deeper pits provided the ore. The Krivoy Rog area on the Dnepr River in the Ukraine still claims nearly half the Soviet iron ore output, even though it is declining in relative importance (Figure 12-2). The second major producing area, the Kursk magnetic anomaly (KMA), lies to the north, roughly halfway between Krivoy Rog and Moscow. It has gained in importance and now claims just under 20 percent of total Soviet iron output.

The Urals producing region, stretching from Serov and Nizhiniy Tagil south to Magnitogorsk, traditionally the second most important Soviet iron ore area, is now in fourth place. Recent declines have meant that ores must now be imported to support local industry. The third major iron ore–producing region in the Soviet Union today lies in Kazakhstan. At Rudnyy in northwest Kazakhstan lies the Sokolovka-Sarbay deposit. About half the regional production

Figure 12-2 Major Iron Ore-Producing Areas in the Soviet Union. *The Krivoy Rog and KMA producing areas dominate iron ore production. The Urals and Kazakhstan areas are secondary production centers.*

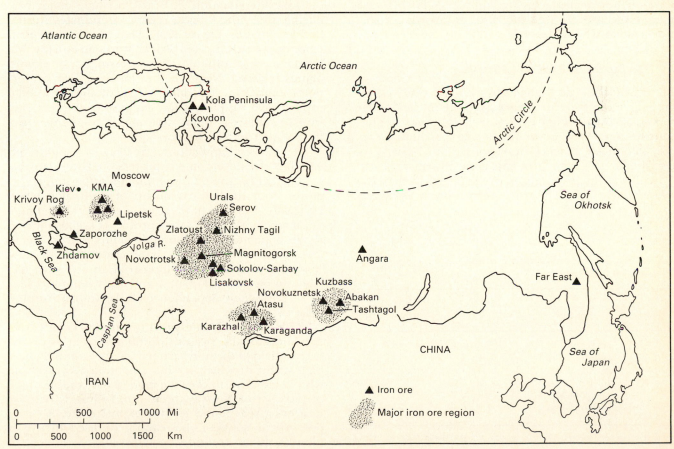

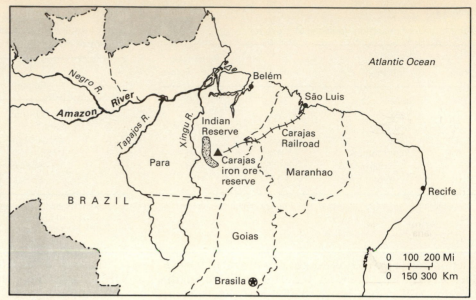

Figure 12-3 Carajas Iron Ore Production Region, Brazil. *The world's largest deposit of high grade iron ore reserves is in this area. Largely due to the output of this region, Brazil now leads the world in iron ore exports.*

occurs there. Other producing sites occur at Karazhal in central Kazakhstan and at Lisakovsk. These shifts in Soviet iron ore production to sites more distant from traditional manufacturing centers have placed additional burdens on the rail transportation system to ship ore to the traditional manufacturing locations as well as providing a motivation for authorities to develop new manufacturing complexes farther east, nearer the sources (see the discussion of the Soviet steel industry in Chapter 15).

The Exporters: Brazil and Australia

The world's largest high-grade iron ore reserves, averaging about two-thirds pure iron, lie in the Carajas region of Brazil (Figure 12-3). Several other nonferrous minerals also occur in the area, including copper, bauxite, and nickel. In response to increased production in this rich iron ore area, Brazil displaced Australia as the leading iron ore exporter in the early 1980s (Table 12-3). The iron reserves occur along two

Table 12-3

Leading Iron Ore–Exporting Nations, 1983
(millions of U.S. dollars)

Rank	Nation	Value	Percent of World Total
1	Brazil	1,744	27
2	Australia	1,419	22
3	Canada	789	12
4	India	474	7
	Subtotal	4,426	68
	World total	6,455	100

Source: United Nations, *1983 Yearbook of International Trade Statistics,* 1985.

sides of the Serra Norte and Serra Sul. Production near the Amazon is emphasized due to cost savings using water transportation. A huge Carajas government-sponsored agricultural and mineral development project which began in the early 1980s will expand export earnings to $15 billion per year by 1990 from this resource complex.

Like the situation in Brazil, a single major producing area—the Pilbara region—dominates Australian production (Figure 12-4). The ore moves to ports from mines at Mt. Tom Price, Mt. Goldsworthy, Mt. Newman, and others for processing and export, mainly to Japan. Nearly half of Japanese iron ore imports come from Australia. Large volumes of foreign capital have supported the expansion of Australian iron ore mining, where five major companies produce 90 percent of the ore.

The Australian economy is heavily resource oriented. Many development professionals feel, for example, that continued development in the country requires even further expansion of mineral exports, especially iron ore, nonferrous metals, and uranium, but many market and labor problems are also linked to this dependence on mineral resources. Iron ore, in fact, ranks second to coal as a mineral export. Because Australia has a history of industrial unrest, especially labor union strikes, Japanese and other importers seek to decrease their dependence on Australian iron ore reserves.

Other major iron-exporting nations include Sweden and Canada. Swedish production centers on the Kiruna field in the northern part of the country. These high-quality ores have a solid reputation. Exports from Kiruna move north over the mountainous spine of the country across Norway to the port of Narvik because it offers a year-round ice-free export route. A closer port to the south, Lulia, cannot be used in winter months, due to freezing.

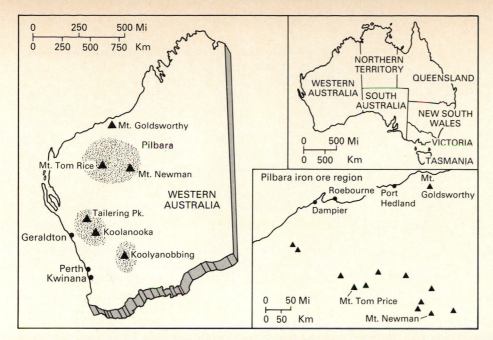

Figure 12-4 Pilbara Iron Ore Production Region, Australia. *The Pilbara region dominates Australian iron ore production. Ore moves to ports, largely for export, from the mines in the interior. Nearly one-half of Japanese iron imports come from Australia.*

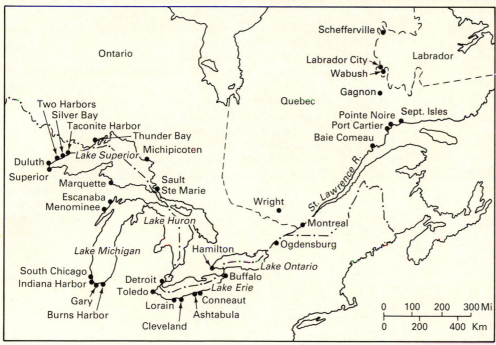

Figure 12-5 Great Lakes/St. Lawrence Seaway Iron Ore Production Areas and Ports in the United States and Canada. *Considerable iron ore movement occurs between the U.S. and Canada due to the proximity of both production and market locations. Both countries reciprocate as the leading trading partner of the other in terms of both imports and exports.*

The Reciprocators: Canada and the United States

Considerable iron ore movement occurs between the United States and Canada, primarily along the Great Lakes and St. Lawrence River waterways (Figure 12-5) due to the proximity of both production and market locations. Both countries reciprocate as the leading trading partner of the other in terms of both imports and exports. This reciprocal relationship occurs because of the economics of international movement along the Great Lakes and St. Lawrence Seaway waterways and location of production such that it is often easier and cheaper to trade ores than to use domestic product in specific manufacturing areas.

The Mesabi range leads output in the United States, most of which occurs in northern Minnesota, accounting for 70 percent of U.S. production in 1980. Taconite represented over 95 percent of this yield. The leading Great Lakes ports, in terms of iron ore tonnage shipments, include Duluth, Minnesota; Superior, Wisconsin; and Two Harbors, Minnesota (Figure 12-5). Since 1950, a downward trend has prevailed in U.S. iron ore production (Figure 12-6 on page 178). Significantly, imports have also shifted downward. Greater use of scrap and a decline in the U.S. steel industry itself account for this pattern.

The focus of Canadian iron ore production shifted away from the traditional Lake Superior producing area of Steep Rock, Ontario, eastward to the Quebec/Labrador boundary region in recent years.

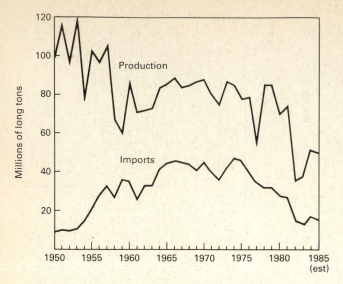

Figure 12-6 Iron Ore Production and Import Trends in the United States. *Since 1950, a downward trend has prevailed in U. S. iron ore production, a trend particularly pronounced in the late 1950s due to a strike, and again in the early 1980s. Imports increased in the first half of this period, but have declined since 1975. Greater use of scrap and a decline in the U.S. steel industry explains this recent trend. (Source: Metal Statistics, various years)*

The leading mines in that area, which now accounts for half the country's output, include Wright, Quebec; Carol Lake, Labrador; Sept-Isles, Quebec; and Schefferville, Quebec (see Figure 12-5).

The Importers: Western Europe and the Far East

With the exception of the production mentioned earlier in Sweden and France, western Europe has few high-quality iron ore reserves. It is not surprising, therefore, that western Europe includes four of the six leading iron-importing nations (Table 12-4). West Germany, France, Italy, and the United Kingdom consume the largest share of this imported ore, which accounted for nearly one-third of world imports in 1980.

Table 12-4

Leading Iron Ore–Importing Nations, 1983
(millions of U.S. dollars)

Rank	Nation	Value	Percent of World Total
1	Japan	3,145	45
2	West Germany	967	14
3	United States	531	8
4	Great Britain	355	5
5	Italy	349	5
6	France	312	5
	Subtotal	5,659	82
	World market economy total	6,925	100

Source: United Nations, *1983 Yearbook of International Trade,* 1985.

The dispersed flow of iron ore to several western European countries does not match the intensity of movement to Japan, which alone purchased 45 percent of the imports among market economies (excluding centrally planned countries) in 1983. Another 2 percent of the flow moved to South Korea. As mentioned earlier, Brazil and Australia provide the largest portion of this supply. The United States now ranks as the third largest iron ore importer, with an 8 percent share.

ALUMINUM INDUSTRY

Although aluminum is an abundant mineral in the earth's crust, heavy concentrations occur infrequently. Bauxite, the raw material for aluminum, occurs most frequently in tropical areas having clay-limestone rocks exposed to weathering. Bauxite can be converted into aluminum by first producing an intermediate product, alumina. To produce alumina, bauxite is crushed, washed, pumped into pressure tanks, heated, and subjected to a precipitation process using caustic soda to remove impurities, such as iron and silica. It is then cooled and dried in large furnaces to drive off moisture and reduce weight. A weight loss of 50 percent occurs in this process. For this reason, concentrating often occurs near the raw material source, but exceptions do occur.

An electrolysis smelting process, typically at a separate location, next converts the white powder alumina substance to aluminum. Alumina is dissolved in a cryolite bath and direct electrical current is passed through it by means of carbon electrodes to complete the operation. Because of the large electrical power needs for this conversion, aluminum producers seek cheap electrical power locations for this third phase in aluminum manufacturing. Aluminum smelting facilities are therefore oriented to the availability of cheap electricity rather than to raw materials or markets.

Several countries with plentiful and cheap elec-

Table 12-5

World Aluminum Production, 1985
(thousands of short tons)

Rank	Nation	Production	Percent of World Total
1	United States	3,857	23
2	Soviet Union	2,535	15
3	Canada	1,414	8
4	Australia	939	6
5	West Germany	820	5
6	Norway	785	5
7	Brazil	600	4
	Subtotal	10,950	66
	World total	16,928	100

Source: Non-ferrous Metal Data, 1985 (New York: American Bureau of Metal Statistics, Inc., 1986).

Table 12-6

World Aluminum Consumption, 1984
(thousands of short tons)

Rank	Nation	Consumption	Percent of World Total
1	United States	5,000	28
2	Japan	2,002	11
3	Soviet Union	1,984	11
4	West Germany	1,277	7
5	China	695	4
6	France	620	4
	Subtotal	11,578	65
	World total	17,637	100

Source: Non-ferrous Metal Data, 1984 (New York: American Bureau of Metal Statistics, Inc., 1985).

Table 12-8

World Alumina Production, 1984
(thousands of short tons)

Rank	Nation	Production	Percent of World Total
1	Australia	9,688	25
2	United States	5,200	14
3	Soviet Union	4,630	12
4	West Germany	1,875	5
5	Jamaica	1,667	4
6	Japan	1,640	4
	Subtotal	24,700	64
	World total	38,193	100

Source: Non-ferrous Metal Data, 1985, (New York: American Bureau of Metal Statistics, Inc., 1986).

trical power have significant aluminum-processing activity even though they possess neither the raw material supplies nor final markets (Table 12-5). Canada and Norway fit this circumstance. Other leading aluminum producers in the world, including the United States, Japan, and West Germany, possess large markets and significant production but little raw material. Historically, some bauxite production occurred in the United States in Arkansas, but the raw material was not plentiful nor of high quality. The Soviet Union stands alone as a major bauxite producer, aluminum manufacturer, and major consumer. Soviet bauxite reserves, however, suffer from poor quality.

Aluminum consumption (Table 12-6) domination by the United States, Japan and the Soviet Union overwhelms the use in other areas of the world, but western Europe also provides a significant market. In addition to the aircraft assembly industry, a wide variety of fabricators use aluminum. U.S. factories now use more aluminum than any other metal except iron.

Bauxite Producers

Australia and Guinea lead the world in bauxite production, while Brazil, Jamaica and the Soviet Union also produce significant quantities (Table 12-7). Most

Table 12-7

World Bauxite Production, 1985 (millions of short tons)

Rank	Nation	Production	Percent of World Total
1	Australia	36	36
2	Guinea	16	16
3	Brazil	7	7
4	Jamaica	7	7
5	Soviet Union	7	7
	Subtotal	73	73
	World total	99	100

Source: Non-ferrous Metal Data, 1985 (New York: American Bureau of Metal Statistics, Inc., 1986).

of the ore extraction occurs in countries that are not major aluminum manufacturers, hence most of the product enters the export trade.

Even though considerable weight loss would occur in the initial conversion to alumina, fully 40 percent of bauxite production at the world scale is now exported prior to processing. An exception occurs in the case of Australia, the world leader in both alumina production and bauxite mining (Table 12-8). Much of the bauxite export flows to the United States, which leads market economy nations in import share, accounting for over one-third of the world movement (Table 12-9). Bauxite produced in the Caribbean, for example, moves to Gulf of Mexico ports in the southern United States for local manufacture into alumina.

Table 12-9

Bauxite Trade, 1983 (millions of U.S. dollars)

Rank	Nation	Total	Percent of World Total
	Imports		
1	United States	324	36
2	West Germany	121	14
3	Japan	97	11
4	Canada	75	8
5	Spain	71	8
6	Italy	56	6
7	France	51	6
	Subtotal	795	89
	World total	899	100
	Exports		
1	Guinea	326	40
2	Jamaica	120	15
3	Brazil	115	14
4	Guyana	63	8
	Subtotal	624	77
	World total	814	100

Source: United Nations, 1983 Yearbook of International Trade Statistics, 1983.

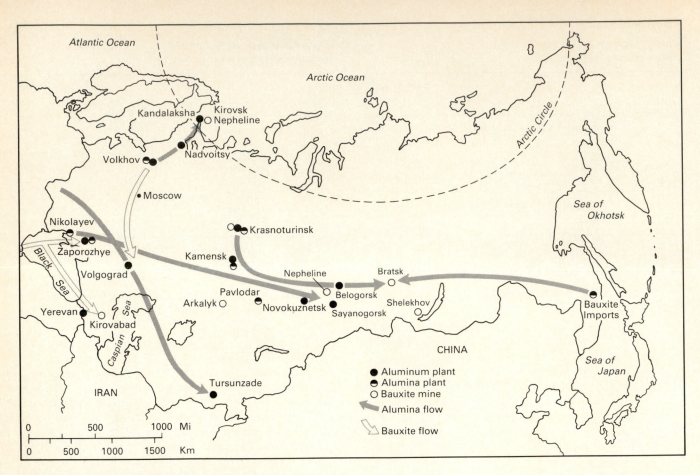

Figure 12-7 Aluminum Manufacturing Locations in the Soviet Union. *Aluminum production occurred exclusively in the Ukraine (Zaporozhye) prior to World War II when facilities were built in the Urals and Novokuznetsk. More recently expansion has occurred in Siberia in the Lake Baykal region which now claims the largest capacity in the country. (Source: Soviet Geography, various issues)*

Today, the Soviets must import half their domestic bauxite need. Bauxite imports accounted for 12 percent of the Soviet use in 1965, 38 percent in 1970, and 45 percent in 1975. Guinea serves as a major bauxite supplier to the Soviet Union, together with Yugoslavia and Greece. Alumina imports come from Hungary as well as from Caribbean sources. In 1980, a new alumina plant was completed on the Black Sea at Nikolayev to process imported bauxite (Figure 12-7).

Multinational Corporations

Together with third-world domination of bauxite production one finds multinational corporate control. Six integrated corporations, all based in developed countries, accounted for over 60 percent of the bauxite output of market economies in 1970. These six firms include the Aluminum Company of America (Alcoa), Pechiney Ugine Kuhlman from France, Swiss Aluminum, the Aluminum Company of Canada (Alcan), and Reynolds Metals Co. (United States). Historically, these corporations became involved in bauxite production due to the large capital and sophisticated technology requirements of the industry, resources not locally available in developing areas. By integrating vertically from mine through refining to fabrication and recycling, these firms also limited their risk exposure.

Using Alcan operations for 1976 as an example, one can observe a typical spatial structure common to the industry. Location shares of Alcan operations appear in parentheses. Bauxite is typically obtained in a developing country (91 percent), alumina processing occurs in both developing (38 percent) and developed countries (62 percent), and aluminum smelting exists almost exclusively in a developed country (90 percent). Alcan operated overseas refineries in Jamaica, Brazil, India, Japan, Australia, Ireland, and Spain in 1979.

Today, cost factors tend to favor developing areas for the manufacturing process, especially if cheap electrical power can be obtained. The market for aluminum is also growing rapidly in these areas. Transnational corporate control continues, although several firms now possess minority partners from the host country. The potential for developing grass roots–based integrated operations exists in only a few countries, such as Brazil and India, but partial integration does occur frequently. Guyana, for example, produces both bauxite and alumina on its own, and

South Korea has developed expertise in the aluminum industry, including plant engineering and construction capabilities.

In order to establish a countervailing power base from which to negotiate more effectively with the multinationals, bauxite producing nations created the International Bauxite Association (IBA) in 1974. Initially 10, and now 11 countries, representing about 75 percent of world production, formed this cartel. Its major weakness revolves around an inability to establish firm price policies applicable to all members. Another stumbling block is that Brazil refuses to join at the same time that it strives to become the world's largest exporter.[1]

U.S. Location Patterns

The alumina factories clustered in the south central part of the United States include six mills strung along the Gulf coast from Texas to Alabama, and two in Ar-

[1]Fillmore C. F. Earney, "The Geopolitics of Minerals," *Focus,* American Geographical Society, 31, No. 5 (May–June 1981), 9.

kansas. These factories consume three types of raw materials: bauxite, fuel, and caustic soda. The weight-loss ratio is high enough—on the average it takes 100 tons of bauxite to make 40 tons of alumina—to discourage long hauls of bauxite by more expensive forms of transportation, such as trucks and trains, while the long shipments of bauxite by water remain economical. The Gulf coast mills feed on bauxite arriving by water, the main source regions being Jamaica, the Dominican Republic, and Surinam. After processing, alumina can be shipped by rail.

From the standpoint of historical development, the East St. Louis alumina mill, constructed in 1902, pioneered production in the United States. In 1938 a second U.S. alumina plant at Mobile, Alabama, began operations. Then, during World War II, three more facilities emerged (Hurricane Creek, Arkansas; Baton Rouge, Louisiana; and Listerhill, Alabama). After the war, six more U.S. plants were built, one in Arkansas, two in Texas, two in Louisiana, and one in St. Croix, Virgin Islands. Since its inception, therefore, alumina manufacturing in the United States has been clustered in the south central portion of the country.

Figure 12-8 Aluminum Manufacturing Patterns in the United States and Canada. *Several clusters of production facilities occur in the United States and Canada, all explained by the availability of cheap electrical power. (Source: Non-ferrous Metal Data, 1985, Secaucus, N.J. 1986)*

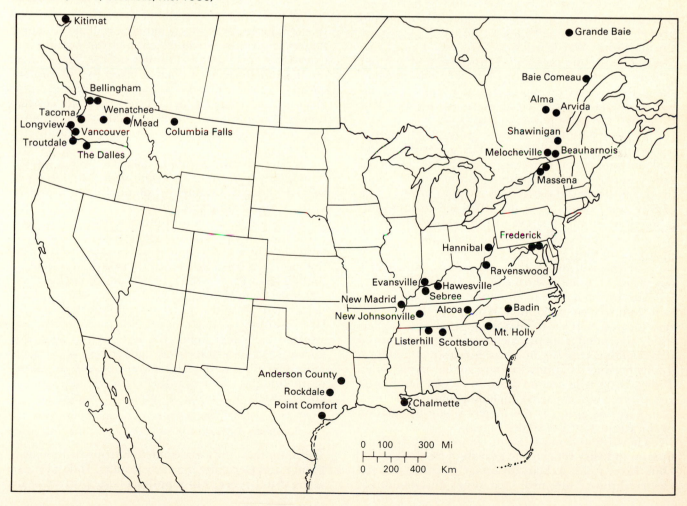

Table 12-10

U.S. Aluminum Production Capacity by Region, 1984
(thousands of short tons)

Rank	Area	Capacity	Percent of U.S. Total
1	TVA	1576	27
2	Pacific northwest	1453	25
3	Gulf coast	779	14
4	Ohio valley	935	16
5	St. Lawrence Seaway	352	6
	Subtotal	5095	88
	U.S. total	5754	100

Source: Non-ferrous Metal Data, 1984 (New York: American Bureau of Metal Statistics, Inc., 1985).

A totally different but rational pattern of historical-spatial development occurred with the U.S. aluminum industry, which has experienced five major developmental phases in various parts of the country (Figure 12-8). Each producing area can be totally or partially explained by access to cheap electricity. A listing of current capacities, by decreasing rank order, in these five areas occurs in Table 12-10. The first aluminum-processing plants in the country occurred in New York State at Niagara Falls (1895) and Massena (1903), where cheap hydropower could be harnessed. A second phase in the industry began in what was to become the Tennessee Valley Authority (TVA) region. During World War I, prior to TVA, two plants came to the region, one in Alcoa, Tennessee, and another in Badin, North Carolina, again to take advantage of cheap hydroelectrical power. Later, a plant at Listerhill, Alabama, constructed at the start of World War II, tapped the same power source.

A wave of four plants built in the Pacific northwest during World War II by the U.S. government also took advantage of relatively inexpensive electrical power in that region. This area has remained a major producer, leading the United States in output. Large energy price increases in the 1970s led to some curtailment of production, but in 1984 about one-fourth of the U.S. output still came from plants in the region, paced by those in Washington, the leading producing state in the country, with seven plants, which produced over 1 million short tons in 1984.

The fourth phase of development in the aluminum industry emerged after World War II with new sites located in Texas to take advantage of cheap electricity generated from petroleum and natural gas-fired power plants. One site in Texas, however, depends on locally produced lignite as a source of electrical power. Today, Texas is in second place in U.S. aluminum ingot production capacity. Included among the four plants in this group is the largest single installation in the country, at Rockdale (Alcoa).

The fifth and most recent trend in the U.S. aluminum industry reflects the growth of smelters in the Ohio River valley to take advantage of cheap thermal electrical power sites near coal mines. Installations at Hannibal, Ohio; Ravenswood, West Virginia; and Evansville, Indiana, illustrate the point. These sites have the further advantages of being near the northeastern market and accessible to Ohio River barge transportation.

Canadian Production

The Canada situation provides a dramatic illustration of a region that has become a dominant world producer even though completely lacking native ore and having scarcely any domestic market (exports largely flow to the United States). Although lacking in bauxite, Canada maintains aluminum mills at the extreme eastern and western parts of the country, near cheap electrical power sites (Figure 12-9). In 1901, the Canadians built a concentrator 60 miles southwest of Quebec City a few miles upstream from its confluence with the St. Lawrence River. Alumina for the plant came from Germany. This mill still operates, but limited water power restricts expansion. In 1926, Aluminum Limited of Canada (Alcan), which operated the Shawinigan Falls concentrator, built a new plant at Arvida in a forested wilderness 120 miles northeast of Quebec on the Saguenay River. The attraction of that remote spot focused on the tremendous hydroelectric potential provided by the 300-foot water drop between Lake St. John and the St. Lawrence River.

At first, imported alumina moved to the Arvida smelter, but in 1928 a local alumina mill also opened. Bauxite arrives by ocean freighters from Jamaica and Guyana at Port Alfred, 40 miles up the Saguenay at the head of navigation, then is taken by rail another 20 miles to the mills. Personnel to operate the port facilities, the railroad, the refinery, and the concentrators all came from elsewhere, and a new community, Arvida, displaced the forest surrounding the mills.

Western British Columbia also claims a large aluminum mill at Kitimat, a development lured to a wet, rugged wilderness, again by potential water power resources. Harnessing the potential of the site required reversing the direction of flow of a stream, causing it to fall down the precipitous western side of the mountains. The hydraulics system developed to overcome this problem involved the construction of a dam on the Nechako River (which flows east) to create a lake 120 miles long (Figure 12-9). A 25-foot-diameter tunnel drilled 10 miles into the mountain drained the lake westward to Kemano, where a powerhouse, constructed 2585 feet below the level of the lake, produced electricity. Electricity generated at Kemano moved 50 miles to the Kitimat mill site (on the shores of the Douglas Channel), which is accessible to ocean-going vessels delivering alumina and bauxite.

European Activity

The aluminum industry came to Europe in 1886, with France and Germany the early leaders. Today, France, along with Greece, still supplies a large portion of the

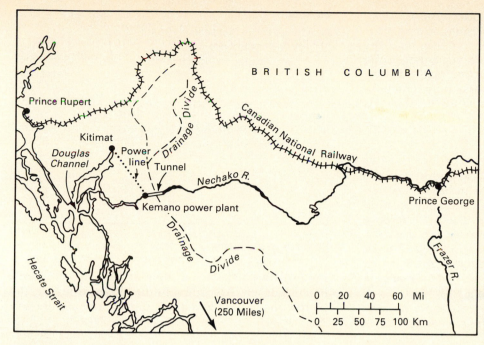

Figure 12-9 **Kitimat, British Columbia Aluminum Production Complex.** *Harnessing the potential of this remote site for aluminum refining required a complex change in the water hydraulics of the area involving reversing the flow of a stream, constructing a dam, and tunneling through a mountain. The mill site is accessible to ocean-going vessels that deliver bauxite and alumina.*

bauxite for western European mills. In the production of alumina and aluminum, Norway, France and West Germany pace the continent. West Germany serves as the largest link in a chain of aluminum manufacturing nations that encircle the Alps. The largest single producer, however, is Norway, which, like Canada, lacks both ore and markets but is blessed with power resources. All told, western Europe accounts for about 12 percent of the world's bauxite output and 20 percent of the refined aluminum.

Locations in the Soviet Union

Before World War II, the Soviet Union's largest aluminum factory was at Zaporozhye, near a large dam in the Dnepr River at Dnepropetrovsk (Figure 12-7). The mill depended almost entirely on Hungarian bauxite. A smaller mill at Volkhov used low-grade ores that came from east of Leningrad. The German army destroyed both of these factories in World War II. But just before the war started, the Soviets built a third plant at Kamensk, on the eastern flanks of the southern Ural Mountains, to take advantage of bauxite in the Urals and local lignite for thermal electricity. Later in the war period the Soviets constructed a fourth mill far to the east, in Stalinsk (Novokuznetsk), with machinery evacuated earlier from Volkhov, which consumed bauxite from the Kuzbass field. After the war, the Russians rebuilt the Volkhov and Zaporozhye mills, using machinery expropriated from Germany, and a large new mill built at Krasnoturinsk tapped the bauxite reserves of the northern Urals.

To take advantage of the hydroelectric capability of dams on the Yenisey and Angora rivers, the Soviets have built new aluminum smelters at Bratsk and Krasnoyarsk. Another plant located near Irkutsk at the southern end of Lake Baykal at Shelekhov, and a

fourth under construction at Sayanogorsk, dramatically increased the capacity in Siberia. These plants will "raise Soviet production to 3 million tons by 1990, and Siberia would account for 60 percent of the country's production of aluminum."[2]

Japanese Production

Japanese aluminum refineries are widely dispersed throughout the country in smaller coastal cities, from Hokkaido Island in the north to Shikoku in the south. Many of these facilities occur near locally produced coal, which provides inexpensive thermal electrical power. In 1980 the largest facility was at Toyama on the Pacific coast of central Honshu. But, as discussed later, aluminum production is a "sunset" industry in Japan today, suffering from higher energy costs, and significant production cutbacks are occurring.

COPPER INDUSTRY

Copper use dates to prehistoric times, but not until the Middle Ages did widespread applications increase, such as those involving brass cannon production. Britain became the center of the copper industry in the eighteenth century and gained a monopoly over production which lasted through the mid-nineteenth century. In the twentieth century copper came into its own with the age of electricity. As demand increased, so did technological production innovations that permitted the profitable use of lower-quality ores. These technological advances also encouraged larger-scale mining operations.

[2]Theodore Shabad, "The Soviet Aluminum Industry: Recent Developments," *Soviet Geography: Review and Translation*, 24 (1983), 96.

Several mines opened in the southwestern United States at Ely, Nevada; Miami, Arizona; and at other locations, placing the U.S. output at over half the world total by 1910. American firms began large-scale mining in South America at that time while British and Belgian firms established production centers in Africa. In the 1930s, Canada became the third largest producer, trailing the United States and Chile.

Copper Ore Production

Copper mining and refining activity exhibits patterns similar to those of the aluminum industry. The terminology and processes differ, but the same location principles apply. The biggest terminology difference is that the terms *concentrate* and *refine* are used dissimilarly in the two industries. Alumina factories first refine bauxite and then aluminum factories smelt alumina. In the copper industry the order is concentrating, smelting, and then refining.

As with most extractive industry, copper process-

Table 12-11

World Copper Ore Production, 1984
(thousands of short tons)

Rank	Nation	Production	Percent of World Total
1	Chile	1495	16
2	United States	1204	13
3	Soviet Union	1140	12
4	Canada	805	9
5	Zambia	573	6
6	Zaire	560	6
	Subtotal	5777	62
	World total	9218	100

Source: Non-ferrous Metal Data, 1985 (New York: American Bureau of Metal Statistics, Inc., 1986).

ing provides a classic example of a raw-materials-oriented industry. According to this classic thinking, the higher the weight-loss ratio during the production

Figure 12-10 Copper Mining and Concentrating in the United States. *Copper mining in the western U.S. extends from the Canadian border, southward to Mexico. Arizona is the principal state. Concentrating also occurs near the mine due to a high weight-loss in this early stage of processing.*

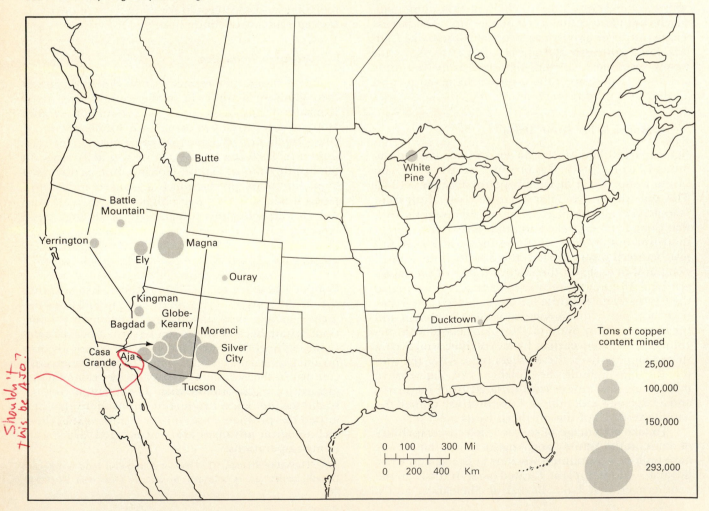

Figure 12-11 Morenci Open Pit Copper Mine in Arizona. *Arizona is the principal copper mining state in the United States. This open-pit mine, one of the largest copper mines in the U. S., is located in a remote area. Ore is hauled by rail. Note the benches or terraces at 50-foot intervals. (Courtesy Phelps Dodge Corporation)*

process, the stronger the tendency for an activity to be located near the raw material. Early stages of the copper manufacturing process, especially concentrating mills, correlate closely with mine locations because of strong weight-loss tendencies. Over 97.5 tons of every 100 tons of ore processed are discarded as waste; only 2.5 tons remain for further processing.

Mining Operations

The world pattern of copper mining activity has not changed drastically in recent decades, with Chile, the United States, and the Soviet Union the major producers (Table 12-11). Output in the United States dropped significantly in the 1980s, however, due to labor problems, strikes, and mine closings. The mining area in the western United States extends from the Canadian border southward to Mexico. Arizona is the principal state, with important mines near Tucson, Morenci, and Globe-Miami (Figures 12-10 and 12-11). The largest single copper mine in North America, at Bingham, Utah, has operated continuously since 1865. The Bingham ores lie near the surface, permitting inexpensive open-pit mining and the use of lower-quality ores.

The most important copper district in the northern United States traditionally occurred at Butte, Montana. The recent Butte story, however, reflects the problems facing the industry in the United States today. The Butte mine was shut down in 1983 when Anaconda could no longer operate it profitably. In December 1985, a local group, the Washington Corporation, bought the mine from Atlantic Richfield, and the mine was reopened in 1986. By avoiding the high overhead costs of a major corporation, employ-

ing a nonunion work force, and taking advantage of cheaper energy costs than existed several years earlier, the new firm aimed to regain a competitive advantage. By also obtaining lower municipal taxes and favorable electrical utility rates, the new management was able to keep costs below industry averages. Finally, the new firm struck a deal to sell ore to Japanese smelters, ensuring a market for the product.

Third-world copper production, paced by the output of Chile, Zambia, and Zaire, remains strong. The interior of southern Africa, especially a zone straddling the Zambia–Zaire border, is a major copper producer. Zaire's Katanga district is also important. In fact, Africa yields about 20 percent of the world's mine production of copper, with Zambia contributing 7 percent and Zaire 6 percent of the total. The South American copper belt extends from northern Peru to central Chile. In most years, Chile mines more copper than any other single nation. The Chuquicamata district in Chile may also contain more ore reserves than any other deposit in the world.

The Urals Mountain region has traditionally provided the bulk of Soviet copper ore. Operations there date back to the seventeenth century. As with iron ore, these resources have now been largely depleted, but significant smelting and refining activity remains. Imports of copper ore to the Urals come mainly from new fields in Kazakhstan. But Kazakhstan has also developed a major copper production complex of its own. The single leading copper center in the country is in Dzhezkazgan. Outlying copper-producing centers occur to the south in Georgia and Armenia between the Black and Caspian Seas; at Uzbekistan in central Asia southwest of Lake Balkash; and in Norilsk to the far north inside the Arctic Circle near the Yenisey River.

Smelting

The first step in transforming ore to metallic copper is concentrating. At concentrating mills the rocky mass goes through a froth-flotation process: the material is crushed, soaked in water, and mixed with oils. During agitation, the copper-bearing materials float to the top and are separated. Much of this activity occurs near the mine. Concentrated ore next goes to a smelter, which is not necessarily aligned with a mining facility. Japanese smelting activity provides a case in point. Whereas Japan ranks fourth in smelting worldwide, the country is no longer on the leading mining country list (Table 12-12). After domestic ores became depleted, Japanese policy emphasized importing raw materials at the least processed stage and refining the product domestically rather than importing refined metals. The movement of product from Butte, Montana, to Japan for smelting provides an example of this situation.

The purpose of smelting is to separate copper from other mineral elements, especially sulfur and oxygen. Chemically, the procedure involves the conversion of copper sulfides into oxides, which are then reduced under great heat to a relatively pure metal. Assuming that the weight-loss ratio for copper smelters is 60 percent, the $2\frac{1}{2}$ tons of material that remain of the 100 tons entering the concentrators are

Table 12-12

World Copper Smelter Production, 1984
(thousands of short tons)

Rank	Nation	Production	Percent of World Total
1	United States	1288	14
2	Soviet Union	1235	13
3	Chile	1200	13
4	Japan	1025	11
5	Zambia	566	6
6	Canada	544	6
7	Zaire	520	6
	Subtotal	6378	69
	World total	9514	100

Source: Non-ferrous Metal Data, 1985 (New York: American Bureau of Metal Statistics, Inc., 1986).

smelted into $1\frac{1}{2}$ tons of waste and 1 ton of *blister copper*, which is 99 percent pure metal. Since the weight-loss ratio is fairly high, smelters are often located rather close to the mines, notwithstanding the exceptions mentioned above (Figure 12-12).

Eight of the 17 copper smelters in the United States in 1984 (Figure 12-13) were located in southern Arizona, the main U.S. copper-producing state. The

Figure 12-12 Copper Smelter. *Some smelters are tied to a single mine, but this one, located in Douglas, Arizona, processes ores from several different mines in southern Arizona. (Courtesy Phelps Dodge Corporation.)*

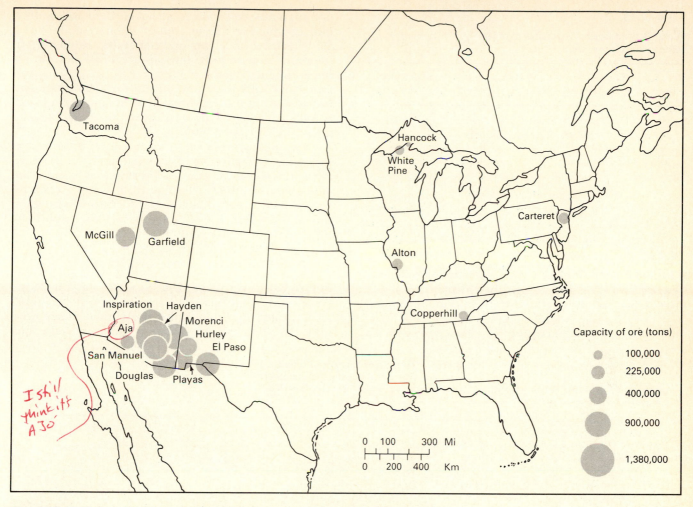

Figure 12-13 Copper Smelting in the United States. *As with concentrating, smelting also occurs near the mine in the United States. Hayden, Arizona claims the largest capacity in the country. Coastal facilities at Carteret, N. J. and Tacoma, Washington depend on foreign supplies and recycled materials.*

largest smelter location in the country is at Hayden, Arizona, where both ASARCO and Kennecott have plants. All told, western copper-mining states contain 12 of the nation's 17 smelters and account for 95 percent of U.S. copper-smelting capacity. Both the Atlantic seaboard (Carteret, N.J.) and the Pacific seaboard (Tacoma, Washington) also possess copper smelters, which depend on foreign supplies of copper concentrate and recycled material.

Refining

Even though blister copper is 99 percent pure metal, it is not suitable for manufacturing electric wires, cooking utensils, and other products. Gold, silver, lead, and zinc impurities must be extracted through a refining process, usually electrolysis. In this process, blister copper is immersed in a bath of copper sulfate through which an electric current is passed. The ratio of weight loss in copper refining is extremely small, scarcely 1 percent of the weight of the refiner's raw material.

Not only is the weight-loss ratio low at the refining stage, but what is lost is also valuable. Consequently, the need for refining activity to locate near copper-mining areas is substantially less than it is for copper for smelters. This explanation weakens at the international scale, where political factors become more influential, especially in third-world areas.

The list of leading copper-refining nations (Table 12-13), for example, closely parallels the smelter roster discussed previously. The U.S. copper output share is slightly higher at the refined than at the smelter stage. This is explained by the greater use of scrap materials today at the refining stage. It takes about one-sixth the power to refine scrap metal than that required by mined ore. At least 15 percent of refined product in the United States today comes from scrap.

The refining capacity of Zambia and Zaire, although significant, is less than it was a decade ago due to problems associated with nationalization of the industry. The rationale for and impact of nationalization of copper production are discussed in a later section.

Table 12-13

World Refined Copper Production, 1984
(thousands of short tons)

Rank	Nation	Production	Percent of World Total
1	United States	1,546	15
2	Soviet Union	1,521	14
3	Japan	1,032	10
4	Chile	975	9
5	Canada	551	5
6	Zambia	528	5
	Subtotal	6,153	58
	World total	10,602	100

Source: Non-ferrous Metal Data, 1985 (New York: American Bureau of Metal Statistics, Inc., 1986).

The United States has 15 copper refineries today (Figure 12-14). A decade ago the capacity was much higher. Six refineries closed in this period. Competition with foreign sources, higher energy costs, outmoded plants, and increasing costs associated with air pollution abatement have all taken their toll. Sulfur dioxide emissions from copper smelter smokestacks are expensive to clean up. Generally, air pollution standards now require the capture of 90 percent of the sulfur entering the smelter (Figure 12-15). Changes associated with the rationalization of multinational investments in the developing world have also played a role in changing domestic activity. These adjustments are discussed later.

Two major areas of refinery concentrations are indicated by Figure 12-14: one centered on the east coast, encompassing the New York and Baltimore regions, and the other extending from western Texas to southern Arizona. Amarillo and El Paso, Texas, have the largest refineries in the country today, with one-third of the total capacity.

Trade Patterns

Before World War II the United States was a net copper exporter but has since become a deficit region. As indicated in Table 12-14, the United States now ranks fourth in refined copper imports and is the sole

Figure 12-14 Copper Refining in the United States. *Two major clusters of refining occur in the U. S., one focused on the New York and Baltimore regions and the other extending from western Texas to southern Arizona, with Amarillo, Texas possessing the largest capacity in the country.*

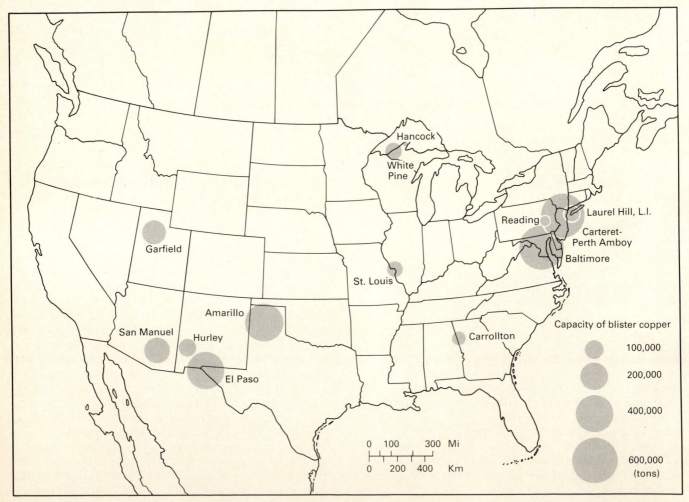

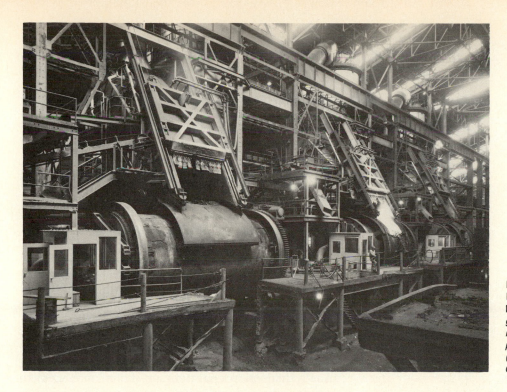

Figure 12-15 Pollution Control Equipment Used in Copper Refineries. *Control of sulfur dioxide gas, generated by converter furnaces, requires water-cooled hoods to prevent sulfur dioxide from escaping. (Courtesy Kennecott Copper Corporation)*

nonwestern European country in the import leader list. The large and growing demand for copper, for which imports supplement the huge domestic output, feeds fabricating companies that produce copper tubing, rods, and wire used in electric motors, electric generators, air conditioners, automobiles, and hardware products. Even though copper faces stiff competition with plastics, aluminum, and fiber optic al-

Table 12-14

Refined Copper Trade, 1983 (millions of U.S. dollars)

Rank	Nation	Total	Percent of World Total
	Imports		
1	United States	758	18
2	West Germany	680	16
3	France	548	13
4	Italy	481	12
5	Great Britain	325	8
6	Japan	300	7
	Subtotal	3,092	74
	Total	4,180	100
	Exports		
1	Chile	1,356	29
2	Zambia	698	15
3	Canada	486	10
4	Belgium/Luxembourg	366	8
5	Zaire	361	8
	Subtotal	3,267	70
	Total	4,681	100

Source: United Nations, *1983 Yearbook of International Trade Statistics,* 1985.

ternatives, specialized applications remain. Electrical uses comprise one-third of the U.S. market; 20 percent goes to construction (wiring and plumbing) and 10 percent to consumer goods.

Copper fabricating traditionally occurred in the so-called *brass valley* of Connecticut. This Naugatuck Valley area also assumed the title of *hardware capital* of the United States. Initially, brass buttons and fasteners established an identity for this industry. Later, clocks, watches, electrical wire, hardware, and artillery shells provided the main thrust. New Britain and Waterbury have been the center of the copper and brass region in recent years. But time has left the area less competitive and the industry has declined.

The leading refined copper-exporting nations include Chile, Zambia, and Canada (Table 12-14). The United States and several western European countries lead as importing nations. In terms of copper ore exports, the leaders include the Philippines, Papua–New Guinea, and Canada (Table 12-15). As mentioned earlier, Japan leads by far in copper imports. Ore imported to the United States comes primarily from the Philippines and Mexico, and blister copper (concentrate) comes from Chile, Canada, and Peru.

World Dynamics

A volatile market and abrupt management changes characterized the international copper industry in the 1960s and 1970s. The changes shocked both the developed and developing economies. Price aberrations, inflation, high interest rates, and changes in demand largely due to the Vietnam war military procurement need caused much of the uncertainty. Nationalization also produced many negative effects.

Table 12-15

Copper Ore Trade, 1983 (millions of U.S. dollars)

Rank	Nation	Total	Percent of World Total
	Imports		
1	Japan	1,475	68
2	West Germany	241	11
3	United States	158	7
	Subtotal	1,874	86
	Total	2,170	100
	Exports		
1	Philippines	423	22
2	Papua–New Guinea	389	20
3	Canada	327	17
4	Chile	274	14
	Subtotal	1,413	73
	Total	1,959	100

Source: United Nations, *1983 Yearbook of International Trade Statistics,* 1985.

Loss of Multinational Control

In 1947 four multinational firms controlled 60 percent of the world copper output, excluding the Soviet Union. In the late 1950s four companies still controlled about half of the world production. By 1974, the four largest companies (Kennecott, Newmont, Phelps Dodge, and Rio Tinto-Zinc) controlled only 19 percent of the copper economy in market countries.

The nationalization process initiated by third-world countries in the late 1960s involved taking control away from multinational firms and returning full sovereignty over resources to domestic producers. An opportunity to increase their share of economic gain and provide more jobs for the local population provided the incentive. The presence of large foreign-controlled vertically integrated firms operating with a profit motive was interpreted as unhealthy for the

Table 12-16

Multinational Firm Losses in Equity Positions in Copper Industry During the 1970s Due to Nationalization (millions of pounds)

Company	1969 Capacity	1969–1974 Capacity Losses	Percent of 1969 Capacity
Kennecott	1,400	400	29
Phelps-Dodge	625	—	—
Asarco	385	—	—
Newmont	385	—	—
Anaconda	1,200	800	67
Amax	500	220	44
Total	4,495	1,420	32

Source: Thomas R. Navin, *Copper Mining and Management* (Tucson, Ariz.: University of Arizona Press, 1978).

local economy. This ownership change typically involved replacing multinational firms with a state-managed company pursuing a social welfare motive to permit more rational economic planning.

Losses experienced by the largest U.S. multinational copper firms in the 1969–1974 period are shown in Table 12-16. These losses ranged from a high of two-thirds for Anaconda to a low of 29 percent of Kennecott among those experiencing takeovers. U.S. losses in copper investments from nationalization occurred in Chile, Peru, Zambia, and Zaire. Similar takeovers in the iron ore field occurred in Chile, Venezuela, and Peru. With bauxite, Guyana and Jamaica nationalized former U.S. interests in this period as well, but multinationals remained strong in the aluminum industry.

In many countries unanticipated negative consequences far overshadowed benefits following this takeover process. Referring to this situation, Shafer coined a phrase—the *myth of nationalization.*[3] While gaining control, developing countries also experienced a *loss of insulation* as a result of the nationalization process. Buffers previously provided by the multinationals disappeared. The new management usually had no preparation or skilled managers to handle increased exposure to economic and political pressures, both domestic and international. The management and custodial functions formerly provided by multinationals disappeared and the vertical integration that formerly cushioned the copper firms from market volatility vanished. Access to capital markets and investment funds also disappeared. Often, production declined, idle capacity increased, and labor unrest rose following nationalization. In Zambia, for example, three strikes disrupted production in 1981.

The Council of Copper Exporting Countries (CIPEC) organization, created in the early 1970s, attempted to provide a stabilizing influence to overcome these weaknesses. But it, too, fell short because to function effectively, a cartel needs control over prices, market, and distribution channels. The copper industry did not have the prerequisites for third-world-cartel success. Threats to substitute other source areas for the product plagued third-world negotiators, as did the prospect of substituting aluminum or glass fibers for copper and recycling scrap material.

The biggest problem facing the copper cartel appeared to be the drying up of private financing formerly available to the producers. *Project financing* rather than *equity financing* increased following nationalization. Project financing is based on anticipated cash flow, not on the credit worthiness of the parent firm. It also frequently entails sharing or *syndicating* the risk, requiring the host country to assume a portion of responsibility of potential failure. It can also entail guarantees extended by the government of

[3]Michael Shafer, "Capturing the Mineral Multinationals: Advantages or Disadvantages?" *International Organization,* 37 (1983), 93–119.

steady markets for the product, a demand that could rarely be met.

The nationalization process that took place in December 1969 in Zambia, a country almost completely dependent on copper exports as a source of revenue, provides a good example of the problems just discussed. In Zambia, nationalization

undermined the government's major development objectives toward the mining industry.... Political leaders apparently thought that they could achieve their developmental goals through their control of the copper companies and could thus avoid making major public investments to achieve those goals. Leaders appeared to see the government's ownership of the copper companies as being a relatively cost free way to develop the country. Not only did this prove impossible, however, but when the copper industry experienced a financial crisis after 1975, the government as the major owner had to bear the financial burden of supporting the multinational corporations. One consequence of Zambia's nationalization of the copper industry, therefore, was to place the country deeply into debt simply to sustain the operations of a financially troubled international industry. This had the effect of forcing the government to suspend all major new development projects and to postpone its development plans for the country.[4]

MINERAL ECONOMY DEPENDENCY PATTERNS

The United States and Japan import tremendous quantities of copper, iron ore, and bauxite to support industrial requirements. These minerals comprise two-thirds of the nonfuel imports of these two countries. The U.S. dependency ranges from 25 percent of the iron ore need to 91 percent of bauxite demand (see Table 12-17). Virtually all of the Japanese mineral requirements are met by imports, but only recently has outside dependence on copper reached such a level. Western Europe also depends strongly on foreign mineral resources.

Strikingly different procurement strategies have been pursued by the United States and Japan in the post–World War II era to obtain these resources. But in the more recent past, a convergence in policies of the two counties has occurred. The United States traditionally relied on investing in foreign mines by purchasing *full ownership rights.* The Japanese, on the other hand, entered into *long-term purchase contracts* for minerals from mines in which they had no financial interest. Roderik indicated that these policies worked well at the time for each nation, but that the Japanese obtained their minerals much more cheaply than did the United States from its approach.[5] Good luck,

[4]Ronald T. Libby and Michael E. Woakes, "Nationalization and the Displacement of Development Policy in Zambia," *The African Studies Review,* 23 (1980), 33.

[5]Dani Roderik, "Managing Resource Dependency: The United States and Japan in Markets for Copper, Iron Ore and Bauxite," *World Development,* 10 (1982), 541–560.

Table 12-17

U.S. and Japan Dependence on Imported Mineral Resources, 1955–1979 (Imports as Percent of Requirements)

Mineral Resource	1955	1960	1970	1979
United States				
Bauxite	71	67	85	91
Copper	34	20	16	36
Iron ore	15	27	31	25
Japan				
Bauxite	100	100	100	100
Copper	31	74	85	96
Iron ore	85	92	99	100

Source: Dani Roderik, "Managing Resource Dependency: The United States and Japan in Markets for Copper, Iron Ore and Bauxite," *World Development,* 10 (1982), 542.

astute bargaining, and the opening of Australian mines all played a part in the success of that policy for the Japanese.

In the 1970s changes in procurement arrangements unfolded in both countries and they now appear quite similar. Nationalization largely ended U.S. multinational ownership patterns. The greater vulnerability facing Japan following the 1973 energy crisis similarly stimulated policy changes. The new strategy emphasized partnerships and joint-venture deals, increasing Japanese equity positions in foreign mines. U.S. firms also began to participate in similar fashion as partners having minority ownership in foreign operations after nationalization ended complete control. An outline of requirements that host countries now mandate for foreign participation in copper mining appears in Table 12-18 on page 192. Notice that foreign ownership is frequently limited to 49 percent and that ownership preferences favor government participation.

Long-Term Purchase Agreements

The former Japanese policy favoring long-term purchase agreements, averaging 10 years in length, can be explained as a natural extension of prevailing Japanese policy in the postwar era.[6] Japanese firms recognized that U.S. multinationals already owned many of the world's resources and that internal rebuilding responsibilities after the war required other uses for scarce capital. Moreover, foreign investment was frowned upon by the Japanese and large import cartels created by the cooperation of the public and private sectors provided considerable leverage when negotiating deals with suppliers. The Japanese experience in setting up long-term iron ore supply agreements illustrates the success of that strategy.

In the early 1950s Japan bought iron supplies on the spot or short-term contract market. Due to high

[6]Ibid., 541–560.

Table 12-18

Selected Requirements for Foreign Investment in Copper-Producing Nations

	Mexico	Zambia	Zaire	Chile	Peru
Maximum percent permitted foreign equity ownership	49% for private lands, 34% for government reserve areas	49	100	100	100
Approval of investment by foreign investment committee	No	Yes	Yes	Yes	Yes
Control of marketing	Domestic buyers have priority	Government agency	Mining company	Mining company	Government agency
Control of export receipts	No government control	Full surrender to central bank	Surrender to central bank but with negotiable arguments for external credits	Negotiable	Surrender to central bank with negotiated provision for portion paid to external creditors
Ownership preference in new investment	Government participation and private domestic	Government with minority foreign participation	Foreign investment with government participation	Private and joint venture for government-owned ore bodies	Government majority ownership

Source: Raymond F. Mikesell, *The World Copper Industry* (Washington, D.C.: Resources for the Future, 1979), pp. 258–259.

transportation costs, premium prices were paid. But things changed with the discovery of Australian iron ore deposits in the late 1950s. Australia relaxed previous export restrictions in 1960 and Japanese commitments to purchase ore helped Australia develop the mines. In 1964 Japanese steel companies formed a cartel that negotiated a long-term deal. Fortunately for Japan, each Australian state tried to negotiate alone and underbid one another to obtain the Japanese business. Australia also realized that Brazil and India competed for the same business as well, leaving them powerless to exact higher prices. The end result was a very favorable deal for Japan.

The low cost of obtaining iron ore resources allowed the Japanese to keep steel prices low. "In 1960, U.S. steel mills had a 16% cost advantage over Japan in their iron ore input; by 1976, this cost advantage had been dissipated into a 43% disadvantage, Japan having obtained its ore at lower cost since 1967. The Japanese steel mills had managed to incur only a 23% increase in iron ore costs between 1960 and 1976, whereas U.S. steelmakers were faced in the same period with a whopping cost increase of 148%."[7]

The Japanese had a similar price advantage in bauxite until the 1973 energy crisis boosted domestic costs for electricity to uncompetitive levels. Today the Japanese appear to be dismantling aluminum production facilities as higher energy prices continue to keep smelting prices high (see Figure 12-16).

[7]Ibid., p. 548.

Transition to Equity Investment

The opportunity the Japanese exercized to negotiate prices for imported aluminum and iron ores never occurred with copper. The London Market Exchange (LME) traditionally exerted more control over world copper prices than with other minerals. It is not surprising, therefore, to learn that pressure on the Japanese to invest directly in foreign copper mines came earlier than with iron ore and aluminum. Japanese investments in the Philippines began in the mid-1950s and by the late 1960s equity positions had been established in mines in Zaire, Malaysia, Canada, Peru, and Chile. In the 1980s these policies even extended to the United States. Mitsubishi and Kennecott, for example, established a joint venture at Chino, New Mexico, to modernize a smelting plant.

The emphasis today in Japanese ore procurement policy focuses on establishing ties that promote long-term security, especially in view of higher energy prices and greater world militancy, such as the unrest in the Middle East. Japan seeks to invest in projects that primarily produce resources for their market. Rather than shopping for a deal, paying a premium price for long-term security receives a higher priority. The Japanese enjoy a reputation as flexible and dependable partners, and elaborate joint-venture projects that go far beyond the mine itself assure continuity. They are also participating in "colossal" multipurpose economic development projects in several third-world countries, including Brazil and Indonesia. In some cases direct government-to-govern-

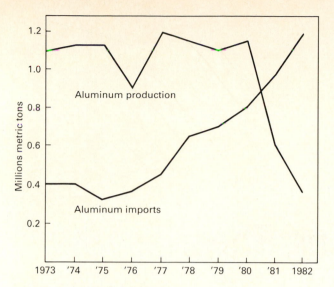

Figure 12-16 Aluminum Production and Import Trends in Japan. *The aluminum industry is a "sunset" business in Japan. Production facilities have been closed in response to higher energy prices. Note the dramatic decrease in production in the 1980–1982 period. Aluminum imports began a rapid increase in 1976 and began exceeding domestic output in 1980. (Source:* The Economist, *July 9–15, 1983, p. 17)*

ment deals involving private-sector banks, trading companies, mineral companies, and government agencies have been negotiated. Such complex collaborative projects between business and government go far beyond the ability of U.S. mineral firms, legally and strategically, because of institutional constraints.

Even though U.S. and Japanese firms now compete in similar ways to gain access to third-world mineral resources, with an emphasis on joint-venture arrangements, it appears that the Japanese investments abroad will increasingly be criticized as being government subsidized and giving entrepreneurs an unfair advantage because of the style and magnitude of joint ventures undertaken. If China becomes a major raw material importer in the future, competition between Japan and the United States and other importers could intensify, unraveling current arrangements.

SUMMARY AND CONCLUSION

Iron ore remains important to industrialized economies but increasingly few major users produce their own high-quality mined product. Only the Soviet Union remains self-sufficient. Japan, western Europe, and the United States, among the major industrial nations, need imports to sustain demand. Similar scenarios occur with copper and bauxite, but in the latter instance the Soviets, too, lack domestic supplies. Given this uneven distribution, international trade has intensified to meet the need, as it has in the petroleum industry. Nationalization of multinational investments also affected these mineral industries, as it did the petroleum producers.

Changes in the location of the mining and refining phases of the mineral industry reflect a fundamental adjustment in the industrial structure of both developed and developing areas. As the former economies retreat from the processing and refining stages of production, developing area economies aim to enhance their share of processing activity, leaving behind the role of a single-dimension mining economy. The contraction of the copper industry in the United States in recent years reflects this transition, as does the retreat of aluminum production in Japan. The direction rather than the magnitude of this shift of production to the third world is the most meaningful to date as considerable capacity still exists in developed areas.

Advanced stages of fabricating activity remain with developed countries because of advantages in technology and markets. Various theories that help account for the location of manufacturing activity that use these raw material resources are examined in Chapter 13.

SUGGESTIONS FOR FURTHER READING

Banks, F. E., *Bauxite and Aluminum: An Introduction to the Economics of Nonfuel Minerals.* Lexington; Mass.: D. C. Heath, 1979.

Bergsten, C. F., et al., *American Multinational and American Interests.* Washington, D.C.: Brookings Institution, 1978.

Brecher, Jeremy, et al., eds., *Brass Valley: The Story of Working People's Lives and Struggles in an American Industrial Region.* Philadelphia: Temple University Press, 1982.

Cobbe, James, *Government and Mining Companies in Developing Countries.* Boulder, Colo.: Westview Press, 1979.

Eckes, Alfred, Jr., *The United States and the Global Struggle for Minerals.* Austin, Tex.: University of Texas Press, 1979.

Graham, Ronald, *The Aluminum Industry and the Third World: Multinational Corporations and Underdevelopment.* London: Zed Books, 1982.

Mikesell, Raymond F., *The World Copper Industry.* Washington, D.C.: Resources for the Future, 1979.

Mizger, Dorothea, *Copper in the World Economy.* New York: Monthly Review Press, 1980.

Navin, Thomas R., *Copper Mining and Management.* Tucson, Ariz.: University of Arizona Press, 1978.

Radetski, Marian, *Mineral Processing in Developing Countries.* New York: United Nations Industrial Development Organization, 1980.

Tsurumi, Yoshi, *The Japanese Are Coming.* Cambridge, Mass.: Ballinger, 1976.

United Nations, *Transnational Corporations in the Bauxite Aluminum Industry.* New York: UN, 1981.

United Nations, *Transnational Corporations in the Copper Industry.* New York: UN, 1981.

Wilkins, Mira, *The Maturing of the Multinational Enterprise: American Business Abroad from 1914 to 1970.* Cambridge, Mass.: Harvard University Press, 1975.

13 Manufacturing Theories and Trends

In Chapters 10 to 12 we have examined several raw materials—coal, iron ore, and copper, among others—that singly or in combination typically experience further processing to create manufactured goods. We have discussed the occurrence of these products and factors associated with their mining and use. Transportation and accessibility factors, costs, markets, and above all the supply and demand for these products at any particular location all play a role in expanding their usefulness in the world economy. Complex political and strategic forces, the scarcity of the item, and the item's relative quality vis-à-vis competing supplies are also important considerations. But there are other factors to consider as well. How much of the product is needed for further processing of a particular item? How much of it is wasted or can be converted to other uses? How does the level of development of a particular area affect the demand for the product? How competitive is it with similar products? What is the likelihood that substitutes can weaken its market in the future?

Clearly, there are many factors to consider when evaluating location patterns and trends in the manufacture of a particular product. In this chapter we turn away from an emphasis on empirical considerations in favor of placing more emphasis on conceptual and theoretical factors that assist in explanations of manufacturing locations. With this framework we will be better able to account for the complex patterns that characterize the industrial landscape today. In the three following chapters we return to an examination

of empirical patterns and processes, using the principles developed here to gain further insights into manufacturing activity.

WHAT IS MANUFACTURING?

Manufacturing activity includes the factories that produce goods as well as the entire production system supporting the factory. The system includes the suppliers and markets for products and transportation facilities that tie together a complex distribution network. The system also includes services and information linkages such as communication and control mechanisms internal to the firm. The external environment surrounding the firm, comprised of various economic, political, and environmental factors, also influences the system. Examples of these factors include transportation rates, tariffs, quotas, and air pollution guidelines. In three accompanying boxes we discuss the definition of manufacturing and several approaches used in measuring its importance and provide a summary of the standard industrial classification system used to systematically group various manufacturing activities on the basis of major commodities produced.

The manufacturing process itself involves changing the form of a good to enhance its value, as the definition in the first box indicates. To undertake the manufacture of a good, inputs in the form of labor, capital, and raw materials are required. The output is

MANUFACTURING DEFINITION

The definition of manufacturing has changed little since the following criteria appeared in the 1904 census:*

1. Process involves changing the form of goods
2. Operations conducted in factories
3. Production process involves division of labor
4. Power-driven machinery used
5. Standard products produced (not custom made)

*Adapted from Harold H. McCarty and James B. Lindberg, *A Preface to Economic Geography* (Englewood Cliffs, N.J.: Prentice-Hall, 1966).

These guidelines exclude cottage or home work-shop types of operations. But it has been pointed out that the word "manufacture" has Latin and French roots which mean "to make by hand." The word "industry" more closely defines manufacturing in a technical sense, but in English-speaking countries we use the term "industry" very loosely to refer to a wide variety of occupations, including many services.†

†F. E. Ian Hamilton and G. J. R. Linge, "Industrial Systems," Chapter 1 in F. E. Ian Hamilton and G. J. R. Linge, eds., *Spatial Analysis, Industry and the Industrial Environment*, Vol. 1, *Industrial Systems* (New York: Wiley, 1979), p. 2.

---

a finished product that can either be used again as a raw material for another manufacturer or consumed in its present form.

THE MANUFACTURING PROCESS AND LOCATION

A manufacturer, when establishing a factory, makes several separate or interrelated decisions about the nature of the business, including the following:

(1) the scale of operation, including how much is to be produced and at what price it is to be offered to the consumer; (2) the technique to be adopted, which involves the selection of the appropriate combination of inputs; and (3) the location of the factory (Figure 13-1 on page 197).

The choice of location cannot be considered in isolation from scale and technique. Different scales of operation may require different locations to give access to markets of different sizes, and if the location decision is made first, this may have an important bearing on the output that the firm can expect to sell. Different techniques will favor different locations, as firms tend to gravitate toward cheap sources of the inputs required in the largest quantities, and location itself can influence the combination of inputs and hence the technique adopted. Scale and technique similarly affect each other, but this is of less direct importance in the present context than the reciprocal relationship between scale and location on the one hand, and technique and location on the other.

The two major sets of economic variables influencing industrial location are thus those relating to technique and those relating to scale. The way these variables affect and are affected by location is very

---

STANDARD INDUSTRIAL CLASSIFICATION

The most comprehensive source of data on manufacturing activity in the United States is the *Census of Manufacturers,* one of the Economic Censuses published by the Bureau of Census each 5 years (i.e., 1977, 1982). Activity reported in this census is recorded by industry groups that conform to the Standard Industrial Classification (SIC) format. The basic SIC classification schema combines all economic activity according to a numerial code of two to four digits. Broad groups are identified at the two-digit level and are further subdivided into three and four digits for more specificity and product identification.

The hierarchical format of the SIC classification is shown in the Appendix. It is conceptually useful to think of 10 one-digit designations as constituting the backbone of the system. The one-digit groups actually appear as SIC subdivisions and are referred to by letters (A through K). The numbers "0" and "1" refer to agricultural, forestry, fishing, mining, and construc-

tion activity. Manufacturing activity is incorporated in categories 2 and 3 (Division D). Groups 4 through 9 refer to various types of services (transportation, retailing, finance, public administration, etc.). The 20 two-digit manufacturing categories and their three- and four-digit derivatives are of concern in this chapter. The activities numbered in the 20s are primarily nondurable industries; those in the 30s are durable manufacture (machinery, electronics, etc.). A breakdown of three industries into three-digit classifications appears in the Appendix to illustrate the subdivision process. Four-digit breakdowns for two groups of activities are also listed as an example of the refinement process. Similar disaggregation can be undertaken for each of the other groups. Specific products are easily identifiable at the four-digit level. The spatial distribution of the production of several of these activities at the four-digit level is examined in this and succeeding chapters.

The level of manufacturing activity can be evaluated with several indices. The measures most often used include (1) the number of employees, (2) the number of plants, (3) the quantity of capital investment or sales, and (4) value added.* Closer scrutiny reveals strengths and weaknesses with each measure for cross-industry comparisons, but within a specific industry, say textiles, any of the measures provides a satisfactory indication of relative strength. As an illustration of problems associated with comparative industrial assessments, consider the relationship of the value-added measure to the number of employees and the amount of capital investment in three industries: (1) petroleum refining, (2) textiles, and (3) electronic computers. Petroleum refining, a highly capitalized industry with a high degree of automation, employs relatively few workers, and yields high-value-added products. In contrast, textile firms are labor-intensive, small-scale, relatively low-value-added operations. The computer apparatus industry lies between these two

extremes in terms of capitalization and employment intensity; it is usually carried on in large-scale facilities and involves very high value-added figures. Comparing these three types of enterprises using the employee measure may lead to overstating the importance of textiles while underestimating the strength of both the petroleum refining and computer industries. Similarly, by just looking at capitalization, one would discriminate against textiles (and other nondurable manufacturing) in deference to petroleum refining and computer machinery.

The nearest index to an ideal measure of manufacturing levels is value added. This index best approximates the importance of a particular activity to the total economy. Since it is expressed in dollar terms, cross-comparisons are easy. Unfortunately, figures on value added and other measures of industrial strength may not be readily available for all industries at the local or state level. The *Census of Manufacturers* gathers these data, but they are often not published, due to a disclosure restriction.† This is particularly a problem at the four-digit product level by small geographic area (i.e., the county). When value-added data figures are not available, the number of plants or another indicator may be published and can be substituted as a measure, but disclosure problems can mean that no indicators are available.

---

*This figure expressed in dollars is the difference between the cost of raw materials, labor, and other costs of production and the final selling price of a product. "Value added data are calculated by subtracting the cost of materials used in the manufacturing process from the value of the finished products. Material costs include all supplies, fuel, electric energy, and resale costs." Truman A. Hartshorn and Frank Drago, "The Manufacturing Map of the South," *Atlanta Economic Review,* 26 (1976), 47. Value added varies from industry to industry fairly consistently, with durable manufacturing having a higher figure than nondurable activity.

†Data are not published if the number of companies or sales figures are so small as to reveal values for individual companies.

---

important, and its introduction here helps to emphasize at the outset the interdependence of the three decisions. Business success may not be determined by choice of location alone, but technique and scale as they influence profits [and] cannot be divorced from location.[1]

Basic manufacturing location principles appear to be simple and straightforward, but the pattern of industry on the landscape today is enormously complicated and not easily explained. When early industrial location theories were developed at the turn of the century, the industrial world was much less diverse than it is today. Manufactured products did not experience as many stages in production, the range of products was much narrower, and regulatory and political factors did not play as important a role in accounting for the patterns. Internationalization of production and the growing interdependence of firms and nations have also contributed to the growing complexity.

In this chapter we discuss the changing world or-

der in manufacturing in the context of the changing body of theory used to explain or account for the distribution of activity. We begin by reviewing classical theory and conclude with a discussion of behavioral and structural approaches to industrial location.

To highlight changes in the approaches to studying manufacturing location, let us introduce two terms: (1) optimizing and (2) satisficing. Classical location theory suggested that manufacturers sought *optimal* or ideal locations in the sense that firms were located in places where production costs were minimized and/or market prices for goods were maximized. Optimizing behavior can be described as *normative* behavior, which means that it assumes that decision makers are completely informed, rational, and choose the best locations. In reality the manufacturing landscape as it actually exists displays a variety of situations, some of which might have represented ideal locations at one time, but not necessarily now, as well as circumstances where firms and businesses are not motivated purely by economic principles, much less in optimizing them.

Some firms choose locations based on family ties, amenities, or by accident. In fact, people and firms often exhibit satisficing behavior rather than the mo-

---

[1]David M. Smith, *Industrial Location: An Economic Geographical Analysis,* 2nd ed. (New York: Wiley, 1981), pp. 23–24.

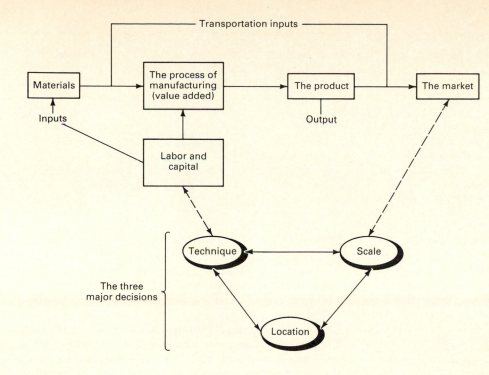

Transportation inputs

Materials

Inputs

The process of
manufacturing
(value added)

The product

Output

The market

Labor and
capital

The three
major decisions

Technique

Scale

Location

Figure 13-1 The Location Factor in the Context of Firm Operations. *Business Success cannot be determined by choice of location alone because technique and scale also influence business performance as shown here.*
(Source: Smith, Industrial Location, *Wiley, 1981, p. 24)*

tivation to optimize their behavior.[2] They accept existing circumstances and make adjustments to overcome any disadvantages. In many cases it may not be possible to identify ideal or optimal locations in any case because of the uncertainties and changing market and supply situations facing firms.

CLASSICAL LOCATION PRINCIPLES

For the greater part of the twentieth century, geographers depended on theory to explain manufacturing patterns developed at a time when industrial processes displayed rather straightforward ties to raw materials and markets. Today, this work still serves us well as a point of departure for understanding manufacturing location, even though the scale and location of operations respond to far more complex forces. In fact, the location of traditional manufacturing processes involved with the refining of mineral resources, such as those discussed in Chapter 12, can still best be accounted for using classical approaches.

Generally speaking, classical industrial location theory assumes *perfect competition* among firms wherein no firms experience monopoly advantages arising from a particular location. Each firm is further assumed to have similar production costs. Some classical theories as discussed below allow for *imperfect competition.* Imperfect competition occurs when an

oligopolistic or monopolistic situation exists, which provides some firms with an overwhelming advantage, thus giving them a chance to drive competitors out of the market. An oligopolistic situation occurs when a few firms significantly influence but do not necessarily control the market. Under a monopoly the demand and supply functions play a minimal role in determining market conditions. Rather, the market is manipulated and controlled by the monopoly firm(s). In addition, the locational constraint that exists in the situation of perfect competition becomes less important under the condition of market imperfection.

Least Cost

Least-cost industrial location theory dates back to 1909 with the publication of a book on the topic by Alfred Weber.[3] Weber simplified real-world situations by making three assumptions in developing his theory. First, he assumed that raw materials occur at only a few locations. Second, he assumed that the market existed only at specific places. Third, he assumed that the labor supply was immobile and available only at several specific locations, but having an unlimited supply at a specific wage level.

Weber distinguished between weight-losing (gross materials) and non-weight-losing (pure materials) industries in order to determine minimum-cost locations. A further distinction between localized materials (restricted availability) and ubiquitous materials

---

[2]In recognition of his work with the concept "satisficing," Herbert Simon from Carnegie–Mellon University won the 1978 Nobel Prize in Economics.

[3]A. Weber, *Theory of the Location of Industries* (Chicago: University of Chicago Press 1929; translation of 1909 original edition in German).

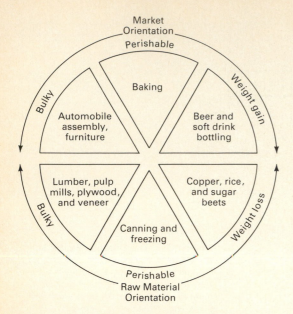

Figure 13-2 Locational Affinities for Selected Industries. *Classical location theory suggests that firms locate either at raw material or market sites. Examples of firms displaying these tendencies are shown here. (Source: Hartshorn,* Interpreting the City, *Wiley, 1980, p. 146)*

(available everywhere) also assisted in the determination of an optimum location.

According to Weber, the least-cost location for industry occurred near the market when manufacturing processes used commonly occurring materials. An example is the use of locally produced limestone for the manufacture of cement. In this situation a market location eliminates unnecessary transportation costs. At the other end of the spectrum, raw material industrial locations predominate when limited alternative sources of material exist such as the mineral-processing activity discussed in Chapter 12.

The Weber least-cost location formulation suggested that firms located either at the *raw material site* or near the *market* in order to minimize distribution costs. Examples of contemporary firms primarily having either raw material or market locations appear in Figure 13-2. Weber assumed that transportation costs varied according to weight and he did not discuss variations according to the stage of production. Later, the transportation cost concept was refined by others to distinguish between raw material and final product transfer costs. Hoover, for example, noted that raw material shipment rates often fell below those of finished goods due to increased fragility and higher packing and hauling costs for manufactured items.[4] Higher final product transportation costs then came to be recognized as important factors encouraging market locations.

Using the Weber approach to plant location, one can argue that intermediate plant locations between the raw material and market locations are undesirable

[4]Edgar Hoover, *The Location of Economic Activity* (New York: McGraw-Hill, 1984).

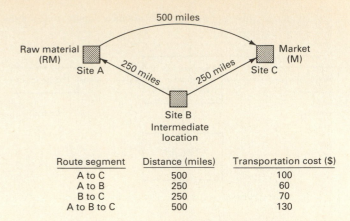

Route segment	Distance (miles)	Transportation cost ($)
A to C	500	100
A to B	250	60
B to C	250	70
A to B to C	500	130

Figure 13-3 Transportation Costs and the Intermediate Manufacturing Location. *Intermediate plant locations between the raw material and market are typically undesirable because they entail higher freight-rate costs as shown here.*

because they induce higher costs of production due to the effect of increased freight-rate charges. It seems that two short hauls, which an intermediate location requires, as shown in Figure 13-3, are more expensive than one long trip, due to the variation in the proportional impact of *fixed costs* and *variable costs* on transportation.

Fixed costs refer specifically to the terminal costs incurred in maintaining such facilities as ports, railroad stations, and airports. These costs are largely fixed since they are not directly related to the level of

Figure 13-4 Freight Rates by Truck, Rail, and Water Transport. *For short distances truck hauls are generally the cheapest, while rail is the least expensive for intermediate shipments, and water for the longest movements.*

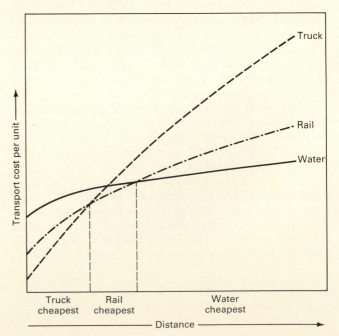

Figure 13-5 Port Newark and Port Elizabeth Freight Terminal. *The New Jersey shoreline in metropolitan New York contains many industrial and transportation facilities. The area serves as a break-of-bulk freight hub for domestic and international business. This view shows the Port Newark and Elizabeth-Port Authority Marine Terminal. (Courtesy the Port Authority of New York and New Jersey.)*

movement and cannot be allocated readily to specific users. An increasing use of terminal facilities as well as of capital equipment (trucks, railroad cars, etc.) results in decreased average fixed costs.

Variable transportation costs are over-the-road costs which increase with the level of movement and can be allocated specifically to users. Some examples include wages, fuel, wear and tear of capital equipment, and tolls, which generally increase with distance. Route maintenance costs may also be regarded as variable costs since these are directly related to traffic volume.

Total transport costs—the combination of fixed and variable costs—are curvilinear rather than linear functions of distance, as is shown in Figure [13-4.] This curvilinearity is caused by scale economies in long-haul transport which make it possible to average fixed costs over longer distances. Figure [13-4] illustrates the differences in movement cost between truck, railroad, and barge. Note that the fixed-cost component for each mode varies greatly: the maintenance of the port

of London is far more expensive than the maintenance of the loading dock of a trucking firm. Note also the different slopes of the three curves and the fact that they intersect. The near linearity of the cost curve for trucking is caused by the lower fixed costs which prevail in this mode. For very long distances, transportation by barge costs less than by either truck or rail. Once the fixed costs are accounted for by barge movement, the increments in variable costs are very slight with increasing distance.[5]

One circumstance that does encourage intermediate manufacturing locations occurs with *break-of-bulk* sites (Figure 13-5). "Break of bulk" refers to the transfer of product from one mode of transportation to another at locations where the mode of transportation changes. Since reloading is required at such places, including ocean port terminals and river/rail

[5]John C. Lowe and S. Moryadas, *The Geography of Movement* (Boston: Houghton Mifflin, 1975), p. 31.

junctions, producers often establish manufacturing facilities at these break-of-bulk locations.

Substitution

Several locational analysts in the Weber tradition have synthesized location theory into a general framework in the face of increasing complexity. Isard, for example, incorporated all relevant locational and spatial factors into a general theory of the space economy.[6] He considered all the costs of inputs and outputs over time and space as well as selling prices in order to create a more general theory.

Isard's *spatial equilibrium through substitution* approach allows for substituting any of the factors in the production process, not just transportation. Cheap coal might be exchanged for higher-cost natural gas in determining an optimum location for one firm, while another might respond to changing costs by adjusting the mix of its output. In general, Isard grouped location factors into three categories: (1) transfer costs; (2) labor, power, tax costs, and so on; and (3) *agglomeration economies* and *diseconomies*. The latter factor reflects the advantage offered by urban areas in the cheaper provision of services and information. These savings are somewhat offset by the increases in cost of other factors, referred to as diseconomies.

Profit Maximization

An alternative to the Weber least-cost location approach was provided by Lösch, who applied the *profit-maximization* approach to the industrial location problem.[7] The central theme of the Lösch theory is that industrial location is characterized by conditions of monopolistic competition, not perfect competition as envisioned by Weber. Lösch indicated that firms locate so as to maximize revenues, not necessarily at locations having the least cost. By avoiding an emphasis on production costs, such firms can concentrate on sales and the market demand.

Market access is the major location consideration for firms according to the maximum-profit theory. When a relatively uniform demand exists for a product, firms locate in the center of this demand. Lösch developed the notion of a *spatial demand cone* to demonstrate this approach (Figure 13-6). The demand cone indicates that the quantity demanded decreases with distance from the center of the market.

One problem with Lösch's approach is that markets for products do not occur in isolation as they overlap. Location equilibrium between a firm and its market therefore rarely occurs using the profit-maximization philosophy because as more firms appear, profits are eroded and optimal location circumstances change.

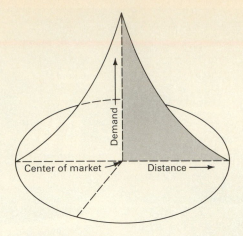

Figure 13-6 Spatial Demand Cone. *According to the profit maximation location principle developed by Lösch firms locate so as to maximize revenues. To maximize profits they locate in the center of their market because the quantity demanded decreases with distance as shown here.*

Spatial Margins

Manufacturing location theory approaches that emphasize least-cost production fail to consider market demand, and similarly, market demand approaches ignore the production side. To overcome these narrow approaches, the *spatial margins* perspective combines features from both the production and revenue sides. Smith and others argue that this approach is more realistic, as it allows for locational flexibility within a range defined by the intersection of space cost curves (SCC) and space revenue curves (SRC), as shown in Figure 13-7.[8] In this example, which evaluates costs and revenues for a steel mill operation in Itabirits, Brazil, distances from Rio de Janeiro, the market, are shown on the horizontal axis, and costs are shown on the vertical axis. At all locations within the shaded area between the SCC and SRC curves, profitable sites occur for the mill. These site possibilities expire at A and B, the spatial margins for profitable steel facility locations. The actual steel plant location at Itabirits, 600 miles from Rio, occurs near the optimal profit location.

A problem with the spatial margins approach is that costs and revenues for a firm are rarely linear as indicated here. It is much more likely that market demand factors are not uniform with distance, due to income and taste and preference variations. The impact of advertising and mass-media exposure further complicates demand factors. But the advantage of this approach is that it combines market demand and production cost variables. The problem is also simplified by the fact that only a linear market is considered.

[6]Walter Isard, *Location and Space-Economy* (Cambridge, Mass.: MIT Press, 1956).

[7]August Lösch, *The Economics of Location* (New Haven, Conn.: Yale University Press, 1954).

[8]David M. Smith, "Modeling Industrial Location: Towards a Broader View of the Space Economy," in F. E. Ian Hamilton and G. J. R. Linge, eds., *Spatial Analysis, Industry and the Industrial Environment*, Vol. 1, *Industrial Systems* (New York: Wiley, 1979).

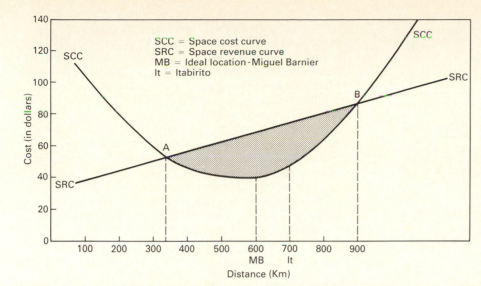

Figure 13-7 Spatial Margins Approach to Location. *Both market demand and production costs are considered by the spatial margins concept. It allows for locational flexibility within a range defined by the intersection of space cost curves (SCC) and space revenue curves (SRC). In this example, which evaluates costs and revenues for a steel mill operation in Itabirito, Brazil. Note that the Itabirito (It) location is very close to the ideal location at Miguel Barnier (MB). (Source: Smith,* Industrial Location. *Wiley, 1981, p. 310)*

Increased competition would change the situation, as could the interaction between supply and demand and economies of scale.

Interdependence

As the scale of industrial production has increased so has *interdependence* among firms. The interaction level of the firm with the external environment has also increased. Such factors include a diverse array of circumstances: the role of labor unions, government, changing trade policies, import–export restrictions, embargoes, growing environmental consciousness, ideological differences and economic factors, including monetary policy, interest rates, and inflation. The interdependence of firms can best be accounted for in a systems framework because it includes the operating units (firms) and their relationships with one another and the external environment.

One manifestation of the increased interdependence characterizing manufacturing activity today is the ever-lengthening chain of processing, fabricating, and assembly between the raw material and marketing stages. In some cases single firms control all the stages in this process through *vertical integration,* but in other instances complex ties have evolved between suppliers (subcontractors) and producers that make them very dependent on one another. An instance of this situation is the large subcontracting industry that has evolved in the automobile industry in the United States and Japan. The greater Detroit area provides this function in the United States, while the Nagoya area of Japan serves a similar function for Toyota and Tokyo for Nissan. In northern Ohio and southern Michigan, highly skilled and sophisticated automotive-linked firms keep in close contact with major automobile corporation design teams to keep abreast of changes and provide rapid adjustments in the component parts required by model specification changes. These information and supply linkages remain very

strong and have helped maintain the dominance of Detroit in the industry.

Forward and *backward* linkages also stabilize manufacturing operations by enhancing marketing (forward linkage) and production (backward linkage) economies. Automobile producers exhibit both of these tendencies. Labor supply linkages are another form of connections that exist in areas requiring a skilled or unskilled labor pool, whichever is the case.

Other types of linkages also encourage interdependence, not among producers, but between firms and their external service providers. Firms using the same support infrastructure, such as that found in an industrial park, illustrate a *commensal linkage,* as has been noted in the retail service industry. In such situations, trucking, warehouse, computer, and financial support service mechanisms, among others, can exert strong locational ties among businesses occupying a common setting.

From a demand or revenue perspective interdependence occurs in the form of *competitive marketing.*[9] Manufacturers in competition with one another often consider the location of other producers because their sales level can be adversely affected by the presence of nearby competitors. Hotelling developed a classic location strategy that two firms might follow when choosing locations. It is presumed that these firms are free to move and sell identical products to a market, providing a consistent demand. A perfectly competitive market is assumed to exist in the area and no firms act in collusion with one another. These circumstances yield a stable location for these firms back to back at the center of the market, with each serving customers to one side (see Figure 13-8). With any other location one firm could move and capture a greater market share than the other. For example, in case B, with each firm having a quartile location, each shares

[9]Harold Hotelling, "Stability in Competition," *The Economic Journal,* 39 (1929).

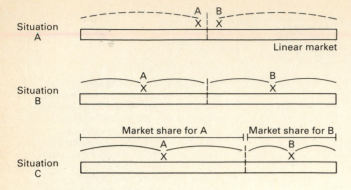

Location strategies for two firms
A = Stable equilibrium situation; back-to-back locations
B = Equal market shares but unstable situation
C = Firm A moves closer to B to claim larger market. Unstable situation such that B can move closer to capture larger share, leading to situation A.

Figure 13-8 Competitive Marketing Location Alternatives. *Firms consider the location of other producers when choosing a location. The dynamics of selecting an equilibrium location when two producers are involved is shown here. Note that neither producer can move to gain advantage when both are located back to back in the center of the market as in Situation A.*

half the market, but one firm could move closer to the other and claim a larger market share because each would dominate all the area nearest it and split the market in between. For situations involving three, four, or more firms, equilibrium locations change, but they do exist, assuming no collusion.

Industrial Complex. Firms supplying raw materials or consuming by-products from one another can also benefit from locations in close proximity. Such a circumstance is illustrated by an *industrial complex* and a *territorial production complex*. The industrial complex is an example of linked firms clustering together to create their own cost-competitive environment.[10] The petrochemical complex on the U.S. Gulf coast illustrates this situation. About 30,000 persons in the New Orleans area are directly employed by these industries, consisting of more than 40 firms producing chemical and oil products in the New Orleans area alone. The Carolina Piedmont industrial crescent, within which numerous textile mills and their suppliers, including synthetic fiber chemical plants and machinery assemblers, have blossomed around a textile core, is another example of an industrial complex.

In developing the notion of the industrial complex Isard linked the Weber least-cost approach to the analytical completeness of *input–output analysis* to show interfirm linkages and monetary flows. Input–output analysis, developed by Leontief in the 1940s, monitors the flows of goods and services in an economy. Sales and purchases among all activities are assumed to be

confined to a closed system in this model such that total purchases (inputs) are equal to total sales (outputs). One of Isard's early studies dealt with the potential for establishing a petrochemical complex in Puerto Rico.[11] It was determined that Puerto Rico had a labor cost disadvantage for the petroleum–chemical combination in comparison with the U.S. Gulf coast. But when a textile industry combine was added to the petrochemical facility, the Puerto Rican disadvantage disappeared, illustrating the fact that a larger integrated facility can be more efficient than a smaller, simpler operation.

The synergism created by the linkages among related activities creates remarkable diversity in output for many manufacturing regions. Petrochemical areas are perhaps the best example. In addition to the affinity of pharmaceutical, chemical, plastics, and rubber industries to such areas, many also have nonferrous metal processing firms, fertilizer plants, and other users of sulfuric acid, clustered together to use by-product materials and gases. Similarly, a tendency exists for synthetic fiber producers, textile firms, and power plant facilities to locate in petrochemical areas. Advantages accrue to these firms from the cogeneration of power and the use of similar production technologies.

Territorial Production Complex. The *territorial production complex* (TPC) is a Soviet development strategy designed to make a broad region self-sufficient and enhance the social welfare of its people. It "is literally a mechanism for spatial organization which optimizes the use of a given area's human and physical resources for a desired economic effect."[12] The TPC represents an attempt to decentralize decision making to the regional level, which is responsible for developing the production and infrastructure facilities. This approach is particularly appropriate in developing frontier areas such as eastern Siberia. In the Soviet Far East several territorial production complexes have been identified and promoted along the Baikal-Amur Mainline (BAM) railway, running eastward from a junction with the Trans-Siberian Railway at Taishet to the Pacific, a distance of more than 3000 kilometers. Opening up access to resources in the region and military considerations motivated the construction of this railway, which traverses very rugged terrain. Thirteen or more territorial production complexes are envisioned for the area, a few of which are identified in Table 13-1. It differs from the industrial complex in that diversification is maximized, with activity not limited to linked products. Economies of

[10]Walter Isard et al., *Methods of Regional Analysis: An Introduction to Regional Science* (Cambridge, Mass.: MIT Press, 1960), pp. 375–412.

[11]Walter Isard et al., *Industrial Complex Analysis and Regional Development with Particular Reference to Puerto Rico* (Cambridge, Mass.: MIT Press, 1959).

[12]Victor Mote, "The Baikal-Amur Mainline and Its Implications for the Pacific Basin," in Robert G. Jensen et al., eds., *Soviet Natural Resources in the World Economy* (Chicago: University of Chicago Press, 1983), p. 170.

Table 13-1

Selected Territorial Production Complexes (TPCs) in the Western BAM Area
of the Soviet Far East

TPC Name	Main Industrial Nodes	Major Activities		
		Nationally Significant	Regionally Significant	Locally Significant
Upper Lena	Ust-Kut	Timber	Oil	
	Kazachinskoye		Gas	
	Magistralny		Hydropower	
North Baikal	Severobaikalsk	Fresh water	Potash	Timber
	Nizhneangarsk	Graphite	Lead	Manganese
	Kholodnaya	Recreation areas	Zinc	Hydropower
Udokan	Udokan	Copper	Iron ore	Coal
	Chara		Titanium	Timber
			Vanadium	
			Nickel	
			Cobalt	
			Gold	

Source: Modified after Victor Mote, ''The Baikal-Amur Mainline and Its Implications for the
Pacific Basin,'' Chapter 7 in Robert G. Jensen et al., eds., *Soviet Natural Resources in the
World Economy* (Chicago: University of Chicago Press, 1983), p. 172.

scale are not as important to the territorial production complex as are diversity and self-sufficiency. Each area typically seeks to include a combination of nationally significant export industries, regionally significant activities, and locally important production, as shown in Table 13-1.

Central Place Theory

Central place theory, developed by Christaller, also provides a framework for explaining the location of industry, even though its major contribution has focused on the explanation of the location of tertiary activities, which will be discussed in Chapter 18.[13] Firms with widely dispersed branch plant locations in various sizes of cities can best be accounted for with central place theory. Many industries deal with products that can be produced in both large- and small-scale plants and are found in all sizes of communities in direct proportion to their population. Typically, these industries have relatively high transportation costs and their products have a low value per unit of weight, or else perishability considerations require dispersed locations. The dispersed locations of soft drink bottlers in various sizes of cities demonstrate this principle.

The concept of *threshold* also assists in the explanation of manufacturing locations. Threshold refers to the minimum market size required to support an activity in terms of sales level. Manufacturers serving a local market are particularly sensitive to this minimum-market-size requirement. Food and bever-

age processors such as bakeries and soft-drink bottlers can operate successfully only in cities where a minimum threshold market size for that activity exists.

Larger cities provide access to larger markets and often support larger-scale operations of similar firms, as noted above, as well as those of firms having a larger threshold level. Beer production, for example, involves a greater entry threshold than that of a soft-drink bottler. Book publishing is an extremely high threshold industry associated with a very few large firms located in only the biggest cities, such as London, Tokyo, or New York.

Critique of Classical Theory

Weber's and Lösch's work, from the early twentieth century, speaks most directly to the location of firms having far simpler locational ties than those experienced by many modern corporations. The emphasis in both approaches is on economics. They are deterministic theories that prescribe a specific, fixed location for firms. They do not anticipate the impact of complicated noneconomic forces—or the large multifunction, multinational conglomerate firms of today that can create their own markets and infrastructure by internalizing many operations. For these reasons classical theory has been supplemented with more realistic behavioral and structural theory alternatives that attempt to identify more closely with the present business climate, recognizing that firms face more uncertainty, incomplete knowledge, and more interdependence with government and foreign competition than acknowledged with the classical approach. These approaches allow for many goals other than minimizing costs or maximizing profits.

[13]Keith E. Beavon, *Central Place Theory: A Reinterpretation* (London: Longman, 1977).

CONTEMPORARY BEHAVIORAL AND STRUCTURAL APPROACHES

Geographers have increasingly turned toward an analysis of the ownership, organizational structure, and management objectives of the firm when studying industrial activity in response to the inadequacy of traditional explanations of firm behavior. This behavioral approach emphasizes the dynamic and uncertain environment in which firms operate.[14] It has grown out of the awareness that today's larger and more diversified multinational firm exerts greater control over its business environment.

The behavioral approach also has limitations. It fails to take into account the impact of the political and social environment in which these decisions are being made. The *structuralist approach* overcomes this weakness by attempting to explain behavior using a historical-political perspective. The structural approach looks at the political economy of economic development in its broader context, encompassing regional development processes, urbanization/suburbanization, and technological innovation. It often fol-

[14]Gunnar Tornqvist, "The Geography of Economic Activities: Some Critical Viewpoints on Theory and Application," *Economic Geography*, 53 (April 1977), 153–162.

lows a Marxist orientation, with its emphasis on location as a consequence of historical and structural conditions governing the utilization of capital allocation or investment over space by firms.

Behavioral Location Theory

By focusing on firm decision-making processes, the behavioral approach provides an alternative perspective on the role of industrial location to the firm. Not surprisingly, the specter of alternative locations often appears to be rather low on the list of options facing a firm, especially in the short run. Many firms are uncertain and defensive in their behavior, which leads them to be very conservative and consider a shift in the location only as a last resort. Other options more frequently utilized include changing the firm's organizational or management structure, changing the product mix, or diversifying output.

Even in the long run, firms often implement de facto locational decisions rather than wholesale shifts. Frequently, this process simply involves on-site expansion (Figure 13-9). In other cases indirect locational strategies involve closures or establishing branch locations. Stafford has indicated that a *spatial increment model* can be useful to account for the behavior of

Figure 13-9 G. Heilman Brewery Expansion in Baltimore. *Firms are much more likely to expand on-site than choose alternative locations. Many firms are uncertain and defensive in their behavior, and consider a shift in location only as a last resort. Unitank construction shown here at the Heilman Brewery. (Courtesy G. Heilman Brewery)*

firms.[15] Typically, firms first locate branches just beyond the previous operating realm. In other words, they place branches no further away than necessary, typically within a 200-mile distance, which is a one-hour trip for a corporate jet.

Uncertainty. In the face of uncertainty, firms often strive to internalize more of their operations—partly as a defense mechanism. This process can lead large multiregional and multiproduct firms to acquire their own material sources and even manipulate the market for their goods. The scale of such operations can broaden to include a multinational presence, allowing firms to control more resources than a host country. By developing vertical integration, firms remove many unknowns but still face coordination and forecasting problems relating to the demand for their products.

Vertical integration in fact reduces a firm's vulnerability to business cycle changes, making the firm more like a diversified conglomerate. Vertically integrated firms tend to stabilize market conditions because output at every stage can be coordinated with market demand conditions. Recall the discussion in Chapter 12 of the role that multinational firms play in smoothing out business-cycle fluctuations in mineral resource industries.

The growing scale of industrial operations, the amount of capital commanded, and the leverage that such firms exert over their *external environment* have become important factors in world industrial geography. The number of firms with more than $1 billion sales in the United States in 1977 was 242, or 15 more than in 1976. From another perspective the concentration is even more dramatic. In the United States 500 firms now control four-fifths of the manufacturing assets.[16]

The increasing scale of firms leads to higher thresholds of entry for competitors. It also means that a greater quantity of industrial activity in a local area is controlled from the outside. Such externally controlled firms may not be as involved in the local economy in terms of making purchases from local suppliers, nor be as responsive to local civic and business needs. One study of industrial firms in Israel, for example, indicated that medium-sized plants (employing about 350 persons) would be better for the growth of a local economy than large ones because the former are not large enough to internalize sources of supply and services.[17] Therefore, medium-sized firms patronized local businesses more often than did larger ones.

Corporate acquisitions provide an excellent means to expand and solidify external control for the firm. Sometimes this control even leaves the country, following acquisition. Foreign control, for example, has long been a concern in third-world countries. Even in highly developed areas such as Canada and Australia, foreign domination is an issue. Frequently, administrative control flows to major metropolitan areas as growth occurs. The case of New York provides a good example of the concentration of international industrial corporate headquarters in a major financial center. Similar processes have increased the importance of Toronto in Canada and Tokyo in Japan.

Business decisions of a firm result from corporate strategies that may or may not be related to profit maximization or least-cost locations or other classical motivating factors. Other goals could be to increase the market share, drive out the competition, enhance product diversity, or simply increase the security of the firm's management by promoting growth.

Some researchers suggest that industrial behavior is governed as much by *risk minimization* as by cost accountability.[18] The uncertainty associated with risk can generally be reduced by concentrating operations in larger urban areas because of greater access to supporting services and professional advice, as well as easier access to markets. Uncertainty normally increases with distance from urban areas because access to the stabilizing factors just described decreases. Uncertainty also encourages firms to maintain smaller plant sizes.[19] In centrally planned economies, uncertainty is minimized by eliminating firm competition. Some observers say that this helps explain the weaker tendency for concentrating activities closer together in planned environments.[20]

As a defensive measure, firms often make "safe" choices by selecting locations near other firms producing similar products, rather than by dispersing. For example, rival cereal manufacturers clustered in Battle Creek, Michigan (Kellogg's and Post). The potential for learning the competitor's secrets and creating an atmosphere of intimidation and bluff is reinforced when firms are in close proximity, which in turn enhances market competitiveness and innovation. A certain *pooling of reserves* also occurs in such locations.[21] This concept refers to the greater likelihood of a successful divestiture of buildings and equipment for firms located in existing urban industrial locations should potentially adverse future business conditions dictate closing the business.

[15]Howard A. Stafford, *Principles of Industrial Facility Location* (Atlanta: Conway Publications, 1979), p. 12.

[16]Rodney Erickson, "Corporations, Branch Plants, and Employment Stability in Nonmetropolitan Areas," Chapter 9 in John Rees et al., eds., *Industrial Location and Regional Systems* (New York: Bergin, 1981), p. 136.

[17]Baruch A. Kipnis, "The Impact of Factory Size on Urban Growth and Development," *Economic Geography*, 53, (July 1977), 295–305.

[18]C. D. Norcliffe, "A Theory of Manufacturing Places," in L. Collins and D. F. Walker, eds., *Locational Dynamics of Manufacturing Activity* (New York: Wiley, 1975).

[19]Michael J. Webber, *Impact of Uncertainty on Location* (Cambridge, Mass.: MIT Press, 1973), pp. 273–282.

[20]Aleksander P. Gorkin and Leonid V. Smirnyogin, "A Structural Approach to Industrial Systems in Different Social and Economic Environments," p. 33.

[21]L. Lefeber, *Location in Space* (Amsterdam: North-Holland, op. cit., (see fn 8) 1958).

Operating Environment. For many industrial firms a core region exists within which most operations are conducted. That area has been described as the *task environment* for the firm.[22] It is the territory the firm calls its home, the territory that it knows well. It is the area "identified with the company." The region can exist at two or more levels. First, at the *executive network* level, the corporate administrative realm deals with management decisions. This network may include other major firms, financial and legal consultants, as well as the board of directors. Research and development operations frequently accompany the headquarters administrative function. Often this interaction occurs around a major national headquarters city, such as New York, Tokyo or Paris.

A second type of linkage is that which exists between the core region and *subsidiary production plants* in the periphery. Branch operations located remotely from the headquarters typically maintain strong communication linkages between the production center and the headquarters office, as well as between the production facility and its suppliers. Once these *contact patterns* are established they reinforce existing locational ties between the core and periphery. If a firm has a regional system of plants, material flows and information ties link them together, as does the more centralized executive network. Such complex overlapping patterns of contact explain the long-term stability of large manufacturing complexes in the northeastern United States, western Europe, and Japan.

Contact patterns obviously vary with the type of firm. Smaller companies generally have more ties with outside supplier firms for products and services, as mentioned earlier. These services include payroll and inventory control as well as sales and advertising services. The more specialized the firm, the greater the demand for services and the closer the connection with top management.

The international arena provides a third level to the firm operating network. Often this "periphery" provides additional production capacity as well as marketing opportunities. Control nevertheless remains in the core. Japanese and European firms have been particularly active in a variety of industries on the international level. The small size of domestic markets in Japan and western Europe literally forced corporations to assume this posture to gain needed economies of scale. Examples of these firms headquartered in Europe include:

Firm	Headquarters	Products	Employees
Unilever	Netherlands	Consumer goods	300,000
Philips	Netherlands	Electrical equipment	377,000
Siemens	West Germany	Electrical equipment	344,000
Nestlé	Switzerland	Food products	140,000

[22]Robert B. McNee, "A Systems Approach to Understanding and Geographic Behavior of Organizations, Especially Large Corporations," in *Spatial Perspectives on Industrial Organization Decision-Making*, F. E. Ian Hamilton, ed., (New York: Wiley, 1974), pp. 47–76.

Unilever operates 400 to 500 subsidiaries in 60 countries, while Nestlé operates 198 factories, only four of which are in Switzerland. All these firms have at least one-fourth of their assets or sales outside their home countries.

The Japanese are also very prominent on the international scene. Leading Japanese firms include Mitsubishi, Mitsui, C. Itoh, Marubeni, and Sumitomo, all large wholesale trade companies called *soga sosha*. Soga sosha traditionally controlled Japanese trade in industrial goods. They remain major heavy industry entrepreneurs but have diversified into high-technology fields. They also participate as investors in third-world mega-projects dealing with minerals and petrochemical projects such as liquefied natural gas (LNG).

The nine largest soga sosha command over one-half trillion dollars in annual sales. Much of their investment occurs in declining industries such as aluminum, petrochemicals, shipbuilding, and textiles. In some ways soga sosha have become the import–export brokers or distributors moving goods in and out of Japan, but they are also major facilitators of trade among third-world countries. Some have developed associations with American or European manufacturers to produce their products as a licensee, while others promote marketing and service networks.

Some Japanese companies have aggressively promoted western-style food and restaurant operations in Japan. Mitsubishi handles Kentucky Fried Chicken and Shakey's Pizza restaurants in Japan as part of an elaborate integrated system. In turn, Mitsubishi controls a Japanese noodles restaurant chain in New York City. Mitsubishi buyers purchase grain in the United States, which is shipped via its ocean fleet to Japan to company-controlled poultry farms as feed for chickens. In turn, chickens are sold to the restaurant chain, completing the cycle from the U.S. grain elevator to a bucket of chicken in Tokyo. American fast-food industry firms have been more successful in penetrating the Japanese market by establishing ties with a soga sosha for that purpose rather than attempting direct marketing. Denny's, McDonald's, Dunkin' Donuts, Shakey's, Dairy Queen, and Wendy's, among others, provide examples of successful U.S. based restaurant chains operating in Japan.

Plant Location Decisions. The location decision process facing firms typically involves top management personnel choosing among alternatives following a thorough review of information available to them, which includes data on firm operations, forecasts, and their own opinions and preferences. Such decisions are necessarily judgmental but they are informed, logical, and rational, if not optimal, given the complexity of the issues involved.[23]

The necessity to relocate to new sites occurs in only about 40 percent of the cases where location decisions are involved, because most decisions (60 to 80

[23]Stafford, op. cit., p. 10.

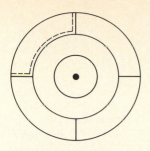 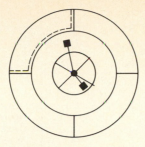 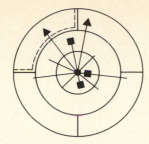

● Headquarters and
production facility

● Headquarters and specialized
equipment production
■ Branch plant for consumer
goods production.

● Headquarters and specialized
equipment production
■ Domestic branch plants
▲ International operations

Figure 13-10 Evolution of Firm Organizational Structure over Time. *From a single operating site, firms often break up operations over time by establishing branch plants while the headquarters and specialized production remains at the original location. This process can eventually lead to international operations as the diagram indicates. (Redrafted after Håkanson, in Hamilton and Linge,* Spatial Analysis, *Wiley, 1979, p. 132)*

percent) involve on-site expansion. When the need to move does occur it is undertaken to solve a problem and needs to be exercised smoothly and efficiently. Top management decision makers like to avoid detailed and difficult analysis and protracted negotiations in these circumstances. Although the process does not necessarily entail crisis management, it is likely that an atmosphere of stress does exist during this process. The pressures facing executives contemplating locational changes are very difficult to evaluate because so many subjective decisions are necessary in terms of both internal operations of the firm and its environment. Smaller firms can potentially do a better job in this assessment process than larger ones because they have simpler external ties with other firms (suppliers and customers).

As mentioned earlier, the location decision can take various forms, ranging from on-site expansion or closure to partial movement of operations, establishing branches, or even moving the entire operations. Quantitative adjustments in the size of the firm can be made relatively frequently. Mergers, acquisitions, and decisions to build branch plants involve location decisions that are also undertaken relatively more frequently than complete firm transfers. When complete firm transfers occur, they normally involve only short-distance shifts. A sample survey of firms in the plastics industry in Great Britain, for example, found that the median distance of a complete transfer move was 13 miles, whereas the median for the establishment of a branch plant was 179 miles, and 128 miles for acquisition.[24] These findings reinforce the notion that *locational inertia* is indeed a strong factor, and that once the roots of a firm become deeply entrenched in an area, complete relocation is rare.

The organizational structure of the firm can affect company expansion policy. Family-owned firms, which accounted for 42 percent of the top 108 U.S. corporations in the early 1970s, often prefer branch plants as an avenue for expansion.[25] In contrast, professionally managed, publicly owned firms, such as General Motors, typically prefer acquisition as a growth prospect. Branches have the advantage of maintaining the family influence factor, while acquisitions appeal to corporations that must maintain a strong profit posture for their stockholders.

Branch plants frequently produce standardized, mass-produced, assembly line items that do not require highly skilled labor. Typically, the production of more specialized products continues in established facilities. Branch operations do not require close top-level communications with headquarters locations and offer lower costs of production, as they can operate in lower-labor-cost locations (Figure 13-10). Dispersed branch facilities might handle mass-marketed consumer products for a firm, while special-order industrial machines and equipment remain competitively produced at traditional sites. The dispersion of home-air-conditioning-unit production away from the manufacturing belt in the northeastern United States to sunbelt plants provides an example of this split.[26] Output of larger, more specialized commercial and industrial air-conditioning units remains tied to traditional production areas in upstate New York, Pennsylvania, and Ohio.

Branch plants often gravitate to rural environments that have lower land and labor costs. Much of the decentralization of the U.S. machinery industry to the south in the 1970s can be explained in this context. This situation helps explain why manufacturing growth in nonmetropolitan areas expanded more rapidly in the United States in recent years than in metropolitan areas, particularly for traditional activity not associated with the high-technology revolution. When choosing new locations, firms typically follow a three-step process that begins with regional assessments, followed by more localized city-by-city comparisons and finally, site level decisions.

[24]D. J. North, "The Process of Locational Change in Different Manufacturing Organizations," op. cit., p. 223.

[25]Philip H. Bunch, Jr., *The Managerial Revolution* (Lexington, Mass.: D. C. Heath, 1972).

[26]The North American manufacturing belt encompasses the majority of manufacturing activity in the United States and Canada. It roughly extends from Boston westward through upstate New York, southern Ontario, southern Michigan, and southeastern Wisconsin, turning south to St. Louis, then extending eastward along the Ohio River valley to Baltimore on the east coast.

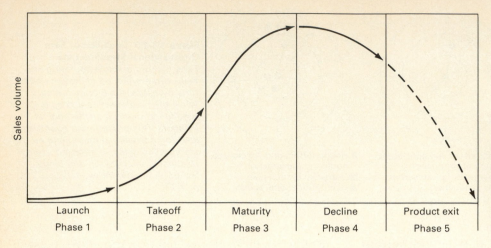

Figure 13-11 The Product Life Cycle. *The birth, growth, and death of products follow a sequence of stages that can be depicted by an s-shaped curve. Studies of product life cycles have relevance for sales and marketing, evaluating economic development processes, and as a barometer of structural changes in economies.*

Product Life Cycle. Changes in manufacturing patterns due to births and deaths of firms can dramatically affect the distribution of activity over time. Using the *product life* of the firm concept, one can observe not only patterns of spatial change in industry, but structural changes as well. The product-life-cycle concept refers to the birth, growth, and death of products.[27] Several stages in the life of a product are depicted in Figure 13-11. An S-shaped curve generally captures the growth and decline of the product, which is not unlike a diffusion process. Phase 1 of a product typically involves slow sales following its commercial introduction into the marketplace. In phase 2 growth increases rapidly, followed by a maturing stage in phase 3, followed by a decline in phase 4, and a withdrawal of the product in phase 5. Each product would obviously have a different experience in the marketplace—some with much longer life spans than others. Among the variables influencing its longevity would be the size and stability of the market served, the degree of "innovativeness" of the product, the rate of technological change that affects its use, and its ability to adapt to new applications.

Studies of product cycles have relevance not only for sales and marketing but also for evaluating economic development processes and structural changes in economies. One aspect of product development on which these studies focus attention is *research and development*. Although the role of development on product life is not well understood, it has become important in innovations in high-technology activity in recent years. We return to this topic in Chapter 16.

In marketing terms, a similar developmental scenario can be observed for a product. In the early or launch phase of the product, strong competition occurs with other products and firms. Many widely dispersed companies often manufacture a product in its formative years. Some producers enter a field as a sideline pursuit. During the takeoff stage a producer gains market share and the market itself grows in size.

Typically, a consolidation in the number of firms producing the product occurs in the stabilization phase. Over time most producers are typically forced out of business or merge, leaving only a few producers in a few locations. Some products pass through these phases in a few years to a decade, while others evolve over a period of 20 to 50 years.

Just as a sifting out of firms and locations occurs in the early phase of product development, another locational dynamic encourages dispersion of facilities in later phases. The emergence of a few dominant manufacturers often leads to the decentralization of plants to many locations to take advantage of production cost and distribution savings. This dynamic arrangement demonstrates an *hourglass effect*, shown in Figure 13-12. The automobile industry in North America provides a case in point.

Early locations of automobile producers exhibited wide dispersion. Some early participants were wagon and buggy makers, while others were machine tool businesses. Gradually, the small, weakly managed, weakly financed, and poorly located firms faltered, especially those not offering cars with gasoline engines (steam, electric), and a centralization gradually occurred. This consolidation of the industry unfolded in the 1910–1940 period, mainly around Detroit. The creation of many branch body assembly plants throughout the country followed World War II. Three parent firms (General Motors, Ford, and Chrysler) fostered this development, whereas dozens of firms accounted for the original dispersed pattern. This dispersion–concentration–dispersion process also characterizes many other industries. It amplifies the importance of branch plants in serving local and regional markets as an industry matures and becomes dominated by a few companies.

One manifestation of product-cycle changes visible on the landscape today is the transition to a post-industrial society. This shift continues to involve significant plant cutbacks in "smokestack" industries in all developed countries. In most countries this shrinkage has involved a downsizing of the workforce, but in the United States, newer jobs are being created faster than the smokestack loss. Small and medium-

[27]Morgan Thomas, "Industry Perspectives on Growth and Change in the Manufacturing Sector," Chapter 3 in Rees et al., op. cit., pp. 41–58.

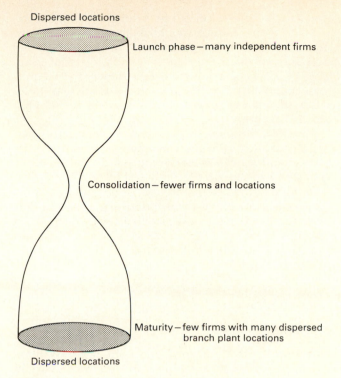

Dispersed locations

Launch phase—many independent firms

Consolidation—fewer firms and locations

Maturity—few firms with many dispersed branch plant locations

Dispersed locations

Figure 13-12 Hour Glass Industrial Location Model. *Many firms typically produce products at widely dispersed locations in the early stages of the industry, followed by periods of consolidation and later, another period of decentralization. These phases are consistent with the product life cycle process.*

sized businesses are leading this expansion, not large firms. Some of the jobs are innovative high-technology activities, but most are low-technology or no-technology businesses. Many new jobs are not occurring in the manufacturing sector at all, but in service-related businesses such as restaurant employment. This trend is discussed in Chapters 17 and 18.

Some observers argue that employment is growing rapidly in the United States because it has successfully shifted its economy to a strong entrepreneurial base that rewards risk and innovation. The plentiful availability of venture capital, a relatively low inflation rate, and stable political environment also favor this expansion. By the mid-1980s this dynamism in the United States exceeded that in western Europe and appeared much more vibrant than the expansion in Japan.

Structural Approach

The structural approach to industrial location focuses attention on the political economy of economic growth and its impact on industrial location. Advocates examine the cause of economic growth and technological innovation starting with a broad top-down view of the economy rather than one focusing on the firm and its environment. A complex organizational structure for industry is assumed, one that strongly influences firm behavior. Conditions in the free-market economy that influence the firm, for example, include the profit motive and a high level of competitiveness.

From a Marxist perspective, the structuralist argument suggests that competitive processes lead "to a fall in the general rate of profit, forcing firms to find whatever means may be available to keep up profits and accumulate capital in the interest of survival."[28] Massey and Meegan, for example, studied problems associated with the electrical engineering industry in Great Britain in the late 1960s from this perspective.[29] That industry, beset with excess capacity, faced eroding profits. This in turn led to cost cutting and plant closures to reinstate higher profit levels. Less productive units shut down to lower overhead costs. Labor costs savings accrued by moving plants to cheap-labor locations and the introduction of more efficient standardized production techniques. This process in turn decreased skill-level requirements in the workplace and led to a disenfranchisement of labor. Many skilled jobs disappeared and those that remained led to the assignment of skilled workers to caretaker work no longer requiring their skills, due to the introduction of computer-controlled machine tools and product standardization. We can use the term *de-skill* to describe this process of downgrading work assignment due to the obsolescence of skilled labor.

The Marxian theory that crisis and cyclical instability such as that discussed in the preceding paragraph operate as normal capitalist processes plays a central theme in the overall structuralist argument. According to this thinking, as capital accumulation occurs, dysfunctions accumulate, creating a crisis of overproduction. Overproduction results from plant expansion promoted by the extension of credit, which in turn lowers profits and encourages management to increase labor productivity. Technological change in this context is viewed as a product of competition and of the desire to control labor and raise productivity. As long as new industry can evolve to replace jobs lost to technological change, some would argue that this process is not only inevitable but healthy. Nevertheless, these short-term and even long-term structural changes must be made to overcome excess capacity in declining industries. The structuralist argument therefore helps us understand problems created by product-cycle theory and the changes in entire economies.

In the British electronic industry case it is apparent that industries responded to higher costs by seeking both cheaper-labor locations and mechanization to reduce labor costs. It is interesting to observe that responses to such changes vary from country to country. In Japan, for example, higher labor production

[28]David M. Smith, *Industrial Location: An Economic Geographical Analysis* (New York: Wiley, 1981), p. 140.

[29]D. B. Massey and R. A. Meegan, "The Geography of Industrial Reorganization: The Spatial Effects of the Restructuring of the Electrical Engineering Sector under the Industrial Reorganization Corporation," *Progress in Planning*, 10 (1979), 155–237.

costs in the 1970s encouraged increased mechanization, such as the introduction of automation and robots. By contrast, in the United States, industry typically moved south and offshore to cheaper labor locations rather than aggressively incorporating automated plants. In Japan, because of its small size and a relatively uniform high cost structure for production throughout, the low-cost-location alternative did not exist. Today, the Japanese pursue offshore locations more aggressively, and U.S. firms have turned to an emphasis on automation, reflecting a convergence of industrial strategies in the two countries. Japanese investments in automobile plants in Europe and the United States illustrate this point, as does the greater use of robots in the U.S. machinery and automobile industry.

In recent years, the greater willingness by the Japanese to accept technology innovations than by their American counterparts reflects contrasting responses to change in the two countries in the past. Cultural and religious differences between the two peoples partially explain these alternative approaches. The Japanese people, for example, do not fear technology as much as do many Americans. They see technology as a natural companion for a people working together for the mutual good. The emphasis in Buddhism, for example, on interconnection and compromise makes it much easier for the Japanese to welcome change than does the more confrontational teaching of Christianity, which emphasizes good and evil, debate, and argument.

Organizational Contrasts. Strong contrasts in the structural organization of manufacturing exist between free-market and centrally planned economies. Although common production and technological processes prevail worldwide, the organizational and spatial characteristics of that activity display differences. Contrasts in the relative strengths of horizontal and vertical linkages between free-market and centrally planned countries provide one means of comparison. In free-market countries such as the United States the competitive work environment yields close horizontal linkages among firms. This situation arises from the strong competition among firms for access to raw materials and the marketing overlap that exists in product distribution. This is not to say that strong vertical linkages do not exist as well. Such vertical ties are typically overshadowed, however, by horizontal ties.

In the planned economy organizational structure, vertical ties similarly overshadow horizontal linkages because competition is minimized. The planned economy organizational structure operates almost exclusively in a top-down command environment, as in the Soviet Union. National economic plans typically guide state executive authorities in planned economies, such as councils of ministers and state planning bodies. The Gosplan in the Soviet Union is an example. Under this umbrella, ministers, directorates (GLAVIK), and associations of enterprises or production unions implement goals. Enterprises form the grass-roots production function, the counterpart of the firm in the capitalist world. There is no segmentation or overlapping of markets in planned economies. Each enterprise maintains a monopoly in its particular field. Often the market is split up into regions, for which a particular enterprise is a sole supplier.

Vertical ties exist in free-market economies when firms establish downstream (raw material) supply sources and upstream marketing arrangements, as occurs with firms involved in mineral resource recovery. Automobile firms frequently develop similar arrangements. As the scale of firms has grown in recent decades, the tendency to intensify vertical linkages has grown. The point being made here is not that these ties are insignificant, but that horizontal linkages are much stronger in comparison to those in planned economies.

SUMMARY AND CONCLUSION

Although many traditional and contemporary manufacturing location principles and theories assist in the explanation of industrial patterns on the landscape, none by themselves are completely satisfactory in accounting for the enormously complicated pattern of activities today. Not only have additional stages in the production process added more complexity but so has the increase in the scale at which firms now operate.

Changes in technology, and the increase in the number of countries seeking expanded manufacturing activity, also come at a time that the entire structure of economies is shifting to less dependence on the manufacturing sector. Political and ideological factors related to the role of capital, planning, and international trade and investment strategies also play a more visible part in manufacturing. We will return to a discussion of the geography of international business in Chapter 20, following a review in Chapters 14 to 16 of several industries that are being shaped by these forces, and a discussion in Chapters 17 to 19 of the service economies and urban development.

The contribution of classical location theory to an understanding of manufacturing location has been downplayed in recent years; nevertheless, raw material costs, transportation, and labor considerations which are emphasized in these theories remain central topics in discussions of manufacturing activity. The growing importance of horizontal and vertical linkages and ownership patterns has changed the focus of analysis to be sure, but the multinational firm which can internalize many factors must still consider the same variables that the small entrepreneur did a century or more ago when making business decisions. Indirect locational decisions such as firm closings, openings, and on-site expansion occur much more frequently than a complete shifting of firm operations. The establishment of branch plants or simply changing the product mix can also be used as a substitute for moving to a new location.

Behavioral studies of the firm increasingly emphasize decision making and management objectives when discussing location strategies. The product-life-cycle concept points out that all manufactured items experience stages in expansion and contraction. Some firms must also endure fixed locations in both the short and long run and must make adjustments other than spatial shifts as changing business conditions dictate.

The structural approach to manufacturing location takes into account the political and social environment in which decisions are made and provides insight into the nature of the instability and dysfunctions created when firms face obsolescence. The severe adjustments affecting labor when technological shifts lead to the substitution of mechanization and robotics for manual labor provides a case in point. Such adjustments are now occurring most visibly in developed countries. We look at the forces that are presently shaping the textile and apparel industry in Chapter 14, the iron and steel industry in Chapter 15, and the high-technology field in Chapter 16.

SUGGESTIONS FOR FURTHER READING

AMIN, A. and J. B. GODDARD, eds., *Technological Change, Industrial Restructuring and Regional Development*. Winchester, Mass.: Allen and Unwin, 1986.

HAMILTON, F. E. IAN, and G. J. R. LINGE, eds., *Spatial Analysis, Industry and the Industrial Environment*, Vol. 1, *Industrial Systems*, Vol. 2, *International Industrial Systems*. New York: Wiley, 1979, 1981.

HEWINGS, GEOFFREY J. D., *Regional Industrial Analysis and Development*. New York: St. Martin's Press, 1977.

HEWINGS, GEOFFREY J. D., *Regional Input-Output Analysis*. Beverly Hills: Sage Publications, Inc., 1985.

LONSDALE, RICHARD E., and H. L. SEYLER, eds., *Nonmetropolitan Industrialization Development*, Berkeley, Calif.: University of California Press, 1980.

McDERMOTT, P. J., and MICHAEL TAYLOR, *Industrial Organization and Location*. New York: Cambridge University Press, 1982.

MILLER, E. WILLARD, *Manufacturing, a Study of Industrial Location*. University Park, Pa.: Pennsylvania State University Press, 1977.

MORIARTY, BARRY M., *Industrial Location and Community Development*. Chapel Hill, N.C.: University of North Carolina Press, 1980.

OUCHI, WILLIAM, *Theory Z: How American Business Can Meet the Japanese Challenge*. Reading, Mass.: Addison-Wesley, 1981.

REICH, ROBERT B., *The Next American Frontier*. New York: Penguin Books, 1984.

RICHARDSON, BRADLEY M., and TAIZO UEDA, *Business and Society in Japan: Fundamentals for Businessmen*. New York: Praeger, 1981.

SCHMENNER, ROGER, *Making Business Location Decisions*. Englewood Cliffs, N.J.: Prentice-Hall, 1982.

SCOTT, ALLEN J. and MICHAEL STORPER, *Production, Work, Territory: The Geographical Anatomy of Industrial Capitalism*. Winchester, Mass.: Allen and Unwin, 1986.

STAFFORD, HOWARD A., *Principles of Industrial Facility Location*. Atlanta: Conway Publications, 1979.

SWEET, MORRIS L., *Industrial Location Policy for Economic Revitalization: National and International Perspective*. New York: Praeger, 1981.

TAYLOR, MICHAEL, and NIGEL THRIFT, eds., *The Geography of Multinationals: Studies in the Spatial Development and Economic Consequences of Multinational Corporations*. New York: St. Martin's Press, 1982.

VOGEL, EZRA F., *Japan as No. 1: Lessons for America*. Rutland, Vt.: Charles E. Tuttle, 1979.

WALKER, RICHARD, and MICHAEL STORPER, *Capital and Industrial Locations*. Berkeley, Calif.: Institute of Urban and Regional Development, University of California, 1980.

WATTS, H. D., *The Large Industrial Enterprise: Some Spatial Perspectives*. London: Croom Helm, 1980.

Changing Order in Textile and Apparel Production

14

The array of manufactured goods having international origins expanded greatly in the 1970s. Today, most manufactured products are truly international, owing to the greater use of raw materials, energy, capital, labor, and/or equipment from more than one nation. This process accelerated after World War II with the diffusion of manufacturing technology to third-world countries and the rapid growth of independent nation-states. The textile industry pioneered the internationalization process that created the new industrial order and associated increases in international trade (Figure 14-1). Over time, various other consumer necessities, such as footwear and apparel, experienced this transition, and later the internationalization phenomenon grew to include more capital-intensive goods. In the 1980s the consumer electronics, automobile, and steel industries became prominent examples of the changing international order. These industries are discussed in later chapters.

In the first segment of this chapter we discuss the diffusion of industry from the traditional core area to the periphery, together with problems associated with this process. The role of multinational firms in fostering this transition, and the rapid emergence of the newly industrialized countries (NICs) as participants, will provide a frame of reference. The dramatic increase in the range and quantity of goods flowing to and from major regions of the world as a result of the changing order is also highlighted.

In the latter part of the present chapter we review recent trends affecting the world pattern of textile and apparel production, including a discussion of the ways that developed countries have modified the structure of their traditional industry mix to respond to the changing world order. The roles of technology applications, economies of scale, market research, strategic planning, and government initiatives are discussed.

DIFFUSION FROM CORE TO PERIPHERY

Typically, the development of new products or processes occurs in technologically advanced nations such as the United States, Canada, Japan, and western Europe. Research and development takes place most intensively in such a setting. Examples of recent innovations resulting from intense research include the laser, fiber optics, holography, and ceramic engines, and before that the silicon chip. As market applications emerge for these items, the demand for them increases and mass-production techniques emerge, as the product-cycle concept suggests.

At the same time as new innovations occur in advanced countries, routine production of traditional goods often becomes too expensive, due to high labor costs, and mass-produced-goods manufacture frequently shifts to lower-cost production areas in the developing world, especially the newly industrialized countries (NICs). This process represents the most re-

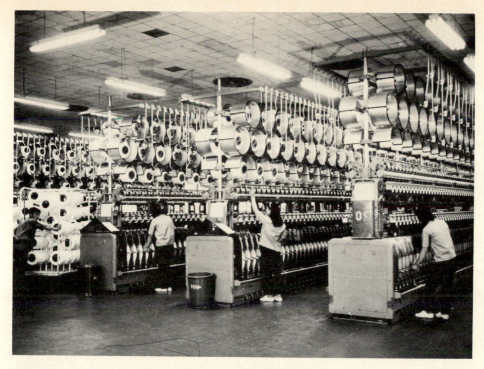

Figure 14-1 Textile Mill in South Korea. *Modern automated thread manufacturing plants using synthetic and cotton fiber have become very competitive in newly industrializing countries such as South Korea. The spooling of thread on cones can be observed in this photograph. (Samsung Company Limited)*

cent phase of the industrial revolution that started in Great Britain with the textile industry itself 200 years ago.[1] Present-day shifts in the location of textile and apparel production reflect this ongoing process involving the expansion of the base of industrial activity.

The age-old concept of *comparative advantage* explains the tendency for advanced nations to specialize in capital-intensive activities, while less developed areas focus on labor-intensive products. By emphasizing activities that yield the greatest return on investment, nations can concentrate on fewer products, relying on international trade to import products in which they have a production cost disadvantage. In this way economies of scale can be maximized and all areas can compete regardless of their stage in economic development. Both products and nations exhibit shifts in output over time as spatial and structural conditions change. The evolution of the Japanese economy provides a good example of these transitions.

Transition in the Japanese Economy

Japanese industry shifted its heavy reliance on labor-intensive textiles to raw materials processing and capital-intensive products, including steel and shipbuilding, in the 1950s. In the 1960s the Japanese moved into capital- and machinery-intensive automobiles and appliance and electronic outputs such as radios, stereos, and televisions. In the 1970s high-technology

[1]Franklin R. Root, "Some Trends in the World Economy and their Implications for International Business Strategy," *Journal of International Business Studies,* 1984, p. 21.

computer products received priority. Knowledge-intensive products related to biotechnology, ceramics, and intelligent fifth-generation computers have the edge in the 1980s. These phases of industrial change illustrate the notion of comparative advantage and the dynamics of a product life cycle discussed in Chapter 13.

The aggressive attempts by the Japanese to move their economy from a heavy dependence on raw materials to a knowledge-intensive base appear successful. Japanese dependence on imported raw materials, discussed in Chapter 12, became even more precarious for the economy following the 1973–1974 energy crisis. Wage rate increases also reduced the competitiveness of traditional activity in Japan as in other developed countries. In addition, growing environmental concerns mitigated against expanded output of smokestack activities (steel, chemicals, etc.) in this period. The federal Ministry of Trade and Industry (MITI) has been particularly active in promoting this transition away from traditional industry.

This involvement of MITI in industrial change is the most recent example of a long history of government and free-enterprise joint involvement in economic matters in Japan. For decades the Ministry has worked hand in glove with the private sector to enhance the business climate of the country. The activity of this agency in guiding industrial development and trade has led to the use of the term "Japan Incorporated" to describe the economy. This agency, together with others, advises, persuades, and insists on many concessions from the private sector. Consultation and consensus typically characterize negotiations between the private and public sectors rather than heavy-handed interventions or direct control.

The Ministry of Trade first emerged in Japan during the reconstruction period following World War II, when it encouraged a few key sectors of the economy to expand rapidly. It has since been deeply involved in establishing import–export guidelines, and an energy policy, as well as deeply influential with specific industries, including automobiles, iron and steel, textiles, and computers. Sometimes industrial consolidations are encouraged; at other times quotas, licensing restrictions or subsidies are meted out to protect specific activities. Whatever the technique, it is clear that this public–private way of doing business is standard operating procedure in Japan. MITI, for example, became deeply involved in the petroleum industry in the 1950s and 1960s, encouraging consolidation of smaller domestic refineries and purchasing shares of companies owned by foreign corporations such that foreign ownership in the industry declined from 70 percent in the mid-1960s to 40 percent in 1978. MITI has also recently directed export restraints for steel, automobile, and shipbuilding sales to the United States.

STAGES OF INDUSTRIAL CHANGE

Based on the Japanese experience and that of several newly industrializing countries (NICs), it is possible to identify at least three stages in the evolution of manufacturing activity as countries try to become first, self-sufficient and later, net exporters of industrial goods. Although the timing of the stages varies from country to country, most NICs have moved through at least two of these three stages in the past 25 years.

Before proceeding, let us define what is meant by the term *newly industrializing country* (NIC). A newly industrializing country is a rapidly expanding third-world country in terms of manufacturing output. The Organization for Economic Cooperation and Development (OECD)[2] identifies 10 NICs.[3] Europe and Asia each have four of these countries, while two are in the Americas. Each had a rapidly expanding industrial economy and increased its export share of manufactured goods significantly in the 1970s. The dramatic increase in exports of seven of these countries in that period is shown in Table 14-1, paced by Taiwan, South Korea, and Hong Kong.

[2]The OECD is a group of 24 market economy developed countries from four continents that work together to solve common problems and seek to promote economic growth and improved standards of living worldwide, including the poorest of the developing countries. Established in 1961, the organization is governed by a council composed of representatives from all member countries, chaired by a secretary general. The agency has an active statistical data-gathering and publications program. Recent emphasis has been placed on publications in the areas of economic policy, energy, environment, agriculture, and industrial policy.

[3]These 10 newly industrialized countries include Taiwan, South Korea, Hong Kong, Spain, Yugoslavia, Singapore, Mexico, Brazil, Greece, and Portugal.

Table 14-1

Manufactured Goods Exports From Newly Industrialized Countries, 1973–1979 (billions of U.S. dollars)

1979 Rank	Nation	Exports 1973	Exports 1979
1	Taiwan	3.8	14.1
2	South Korea	2.7	13.4
3	Hong Kong	3.6	13.2
4	Singapore	1.6	6.4
5	Brazil	1.2	5.6
6	Mexico	1.5	3.2 (1977)
7	Argentina	0.7	1.6 (1978)
	Total	18.0	57.5

Source: Colin Bradford, Jr., ''The Rise of the NIC's as Exporters on a Global Scale,'' Chapter 2 in Louis Turner and Neil McMullen, eds., *The Newly Industrializing Countries: Trade and Adjustment* (London: Allen & Unwin, 1982).

Stage 1 of manufacturing growth, the resource export/finished goods import stage, represents the traditional preindustrial role of third-world countries as raw materials supplier to the developed world, income from which financed the import of finished goods from developed countries. North/south and south/north flows predominate in this stage. Often, this export–import relationship produces unsatisfactory income flows to developing or colonized countries, as often happens with the mercantilist process described in Chapter 2. This imbalance often provides the motivation to move to stage 2.

In stage 2 of the manufacturing process, *import substitution,* developing countries strive to replace foreign imports with domestically manufactured products. Import substitution as an economic development policy appeals to developing countries, as domestic production can be substituted for imports, saving money otherwise lost in foreign exchange. It also fosters the industrialization process by assisting the transformation of the traditional bazaar economy into an industrial state. Often, emphasis is placed on producing sophisticated consumer goods, which requires imported raw materials and capital to build industrial plants. Frequently, such facilities also require skilled labor.

The import substitution process frequently proves unsatisfactory for developing countries because these nations typically have an abundance of unskilled labor, which may not benefit. The high-skill requirement of labor needed by these highly capitalized industries excludes most of the labor force and does little to foster the growth of a mass-consumption middle-class market required for the long-term success of businesses producing specialized products. The premature expansion of specialized manufacturing can also deplete foreign exchange reserves and create balance-of-payment problems. Not only must the plant and technology be purchased abroad, but the need for imported raw materials also increases, further taxing the fragile local economy.

Domestic demand in developing countries is typically not large enough to make manufacturing facilities efficient, resulting in higher prices for locally produced products than for similar imported goods. This problem can lead to another—the imposition of tariffs and other forms of protection to enhance local products in relation to competing foreign goods—sheltering the domestic industrial sector. Local producers in turn can raise prices, which further restricts the market for the product, and often negatively affects other sectors of the economy. Moreover, this uncompetitive price structure limits the ability of producers to tap export markets. These situations have led some countries to rethink the import substitution policy and place more emphasis on products for which greater potential exists for generating currency from exports.

A country electing to change priorities and emphasize exports enters stage 2A, *labor-intensive activity substitution.* This adjustment involves placing greater emphasis on labor-intensive, low-capital-investment activities rather than more capital-intensive products. It is often much easier for developing countries to create a comparative advantage in these low-capital-investment activities, such as textiles, footwear, and other entry-level manufacturing operations such as toy and game production. In the early 1960s, South Korea and Taiwan made adjustments to their economies in this direction that proved very worthwhile. Rather than promote the expansion of machinery and consumer electronics, emphasis shifted to textiles, apparel, and footwear production. In addition to producing products with a price advantage, such a strategy also built up a skilled labor pool and middle-management group. As these industries matured they became competitive worldwide and set the stage for a return to more skilled-labor and higher-capitalized operations by the late 1960s and early 1970s.

Table 14-2 indicates that four waves of change characterized the manufacturing change process in South Korea from 1961 to 1981. By the late 1970s the country became a major world competitor in electronics, shipbuilding, fertilizers, and steel, in addition to the textile, clothing, and footwear base developed earlier. In the early 1980s it appeared that motor vehicle assembly, consumer electronics, specialty steel, and precision goods had also joined the group of self-sustaining industries.

Stage 3, the *export stage,* wherein worldwide marketing becomes a feasible option for developing nations, occurs when the quality, price, and volume of output justifies developing overseas distribution networks. In the Korean case, textiles and clothing achieved this status by the early 1970s, followed by electronics, shipbuilding, and steel by the late 1970s. More sophisticated high-technology products reached the export stage in the 1980s, as just mentioned.

Some nations elect to short-circuit this three-stage process by purchasing advanced technology, and even import operational plants as a *turnkey* project. The Japanese have, in fact, converted former shipbuilding facilities into large-scale production facilities to make offshore drilling rigs, power plants, aluminum refineries, and other facilities that can be "floated" to their destinations in third-world countries. Often, management and maintenance services are provided as a part of the package. Typically, a large multinational firm provides this service. Recent sales of these

Table 14-2

Stages in Manufacturing Change in South Korea, 1961–1981

Stage	1961–1966	1966–1971	1971–1976	1976–1981	1981–
Infant industries	Textiles Clothing Footwear	Electronic assemblies Shipbuilding Fertilizers Steel	Motor vehicle assembly Consumer electronics Specialty steel Precision goods (watches, cameras)	Automotive parts Machine tools Simple instruments Heavy electrical machinery assembly Semiconductors	
Industries becoming competitive		Textiles Clothing Footwear	Electronic assemblies Shipbuilding Fertilizers Steel	Motor vehicle assembly Consumer electronics Specialty steel Precision goods	Automotive parts Machine tools Simple instruments Heavy electrical machinery assembly Semiconductors
Self-sustaining industries			Textiles Clothing Footwear	Electronics assemblies Shipbuilding Fertilizers Steel	Motor vehicle assembly Consumer electronics Specialty steel Precision goods

Source: G. J. R. Linge and F. E. Ian Hamilton, "International Industrial Systems," Chapter 1 in F. E. Ian Hamilton and G. J. R. Linge, eds., *Spatial Analysis, Industry and the Industrial Environment,* Vol. 2. *International Industrial Systems.* New York: John Wiley and Sons, 1981, p. 33. Adapted from Boston Consulting Group, *A Framework for Swedish Industrial Policy.* Stockholm: Ministry of Industry, 1978, Appendix 5, p. 30.

plants to Middle Eastern republics and Latin American countries illustrate examples of turnkey projects (Figure 14-2). The Soviet Union has also placed many orders for turnkey projects from western nations—for petrochemical plants, automobile assembly plants, and aluminum refineries.

East-west cooperative agreements involving technology transfer rather than turnkey plants include such techniques as *specialization* and *co-production* to foster development. Specialization refers to splitting up the production of complementary goods, which are then exchanged to give each country a full line of products. For example, an arrangement for the production of diesel engines between a West German firm and a Yugoslav enterprise led to the exclusive production of specific models in each country, which could then be exchanged to create a full line of motors. Co-production refers to sharing in the production of component parts and/or designs for products. The transnationalization of the motor vehicle industry, with production facilities for various component parts located in different countries, illustrates such a process and is discussed in Chapter 15.

The dramatic changes affecting the location of production of low-skill, low-technology consumer goods, such as apparel and footwear, have not yet had the same locational impact on higher-technology goods as completely but appears to be under way. The largest share of imported manufactured goods by developed market economies, represented by OECD countries, still comes from other OECD countries, but NICs are gaining in competitiveness as suppliers of apparel and footwear (Table 14-3). It appears that parity may soon emerge in clothing sources, as will be discussed later. High-technology items such as transportation equipment and chemicals production remain most competitive in more advanced economies.

TEXTILE AND APPAREL PRODUCTION

Textile and apparel activities, owing to their roles as entry-level industries in developing economies, occur in a more dispersed pattern throughout the world than any other major industrial enterprises today. This expansion has occurred as an evolutionary process, as discussed in the preceding sections. But what becomes of the industry in mature economies as competition increases for newer, lower-cost producing areas? How can traditional producers compete? Sev-

Figure 14-2 Turnkey Floating Factory, Bahia Blanca, Argentina. *A Dutch ship is bringing a floating Japanese-built factory into dock. Some former petroleum supertanker construction yards in Japan have been converted to turnkey floating factory construction sites. (Courtesy Will McIntyre, copyright 1981)*

Table 14-3

Imports of Manufactured Goods by OECD Nations from NICs and Member Nations, 1977

Product	Percent of Imports from OECD Nations	Imports from Developing Countries as Percent of OECD Total
Clothing	9	47
Leather, footwear, and travel goods	5	39
Textiles	15	20
Electrical machines and appliances	34	15
Transportation equipment	69	3
Chemicals	46	5
Iron and steel	22	7
Average	29	19

Source: See table 14-2, p. 43. Derived from Secretary-General OECD, *The Impact of Newly Industrialising Countries on Production and Trade in Manufactures.* Paris: OECD, 1979, pp. 24, 60–7.

eral decades ago the industry in the United States shifted from north to south to find a lower-cost working environment in which to operate. Pressures also existed at that time to develop new technology to sustain productivity advances. In reality, introducing new technology to the textile and apparel industries has been a continuous process (see Table 14-3). Attempts are now being made to make them more "high tech" in nature (Figure 14-3).

Three stages in textile technology development are shown in Table 14-4. Stage 1 spans the two-century period prior to World War II. During that era the technology remained essentially that associated with the industral revolution itself—focused on labor-intensive mechanized spinning and weaving processes. This technology, originally from England in the eighteenth century, later spread to the United States, western Europe, and Japan.

The second textile technology stage began in the 1930s and lasted about 30 years. The primary change in this period revolved around the introduction of new synthetic and man-made fibers. The technology emanated from the United States and diffused swiftly to other developed and developing countries.

The third stage in technology development began in the 1960s and continues to the present. This time the technology innovation came from Japan and western Europe, primarily West Germany, and involved the introduction of highly engineered and automated machines that could produce many different products as market needs dictated. The shuttleless loom is an example. Higher-skilled workers displaced low-skill

Table 14-4

Stages in Textile Technology Development

Stage	Time	Characteristics
1	1750s–1920s	Originated in England and spread to U.S.; labor-intensive spinning and weaving technology
2	1930s–1960s	Synthetic and blended fiber innovations predominantly originated in U.S.; spread to other countries
3	1960s–1980s	Product differentiation and automation dominate in response to increased competition; required upgrading of worker skills; Japanese and European technology

Figure 14-3 Modern Textile Factory in Bayreuth, Germany. *The textile industry pioneered the internationalization process that created the new industrial order. Automation and high technology applications have kept the textile industry competitive in highly developed areas as this photograph suggests. (Courtesy German Information Center)*

operatives as a part of this technology adaptation, and the industry became more capital than labor intensive. The terminology used to describe the new generation of looms itself speaks of their "high-tech" nature: missile looms, rapier looms, water-jet looms, and air-jet looms. See the accompanying box for a further explanation.

Many textile plants now require a mix of various loom types; the day of all-purpose weaving has passed in an age when multiple fabrics and multiple technologies must be utilized to remain competitive. Only the largest plants owned by the largest corporations in highly developed areas can compete effectively in this environment, and many of these businesses remain headquartered in developed nations. The production of generic products displaying few points of design and styling uniqueness shifted to new loom technologies the most rapidly. Bed sheeting and terry cloth are examples. The broad width and simple design requirements of sheeting make it economical for production on large-scale machines.

As technology applications intensify, the risk of loss to the firm also increases and firms become more management intensive. Typically, firm sizes also increase. By changing in these ways the industry can remain competitive in highly developed areas if costs remain competitive, which often means shedding generic products in favor of more specialized high-fashion merchandise. In succeeding sections we review how various segments of these industries have fared in the face of heightened international competition.

ANATOMY OF THE TEXTILE INDUSTRY

To understand the competitive dynamics of international textile production, it is necessary to subdivide the industry into several segments representing different stages of product development, including fiber supply, textile and apparel production, and marketed products (Figure 14-4). Each stage of product development responds to different technology, scale, and labor factors, as well as exhibiting a different spatial structure.

Fiber and filament production typically occurs in highly capitalized, automated, large-scale factories. As dependence on cotton fiber gave way to an emphasis on man-made fibers, linkages to the petrochemical in-

dustry increased. Over two-thirds of the fiber consumed in the United States is man-made, primarily polyester and nylon. The man-made fiber production process involves extruding, under pressure, a sticky, viscous chemical solution called viscose through tiny holes in thimblelike objects called spinnerets. Synthetic fibers include nylon and acetate, which are cellulose-based products derived from coal.

The United States easily dominates world synthetic fiber output, claiming nearly one-fourth of total production (see Table 14-5). Japan ranks second, followed by South Korea, the Soviet Union, West Germany, and China. U.S.-based firms have also traditionally led in fiber production (see Table 14-6). Dupont alone claimed 14 percent of the market in 1979, followed by Celanese and Monsanto, with 6 percent each. The largest foreign firm, Akzo, from the Netherlands has a 6 percent share of output. The fifth- and sixth-leading producers, both from Japan, produce 4 percent of world synthetic fibers, as does Hoechst from West Germany.

Yarn and fabric can be created with four distinct processes: (1) weaving, (2) knitting, (3) tufting, and (4) nonwoven techniques. Weaving involves interlacing threads or yarn on a machine called a loom to create a fabric. Knitting involves looping thread or yarn together with needles to make a cloth fabric. Weaving predominated until the 1950s when knitting became more important in fabric production. By the 1970s a new process, double knitting, represented about one-third of fabric output.

Tufting, which occurs today primarily in the carpet industry, was, until the 1950s, limited to decorator items such as candlewicking. Tufting involves thrusting needles with a yarn attached through a backing material to create a loop. The resulting loops form the carpet surface, which can be sheared to create a cut pile or left as loops. Nonwoven fabrics include a variety of products made directly from fibers by pressing the material without first passing it through the yarn stage. Examples of products made this way include padding, disposable napkins and diapers, insulation, and filters. All four of these operations are mechanical processes, highly adaptable to mass-production techniques, including computer and robot-driven operations.

Finishing processes cannot be as highly automated as fabric production itself because of the com-

TYPES OF LOOMS

Shuttle loom: moves a bobbin of yarn across a field of parallel yarns mechanically, performing an interlacing function; different patterns of interlacing create various weaves

Shuttleless loom: faster speed, less noise, and fewer auxiliary operations than shuttle looms; $2\frac{1}{2}$ times as much cloth produced per hour

Rapier loom: first-generation shuttleless loom
Water- and air-jet looms: current generation of shuttleless looms; moves yarn across weaving loom by air or water pulses from a nozzle; pressures too great for natural fibers but works well with synthetics

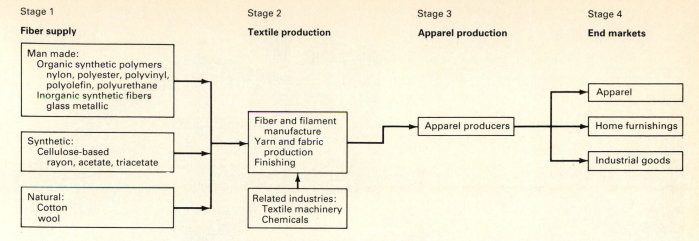

Stage 1	Stage 2	Stage 3	Stage 4
Fiber supply	**Textile production**	**Apparel production**	**End markets**

Figure 14-4 Textile Product Development Stages. *Each stage responds to different technology, scale, and labor factors, as well as processing a different spatial structure. Textiles have remained more competitive in highly developed areas than apparel.*

plexity and variety of chemical processes involved, but it is a technology-intensive field. These processes include color dye treatments, bleaching to remove stains and color, singeing to remove lint, mercerizing to improve appearance, shrinking, and surface adjustments such as glazing and napping, among others. New frontiers of development in dyeing offer considerable promise for the future, including the greater use of foams.

STRUCTURAL CHANGES IN THE U.S. TEXTILE INDUSTRY

Structural and locational changes in the U.S. textile industry reflect the stages in evolution of the industry itself (Table 14-4). New England's textile mills became the first and foremost manufacturing region in the United States in the late eighteenth century. Originally associated with larger port cities, such as Prov-

idence and Boston, as financing became available, these mills gradually spread inland to small towns along major waterways. The first textile mills were two-story wood-frame buildings, but they gradually gave way to larger three- or four-story brick buildings at major water power sites. Lowell, Massachusetts, located at Pawtucket Falls on the Merrimac River, typified the structure of the early mill town. Over 30 factories lined the river and the man-made canals paralleling the river in 1840 (Figure 14-5). By 1850, Lowell became the second-largest city in Massachusetts, and the largest textile center in the country.

At first textile makers recruited young women from surrounding rural communities to work in the mills, accommodating them in rows of boarding houses near the mills. Later, in the 1840 and 1850s, waves of foreign immigrants displaced the native female labor force. Thus plentiful, cheap labor became a major factor in the growth of the industry, together with cheap power locations, and growing markets in the region.

Table 14-5

World Production of Synthetic Fibers, 1983*
(thousands of metric tons)

Rank	Nation	Quantity	Percent of World Total
1	United States	1885	23
2	Japan	1058	13
3	South Korea	694	9
4	Soviet Union	670	8
5	West Germany	568	7
6	China	457	5
	Subtotal	5312	66
	World total	8054	100

Source: United Nations, *1983 Industrial Statistics Yearbook,* 1985.
*Non-cellulosic (synthetic) and artificial (rayon and acetate) fibers.

Table 14-6

Leading World Synthetic Fiber Production Firms, 1979

Rank	Firm	Home Nation	Percent of Market Share
1	Dupont	United States	14
2	Celanese USA	United States	6
3	Monsanto	United States	6
4	Akzo	Netherlands	6
5	Toray	Japan	4
6	Teijin	Japan	4
7	Hoechst	West Germany	4
8	Rhône-Poulenc	France	3
9	ICI	Great Britain	3

Source: United Nations, *Fibres and Textiles.* New York: U.N. Conference on Trade and Development, 1981, p. 20.

Figure 14-5 Historic Boott Textile Mill Complex in Lowell, Massachusetts. *This mill, named for Kirk Boott an early investor and the first agent in the city, consists of 10 buildings that are now being rehabilitated for a variety of uses, including private business, housing and a National Park Service-sponsored museum. The Mill, located on the Eastern Canal which was built in 1835, still provides energy for electrical power generation in the city. Note canal in foreground of the 5-story mill number 6 shown here. The low building in foreground, built in 1835, served as the Counting House for the firm and housed the agent (general manager), clerical staff, and the business end of the operation. The National Park Service museum will include 140 operating weaving looms and exhibit space. (Courtesy of National Park Service.)*

Time took its toll on the New England textile industry by the beginning of the present century. It ceased to grow, labor productivity declined, plants became obsolete, and costs of operations increased as labor costs grew, electric power replaced water power, and tax levels rose. In short, the product cycle reached a sunset phase and a new era dawned. Plants in the south offered an attractive alternative in the 1920s and thereafter. Many of the factors favoring the south in the early twentieth century were the same as those that favored New England locations a century earlier—a plentiful low-wage labor pool, cheap hydroelectric and coal power, low taxes, and low living costs.

Expansion in the South

With the coming of air conditioning after World War II, textile mills could easily replicate the more humid conditions of New England that once favored that area for spinning. The newer southern mills utilized modern machinery and single-story plants rather than vertical operations, facilitating an efficient assembly line product flow.

Cotton producers, the first textile firms to leave New England, operated more cost-sensitive facilities than wool firms, which remained in New England. Woolen mills usually employed more skilled workers and their market was more specialized—primarily men's clothing—allowing them to compete more successfully in a higher-cost environment.

Today, the south totally dominates the domestic textile industry. Small-town locations are particularly attractive to the industry, especially in the Piedmont region of North and South Carolina and northwestern Georgia. The greater use of synthetic fibers today also favors that region, as many synthetics are also produced in the Carolinas Piedmont area. Other factors continuing to favor the area for the textile manufacture include the low level of labor union activity and the plentiful supply of low-cost labor in rural areas and small towns.

From Trade Surplus to Deficit

Increasing world competition and an aggressive automation strategy continued to serve the U.S. textile industry well in the 1980s. To be sure, layoffs continued as surplus labor was shed, but overall the country enjoyed a trade surplus in the early 1980s (see Figure 14-6). North and South Carolina, for example, lost

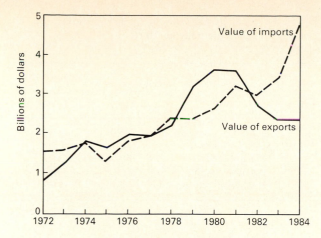

Figure 14-6 Textile Trade Trends in the U.S., 1972–1984. *The value of textile imports rose in roughly comparable fashion until 1982, when the value of exports dropped precipitously while the value of imports rose dramatically. This adjustment created a vast trade deficit which exceeded $2 billion in 1984. (Source: David Avery and Gene D. Sullivan, "Changing Patterns: Reshaping the Southeastern Textile—Apparel Complex," Economic Review, November 1985, p. 42)*

60,000 jobs in the industry in the 1980–1985 period, but the benefit gained was that the industry remained the most productive in the world as a result of this retrenchment. This competitive posture, together with trade restrictions, slowed the rate of textile imports into the country in the 1970s and early 1980s as the value of exports rose slightly faster. "In 1980, the U.S. textile industry had in excess of $1 billion trade surplus."[4]

By 1982, the situation had changed dramatically, as the value of textile imports surged upward at the same time as the value of exports plummeted, creating an annual trade deficit of over $1 billion in 1983, and $2 billion in 1984 (Figure 14-6). Why this abrupt change? The continued labor-cost differential between the United States and textile competitors such as Taiwan, South Korea, and China obviously played a part, but so did the extraordinarily high value of the U.S. dollar at that time, which placed U.S. products at a comparative disadvantage in world trade.

U.S. production of textile items remains strong (Table 14-6), and the competitiveness of products began stabilizing as the dollar declined considerably in value against foreign currencies in 1986 and 1987. In 1986, a farm bill also became effective, dropping the price of U.S.-produced cotton about 30 percent against the world market price. This action benefited the textile producer as well as the apparel industry. Years of consolidation also left a leaner, more market-oriented industry, which also enhanced competition. Nevertheless, the U.S. textile industry remained wary about the future, given the greater intensity of foreign competition. Pressure also accelerated in Congress for more protection of the domestic producer, but the

[4]Joseph Pelzman, "The Textile Industry," *Annals, American Academy of Political and Social Science,* 460 (March 1982), 96.

administration remained committed to a free-trade environment.

Other hazards also faced textile manufacturers in the 1980s. The enormous cost of new equipment and unstable market prospects thwarted new investments by all but the largest firms. Water pollution, emanating from dyes dissolved in wastewater, required attention. Rapidly escalating fiber costs, due to higher petroleum prices, also joined the litany of negative impacts facing the industry. Adverse product-cycle developments added to the problems of smaller firms, which found modernization out of reach financially.

THE JAPANESE INDUSTRY

The Japanese textile industry rebounded rapidly after the World War II–related devastation, aided by U.S. capital, but today faces the same cloudy outlook as it does in the United States. By 1956 Japanese textile exports to the United States reached their prewar levels. Although this flow represented less than 2 percent of U.S. domestic output, a clamor began in the United States for trade restrictions. The Japanese voluntarily curtailed exports in 1957, fearing retaliatory measures in the United States. At that time 60 percent of U.S. textile imports came from Japan. The impact of the Japanese cutback then shifted import sources for the U.S. market to other countries, such as Hong Kong and India.

By 1961 the United States took action to initiate multilateral trade discussions using the General Agreement on Tariffs and Trade (GATT) umbrella. A one-year Short Term Cotton Textile Arrangement (STA) resulted, followed by a Long Term Arrangement on Cotton Textiles (LTA) in 1962. The latter agreement specified a 5-year import control arrangement, item by item. In the late 1960s restrictions on textile imports to the United States affected 17 nations; nevertheless, the total import share of textiles increased from 6 percent of consumption in 1960 to 9 percent in 1967. Thirteen additional countries faced restrictions by 1970, bringing the total number to 30. Imports continued to rise, reaching 15 percent in 1973. In the face of pressures of growing imports, restrictive bilateral agreements were worked out with many countries, but the major undertaking of the time (1973) was a multilateral agreement, the Multi-Fiber Arrangement (MFA). It extended agreements to cover all fibers, not just cotton, and established a surveillance body to mediate disputes. In 1977 this agreement was renewed. Bilateral agreements also became more effective, and these controlled over 80 percent of U.S. imports by 1980.

The Japanese textile industry experienced the same competitive pressures as those experienced in the United States in the 1970s, facing growing imports from South Korea, Hong Kong, and Taiwan. The Japanese responded by promoting exports through more aggressive marketing and by shifting operations offshore to lower-cost production settings. In the mid-

1980s these initiatives largely ceased. Cultivating the smaller, high-quality domestic fashion market became a new priority. Some outmoded mills now produce electronics parts for high-technology industries. Other companies have diversified into cosmetics production or biotechnology.

Osaka traditionally served as the center of the Japanese textile industry, and that region now serves as the innovator in new techniques to keep the industry competitive. The overriding goal in the industry remains to reduce labor costs and maintain a reputation for a high-quality product. The Ministry of Trade (MITI) began a 7-year automation program in 1981 involving such initiatives as an automated three-dimensional sewing machine, a flexible manufacturing system (FMS) that can handle the production of a variety of styles and fabrics, and laser cutting machines. Japanese textile machines are now enjoying a larger share of the world market due to their superior engineering. West German and Swiss textile machines continue as the leading sellers, but this dominance may be broken in the 1980s by the Japanese.

The Japanese textile industry has had a long tradition of cooperation among the many small independent firms that provide the bulk of the output. A series of laws dating to 1956 encouraged this cooperation, along with modernization and integration of production and marketing activities. Subsidies and low-interest government loans encouraged groups of firms to establish "product development centers" to provide research and development services. Cooperatives provided raw materials, design, and marketing support. Wholesale houses aided access to overseas markets.

Despite these initiatives, the Japanese textile industry remains depressed and in retreat. The Japanese deficit in textile and fiber trade exceeded $1 billion in 1979 and contributed the biggest single source of the trade deficit facing the nation.

APPAREL ACTIVITY

The structure and economics of the apparel industry are quite different from those for textiles. Apparel manufacture is much more labor intensive, as it involves the cutting, sewing, and pressing of individual garments by the piece rather than through mass production. Automation advances, such as the use of laser cutting, substituting fusing techniques for sewing with thread, and holding fabrics in place by suction to reduce the need for pressing, will eventually have a big impact on the industry, but advances in apparel technology have been slower to emerge than those in the textile industry. As a result, the product cycle for apparel has worked more consistently than with textiles to shift production away from developed countries to developing areas. Apparel manufacturers, in contrast to textile producers, largely remain small independent operations in both the developed and the developing worlds. Nevertheless, there are examples of

Table 14-7

Major U.S. Apparel Corporations, 1978
(millions of dollars)

Rank	Firm	Total Sales	Apparel Sales as Percent of Total Sales
1	Interco	731	40
2	Gulf & Western	725	17
3	General Mills	534	17
4	Cluett-Peabody	495	86
5	Northwest Industries	454	19

Source: United Nations, *Fibres and Textiles,* New York: U.N. Conference on Trade and Development, 1981, p. 24.

giant firms operating in the industry dominated by midgets (see Table 14-7). Then, too, the industrial apparel market is growing, and this is a high-technology field also dominated by larger firms.

Since economies of scale have had far less impact than with textile operations, apparel plants remain more like handicraft than like mass-production enterprises. Part-time seasonal labor typifies apparel businesses, with little multinational firm involvement. Apparel producers have the reputation of being "back street" operations employing low-skilled minorities and immigrants (sometimes illegal aliens) for low wages in a poor working environment—hence the label *sweat-shop* operation.

The shift in the apparel industry away from developed market economies to state-run enterprises in eastern Europe and lesser developed countries has accelerated in the past 25 years. The world map of apparel employment, however, highlights the continued dominance of this industry by western Europe, the Soviet Union, the United States, and Japan. Third-world concentrations occur in Latin America, South Africa, and southeastern Asia (Figure 14-7).

High-fashion apparel remains competitive in developed countries, as do mass-produced goods such as hosiery, underwear items, and denim jeans. Materials, primarily man-made fibers, represent about one-half of apparel production costs. Most of this man-made fiber output occurs in developed countries, somewhat reinforcing the competitiveness of their domestic apparel industry. But most highly developed countries today import a large share of their apparel.

Apparel consumption, especially fashion merchandise, increases with income. This demand factor also works in favor of the expansion of textile output in developed market economies. Consumption in the United States alone equals the total of that in all other advanced economies combined. The list of the largest textile and apparel retailers reflects this U.S. dominance (see Table 14-8). Seven of the eight largest firms worldwide are U.S. corporations. West Germany has the second largest number of large retailers.

Three distinctive types of operations maintain locational inertia in the apparel production industry.

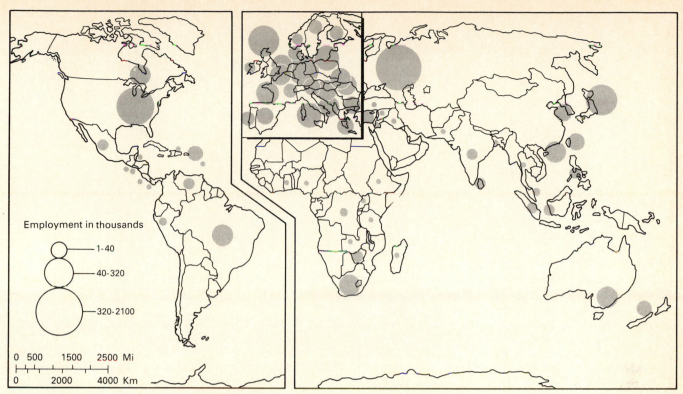

Figure 14-7 World Apparel Production. *The world map of apparel employment is dominated by western Europe, the Soviet Union, the United States, and Japan. Brazil, South Africa, Hong Kong, and Australia are other production leaders. (Source: Redrafted after Steed, in F. E. Ian Hamilton and G. J. R. Linge, eds.* Spatial Analysis, Industry and the Industrial Environment, *vol II,* International Industrial Systems, *Wiley, 1981, p. 267)*

The standard producer, the *manufacturer,* as in other industries, creates a completed product from purchased fabric. The *jobber* performs specialized functions for a manufacturer, serving as a buyer, a designer, and/or a marketer. The third type of operation, the *contractor,* employs workers to make items on consignment from another firm. Because of the communication and material linkages among these operations, physical proximity provides an advantage. These linkages become more important with high-fashion products. The need for quick reactions to style changes and shifts in demand require such close ties. An exception to this generalization occurs in the situation with South Korea, which produces high-fashion name-brand men's apparel under contract to U.S. firms under the label of Christian Dior, Calvin Klein, and others, in the designer clothing field. The advantages of lower labor costs clearly outweigh local ties in this situation. But constant contact must be maintained between couriers and sales representatives to streamline the order and production process.

LOCATION OF THE U.S. APPAREL INDUSTRY

The manufacture of clothing in the United States occurs primarily in major metropolitan areas (see Figure 14-8). The largest concentration lies in a corridor between metropolitan New York and Philadelphia,

with New York city serving as the nucleus. Measured in terms of value added, about 40 percent of the nation's apparel industry clusters in this small zone. Other conspicuous clothing centers include Los An-

Table 14-8

World's Largest Textile and Apparel Retailers, 1979
(billions of U.S. dollars)

Rank	Retailer	Total Sales
1	Sears, Roebuck (U.S.)	17
2	K Mart (U.S.)	13
3	J.C. Penney (U.S.)	11
4	F.W. Woolworth (U.S.)	7
5	Karstadt (W. Germany)	6
6	Federated Department Stores (U.S.)	6
7	Household Finance (U.S.)	5
8	Montgomery Ward (U.S.)	5
9	Kaufhof (W. Germany)	5
10	Schickedang (W. Germany)	4
11	Hertie (W. Germany)	4
12	Marks and Spencer (Britain)	4
13	Coop (Switzerland)	4
14	Carrefour (France)	4
15	Vroom and Dressman (Netherlands)	3
16	Dayton-Hudson (U.S.)	3
17	G.B.-Inno (Belgium)	3
18	C & A Brenninkmeyer (W. Germany)	3

Source: United Nations, *1983 Industrial Statistics Yearbook,* 1985, p. 219.

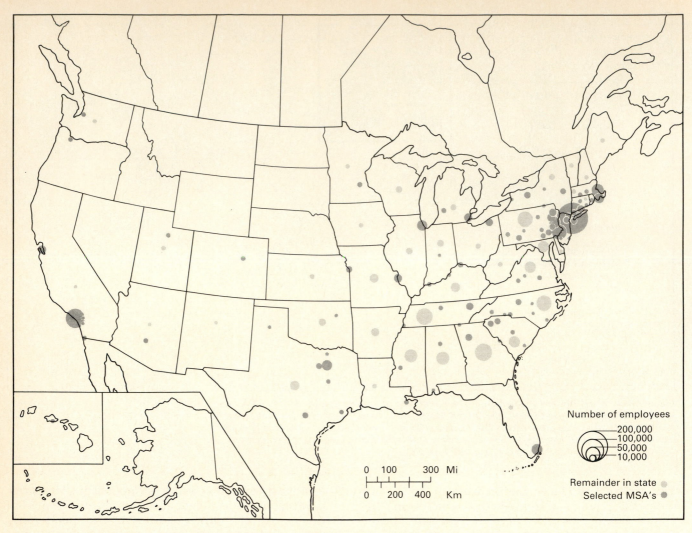

Figure 14-8 Apparel Employment in the United States. *The manufacture of clothing occurs primarily in major metropolitan areas, with the largest concentration occurring in the northeast in an area centered around New York City. The southeastern U.S. specializes in men's and boy's clothing and children's outerwear.*

Number of employees

200,000
100,000
50,000
10,000

Remainder in state
Selected MSA's

geles, Chicago, Miami, Boston, Providence, and a string of centers in the Carolinas. Nearly all southeastern states have strong apparel production levels in smaller towns, as indicated by the lightly shaded circles in Figure 14-8.

The apparel industry remains largely a small-scale industry with an average number of 56 employees in a clothing factory compared with 132 in a typical textile mill. The smallest apparel operations occur in metropolitan area shops that create high-fashion merchandise. Factories that make women's clothing are much smaller than those that turn out men's clothing. The larger firms in the United States occur predominantly in the southeast, producing low-style generic items such as hosiery, underwear, and work clothes. In the early 1970s, for example, 242 of the 312 women's hosiery plants in the United States were in the south. The southeastern United States also specializes in men's and boy's clothing and children's outerwear.

STYLE CENTERS

Designers of men's, women's, and children's clothes locate in larger metropolitan areas, principally New York. Since apparel manufacturers maintain constant contact with these designers, they are a powerful force in keeping clothing factories in New York. The garment business is one of the rare industries still located in the downtown sections of cities. In Manhattan alone 170,000 apparel workers run scores of small factories. Thousands of garment workers similarly operate shops in the Loop of downtown Chicago. Land in central business districts is expensive, and these factories cannot afford first-floor space. Often, the ground floor in a building is occupied by a clothing retailer, with wholesalers on the second floor, while the clothing manufacturers occupy the upper stories, as the label *loft* industries suggests.

The main reasons for the downtown concentration of garment manufacturing include labor supply availability and historical inertia factors. The downtown area, having a central location to the local population, provides an ideal nexus for assembling a work force, including access to recent immigrants, such as Puerto Ricans and others of Hispanic origin in the case of New York. More workers can gather with less effort in Manhattan than in any other part of the city because mass transportation arteries converge upon the island. This one urban area contains 40 percent of the nation's garment factories, supports 20 percent of the industry's employees, and accounts for 26 percent of the value added.

The Manhattan Garment Center encompasses an area of 150 acres, bounded by the Pennsylvania Rail-road station on the south, Times Square on the north, Seventh Avenue on the east, and Ninth Avenue on the west (see Figure 14-9). Here are the lofts that house the hundreds of small enterprises that constitute the garment business. In the neighboring streets sellers maintain showrooms where buyers, staying in nearby hotels, converge to make their deals.

At the world scale, several other major retail and wholesale centers provide a role in the apparel industry comparable to that of New York, including Montreal, Osaka, Tokyo, and Hong Kong.

INDUSTRIAL APPAREL

Notwithstanding the changes in the traditional apparel market, the most dramatic growth in the industry today encompasses the use of apparel for home finishing, floor coverings, and various industrial uses. In fact, the traditional clothing industry accounts for only 40 percent of the textile fiber market today. Home and industrial uses for fibers now claim about 60 percent of the market (Figure 14-10). Long-established uses still dominate this field, such as automobile upholstery, vinyl roofing for cars, tires, and tarpaulins, even as new applications emerge in the market for shoes, luggage, tents, awnings, and arena and shopping center coverings. The outer-space exploration program in the 1960s and 1970s stimulated a technology transfer in firefighting suits and aircraft interiors, greatly benefiting the industrial apparel market. But the most exciting innovations may be around the corner as a convergence of technologies bridging such diverse fields as carbon fibers, fiberglass-reinforced plastics, stainless steel, and textile glass leads to new applications in automobile and aircraft bodies, among others.

The home furnishing market accounts for 13 percent of the textile fiber market and floor coverings,

Figure 14-9 Manhattan Garment Center. *Designer clothes production continues as a major industry in mid-town Manhattan. The Manhattan Garment Center encompasses an area of 150 acres. Near the lofts that house this business, sellers maintain showrooms where buyers make deals. (Redrafted after Kenyon, in R. S. Thoman and D. J. Patton, eds.,* Focus on Geographic Activity, *McGraw-Hill, 1964, p. 160)*

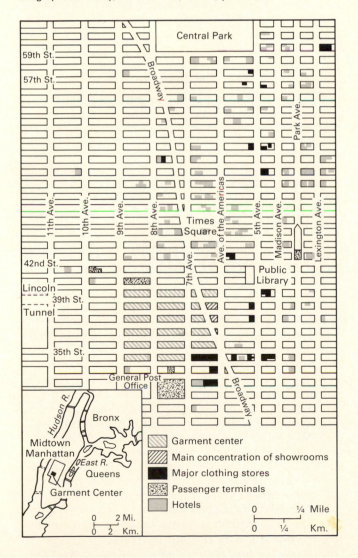

Figure 14-10 Textile Fiber Markets. *The traditional clothing industry accounts for only 40 percent of the textile fiber market today. Home and industrial uses now claim about 60 percent of the market. (Source: Seidel,* Textile Industries, *November, 1982).*

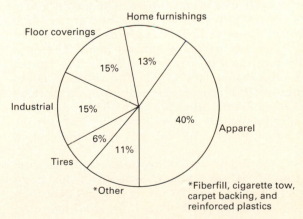

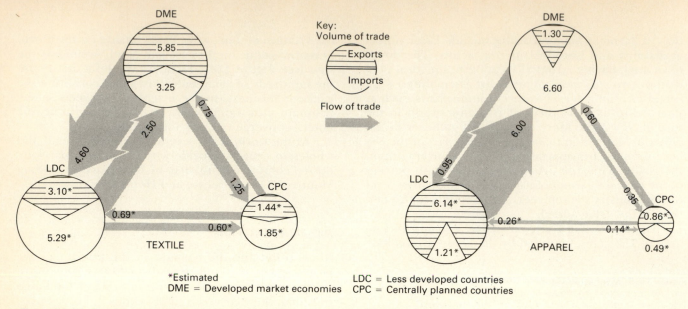

Key:
Volume of trade

Exports
Imports

Flow of trade

*Estimated
DME = Developed market economies

LDC = Less developed countries
CPC = Centrally planned countries

Figure 14-11 Textile and Apparel Trade. *World trade in textiles far exceeds apparel flows. Apparel is more market-oriented and more nations produce a greater share of their own domestic need. Whereas textiles predominantly flow from north to south, apparel flows from south to north. (Redrafted from United Nations,* Fibres and textiles: Dimensions of Corporate Marketing Structures, *New York, 1980, p. 205)*

primarily carpet, another 15 percent. Demand for these products is especially cyclical, tied as they are to the new home building industry, a market that exists primarily in the highly developed world.

TEXTILE AND APPAREL TRADE

World trade in textiles far exceeds apparel flows (Figure 14-11), even though the latter are higher-value products. The explanation of this inconsistency is that apparel is more market oriented and more nations produce a greater share of their own domestic need. Whereas textiles traditionally flowed from developed to developing countries, north to south, apparel overwhelmingly flows from developing countries to the developed market economies, or from south to north. This trend first emerged in the 1960s and is particularly evident in Figure 14-12. Trade between the centrally planned economies and the other two trading blocs—developing and developed countries—is much lower for both textiles and apparel. The textile flow in and out of centrally planned countries is greater than the apparel movement, but as a group these countries are only net exporters for apparel items. The most rapidly growing exporter of textiles to the United States today is mainland China. This situation led to the imposition of unilateral restraints on textile flows from China by the United States in 1982, resulting in a counter-retaliation by the Chinese by banning the purchase of fibers and soybeans from the United States.

A map of bilateral clothing flows emphasizes the flow into the United States, primarily from Hong Kong and Taiwan, and the strong linkages occurring in western Europe, particularly flows from the south (France, Italy, Yugoslavia) to the north (particularly

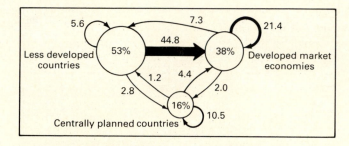

Figure 14-12 World Trade in Clothing, 1975. *The trade between developed and developing areas far exceeds the flow between the centrally planned economies and the other two trading blocs. Mainland China is the most rapidly growing exporter of textiles to the U.S. today. (Redrafted after Steed, "International Location. . . ." in Hamilton & Linge,* Industry and the Industrial Environment, *Wiley, 1981, p. 270)*

West Germany). Figure 14-13 graphically portrays the importance of east Asia as the primary clothing trading partner with the United States.

Since clothing shipments involve low transportation costs in relation to the retail value of garments, apparel goods can easily absorb the cost of international trade. The lower labor costs of the third world also create marketing advantages for producers in developing areas. But there are noneconomic factors at work in the industry as well which affect the production map. Trade barriers such as quotas, tariffs, and subsidies increasingly affect the international pattern of production. The GATT system and trade barriers are discussed in Chapter 20.

The U.S. apparel industry can expect greater foreign competition in the future. While 30 percent of the product came from offshore in the early 1980s,

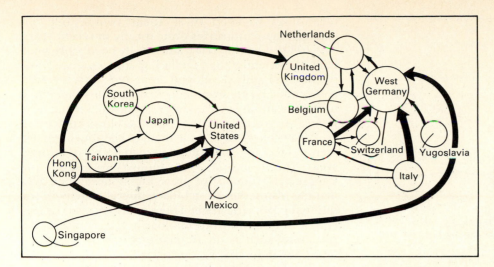

Figure 14-13 Bilateral Clothing Flows, 1973. *Hong Kong and Taiwan dominate bilateral flows to the U.S. The strong linkages among western European countries highlight the flows into West Germany, especially from Italy and France. (Source: as in Figure 14-12, p. 282)*

this level could double by the end of the decade if domestic labor costs continue to escalate and foreign competition becomes more efficient. In the mid-1980s, "while U.S. apparel workers earn $6.52 per hour on average, their counterparts in Taiwan and Korea earn $1.43 and $1.00, respectively. Chinese workers earn about 26 cents an hour, which suggests why American manufacturers are so apprehensive about unrestrained imports from that nation."[5]

As indicated in Table 14-9, the list of U.S.-produced apparel items subject to extreme sensitivity to imports will probably grow rapidly in the late 1980s. Whereas only men's shirts, coats, and sweaters experienced extreme sensitivity in 1982, the list may include suits, sportcoats, and pajamas by the end of the decade. A parallel growth in sensitivity levels will probably affect women's clothing items in the future.

For domestic apparel manufacturers to compete more effectively in the international market, many are exercising a provision 807 of the U.S. tariff code that permits "domestically cut material to be shipped for sewing to countries with lower labor costs, then reimported. Duty is paid only on the value-added portion of the commodity, and that value is modest because of the low-cost foreign labor used in the sewing operation."[6]

The overall deficit created by the difference in the value of apparel imports over exports in fact reached unprecedented levels in the United States in 1984 when it approached $12 billion, as Figure 14-14 indicates. "Textiles and apparel accounted for 13 percent of the nation's overall trade deficit in 1984, as imports continued to erode the competitive position of the U.S. industry."[7] But as mentioned earlier, the situation facing the industry stabilized by the middle of 1985, when the U.S. dollar began falling against for-

eign currencies and consumer spending on apparel remained strong.

It appears that in the future the apparel and textile industries will have to place as much emphasis on market research and strategic planning as on investments in technology. Although the latter may assist in cost containment, it does not necessarily assist in market penetration because many buyers are looking for style, quality, and snob appeal of products. Some observers also argue that U.S. producers should work together to create an export trading company such as exists in Japan to enhance market penetration.

SUMMARY AND CONCLUSION

As former agrarian economies climb the technological ladder, they typically experience initial growth in textile and apparel industries and related goods. Great

Table 14-9

U.S. Apparel Import Sensitivity, 1982 and 1987

	Extreme Sensitivity	
Category	1982	1987
Men's and boys		
Suits		×
Sportcoats		×
Shirts	×	×
Sweaters	×	×
Coats and jackets	×	×
Pajamas		×
Women's and girls		
Dresses		×
Blouses and shirts		×
Knit tops		×
Skirts		×
Sweaters	×	×
Coats and jackets	×	×

Source: Leon Seidel, "The Tokyo Round," *Textile Industries,* April 1983, p. 36.

---

[5]David Avery and Gene D. Sullivan, "Changing Patterns: Reshaping the Southeastern Textile-Apparel Complex," *Economic Review,* Federal Reserve Bank of Atlanta, November 1985, p. 40.

[6]Ibid., p. 39.

[7]Ibid., p. 41.

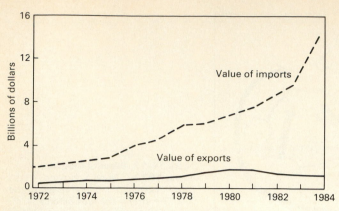

Figure 14-14 Apparel Trade Trends in the U. S., 1972–1984. *While trade deficits have occurred in the apparel industry for years, they reached unprecedented levels in 1984, approaching $12 billion. Some observers argue that, to enhance market penetration, U.S. producers should work together to create an export trading company such as exists in Japan. (Source: as in Figure 14-6.)*

technology to keep these firms competitive has changed these industries dramatically in recent years. Apparel production has traditionally been more manual labor oriented than textiles, so it experienced contraction in highly developed areas and shifts to lower-cost producing environments first. On the other hand, new innovations such as laser cutting and sewing techniques are revolutionizing the apparel workplace and creating economies of scale that are keeping large segments of the industry competitive in highly developed areas. The strong tie of the industry to fashion and the greater affluence of more highly developed area markets also serves to stabilize apparel activity in traditional producing areas. These areas are also the centers of innovation in technology and new market applications, such as the dramatic growth in the use of fabric for industrial applications.

The United States nevertheless experiences a tremendous trade deficit in the apparel industry, as imports appear to exceed exports at a higher level each year. The deficit with textiles is far less, as that activity has adjusted more readily to "high-tech" production techniques.

Pressures to increase tariffs to protect domestic producers in the United States and other highly developed areas have increased as a response to competitive pressures. Free-trade proponents have argued against such a move. The United States has established bilateral trade agreements with many producing nations but has also subscribed to the international GATT arrangements that seek to reduce trade barriers.

In the next chapter attention will focus on spatial and technological forces affecting highly capitalized industries such as steel and automobiles, industries that have also faced increasing pressure from expanding production levels in newly industrializing countries.

Britain and the United States experienced this same transition at the beginning of the industrial revolution. But today the world's much more complex. Some nations try to bypass the labor-intensive phase of growth and introduce capital intensive production from the start. Drawbacks to this approach include a lack of a skilled labor force and small domestic markets for specialized goods. Textile and apparel activity itself is also more capital intensive today.

The textile and apparel industry remains strong, if leaner, in highly developed areas as increased international competition with lower-wage countries has necessitated retrenchment. Dramatic increases in capital expenditure through the introduction of new

SUGGESTIONS FOR FURTHER READING

CLAIRMONTE, FREDERICK, and JOHN CAVANAGH, *The World in Their Webb: Dynamics of Textile Multinationals.* London: Zed Books, 1981.

LINGE, G. J. R., and F. E. IAN HAMILTON, "International Industrial Systems," in F. E. Ian Hamilton and G. J. R. Linge, eds., *Spatial Analysis, Industry and Industrial Environment,* Vol. 2, *International Industrial Systems.* New York: Wiley, 1981.

OKOCHI, AKIO, and SHIN-ICHI YONEKAWA, eds., *The Textile Industry and Its Business Climate.* Tokyo: University of Tokyo Press, 1982.

OLSEN, RICHARD P., *The Textile Industry.* Lexington, Mass.: Lexington Books, 1978.

STEED, GUY P. F., "International Location and Comparative Advantage: The Clothing Industries and Developing Countries," in F. E. Ian Hamilton and G. J. R. Linge, eds., *Spatial Analysis, Industry and Industrial Environment,* Vol. 2, *International Industrial Systems.* New York: Wiley, 1981.

TOYNE, BRIAN, ET AL., *The Textile Mill Products Industry: Strategies for the 1980s and Beyond.* Columbia, S.C.: University of South Carolina Press, 1983.

UNITED NATIONS, *Fibres and Textiles.* New York: United Nations Conference on Trade and Development, 1981.

UNITED NATIONS, *The Textile Industry.* Vienna: United Nations Industrial Development Organization, 1971.

Capital-Intensive Steel and Motor Vehicle Production

15

While the textile industry was in the vanguard of the industrial revolution, iron and steel later came to symbolize the prowess of the industrial state. But today steel no longer claims the lofty position it once held, due to the movement of world economies away from dependence on capital- and raw material-intensive manufacturing activity. In developed economies such industry remains strong but has entered a mature and possibly sunset phase in the product life cycle.

In the mid-1970s crisis overtook the steel industry worldwide and it continued to reel from this malaise well into the 1980s, following decades of expansion after World War II (Figure 15-1). A commitment to maintain excessive steel production capacity in both developing and developed areas over a 25-year period planted the seeds of decline, together with structural changes in the world economy. The 1973–1974 energy crisis and ensuing recession dashed the expansionary approach permanently. But the outlook for steel did not remain negative everywhere. In developing countries, especially the NICs, the outlook remained bright. Advanced countries, on the other hand, continue to experience uncertainty and the prospect of further retrenchment.

In this chapter we review the shift of the iron and steel industry into a mature product-cycle stage in advanced industrial nations and its emergence in the NICs. Nowhere, for example, is the decline of the industry more evident today than in Great Britain, and nowhere is it more prosperous and vital than in South

Korea—two nations on the opposite ends of the development spectrum.

Transitions occurring today in the automobile assembly industry, the biggest steel consumer, will also be discussed. In this case transnationalization and standardization of product has reached advanced levels, providing a situation wherein competition among producers from many nations has intensified as firms seek greater worldwide market shares. It appears that standardized world-class cars produced by a dozen or fewer major transnational firms, with operations in many countries, will be the end result of processes well under way in that industry.

IRON AND STEEL INDUSTRY

Iron and steel once commanded the envious role as the barometer of economic health of a country, but with the structural changes that altered world economies in the 1970s this bellweather function shifted to more knowledge-intensive industries such as computers and other high-technology activities. Once the decline of steel began, the drop was precipitous because many industries depending heavily on steel as a raw material simultaneously faced declining markets. Shipbuilding, automobile production, engineering, and construction all suffered severe reversals in highly developed countries in the 1970s and early 1980s. To make matters worse, the transition away

Figure 15-1 Demolition of U.S. Steel Mill, Youngstown, Ohio Works. *The malaise affecting the U.S. steel industry in the 1980s is dramatically portrayed by this photograph showing the demolition of the four abandoned blast furnaces at the Youngstown Works on April 28, 1982. This turn-of-the-century facility was replaced by an industrial park. (Source: THE VINDICATOR, Youngstown, Ohio)*

from this heavy industrial base was aggravated by two severe recessions in the decade between 1973 and 1983.

Actually, world iron and steel production peaked in 1974 at the time of the energy crisis, which brought into focus many unstable situations already facing the industry. The output of steel in most advanced nations increased continuously for over two decades prior to this time. Developing nations, too, joined the producing ranks during this period. World trade increased as domestic production met a smaller share of the need in many countries, including the United States, which became a net importer of steel in 1959. The overheated industry did not keep its production costs in line in the 1960s nor adopt new technology innovations aggressively. This is particularly true with the older producing areas such as the United States and western European countries.

As the steel boom continued in the early 1970s overordering became commonplace and inventories expanded to counteract the prospect of inventory shortages. Suddenly, as energy prices skyrocketed in 1974, orders ceased and firms began whittling down inventories. Price discounting became an all-too-common technique to capture remaining markets. These adjustments led to a collapse in steel prices and as a result many nations, especially in western Europe, faced a severe steel industry recession. Japan and the United States also suffered. Only the Soviet Union and eastern Europe, among the major producers, managed to continue production increases in the late 1970s. Production expansion also increased in the third world as output rose 67 percent between 1974 and 1980.

With the collapse of world steel prices, inefficient producers closed down and the call went out for subsidies, price controls, and production quotas to save other steel producers in country after country. Even the very survival of the industry in Europe became an issue. In the face of this crisis the need for modernization and adoption of new technology took on more urgency.

Technology Trends

The primary technology innovation affecting the steel production process in the post–World War II era was the conversion from the less efficient open-hearth process to the basic-oxygen production method. While the oxygen process did not become a feasible alternative until the mid-1950s, the switch to this process became very appealing in the 1960s, due to the shorter time it required to make steel, not to mention energy savings. Europe and Japan generally converted to this oxygen process more rapidly than the United States, which lagged because its plants expanded in the early 1950s during the Korean conflict, before it became a workable alternative. The Soviet Union, also a slow adopter, converted to the basic-oxygen process in the late 1970s.

Another technology innovation that gained acceptance in this period due to the potential for significant energy savings was *continuous casting*. This technique eliminated the intermediate step of producing ingots to produce specialty steel products. By linking the production of raw steel directly to that of the final product in a single process, firms avoided reheating the product, creating obvious energy savings.

Other innovations involved the greater use of scrap metal in *electric furnaces* and the direct reduction of iron, which involved removing oxygen from iron ore without using coking coal in the blast furnace. *Direct reduced iron* (DRI) production typically uses natural gas to remove the oxygen in a kiln, yielding a very desirable low-carbon product especially suitable for electric furnace processing. Mexico, Venezuela, and the United States led in DRI output in the late 1970s. Unfortunately, western Europe, having a shortage of natural gas, has been unable to use this process effectively.

Table 15-1

Leading Crude Steel–Producing Nations, 1984 (millions of metric tons)

1980 Rank	Nation	Production	Percent of World Total
1	Soviet Union	155	22
2	Japan	106	15
3	United States	85	12
4	China	44	6
5	West Germany	39	6
	Subtotal	429	60
	World total	710	100

Source: Metal Statistics, 1985 (Fairchild Publications, N.Y.: 1986)

World Production Levels

In the 1980s the Soviet Union led the world in steel production, claiming a substantial lead with 22 percent of total output, followed by Japan with 15 percent and the United States (Table 15-1) with a 12 percent share. An intermediate group of producers included several western European countries and China. A decade earlier the same major producers dominated production, but the United States still outproduced Japan.

The steel industry has lost considerable ground in western Europe in the past two decades as it has in the United States and to a lesser degree in Japan (Figure 15-2). Steel output of the Soviet Union in this period experienced a steady upward trend. We take a closer look at the Soviet industry following a discussion of retrenchment in western Europe.

European Economic Community Steel

The European Economic Community (EEC) accounted for just over half of the steel on the world market in 1966 but declined to about one-fourth of

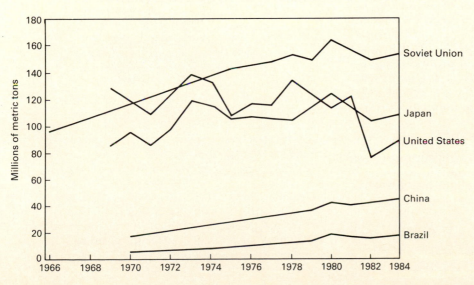

Figure 15-2 Steel Production Trends in Selected Nations, 1966–1984.
Unlike the trend in the United States and Japan, steel output experienced a steady expansion in the Soviet Union. Output in China and Brazil has also grown but remains one-third or less than that in leading producing areas. (Source: Metal Statistics, various years)

Table 15-2

European Economic Community Steel Employment,
1974–1981 (thousands of employees)

Year	Total	West Germany	France	Italy	Benelux	Great Britain
1974	796	232	158	96	112	194
1978	685	203	132	96	87	165
1981	550	187	98	98	79	88

Source: Kevin Morgan, "Restructuring Steel: The Case of Labor and Locality in Britain," *International Journal of Urban and Regional Research,* 7 (1983), 176.

world production by 1976. To capture a greater market share, price cutting became popular in the EEC, but this led to increased losses and diverted resources away from needed plant modernization. Excess capacity and inefficient processes further affected the industry. Unfortunately, the solution often chosen to remedy these problems—subsidies—led to more inefficiencies rather than greater competitiveness and further delayed needed restructuring.

One author indicated five major problems with the EEC steel industry:[1]

1. Chronic overcapacity
2. Depressed prices
3. Astronomical deficits
4. Massive labor displacement
5. Vast subsidies

Coping with these problems proved difficult because the importance of steel in national and regional economies varied widely in the EEC and no supranational political authority existed to enact needed changes. The voluntary *Davignon Plan* developed in 1977, for example, attempted to reduce capacity, increase prices in a coordinated fashion, and abolish subsidies, but violations in several countries continued. This situation resulted in mandatory EEC regulations in 1980, which did not work either. Labor unrest in France and the Benelux countries worked against reforms, as did the lack of transnationalization of ownership. In fact, the trend in recent years has been away from multinational ownership in steel in western Europe toward more independent status. Only Great Britain among the major western European producers significantly decreased its steel capacity in the late 1970s and early 1980s (see Table 15-2). On the other hand, employment in the steel industry in Italy actually rose in this period of uncertainty.

Great Britain. Steel employment dropped by half in Great Britain in the 1970s. Nationwide, 30 enterprises still produced steel in Britain in the 1980s, but one firm, British Steel Corporation (BSC),

[1]Kevin Morgan, "Restructuring Steel: The Crises of Labour and Locality in Britain," *International Journal of Urban and Regional Research,* 7 (1983), 176.

founded in 1967 by the nationalization of over 90 percent of the capacity in the country, remained the dominant producer. Unlike the rest of the British steel firms, it is government owned. Smaller privately owned firms mainly produce specialty steel, and BSC increasingly finds itself producing less profitable bulk steel. The process of selling off specialty production facilities accelerated in the 1980s as the conservative Thatcher government began a process to privatize the industry, reversing the earlier nationalization trend.

Over 70,000 employees lost their jobs in the British steel industry in 70 weeks in 1979–1980. Mills located in the interior of the country suffered most severely during this downturn, because larger coastal installations using imported iron ore produced steel

Figure 15-3 Employment Trends for British Steel Corporation, 1973–1981. *Over 70,000 employees lost their jobs in the British Steel industry in 70 weeks in 1979–1980. Mills located in the interior of the country suffered most severely during this downturn. Coastal installations processing ore in efficient plants experienced less contraction. (Redrafted after Kevin Morgan, "Restructuring Steel. . . ." International Journal of Urban and Regional Research, 1983, p. 195.)*

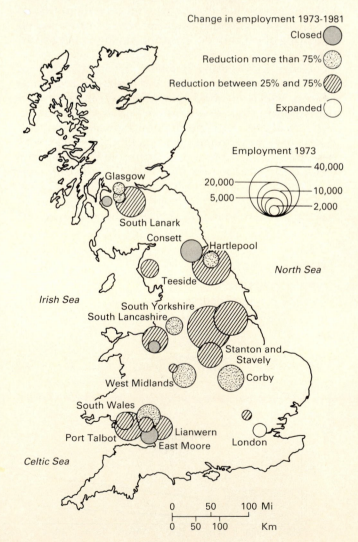

more efficiently (see Figure 15-3). Older open-hearth manufacturing processes requiring more time and energy succumbed to larger-scale basic-oxygen processing plants during this downturn. A growing market share for imported steel also hampered British steel producers in the 1970s as in other advanced economies. The import share of the market rose from 5 percent to 25 percent of total consumption in this period in Great Britain.

The downsizing of British steel capacity occurred in several phases. The first major reduction called for in the 1979 *Business Proposal* endorsed a drop in capacity from 21.5 million tons to 15 million tons. This was followed in December 1980 by the *Corporate Plan,* which reduced capacity to 14.4 million tons. The latter reform reorganized the industry from a regional network into a decentralized product-based series of profit centers. A third call for reduction in July 1982 involved the loss of 17,000 more jobs.

Unlike the situation elsewhere in western Europe, British steel producers do not enjoy the benefits of indirect subsidies, creating a source of contention. Outside the British realm, energy prices, rail transportation rates, and research and development and labor costs are frequently subsidized in western Europe. Unions have also been more cooperative in Britain, as they are fragmented, and the strong populist pressure exerted by the citizenry weakens the movement. Social crises created by the massive displacements in such a short time have been severe in Britain, but the process has created a more competitive industry. Job losses in the late 1970s were particularly acute in manufacturing centers in Great Britain, where unemployment rates skyrocketed at a time when national unemployment rates already exceeded 12 percent, compared to previous postdepression high unemployment levels of 5 to 6 percent.

Other European Steel Powers. West Germany leads all western European steel producers, turning out about twice the quantity of its nearest rivals, Italy and France (Table 15-3). Unlike the situation with most of their neighbors in western Europe, West German steelmakers remain essentially free-enterprise operations, but subsidies from the government do exist.

Table 15-3

European Economic Community Steel Production, by Process, 1980 (millions of metric tons)

Rank	Nation	Total Production	Percent Oxygen	Percent Electric
1	West Germany	44	78	15
2	France	23	82	16
3	Italy	27	45	53
4	Belgium	12	95	5
5	Great Britain	11	59	41

Source: William T. Hogan, *World Steel in the 1980's: A Case of Survival* (Lexington, Mass.: Lexington Books, 1983), p. 42.

Company mergers have also continued in West Germany, increasing their market competitiveness.

Italian steel plants emphasize electric furnace technology, many of which are minimills with a high cost structure. Italsider, a large government-owned firm, has three fully integrated plants accounting for over half of the country's steel capacity, most of which is produced by the basic-oxygen process. The two largest integrated companies in France, Unisor and Scailor, were nationalized in the late 1970s. Belgian firms remain small and produce high-cost steel. Considering that steel is a mainstay of the Belgian economy,

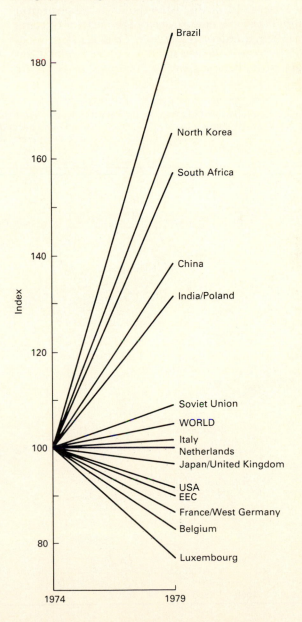

Figure 15-4 Changes in Crude Steel Production in Selected Nations, 1974–1979. *Using output in 1974 as a base, only the Soviet Union, China, and several third world nations expanded output faster than the world average. Note the declining index values in highly developed areas. (Redrafted after Evans, "Europe's Steel Industry. . . . ," Geographical Magazine, July 1982, p. 385)*

recent studies suggest that reforms are needed to avoid further catastrophe.

Soviet Iron and Steel

Unlike the situation facing advanced market economy nations whose iron and steel industry is in decline, the Soviets have experienced expansion until quite recently, as production continued to grow in the 1980s. Only the Soviet Union among major producers expanded its output at a faster pace than the world average following the energy crisis of the early 1970s (see Figure 15-4 on page 233). The growth of heavy industry remains an important goal in the USSR, and steel plays an important part in this policy. Although the country has many steel-producing districts, two-thirds of the output comes from the eastern Ukraine and Urals (see Figure 15-5). These regions are often referred to as the first and second metallurgical bases, respectively.

In Chapter 12 we discussed the importance of iron ore and coal in the emergence of the eastern Ukraine as a major manufacturing center. That area produces over 35 percent of Soviet steel. A similar explanation has been given for the growth of manufacturing in the Urals region. Three other steel-producing areas of lesser importance include Karaganda, the Kuzbass, and KMA, where both iron ore and coal are both readily available. The final area of importance is the Moscow area, which relies on domestic imports of iron and coal.

Despite the enormity of its output, a technology deficiency in Soviet steel processing is demonstrated by the inferiority of specialty products such as rolled sheets used by the automobile industry and large-diameter seamless pipe for the petroleum and gas industries. Most pipe for the petroleum industry, both well and pipeline stock, is imported due to this weakness. Modernization of the steel industry accelerated in the 1980s and technology was sold abroad, mainly to India, Nigeria, and Algeria. Exports flowed primarily to eastern Europe, becoming more important as production glitches occurred in Poland due to increased domestic unrest.

Japanese Steel

Japanese steel production, together with all other manufacturing activities, began again from scratch after World War II, growing from a production level

Figure 15-5 Iron and Steel Industry in the Soviet Union. *Two thirds of the Soviet output occurs in the eastern Ukraine and Urals, the first and second metallurgical bases, respectively. Note secondary concentrations in Karaganda, the Kuzbass and KMA areas.*

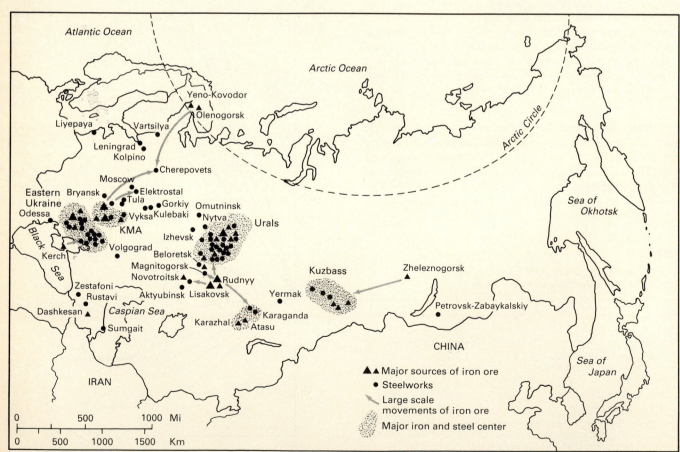

of 1 million tons in 1947 to a peak of 119 million tons in 1973 (see Figure 15-1). This tremendous expansion occurred in very large modern plants. The Japanese in fact now lead the world in the tonnage of continuous cast steel production. The Japanese have more large integrated plants than any other country. Six integrated companies operate 16 facilities, 10 of which have a capacity of 8 million tons or more on an annual basis. The Nippon Kokan facility in the Okayama area claims the distinction of being the world's largest plant with a capacity in excess of 15 million tons annually. The enormity of this facility is demonstrated by the fact that it alone exceeds the steel-producing capacity of Great Britain.

All the Japanese steel capacity has a coastal exposure and direct access to deep-water ports (Figure 15-6), which serve as entry points for iron and coal imports and the exit point for the final product destined for overseas markets. Fully 40 percent of Japanese steel is exported. If automobile exports are included, over half the steel product is exported.

Figure 15-6 Japanese Steel Production Centers. *All Japanese steel production occurs at coastal locations which serve as entry points for iron and coal imports and the exit point for exports. Note concentrations in Chiba (Tokyo suburb) Nagoya, Osaka, and Kobe areas. (Source: Japan Statistical Yearbook, 1986)*

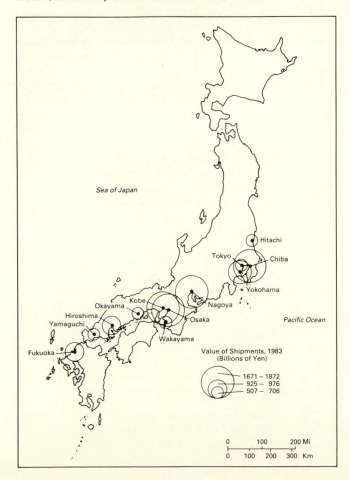

The origin of the steel industry in Japan dates to the period following the Meiji Restoration in 1868. In the early 1900s steel production facilities were developed in the Osaka, Kobe, and Tokyo Bay areas. As recently as 1930 output was only 2 million tons of steel a year. But the demands of rearmament in the 1930s fostered considerable expansion. In the late 1930s the Japanese began importing large quantities of scrap primarily from the United States. Foreign mineral imports also increased, as did flows of pig iron and steel from colonial possessions, especially Manchuria. Production peaked in the war years at about 8 million tons in 1943.

Post–World War II recovery of the steel industry in Japan received a boost during the Korean War. Those years were the beginning of the *Japanese miracle,* the rapid-growth era of the 1950s and 1960s. "In 1970 industrial production in Britain was 10 percent higher than in 1966, in the USA 9 percent higher and in West Germany 32 percent higher. In Japan the increase was 91 percent."[2] In this period Japan became the world's leading shipbuilding nation and the automobile industry came of age. The booming economy also spawned a major machinery industry the output of which in 1963 was six times that in 1955. In 1971 only 2 percent of Japanese steel was produced by the traditional open-hearth process, while nearly one-third was produced this way in the United States.

After the mid-1950s Japan's steel exports grew faster than domestic consumption. Lower labor costs and aggressive adoption of the technology available made prices very competitive on the world market. Although the Japanese borrowed this technology originally, they worked hard to improve it. The Japanese have the reputation in industrial circles as being workaholics as problem solvers, always fine-tuning production processes, giving their engineering industries unparalleled excellence in refining and perfecting new applications.

Nagoya earned the reputation of the "Detroit of Japan" in the 1960s. Fuji and Yawata Steel created large facilities in the area, as did Toyota. In 1970, Yawata and Fuji merged to create Nippon Steel. Other large Japanese producers, all private-sector firms, include Kawasaki, Sumitomo, Kobe, and Nisshin Steel.

Third-World Steel

The biggest growth story for steel in recent years belongs to the third world (Figure 15-4). Nine third-world countries now produce 2 million tons or more annually. The leading newly industrializing producers, paced by Brazil, are shown in Table 15-4. Three large government-owned plants produced over half the Brazilian output in 1980. Brazil shifted status from

[2]Kenneth Warren, *World Steel: An Economic Geography* (New York: Crane, Russak, 1975), p. 102.

Table 15-4

Leading Newly Industrialized Country Steel Producers,
1984 (millions of metric tons)

Rank	Nation	Production	Percent of World Total
1	Brazil	18	3
2	Spain	14	2
3	South Korea	13	2
4	India	11	2
5	South Africa	8	1
6	Mexico	8	1
7	Taiwan	5	1
	Subtotal	77	11
	World total	710	100

Source: *Metal Statistics, 1985* (Fairchild Publications, N.Y.: 1986)

a net importer to a net exporter of steel in 1978. In 1979, and again in 1980, exports of steel in excess of 1 million tons occurred for the first time. The very high quality of iron reserves in Brazil, as discussed in Chapter 12, bodes well for the industry in the future, but coking coal is not as plentiful and must be imported. The country is expected to continue building exports to alleviate foreign balance-of-payment problems.

The newest major world producer is South Korea, with a 1970 capacity of only 500,000 tons; production rose to 10.7 million tons in 1981 and 13 million in 1984. As in Japan, coastal sites are favored. Iron ore and coking coal are also imported. The leading South Korean steel company, Pohang Steel, has a production capacity of over 9 million tons. The South Koreans have also accelerated exports, which now account for about one-third of production, but the country remains a net importer because the output of specialty steel such as flat-rolled product lags. Exports flow mainly to the United States, while imports come primarily from Japan.

Another success story among newly industrializing countries occurs with Spain, the second-largest NIC producer (see Table 15-4). Spain now produces as much steel as Great Britain and Canada, and along with Portugal is the newest member of the European Economic Community, having joined on January 1, 1986. As with most NIC producers, Spain exports a large share of its product to earn hard currency to pay for imported technology.

In some third-world countries external sales involve *buy-back arrangements* to ensure that adequate sales levels exist to make production feasible. Since very large scale plants are preferred by financiers, production frequently exceeds the domestic market demand during early years of production in these areas. Buy-back arrangements involve export sales to creditor nations, which provide the sales cushion to enhance manufacturing operations in third-world areas.

U.S. Steel

The U.S. steel industry, although performing better than its competitors in western Europe, has failed to keep pace with the Japanese or eastern bloc in the past two decades. The reasons for its poor performance run deep. The fact that the industry operated independently for so many decades in an expanding and unencumbered market unfettered by foreign competition allowed uneconomic practices to develop. Unfair marketing arrangements also occurred. The industry was reprimanded in the 1940s, for example, for collusion practices, including the use of a basing-point pricing system called *Pittsburgh Plus.*

This system had the effect of prolonging the dominance of Pittsburgh and keeping prices to customers unnecessarily high. Consumers paid a price for steel based on production and shipment from Pittsburgh, even though the product usually moved to the buyer from a much closer plant.

Signs of greater challenges ahead for the U.S. steel industry emerged in 1959 when a prolonged strike opened the door for larger quantities of cheaper imported products. Prior to that time the industry accommodated growing wage demands from labor simply by passing increased prices on to the consumer. In the 1950s, for example, wages in the steel industry increased twice as fast as overall U.S. wholesale prices. Bad management practices also contributed to the problem of growing obsolescence in the industry. No comprehensive replacement policy emerged for modernizing plants or adopting new technology. Several steel firms, in fact, diverted funds that could be used for modernization to diversify holdings into nonsteel investments.

Mergers, bankruptcies, and changing corporate management strategies cut the number of fully integrated steel mills in the United States nearly in half from the late 1960s to the late 1970s, dropping from 23 to 14 in number. Industry leaders often turned to political solutions for the growing problems facing the industry as the crisis worsened. One author noted that the industry's political response to greater international competition was far more innovative than its economic policy.[3]

In 1980 about 15 percent of total U.S. steel consumption came from imports, a level comparable to that held by foreign producers since the late 1960s, but that level increased to 26 percent in 1984. Nearly one-half of the imports came from Japan in the early 1980s and about one-fourth from the EEC. In addition to the failure of plant modernization policies, these other reasons also explain the growing market share of imports:[4]

1. [F]ailure of the large American producers to build modern steel mills in the fast-growing markets of

[3]Hans G. Mueller, "The Steel Industry," *Annals,* American Academy of Political and Social Science, 460 (1982), 77–78.
[4]Ibid.

the Pacific and Gulf states caused a supply vacuum in those regions. As the cost of overland transportation rose in relation to ocean haulage, it became cheaper to ship steel from new coastal steelworks abroad than from the big American mills located in the Midwestern states.

2. [F]oreign producers, especially the large Japanese companies, have taken the lead in achieving better quality control of their production. This can be attributed in part to the vintage effect; that is, the relative young age of their rolling facilities, but also to the high quality that management has placed on product quality.

3. Continued threats of strikes in the U.S. led steelusers to turn to imports as a more stable source of supply.

4. European and Japanese integrated steel firms began to have lower costs per ton of steel than their American counterparts in the early 1960s.

Lower wage rates, cheaper foreign sources of raw materials, and falling freight rates also contributed to this lower cost structure. These factors were discussed in Chapter 13.

Voluntary import restrictions on steel shipments to the United States began in 1969, but had little impact because worldwide market expansion diverted to other countries flows previously directed to the United States. Uncertainty followed the renegotiation of quotas in 1971. Suits filed in 1972 led to the abandonment of quotas in 1973 as the industry again experienced a worldwide boom. But the 1973–1974 energy crisis and recession abruptly changed the circumstances facing the industry.

Dramatic production cutbacks occurred in 1975, and price cutting accelerated, as mentioned earlier. By 1977 imports to the United States rose again and domestic plant closings eliminated 5 million tons of capacity. In turn, the Japanese were accused of dumping steel and the United States invoked the controversial *trigger price* mechanism. It developed a "system of trigger prices based on the full cost of steel pro-

duction, including appropriate capital charges, of steel mill products by the most efficient foreign producers [then considered the Japanese] . . . plus freight to the U.S., plus a minimal profit."[5] The U.S. situation stabilized in 1978 and 1979, but in 1980 antidumping suits filed by U.S. Steel led to the temporary suspension of the pricing system, only for its restoration later in the year. Several more suits against foreign producers came in 1981, followed by over 80 more in early 1982.

Regional Production Patterns. The American iron and steel industry, with origins in the Pittsburgh area, migrated westward with growing markets in the late 1800s. The expansion of the industry in the greater Chicago area illustrates this tendency, an area that now has the largest capacity in the country, with six of the seven largest integrated companies (Figure 15-7). Steel producers also expanded on the east coast in response to a growing reliance on imports. The classic heartland of the industry, the Pittsburgh–Youngstown–Cleveland corridor, has been stressed by this newer competition. Core-area plants were slow to modernize and became less competitive. Many of these facilities merged or closed in the past few years. Closures of the Youngstown Steel and Tube plant and the U.S. Steel plant in Youngstown around 1980 are cases in point.

The steel industry in the Pittsburgh area traditionally enjoyed easy access to coal fields. Iron ore itself moved southward by Great Lakes barge and rail from the Mesabi, Cuyuna, and Marque mines in northern Minnesota. Gradually, the industry migrated northward to the Great Lakes shores. Coal moving north by rail met the iron ore at this *break-of-bulk* location. Notwithstanding the break-of-bulk advantage,

[5]William T. Hogan, *World Steel in the 1980s: A Case of Survival* (Lexington, Mass.: Lexington Books, 1983).

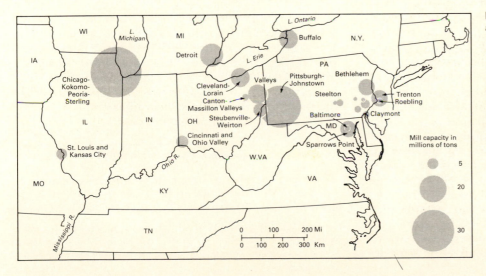

Figure 15-7 Major Steel Production areas in the United States, 1970. *Originally concentrated in the Pittsburgh area, the steel industry migrated westward to the greater Chicago area in the late 1800s, and in the twentieth-century has moved eastward again to the east coast, relying on imported iron ore. (Redrafted from Warren, The American Steel Industry, 1850–1970, Clarendon Press, 1973)*

the primary reason for expansion in Great Lakes locations was the growing markets in the area. The industry soon expanded to the east and west in the late nineteenth century, extending from Buffalo to Detroit and Chicago, on the Great Lakes system. Some peripheral production developed over the years at Birmingham in the south and in western states. Before World War II only two fully integrated iron and steel mills existed outside the northeastern manufacturing belt, one at Birmingham, relying on local coal and ore, and the other at Pueblo, Colorado. During World War II two new major plants opened: at Geneva, Utah (south of Salt Lake) and at Fontana, California (east of Los Angeles).

Most iron and steel production today remains in the manufacturing belt, but Pittsburgh and Youngstown have lost their dominance. Fuel needs are not as high as they once were, due to technological advances which have reduced the need for proximity to coal reserves. Pittsburgh and Youngstown are not central to the markets and their hilly and constricted sites are not as ideal as the flats along the Lakes. It is not surprising to learn that Pennsylvania still ranks as the largest steel-producing state, but this prominence is not now due to Pittsburgh as much as it is to the greater Philadelphia area (Figure 15-7). Fully one-fifth of total U.S. production occurred in Pennsylvania in 1981, followed in ranking by Indiana, Ohio, Illinois, and Michigan.

Capacity Shedding. Mergers and further plant closings continued to cast a pall over the U.S. steel industry in the mid-1980s. From 1979 to 1984, for example, U.S. Steel sold over $3 billion in assets, including coal, real estate, and timber interests, and cut production from 34 million tons to 26 million tons annually and closed 72 steel-centered operations, as well as reducing management by 45 percent. In 1984, Nippon Kokan (NKK), Japan's second-largest steel producer, purchased half of the National Steel company, the fourth leading U.S. producer, to gain greater access to U.S. markets. Another Japanese firm, Nisshin Steel, bought into Wheeling-Pittsburgh in 1984. Republic and LTV also contemplated merger at that time. Other firms discussed plant swapping to gain more efficiency.

The anticipated enactment of voluntary import controls in the United States provided one motivation for Japanese buy-ins of U.S. companies. Steel jointly produced in the United States by such partnerships bypassed import restrictions and cushioned Japanese producers from the declining domestic demand for steel by the automobile industry. Japanese automobile producers also began shifting operations to the United States in the mid-1980s to avoid potential future import quota restrictions. Foreign investment in U.S. steel companies also provided U.S. firms with critically needed capital to modernize operations and develop new products, such as rustproof steel for automobiles.

In the fall of 1984 the Reagan administration announced its "comprehensive fair trade in steel program," which did indeed establish a voluntary import quota system. The import share thereafter was to remain at 18.5 percent of the U.S. market for finished steel for 5 years. Specialty products had tighter restrictions. In mid-1985 Japan announced that it would adhere to this policy.

Massive debt accumulation and declining markets encouraged firms in the United States and Japan to diversify in the 1980s. U.S. Steel, for example, purchased Marathon Oil in 1982. *Fortune* magazine now classifies U.S. Steel as a "petroleum refining" company in view of the fact that steel accounts for 30 percent of revenues and oil and gas 50 percent. A corporate name change to USX Corporation followed this restructuring. Production in 1983 in Japan failed to meet 1982 levels and many firms sought to diversify to preserve their strength. "Nippon Steel, the world's largest steelmaker, has an eye on the industrial ceramics industry. Sumitomo Metal Industries and Kobe Steel are moving into production of special metals like titanium."[6] Steel has been called "Japan's next sunset industry."

At the start of 1984, *Time* magazine made the following assessment of the U.S. steel industry in reaction to the U.S. Steel announcement of pending closures:

> Last week's surgery was by far the most drastic by a single steel company in an industry battered mercilessly during the past decade by rapidly changing economic forces. Steel has been the hardest hit of America's once proud smokestack industries....
>
> American steel has been pounded by cheaper imports from Japan, South Korea and Brazil, crippled by high wages and inefficient plants, and stunted by management that sometimes seems to have just given up on the industry. Employment has plunged from a post-war high of 620,000 in 1953 to about 250,000 last year [1983]; half of that loss has come since 1970. Use of American steelmaking capacity has shrunk from 90% in 1979 to 50% in 1983.[7]

Other large steel firms in the United States faced similar hardships as the USX Corporation in the 1980s. From 1982 to mid-1986, the second-largest steel firm in the United States, LTV, lost more than $1.5 billion. Unable to meet debt obligations, the firm filed for bankruptcy in July 1986. One observer noted at the time that "the whole industry has one foot in the grave and another on a banana peel."

Minimill Competition

There is a bright spot in the steel industry in another market segment. Small mills producing less than 1 million tons of steel a year, called *minimills*, experi-

[6]"Japan's Next Sunset Industry," *The Economist*, January 29, 1983, p. 62.
[7]"Grim Tradition: More U.S. Steel Layoffs," *Time*, January 9, 1984, p. 48.

Figure 15-8 Minimill Operation. *Minimills typically use a nonunion labor force to produce steel for local markets using electric furnaces which consume scrap metal. Their small size permits rapid adjustments in product mix which emphasizes light structural products and consumer goods such as fencing, wire, nails, and staples. (NuCor Corporation)*

enced growth and profitability in the United States and elsewhere in recent years, unlike their giant counterparts (Figure 15-8). The greater decentralization of the minimill industry closer to local markets, the increased use of scrap, and the tendency for more dependence on electric furnaces for such facilities all account for this growth. Minimill producers avoided union labor and their small size permitted rapid adjustments to market conditions, making them ideal producers for light structural products and consumer goods such as fencing, wire, nails, and staples. While they now account for 20 percent of U.S. steel production, having experienced a capacity increase from 9 million tons in 1979 to 13 million in 1984, the market now appears saturated. Stable prices, overcapacity, and foreign competition may give minimills the same headaches that big steel experiences in the future.

International Trade

After World War II barely 10 percent of total steel production entered world trade. At that time the United States was the leading producer. In the 1980s over one-fourth of world iron and steel output moved in international trade. Japan led the exporters, accounting for 18 percent of the total movement in 1983 (Table 15-5). As a trading block, the EEC overshadowed the Japanese export level, paced by West Germany with 16 percent of world export; Belgium/Luxembourg and France, 10 percent; the Netherlands, 7 percent; and Italy, 5 percent. Western Europe as a region also paced the world in imports, due to the high level of flows among EEC member countries. The U.S. share of imports ranked seventh, with a 5 percent share.

Dual Steel Market

In 1983, U.S. Steel began negotiations to import semi-finished steel slabs from Britain as an alternative to completely closing one of its aging domestic plants. This discussion signaled the dawn of a new era of international cooperation in order to save as much steel capacity as possible in developed nations. This plan involved closing the terminally obsolete open-hearth production plant at the Fairless Works north of Philadelphia in Bucks county, but keeping the finishing

Table 15-5

Leading Iron and Steel Trading Nations, 1983
(millions of U.S. dollars)

Rank	Nation	Value	Percent of World Total
	Imports		
1	France	608	11
2	Italy	608	11
3	West Germany	571	11
4	Japan	345	6
5	Belgium/Luxembourg	325	6
6	South Korea	316	6
7	United States	266	5
	Subtotal	3,039	44
	World total	5,440	100
	Exports		
1	Japan	970	18
2	West Germany	892	16
3	France	740	13
4	Belgium/Luxembourg	730	13
5	Netherlands	380	7
6	South Korea	330	6
7	Italy	267	5
	Subtotal	4,309	76
	World total	5,539	100

Source: United Nations, *1983 International Trade Statistics Yearbook,* 1985.

plant in operation. By using imported semifinished slabs for further refining the firm hoped to preserve many jobs. But at year's end the Fairless works joined the list of closing plants when no agreement was reached. It appears inevitable, despite this failure, that as basic steel production increases in the third world in modern, efficient plants, developed nations will

Table 15-6

Car, Bus, and Truck Production, by Nation, 1985
(millions of units)

Rank	Nation	Production	Percent of World Total
1	Japan	12.3	27
2	United States	11.7	26
3	West Germany	4.5	10
4	France	3.0	7
5	Soviet Union	2.2	5
6	Canada	1.9	4
7	Italy	1.6	4
8	Spain	1.4	3
9	Great Britain	1.3	3
10	Belgium	1.0	2
11	Brazil	1.0	2
	Subtotal	41.9	93
	World total	45.0	100

Source: Ward's Automotive Yearbook (Detroit, Mich., Ward's Communications, Inc., 1986).

import semifinished goods rather than raw iron ore as in the past.

In summary, there is evidence that a two-tier world steel industry now exists, with raw steel increasingly produced in the lower-cost environment of the third world and higher-cost production of specialty steel occurring in developed areas. Examples of the latter include seamless pipe, alloy steel for aerospace, and flat-rolled steel for the motor vehicle industry. The third-world share of crude steel production in fact increased from 7 percent in 1970 to 17 percent in 1982. This dual steel market thus reflects changing patterns of comparative advantage in producing steel and means that the industry will continue facing worldwide production adjustments for the foreseeable future. Shedding excess raw steel capacity will remain a significant problem for Japan, the United States, and western Europe, even as the conversion to making efficient specialty products proceeds.

MOTOR VEHICLE PRODUCTION

The world automobile industry, perhaps the glamor industry of the twentieth century, encompasses the largest corporate business entities, in both assets and employment, in the world. Even though the manufacture of automobiles has received priority in developing countries, the production of motor vehicles remains highly concentrated in developed nations, with a triad consisting of Japan, the United States, and western Europe accounting for over 80 percent of world output in 1985 (see Table 15-6). The leading producer outside this bloc, the Soviet Union, manufactures about 6 percent of the world total. The other areas producing more than 1 million motor vehicles annually, Canada, Spain, and Brazil, have no domestically based firms.

The biggest transition occurring in automobile production in the past 40 years relates to the displacement of the United States as the dominant producer and the substitution of Japanese leadership. In 1950, the United States alone claimed 75 percent of world motor vehicle output (see Figure 15-9). By 1965, the U.S. share declined to 45 percent, dropping to 28 percent in 1970 and 21 percent in 1980, and subsequently rising to 26 percent in 1985. In the early postwar period, expansion of the industry in western Europe mainly explained the decline in American dominance, but since 1960, and especially in the 1970s, growing Japanese output provided the most competition.

Japan produced more cars than the United States for the first time in 1980, capturing 28 percent of the world output. The emphasis in Japan on producing competitively priced smaller cars and developing a reputation for quality explains much of this growth, especially the larger export share. Exports to the United States and western Europe grew explosively in the late 1970s and early 1980s as the overseas Japanese distribution and marketing matured.

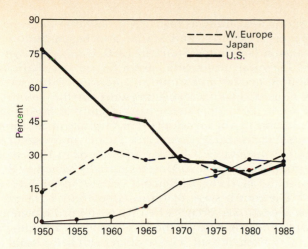

Figure 15-9 Changing Shares of World Automobile Production, 1950–1985. *The biggest transition in automobile production in the past 40 years involved the displacement of the United States as the dominate producer. Note the virtual parity of production among the Western Europeans, Japanese, and United States' in the mid-1980s. (Source:* Ward's Automotive Yearbook, *various years)*

The United States ceased to be a net exporter of automobiles in the 1950s in terms of numbers of units, but continued to maintain favorable balance of trade, because more higher-priced vehicles were exported than imported until the late 1960s. By the early 1980s, the United States registered a massive trade deficit in automobile trade. While the country exported about $16 billion worth of cars, parts, and commercial vehicles, the import bill for autos alone exceeded $27 billion.

The concept of *surplus capacity* assists in the understanding of the changing world order in automobile production, especially the decline in market shares of western European and U.S. producers. The notion of surplus capacity also provides insights into the dynamics of international trade. Surplus capacity refers to a situation wherein excess production capacity results in dramatic changes in industrial policy following a sales slump or loss of markets. To cope with these changes, trade policies and government responses typically become more protectionist. At the same time, competition from the greater productive capacity in developing areas increases and the share of product involved in international trade rises significantly.

After World War II the prevailing attitude toward industrial expansion worldwide suggested that all producers could expand their output, developed and developing countries alike. The presumption that expanding market demand would absorb increased production and all countries would maintain a liberal trade policy allowing for unlimited import and export flows accompanied this philosophy. Essentially this is the way the automobile industry operated. Expansion occurred in all regions and trading became easier. In 1965, Canada and the United States worked out arrangements to eliminate automobile trade re-

straints. Similar agreements among U.S. and Japanese producers also occurred. Chrysler and Mitsubishi, for example, completed negotiations whereby Chrysler would import Mitsubishi-made cars from Japan under the Chrysler name beginning in 1971.

But the tide turned later in the 1970s. By mid-decade, western European automobile producers began to show symptoms of surplus capacity and several countries invoked voluntary restraints designed to limit Japanese imports. By the end of the decade the same surplus capacity problem faced the United States and a similar response occurred. In May 1981 a voluntary import restriction directed at Japanese producers went into effect, and Congress began discussing more restrictive measures, such as a 90 percent local content bill which would require the use of domestic parts in cars marketed in the United States. Government involvement to protect the U.S. auto industry at the time became most visible with the massive loan guarantee program for Chrysler in 1980. Another response to the surplus capacity problem involved complex international corporate arrangements, such as joint ventures, licensing agreements, technology trades, and research pooling programs.

The following examples illustrate this process of corporate consolidation. In Europe, Peugeot took over Citroen in 1974. In 1980, Renault acquired 25 percent of American Motors, providing greater access to the U.S. market and the opportunity to assemble cars jointly in the United States. In 1979, Ford acquired a 25 percent equity interest in the Japanese firm Toyo Kogyo, manufacturer of Mazda. At the same time Nissan bought a one-third share of Motor Iberia in Spain. British Leyland began producing cars with Honda, and VW agreed to a venture agreement with Honda.

In summary, the surplus capacity concept accounts for the responses of companies and governments when overproduction causes problems for a particular industry. Changing attitudes toward trade, the timing of responses, and growing interests in greater levels of transnationalization can all be accounted for with this concept. Although the level of acrimonious debate that occurred with international textile negotiations has generally been avoided in the automobile industry, domestic automobile production futures can spark nationalistic responses. Before reviewing trends in each major producing bloc it will be useful to elaborate on the extensive transnationalization process under way in the industry today.

Global Operations

Automobile producers operate much more intensively at the international scale than most other manufacturers. The leading corporations in Japan, the United States, and Europe all engage in worldwide operations. The scope and sheer size of the leading firms in the industry, including General Motors and Toyota, is awesome. General Motors, for example, maintained a global work force of 690,000 employees in 1983. These producers compete head-on globally, not just within the borders of one or two countries. Such world-class competitors know no political boundaries because they operate at a different level.

In some cases the roots of this global involvement date back two generations or more, but in other cases it represents a recent development. The drive to increase productivity through automation and robotization strongly motivates firms to expand their operations today. As more capital is invested and the industry becomes less labor intensive, giant firms can solidify their position as innovators and efficient producers. In the mid-1980s the Japanese gained a cost advantage over U.S. and European producers in the small-car market (1000 to 2000 cubic centimeters). This competitive edge gave Japanese firms additional incentives to penetrate overseas markets and develop strategies to produce automobiles in other countries. At the same time European and American producers welcomed access to Japanese management and technology advances.

Transnationalization Process. Five stages in the evolution of the motor vehicle industry can be identified, leading to the present global production system:

1. Domestic expansion
2. Direct exports
3. Local assembly
4. Integration of production
5. Export from host country

Efficient mass production and marketing in the automobile assembly industry can best be achieved by specialized, heavily automated plants that provide scale economies. As the volume of production increases, costs decrease, and domestic market saturation may result. The first phase in the growth of motor vehicle production, *domestic expansion,* then leads to the second, *direct export,* stage. European, American, and Japanese firms all experienced this transition from domestic to international marketing. While American-produced vehicles are no longer exported in large numbers, European, and especially Japanese, producers represent very active stage 2 producers.

American- and European-based firms moved to later stages in the transnationalization process more aggressively than the Japanese until recently. Stage 3, *local assembly* in foreign countries, occurred among American firms in the immediate post–World War II era, followed by European company initiatives. The Japanese, heretofore more cautious, began investing in U.S. plants in the early 1980s to avoid import restrictions. Honda began production of the Accord model car in Ohio in 1982, Nissan began producing pickup trucks in Tennessee in 1983 and cars in 1985. Toyota and GM began joint production of automobiles in Fremont, California, in 1985. By far, the majority of the Japanese automobiles in America today continue to be imported directly from Japan, but the trend is weakening as overseas production increases. Toyota, Mitsubishi, and Mazda also began building plants in the United States in 1986.

European producers have a far smaller share of the U.S. market and would not be expected to have as large a manufacturing presence in the country. Mercedes-Benz operates a truck facility in West Virginia, and Volvo, a truck plant in Virginia, in addition to the Volkswagen of America facility in Westmoreland, Pennsylvania. Chrysler announced in 1987 that it would buy American Motors, partially owned by Renault and marketer of Renault vehicles in the U.S.

The fourth stage of transnationalization, *integration of production,* occurs when parts and procurement purchases originate in the host country rather than coming from abroad. This arrangement helps build a base of production indigenous to the host country and assists in the development of a skilled labor force. Psychologically, this process also fosters pride and prestige. Some countries require local content in the manufacturing process, while others, such as Brazil, have promoted integration of production by excluding imports totally.

Often, the producer is obliged only to purchase nonautomotive products initially, developing subsidiary auto-parts industries later. Cooperative licensing agreements with local firms, subcontracting or the development of a subsidiary firm, can be used to increase the local production of parts. Some corporations, such as Fiat, practice joint-enterprise involvement using local capital participation in projects. Countries reluctant to surrender control to multinationals find the latter approach particularly appealing. The trend to expand "just in time" parts procurement programs rather than stockpiling parts in the automobile industry also favors reliance on local suppliers.

The fifth stage in this process, *export from the host country,* receives more attention with the maturing of the industry and growing cost structure diseconomies in the parent country. Western European firms, for example, have been actively involved in developing production capacity in Spain, Portugal, and Latin America for the export market, to take advantage of lower labor costs and the greater flexibility offered by less organized labor markets.

Other scenarios also explain the expansion of world production. Some countries purchase technology abroad to gain a competitive edge. Fiat sold a large production facility under a licensing agreement to the Soviet Union in 1970. Although that arrangement does not fit the stages under discussion here, it gave the Soviets more competitive muscle in motor vehicle production. As a result, Fiat now faces greater international competition for its automobiles with the greater level of Soviet exports.

The most dramatic symbol of the growing internationalization process in the automobile industry has been the emergence of the "world car." Not only do cars from all producers look quite similar today for a particular market niche, but the same product, except perhaps for the nameplate, is being marketed in several countries simultaneously. For example, for several years Volkswagen called its popular subcompact front-wheel-drive vehicle Rabbit in the United States and Golf in Europe.

The internationalization of automobiles is more than just a marketing phenomenon. Specialized production of component parts now occurs in many countries, each of which may send items for assembly to several other nations. The Ford Fiesta, for example, was conceived on a global basis, with production facilities located in several countries and component-parts production in several more. Ford now has a Europe strategy that entails marketing the same car in Britain and Germany. The Iveco truck, discussed later, is a joint product developed by the Italians, French, and Germans, with eight plants in Italy, three in France, and two in Germany. "GM's philosophy is to produce a world-wide concept design which individual countries can tailor to their individual needs. An example of this is the 'T' car which was marketed as the Chevette in the U.S. until 1986, Opel Kadett in Germany, Chevette in the U.K., and Isuzu in Japan."[8]

Japanese Automobile Industry

Some observers have noted that a successful automobile producer today must market at least 1 million vehicles a year, and others suggest a threshold of at least 2 million. By the latter criterion, Japan has two strong automobile producers, Toyota and Nissan (see Table 15-7). By the former standard, Toyo Kogyo, Mitsubishi, and Honda should also be counted as strong competitors. In the mid-1980s Japan possessed nine automobile manufacturers out of a total of eleven motor vehicle producers. These smaller firms included Isuzu, Daihatsu, Subaru, and Suzuki. Along with the five leaders, these four firms could not prosper if it were not for the incredible size of the Japanese export market. Over 6 million units or 54 percent of total production in Japan went abroad in 1981.

History. Two phases of U.S. involvement account for the early history of the Japanese automobile industry. First, in the 1920s the big three U.S. producers, General Motors, Ford, and Chrysler, all established operations in Japan, producing Plymouths, Fords, and Chevrolets, but the industry remained relatively weak. After the war a second phase of development occurred as Japanese companies entered into agreements with American and British firms, such as Mitsubishi-Willys (Jeep) and Nissan-Austin. Once facilities were fully integrated, purely Japanese designed and engineered vehicle production began in the early 1950s.

The tremendous demand of the domestic Japanese market fueled the initial expansion drive in the

Table 15-7

Leading Japanese Motor Vehicle Producers (Including Output for Export), 1985 (millions of units)

Rank	Company	Production	Percent of Japanese Total
1	Toyota	3.7	30
2	Nissan	2.5	20
3	Toyo Kogyo (Mazda)	1.2	10
4	Mitsubishi	1.2	10
5	Honda	1.1	9
	Subtotal	9.7	79
	Japanese total	12.3	100

Source: Ward's Automotive Yearbook (Detroit, Mich., Ward's Communications, Inc., 1986).

industry. In 1960, for example, there were just over 1 million motor vehicles in Japan, a number that rose to a phenomenal 40 million units in 1976. To protect domestic producers, import quotas and tariffs were established, but later phased out in the 1970s. Nevertheless, licensing agreements and strict inspections served to continue limiting imports.

The leading Japanese manufacturer, Toyota, has roots in the textile field. Sakichi Toyoda, the founder of Toyota, invented an automatic weaving machine, and later began producing cars in the Nagoya area in 1936. Production by that firm remains localized in the same region today, now called Toyota City. A major integrated steel plant provides body and forge products locally. About half of the 40,000 plus Toyota workers in the Toyota City region work at the headquarters plant, opened in 1938. It was a major military supplier during the war, outproducing Nissan and Isuzu, both located in the Tokyo area. The headquarters facility was joined by seven additional plants from 1959 to 1975 to produce cars, trucks, engines, transmissions, and chassis parts. Many subcontracting firms are also closely tied to the area as parts manufacturers, a business that has also spread to the Tokyo area.

The second major Japanese motor vehicle producer, Nissan, identifies with the Tokyo region. Unlike Toyota, which marketed cars both at home and abroad under the same name, Nissan traditionally used the corporate name only for domestic sales and the Datsun label abroad until the early 1980s, when the Datsun name was gradually phased out. The label Datsun has a fascinating origin, dating back to the early 1900s. The financiers of Kwaishinsha Motor Carworks, founded in 1911, were three businessmen named Den, Aoyama, and Takeuchi, and the car produced by the firm took its name from their initials, D.A.T. The company failed and the successor firm produced a car known as "son of DAT," or Datson. The spelling was changed to Datsun because the "son" expression "as pronounced in Japanese sounds like an expression for losing money. . . ."[9] In the 1920s the

[8]Krish Bhaskar, *The Future of the World Motor Industry* (New York: Nichols Publishing, 1982), p. 98.

[9]John B. Rae, *Nissan/Datsun: A History of Nissan Motor Corporation in U.S.A., 1960–1980* (New York: McGraw-Hill, 1982).

DAT firm merged, followed by another reorganization in 1934 that resulted in the formation of the Nissan Motor Company, Ltd. During the war the company produced military vehicles and aircraft engines, resuming civilian truck production in 1946 and cars in 1948. In 1952 an agreement with the Austin Motor Company of Great Britain led to the production of Austin cars by Nissan. This involvement provided needed technical expertise. The demands of the Korean War also provided a needed stimulus through sales to the U.S. forces. Nissan production first reached the 100,000-unit threshold in 1956.

Significant exporting of Datsun automobiles began in 1958, primarily to southeast Asia, but sales to the United States also began at that time, following an automobile show in Los Angeles, where the first Japanese automobile was introduced to the U.S. market. Datsun sales increased to 1300 units in the United States in 1959 and in 1960, Nissan Motor Corporation-USA was chartered. At that time the leading imports to the United States were European products, dominated by Volkswagen, which enjoyed a 50 percent share of all imports in the early 1960s, a share that grew to two-thirds by the mid-1960s. In 1965, for

Figure 15-10 Imported Japanese Automobiles on Dock in Los Angeles. *Japanese dominance in the U. S. imported car market began in the 1970s. (Courtesy Nissan Motor Corporation in U.S.A.)*

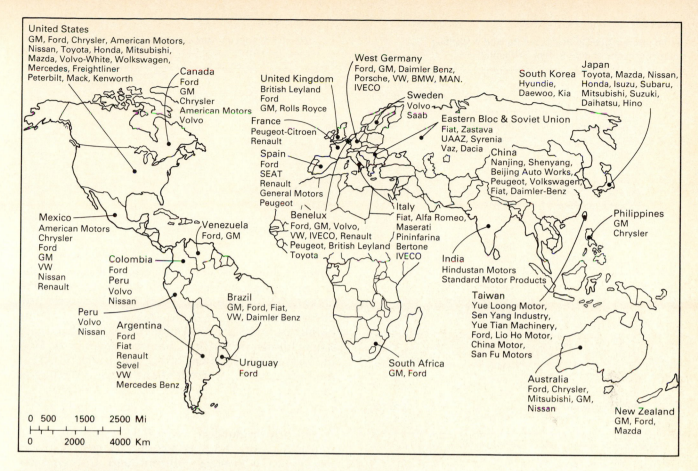

Figure 15-11 Major Motor Vehicle Manufacturers by Nation. *Unlike the situation in the United States and Japan, only one-third of European production occurs with domestic firms. General Motors, Ford, Chrysler, Toyota, Nissan, Volkswagen-Audi, and Renault are the world's seven leading producers, owing to their extensive global operations. Domestically-based firms dominate in the Soviet Union, India, China, Taiwan, and South Korea, but some foreign joint ventures occur. (Redrafted and updated from* Ward's Automotive Yearbook, *1980, p. 95)*

example, there were 16,000 Datsuns (2 percent of U.S. imports) sold in the United States, compared with over 380,000 Volkswagens (67 percent of U.S. imports).

Toyota overtook Datsun as the leading Japanese exporter to the United States in 1967, when each company shipped in excess of 30,000 units. Volkswagen exports also climbed, topping 500,000 units in 1968, while Toyota shipped 68,000 units and Datsun 40,000. In that year imports accounted for 10 percent of U.S. sales for the first time. In 1969, Toyota displaced Opel as the number two imported car in the U.S. market, while Datsun remained in fourth place, only to overtake Opel itself in 1970, giving the Japanese two of the top three import positions.

Toyota surpassed Volkswagen in U.S. import car sales in 1975 for the first time, and Datsun, which shipped more trucks to the United States than Toyota at that time, passed them both, counting car and truck imports (Figure 15-10). Honda captured fourth place at that time, further establishing Japanese dominance in the import market. In the mid-1980s Japanese firms gained greater dominance as major suppliers to the

U.S. market, as several more Japanese firms gained in prominence in U.S. sales, including Mitsubishi, Izusu, and Mazda. In 1986, Honda for the first time exported more cars to the U.S. than Toyota, including the cars it produced domestically.

European Automobile Production

Unlike the situation in the United States and Japan, where the overwhelming share of automobile sales come from domestically headquartered firms, only one-third of European sales have traditionally originated from domestically based firms (Figure 15-11). This occurs because of the large presence of U.S. firms in Europe and because many European firms maintain continent-wide operations. Overseas involvement, in addition to dominance at home in the United States, make General Motors, Ford, and Chrysler three of the world's seven leading motor vehicle producers, along with Toyota and Nissan from Japan, and Volkswagen-Audi and Renault, based in Europe (Table 15-8).

Table 15-8

Leading Motor Vehicle Producers Ranked According
to Production by Nation (millions of units)

Rank	Producer	Nation	1985 Production
1	General Motors	United States	6.4
2	Toyota	Japan	3.7
3	Ford	United States	2.8
4	Nissan	Japan	2.5
5	Renault	France	1.5
6	Chrysler	United States	1.5
7	Volkswagen-Audi	West Germany	1.5
8	Peugeot S.A.	France	1.5
9	General Motors	Canada	1.4
10	Fiat Group	Italy	1.2
11	Toyo Kogyo (Mazda)	Japan	1.2
12	Mitsubishi	Japan	1.2
13	Honda	Japan	1.1
14	Opel	West Germany	0.9
15	Lada	Soviet Union	0.8
16	Suzuki	Japan	0.8
17	Daimler-Benz	West Germany	0.7
18	Ford	Canada	0.7
19	Isuzu	Japan	0.6
20	Fuji (Subaru)	Japan	0.6
21	Daihatsu	Japan	0.6
22	British Leyland	Great Britain	0.6
23	Ford	West Germany	0.5

Source: Ward's Automotive Yearbook (Detroit, Mich., Ward's Communications, Inc., 1986).

Table 15-9

Western European Automobile Sales
Leaders, 1983

Firm	Percent of European Total
Renault	13
Ford of Europe	13
Fiat	12
Volkswagen	12
Peugeot	12
General Motors–Europe	12
Subtotal	74

Source: Forbes, February 13, 1984.

wagen is the largest firm operating in Brazil and the largest auto producer in Latin America. Unlike the decentralized domestic network of Volkswagen producers, Daimler-Benz continues to concentrate its output around Stuttgart. Daimler-Benz also has a very strong foreign presence in many countries in the Middle East, South America, Africa, and India, as well as the United States. Mercedes-Benz Brazil is the largest producer of trucks and buses in Latin America.

French Production. In 1960, France supported four major car producers: Renault, Peugeot, Citroen, and Simca (Chrysler). Peugeot and Citroen later merged and Chrysler sold Simca to Peugeot-Citroen in 1978. The two remaining firms, Renault (51 percent of French production in 1981) and Peugeot-Citroen (42 percent of production), possess significantly different organizational structures. Renault was nationalized in 1945. It has plants decentralized throughout the country and accounts for 40 percent of French motor vehicle exports. The company also has big operations in Spain (eight plants), Portugal, Mexico, Argentina, Venezuela, eastern Europe, and Africa (Figure 15-12). Renault began joint marketing with American Motors in the United States in the late 1970s.

Peugeot-Citroen is privately owned. Peugeot production has traditionally been localized in Sochaux, while Citroen operated in the Paris region. Citroen's nondomestic operations concentrate on Europe, while Peugeot is represented in the market in Africa, Indonesia, Thailand, Australia, and New Zealand, among other areas, in addition to its European facilities.

No automobile manufacturer producing cars in Europe, domestic or foreign, has a dominant sales position. In fact, six firms traditionally split the market equally, each with a 12 to 13 percent market share (Table 15-9). Some leading European producing countries, in fact, possess no domestically based firms, but serve as premier auto assemblers. Belgium, for example, produces over 1 million automobiles annually for foreign firms.

The largest GM assembly plant outside the United States occurs in Antwerp, as does a Ford plant, while Volkswagen, Renault, and Peugeot-Citroen all operate production facilities near Brussels, and Volvo, British Leyland, and Toyota (Daihatsu) are also represented in the country. Belgian assembly plants rely on imported component parts and the majority of the units are exported.

West Germany. In 1985, West Germany, Europe's major motor vehicle producer, had 10 percent of world production, with an output of 4.5 million units (Table 15-6). The leading producers include Volkswagen with a 40 percent market share; Opel (General Motors), with a 21 percent share; and Daimler-Benz, with 16 percent. Volkswagen alone operates seven assembly plants in West Germany and Belgium. Foreign operations of Volkswagen occur primarily in South Africa, Brazil, and the United States. Volks-

Italian Production. Fiat dominates Italian motor vehicle output, representing 78 percent of the total in 1981. Alfa Romeo was a distant second, with 14 percent of total production. In 1986 Fiat bought out financially ailing Alfa Romeo from the government-owned holding company that controlled the firm. The majority of Fiat plants form a cluster in the Turin area

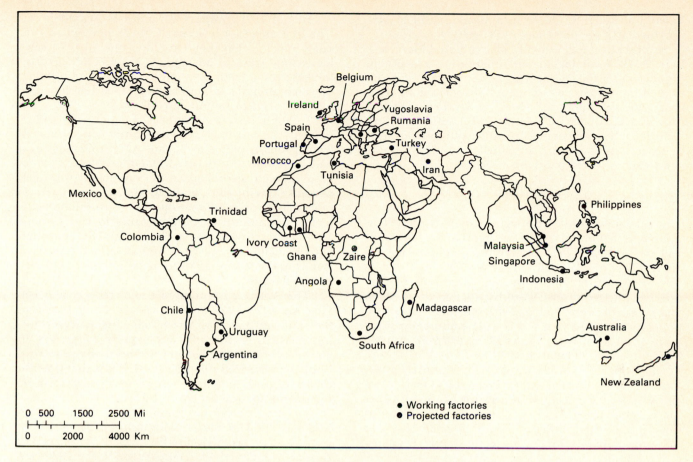

Figure 15-12 Renault Global Manufacturing Operations. *Renault produces about half the French automobile output, and maintains operations in about 28 other nations. Note the widely dispersed involvement in Third World areas in addition to Western Europe, Australia, and New Zealand. The U.S. linkup with American Motors is not shown on the map. (Redrafted from Sékaly,* Transnationalization, of the Automobile Industry, *University of Ottawa Press, 1981, p. 98.)*

of northwestern Italy. Foreign operations occur in eastern Europe, the Mediterranean basin, and South America. Exports account for nearly one-half of Italian vehicle assembly.

Fiat modernized its line of very competitive automobiles in the early 1980s and anticipated a strong performance for the remainder of the decade. The corporation jointly produces the Iveco brand of trucks with France and West Germany. The company withdrew its distribution network in the United States in 1983, after several years of declining sales. The company has also diversified into aircraft engine production and telecommunications, but the firm's future clearly depends on automobiles.

British Auto Industry. The British automobile industry retreated significantly in the past 20 years. It is the smallest among the major western European countries and the leading British producer, British Leyland, ranks twenty-second in the field of major world producers (Table 15-8). British Leyland, created by merging several weak producers in 1968, was nationalized in 1974. About 40 percent of domestic pro-

duction comes from this firm, while Ford accounts for 37 percent of output and General Motors (Vauxhall), 12 percent. Chrysler withdrew from the British market in the 1970s. British Leyland has over 25 foreign plants but has yet to regain its strength at home or abroad following years of restructuring.

European Outlook. More consolidation among European producers will probably occur in the future. Volkswagen, Renault, and Fiat appear to be the strongest competitors. The future is somewhat clouded by the weak presence of European producers in the Pacific basin, where Japan appears strong. Excessive exposure of European firms in the sluggish European and uncertain Latin American markets will probably dampen future growth prospects. The industry remains very fragmented in Europe and the continuing protectionist philosophy, undergirded by rather high import duties, artificially protects domestic producers and will probably prove detrimental. In the mid-1980s more joint ventures with Japanese and American producers gave several firms more strength. Nissan, for example, established ties with Volkswagen,

Volvo and Renault bought into Suzuki, and Chrysler obtained a 15 percent share of Peugeot.

Several European vehicles, such as Volvo, Mercedes-Benz, BMW, Saab, Jaguar, and others enjoy a favorable reputation for the engineering and styling of their top-of-the-line models, which generally set the international design standard. These and other high-performance European vehicles are favorites for affluent American drivers seeking a prestige automobile. Most of these manufacturers are small producers of specialty cars by world standards.

U.S. Production

The U.S. automobile production map remained as volatile as that of the rest of the world in the late 1970s and early 1980s. The U.S. automobile giant, General Motors, lost nearly $800 million in 1980, its first loss since 1920. At the end of the year it laid off 140,000 workers, merged several subsidiaries, and closed some plants. Profits returned in 1981 but sales continued to sag for the third year in a row. The number of company employees worldwide fell to 657,000 in 1982 from a level of 853,000 in 1979.

The poor performance of GM in the early 1980s was shared by the two other major producers, Ford and Chrysler. It occurred as a result of sluggish new car sales due to escalating gas prices, worldwide recession, double-digit inflation, poor management and planning, and the increasing competition from imports. Rapidly rising car prices created another phenomenon—*sticker shock*—which refers to negative customer reaction to posted prices on cars. As the U.S. economy rebounded in 1983 and 1984, however, so did car sales. Not only did General Motors begin generating record profits as a result, but so did Chrysler and Ford. Throughout this period, General Motors nevertheless led in worldwide motor vehicle production by a wide margin (Table 15-10).

Table 15-10

World's 12 Leading Motor Vehicle Producers
(Cars and Trucks), 1985 (millions of units)

Rank	Company	World Production
1	General Motors	8.9
2	Ford	4.6
3	Toyota	3.7
4	Nissan	2.5
5	Volkswagen-Audi	2.1
6	Chrysler	1.9
7	Renault	1.9
8	Peugeot	1.5
9	Fiat	1.4
10	Toyo Kogyo (Mazda)	1.2
11	Mitsubishi	1.2
12	Honda	1.1

Source: Ward's Automotive (Detroit, Mich., Ward's Communications, Inc., 1986).

In the 1970s import car sales increased in the United States to a level of 20 to 25 percent of total sales, and in 1980 they exceeded one-third of sales, increasing again to 38 percent in 1986. Most of the increase came from Japanese vehicles, the producers of which offered a better engineered and designed vehicle in terms of public perception.

It should be noted, however, that one-third of U.S. imports came from U.S.-owned producers, primarily from assembly plants in Canada. Over one-half of Canadian production occurs in General Motors plants. Ford claimed second place in production with a 27 percent share in 1981, and Chrysler was third with 11 percent of the Canadian output. This production clusters predominantly in southern Ontario, with General Motors operating plants in Oshawa and Scarborough, Ford at Oakville and St. Thomas, and Chrysler, in Windsor. One General Motors plant exists in Quebec, at St. Therese.

The American automobile changed substantially after the 1973–1974 energy crisis. Prior to that time the full-sized U.S. car was much longer, wider, and heavier than its foreign counterparts. Large V-8 engines, automatic transmissions, and many power options characterized the typical car. Since that time the car has shrunk to match the international preference for smaller, more efficient vehicles.

General Motors has maintained its status as the leader in the U.S. automobile industry for nearly 50 years. In the past 25 years the market share for GM products remained 40 to 50 percent. In the 1920s and 1930s Ford led the industry, but its share declined over the years. Chrysler remained a strong competitor until the late 1970s, when its share declined to about 10 percent of the market, only to rebound again with innovative products in the early 1980s.

U.S. producers developed a reputation for *badge engineering* in recent years because each of the big three producers marketed many nearly identical products as different brand-name products. Each of the five divisions of General Motors, for example, marketed a full range of parallel models, except for the luxury Cadillac division at the top end of the market. This situation led to customer confusion and an identity crisis for the models. U.S. producers have also had a difficult time building their smaller-car market mix. Chrysler was very slow to enter this market in the late 1970s, but was able to recover strongly in the early 1980s with a new line of well-engineered front-wheel-drive vehicles. Ford also badly misjudged the market as it entered the 1980s but gradually became stronger well into the decade, when it belatedly introduced downsized front-wheel-drive vehicles with European-inspired designs. The payoff for this effort came in 1986 when much smaller Ford generated larger profits than General Motors.

In early 1984 General Motors made its first major organizational change since 1920 when it announced that two major engineering and design divisions would be created while retaining five brand-name cars. At the lower end, the new Chevrolet–Pontiac division would emphasize small cars, while the Oldsmobile–Buick–Cadillac division would produce large cars. This shift promised more efficient planning, engineering, and marketing with a leaner management superstructure.

Ford's international involvement grew considerably in the postwar years, making it the most international of U.S. automobile manufacturers. Foreign production accounted for over 40 percent of total output in 1980. Wholly owned subsidiaries manufacture vehicles in Argentina, Australia, Brazil, West Germany, Great Britain, Mexico, and South Africa. In recent years these foreign Ford plants have proven more profitable than U.S. operations.

Fearing a restrictive quota barring greater access to U.S. markets, the Japanese agreed to a voluntary limit of exports to not exceed 1.68 million units for the year ending March 31, 1982, 7.7 percent below the 1980 level. The following year exports were to expand to no more than 16.5 percent of the growth of the U.S. market. The purpose of this restriction was rationalized as providing a *protective shield* for U.S. producers, to give them time to restructure their plants. Critics of this arrangement pointed out that this restriction would have the effect of increasing the price of cars, giving a greater market share to more expensive U.S. cars. Adroitly, the Japanese upscaled the mix of exported units, giving a greater share to higher-priced models, which were more profitable. Toyota, for example, exported a richer mix of cars to the United States in the first half of 1982 than in 1981. In 1981 about 7 percent of its exports to the United States had prices over $10,000, while 12 percent were priced at that level or above in 1982. These voluntary export restraints lapsed in 1985, but the Japanese continued to observe export restraint even as they moved aggressively to build production facilities in the United States.

Location Trends. The manufacture of automobiles, buses, and trucks has traditionally clustered in a portion of the Middle West, a region sometimes called the *automotive triangle* (Figure 15-13). If we drew an imaginary line from Buffalo to Cincinnati to Janesville (Wisconsin) and back to Buffalo, we would delimit an area containing some two-thirds of the nation's automotive employees. The southern half of Michigan alone accounts for almost 40 percent of the employees in this industry. Few other large industries are so highly concentrated in a single state. On an even finer scale, Detroit and its suburbs can boast 20 percent of the automotive jobs in the country.

In addition to the Detroit cluster, the automotive triangle includes these other centers: Flint, Lansing, Saginaw, and Grand Rapids in Michigan; and Cleveland, Buffalo, Toledo, South Bend, Chicago, Milwaukee, Indianapolis, and Cincinnati in neighboring states. This triangular area claims the majority of automotive manufacture today except for the assembly function, which is widely dispersed (Table 15-11). Automotive assembly now extends from coast to coast in eight states, with major voids in the interior west. These dispersed locations save transportation costs incurred by shipping bulky motor vehicles.

The production of American cars declined the most significantly in California in recent years. The California share of U.S. output dropped from 7 percent to 2 percent from 1982 to 1985 (see Table 15-11). The strong competition of Japanese imports, for example, led General Motors to close its Fremont plant in this period. As noted elsewhere, this plant has now reopened as a joint Toyota/GM operation, the New United Motor Manufacturing Inc. (NUMMI), building small cars, which will again raise the share of California output.

Japanese producers entering the U.S. market have selected several sites in both the midwest and southeast to construct their new assembly plants, including the Mitsubishi facility in Illinois, Mazda in Michigan, and Honda in Ohio. Nissan chose Tennessee for its first U.S. plant, and Toyota built in Kentucky. The new General Motors Saturn small-car plant will also build vehicles in Tennessee.

Nearly all production in the automotive parts industry occurs within the automotive triangle or the Middle Atlantic seaboard. This region contains about 85 percent of those engaged in the manufacture of motor vehicle parts and accessories.

This automobile "heartland" now houses the oldest, least efficient plants in the country. With increasing competitiveness in the industry in the 1980s, attempts at streamlining production and containing costs have led to the closure of these older assembly plants and company-owned parts manufacturers. Chrysler, for example, now buys up to 70 percent of its parts from outside suppliers, including a growing quantity from foreign sources. General Motors announced in late 1986 that it would close 11 plants in 4 states by the end of the decade, including both assembly and parts operations. This cutback will involve a workforce reduction of 29,000 persons. Industry observers have suggested that General Motors was forced to trim its surplus production capacity and excessive manufacturing costs to remain competitive. Prior to cutbacks, for example, it cost General Motors about $500 more per unit to produce an automobile than Chrysler, not to mention the additional disadvantage it faced vis-à-vis Japanese producers.

Automobile Origins. The birth of the automotive industry occurred in 1886 with the invention of the first internal combustion engine using petroleum by Gottlieb Daimler in Germany, but several years elapsed before anyone successfully harnessed the engine to a wheeled vehicle. By 1895 several inventors had built experimental vehicles, with Ford, Hanes, and Duryea each claiming to have built the first successful one in the United States.

In 1900 a large number of companies made cars,

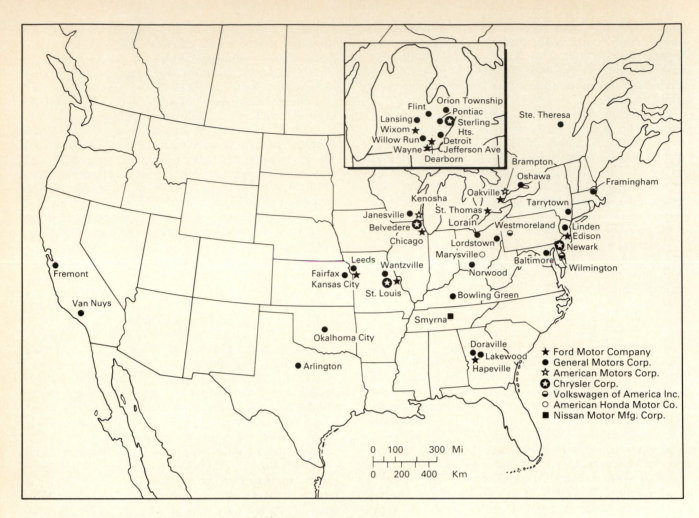

Figure 15-13 Major Automobile Production Facilities in the United States and Canada. *Manufacture traditionally clustered in the* **automobile triangle** *connecting Buffalo, Jamesville, Wisconsin, and Cincinnati, with nearly 40 percent of the employees in the industry clustered in south southern Michigan. Assembly of automobiles occurs widely, with major voids in the interior west. Several newer Japanese-based firms have recently established production facilities in Tennessee, Kentucky, Ohio, and Illinois. (Redrafted from* 1986 Ward's Automotive Yearbook, *p. 80)*

primarily buggy makers, resulting in the label *horseless carriage* for the motorized vehicle. Then Henry Ford put together three technologies that catapulted his company to fame and gave his home town, Detroit, a tremendous lead over all other places in the manufacture of automobiles (Figure 15-14). First, he designed an extremely simple, durable internal combustion engine that was easy to operate. This "Model T" vehicle immediately caught the fancy of the American buyer. Ford's second contribution was to abandon the strategy of producing many models at many prices and concentrate on one model in order to produce the most efficient car at the lowest price. The third idea was to manufacture these cars, not piece by piece as one builds a house, but by using mass-production methods developed by Eli Whitney a century before, whereby all parts of the car were standardized and produced by extremely specialized machine tools and then fitted together on a moving *assembly line.* The

assembly line technical innovation, combined with the marketing of a single model, was so successful that in 1909 Ford turned out 10,000 vehicles. By 1923 production passed 2,000,000. At the same time, selling prices dropped steadily, in response to mass production system efficiencies, from $950 per car in 1909 to $295 in 1922.

Ford was not the only successful automobile producer, of course. Numerous other companies followed Ford's lead and most of them started out in Michigan. Lansing was the home town of R. E. Olds, a young mechanic who launched the Oldsmobile Company in 1901. Flint, the hometown of carriage producer W. C. Durant, became the center of the Buick Motor Co.

Why Detroit? Why did the automobile industry concentrate in Detroit? Part of the answer lies in Detroit's early start. Even before the day of the large cor-

Table 15-11

U.S. Automobile Assembly, By State, 1985 Model Year (thousands of units)

Rank	State	Units	Percent
1	Michigan	2,269	29
2	Missouri	1,088	14
3	Ohio	767	10
4	Georgia	545	7
5	Illinois	505	6
6	Delaware	460	6
	Subtotal	5,634	72
	U.S. total	7,817	100

Source: Ward's Automotive Yearbook (Detroit, Mich., Ward's Communications, Inc., 1986).

poration, in the late 1890s and early 1900s when cars were made by numerous small companies, Detroit was the leading center. This statement applies only to gasoline-burning vehicles; in 1900 the United States produced 4000 horseless carriages, about half of which came from New England shops. But over 3000 of these vehicles were powered by either steam or electricity—types of motive power that soon disappeared. When gasoline burners won the competition as a power source, therefore, southern Michigan was ahead of all other regions in this phase of the industry.

The concept of "early start" is only a partial answer, as many authorities point out that horseless carriages did not encounter problems in navigating up hills in southern Michigan as they did in New England. Moreover, the glacial soil in Michigan, with a high proportion of sand and gravel, made for more "navigable" roads in inclement weather; although horse-drawn vehicles could make their way through mud roads, horseless ones had more difficulty. Then, too, the Detroit River, a major link in the Great Lakes water system, provided a market for internal combustion engines for the boat industry. All these factors played a part in the early start of automobile production in Michigan. But these conditions are equally applicable for locations in eastern Wisconsin and northern Ohio as well. What did Michigan have that these other states lacked?

Figure 15-14 Original Model T assembly Line in Highland Park, Michigan, in 1913.
This photograph, circa 1913–14, shows the exterior use of the Ford building, located in suburban Detroit, for lowering the auto body carriage onto the chassis. By 1924, the automobile chassis traveled on an elevated, powered conveyor system. Unlike modern facilities, this plant boasted four stories. The building still stands, but is no longer used by the firm. (Source: Ford Motor Company.)

The concepts of *spatial limits* and *satisficing* help explain why Detroit remained so strong. Although it may not be the ideal, optimal location, neither is it uneconomical region for the industry. Moreover, the specialized labor force that evolved in the closely tied machine tool and subcontractor industries stabilized and reinforced the pattern. In the final analysis, then, the industry remained strong in Detroit due to *historical inertia*. There is good reason at times for some assembly operations to disperse and decentralize, but the engineering, design, and management decision making for the industry has had less reason to relocate. It should be pointed out, however, that Japanese firms have made California the base of many of their engineering and sales operations in North America, and that California now houses design studios for eleven major firms, including U.S., Japanese, and European producers.

Soviet Production

Soviet motor vehicle production exceeded 1 million units for the first time in 1971. That year also marked the first time that passenger car output exceeded truck production, a traditional emphasis in the Soviet motor vehicle industry. Car production alone topped 1 million units in 1974 and continued to grow thereafter, reaching 1.3 million units in 1980, while total motor vehicle production exceeded 2.2 million vehicles.

The increase in Soviet automobile production in the 1970s coincided with the opening of the Togliattia facility in the Volga River valley, which produced Lada cars. That plant, near Kuybyshev, was engineered and built in collaboration with Fiat and is the largest car-producing facility in the country. A plant in Moscow producing Moskvich automobiles formerly led output. Truck plants in Moscow and Gork'iy produced the majority of Soviet trucks, but the largest single plant is located on the Kama River at Tatariya. Other motor vehicle production facilities occur in the Urals, Transcaucasus, Byelorussia, and the Ukraine. Soviet automotive engineering lags by western standards. Vehicles are not reliable and lack the crisp styling and creature comforts found in Japanese, American, and western European models.

Third-World Production

Brazil, Mexico, and South Korea are among the most active third-world motor vehicle producers. South Korea has ambitious expansion plans, including a massive export business. The Hyundai firm doubled its production capacity to 100,000 units in 1979 and completed another doubling in the mid-1980s when sales began in the United States and Canada. The most popular Hyundai model, the Pony (Excel in the United States), sells mainly in Saudi Arabia, but it is marketed in over 40 countries, predominantly in the third world. Daewoo and Kia are other major producers. A goal of South Korean producers to develop

a capacity to rival Japanese production may not be attainable, but the country does have a distinct labor cost advantage, which provides a competitive advantage. Plant capacity nevertheless remains low by Japanese standards, with output of 265,000 vehicles in 1984.

Although significant motor vehicle production levels occur in several Latin American countries, including Mexico, Argentina, and Venezuela, Brazil serves as the dominant producer. The industry is fragmented in these countries much as it is in Europe, with the presence of American, European, and Japanese companies. European (VW, Fiat) and American (GM, Ford) producers dominate.

In Brazil, production is shared among six firms, dominated by Volkswagen. Brazil bans all imports and doubled its output in the 1970s. From the 1920s through 1962 the Mexican automobile industry was "an assembly operation for completely knocked down (CKD) vehicle kits imported from the United States and Europe. The industry's development accelerated and changed dramatically in 1962 with the promulgation by the Mexican government of the first of several automotive policy statements."[10] These guidelines called for an *import substitution* policy which was reinforced with high tariffs on imports, licenses, and local content legislation. Investment and fiscal incentives such as tax credits were also provided by the government, as well as preferential energy pricing. Amendments to those guidelines over the years further protected the domestic industry.

Output in the Mexican automotive industry grew steadily in the 1960s and 1970s, and in 1981 more than 500,000 units were produced for the first time. Nevertheless, Mexico still has an export trade deficit with the United States, but more balance is anticipated by the late 1980s. "Seven international producers now operate in Mexico: Renault, AMC, Chrysler, General Motors, Ford, Volkswagen, and Nissan."[11] Volkswagen, Chrysler, and Ford lead production.

Future Prospects

If future output levels and surviving producers for the long term in the motor vehicle industry remain unknown, the number of competitors in the future will be quite few. Each will have global operations, marketing similar cars in all major world markets outside the Soviet bloc (Figure 15-15). The vehicles themselves will also evolve rapidly in terms of design and performance.

The future automobile will be lighter in weight, have a more durable body, and be more technology intensive in terms of electronics. The use of aluminum and plastic in engines and bodies will increase.

[10]Charles A. Ford and Alejandro Morlock, "Policy Concerns over the Impact of Trade Related Performance Requirements and Investment Incentives on the International Economy: U.S.-Mexican Relations," *Inter-American Economic Affairs* (Autumn 1982), 12.
[11]Ibid., p. 25.

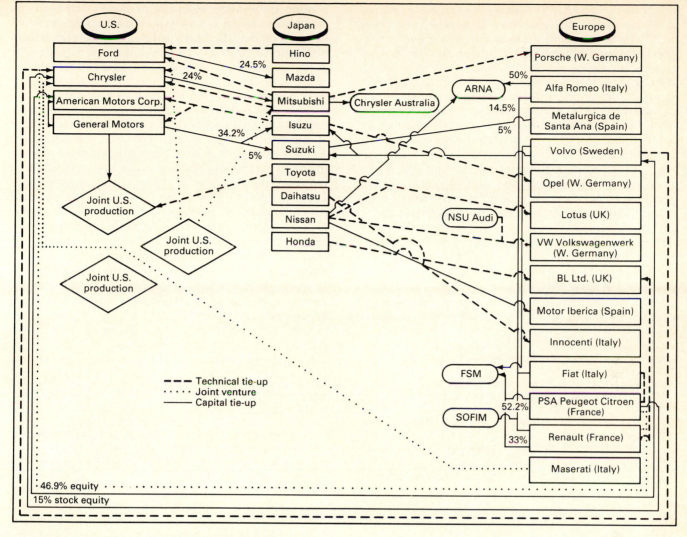

Figure 15-15 Global Linkups of Major Automotive Firms. *All major producers now maintain extensive world-wide operations, marketing similar vehicles in several markets. Chart shows the complex global linkups of technology, joint ventures, and capital tie-ups among automobile firms. (Redrafted after Kenichi Ohmae,* Triad Power, *Free Press, 1985, p. 98 and* Detroit News, *April 21, 1985.)*

A ceramic engine that burns fuel more efficiently may also be around the corner. The use of robots and computers in production lines will increase, as will quality control standards. Vehicles themselves will also have more electronic features, including greater use of fuel injection, fiber optic wiring, and video monitors that will be used as readout devices and navigational aids.

SUMMARY AND CONCLUSION

Both the steel and motor vehicle industries are capital- and material-intensive activities. Both are also mature industries. Although production remains at a high level in both industries in highly developed countries, production dominance has and will continue to decline. The Soviet Union now leads in world steel output and third-world producers are rapidly gaining a comparative advantage for the production

of steel, largely at the expense of highly developed nations. The outlook for specialty steel in developed areas does, however, look bright.

Shedding excess steel capacity has been a particularly troublesome process in the United States and western Europe. The concept of surplus capacity assists in an understanding of the implications of this adjustment. Quotas, import restrictions, and poor management have all impeded this transition. The quality of specialty steel products manufactured in the Soviet Union remains inferior despite its position as the production leader. The market for quality seamless pipe used in the petroleum industry and sheet steel for automotive production suffers from this weakness.

While the prospect for minimills as efficient alternative producers for high-quality steel once appeared bright in highly developed countries, the excess capacity in the industry and high production costs in traditional production areas does not bode well for them. Japanese companies have increasingly invested in U.S. firms in recent years as a hedge against anticipated import restrictions. The transnationalization process has been much slower in western Europe, where independent producers still dominate.

The automobile industry has adjusted more aggressively to the increasing competitiveness of foreign producers by shifting production to more efficient assembly plants jointly owned and/or operated in host countries under a transnational umbrella. Accompanying this process of global involvement of the major producers has been the emergence of the world car. Developed countries have fared much better in automotive production than with steel in recent years, due to this global presence. A few larger producers account for the majority of world automobile production today.

Although Japan remains the leading automobile-producing nation, a U.S. producer, General Motors, leads in worldwide output, owing to its extensive operating environment. The United States is also the leading automobile importer, because U.S. producers responded slowly to changing consumer preferences in the past decade, and the quality and relative cost of imported products from the leading exporter, Japan, is perceived to be much better. As in the case of steel, growing protectionism pressure for automobiles has also increased in mature producing regions such as the United States and western Europe. Western European producers are not positioned in the world market as well as their Japanese and U.S. counterparts because they have a greater relative exposure in the sluggish Latin American market, in contrast to the more rapidly growing Asian realm, in which Japanese producers have better prospects. Third-world producers such as Brazil and South Korea have also become more competitive. As with steel, the traditional producing countries still retain a competitive edge in developing new design and production technologies. In the next chapter we examine several high-technology activities that represent yet another dimension of economic development—that of knowledge-intensive activity.

SUGGESTIONS FOR FURTHER READING

ALTSHULER, ALAN, *The Future of the Automobile: The Report of MIT's International Automobile Program.* Cambridge, Mass.: The MIT Press, 1984.

BHASKAR, KRISH, *The Future of the World Motor Industry.* New York: Nichols, 1980.

BLOOMFIELD, GERALD, *The World Automotive Industry.* North Pomfret, Vt.: David & Charles, 1978.

HARRIS, A. W., *U.S. Trade Problems in Steel.* New York: Praeger, 1983.

HOGAN, WILLIAM T., *World Steel in the 1980's: A Case of Survival.* Lexington, Mass.: Lexington Books, 1983.

OHMAE, KENICHI, *Triad Power: The Coming Shape of Global Competition.* New York: Free Press, 1985.

RAE, JOHN B., *Nissan/Datsun: A History of the Nissan Motor Corporation in USA, 1960–1980.* New York: McGraw-Hill, 1982.

SAKIYA, TETSUO, *Honda Motor: The Men, the Management, the Machines.* New York: Kodansha International, 1982.

SÉKALY, RAYMOND, *Transnationalization of the Automotive Industry:* Ottawa: University of Ottawa Press, 1981.

WARREN, KENNETH, *World Steel: An Economic Geography.* New York: Crane, Russak, 1975.

ZUMBRUNNEN, CRAIG, and JEFFREY P. OSLEEB. *The Soviet Iron and Steel Industry.* Totowa, N.J.: Rowman & Allanheld, 1986.

16 Knowledge-Intensive High Technology Activity

What is high technology? Although no precise definition exists, high-technology companies place a great deal of emphasis on research and development and create products in the forefront of science (see the accompanying box). Professional white-collar workers comprise a large share of the high-technology labor force. This unique labor force mix includes a very high proportion of nonproduction specialists—up to 40 percent in many instances, including engineers, scientists, skilled technicians, and other professionals, not counting executives, administrators, marketing, and supervisory personnel. Such a large share of professional and technical workers does not occur with traditional capital-intensive manufacturing activities as those discussed in Chapter 15. These traditional activities are much more labor intensive on the production line rather than in research and development departments.

High-technology research and development occurs primarily in North America, western Europe, and Japan. The Soviet Union has been a notable underachiever in this field. The rapid pace of product innovation in the microelectronics field drove the high-technology development wave in the 1960s and 1970s. Miniaturization of component parts played a key role in this process. Expanding markets and lower prices also contributed to the expansion.

High-technology industries include many more activities than computers. In fact, there are high-technology aspects of nearly all industrial groupings today. Recall, for example, the trends in the textile and apparel industry as well as those in the steel and automotive industry examined previously. Nevertheless, several other fields not yet discussed fall more exclusively in the high-technology arena, including pharmaceutical, chemical, biotechnology, medicine, genetic engineering, military weaponry, telecommunications, robotics, space technology, and laboratory equipment. In addition to these "hardware" components, the high-technology field includes service firms providing data processing and software support.

In this chapter we discuss the computer and electronics industry, the impact of robotics on the electrical machinery industry, biotechnology, and the telecommunications revolution, including the impact of fiber optics. Particular attention will be placed on U.S./Japanese rivalry in developing new products and the profound impact of these new fields on the structure of the labor force. Indeed, the change appears to be as fundamental and far reaching as the emergence of the industrial age over 250 years ago.

THE ELECTRONICS AGE

Traditional classifications divide the electronics industry into three parts: (1) consumer products, (2) military applications, and (3) computers. The electronics industry owes its start to the inventive genius of several American and British scientists in the late nineteenth century. The radio, and later television, represented great strides in consumer electronics

Although a precise definition of high-technology activity is not possible, it is possible to distinguish activities on the basis of their employment profile. The U.S. Department of Labor has developed three categories:

1. Research and development (R&D) spending in relation to sales is twice the industry average. This definition identifies theory-driven firms with a heavy emphasis on R&D, such as biotech and computer development.

2. Firms employing a greater-than-average proportion of engineers and scientists and a ratio of R&D spending equal to or above average. This definition is not as restrictive as above and identifies firms that manufacture products as opposed to R&D.

3. Firms with a science and engineer work force proportion 1.5 times the industry average. This definition is the broadest of the three and is general enough to include firms producing high-technology items for the mass market.

applications, but military procurement needs propelled the electronics industry into an advanced development stage during and after World War II. The computer industry in particular benefited strongly from military initiatives. Radar, submarine and rocket guidance, and smart bombs provided much of the early incentive, just as missiles, satellites, sophisticated warheads, and electronic intelligence devices have pushed back computer miniaturization frontiers in recent years.

Although the electron was discovered in the late 1800s by J. J. Thomson, most observers suggest the role of "father of electronics" falls on the shoulders of J. A. Fleming, the inventor of the thermionic tube, which put electrons to work. A Britisher, Fleming studied Thomas A. Edison's work with electric lights and became a consultant on developing an improved wireless signal technology for the radio. Improvements to the Fleming thermionic tube came with refinements to the diode and the triode valve in the early twentieth century.

In the 1912–1925 period, AT&T and GE applied for vacuum-tube patents, providing a breakthrough for the radio. Photoelectric cells were perfected in this era as well, greatly assisting the sound-moving-picture industry. In this period the term "electronic" was generally limited to the properties of the electron itself. The association of electronics with applications and instruments awaited the post–World War II era, even though radio, telephone, and television technology had been perfected by the beginning of World War II (see Table 16-1).

Electronic advances developed steadily in the immediate prewar period, with the British having a distinctive lead, especially in communications and radar. During the war the need for lighter, more reliable equipment created pressures for miniaturization for aircraft guidance systems. By the end of the war, the first-generation electronic computers were available, but their commercial debut did not occur until the 1950s. Again the British stood in the forefront of computer development in 1950, but by 1960 the United States captured the lead it still maintains today, even if the Japanese have cut the margin to a slim advantage.

Notable achievements in the electronics and communications field occurring after World War II included the introduction of holography in 1947 and the transistor in 1948 (see Table 16-1). The latter achievement signaled the beginning of the microelectronics field, which led directly into the *information age* 25 years later. Two achievements in the 1950s had a comparable impact: the silicon transistor/semiconductor in 1954 and the integrated circuit in 1959, both attributed to Texas Instruments (see Figure 16-1). The 1960s witnessed the perfection of laser and electrostatic copiers. Several major advances occurring in the 1970s, again in the United States, involved the first

Table 16-1

Selected Communications and Electronics Milestones

Innovation	Inventor	Year
Telegraph	Morse	1839
Telephone	Bell	1876
Census Tabulator	Hollerith	1890
Radio	Marconi	1898
International Business Machines Corporation (IBM)	Watson	1924
FM radio	Armstrong	1935
Xerography (electrostatic copier)	Carlson	1937
Television network broadcasts	Great Britain: BBC; U.S.: NBC, Zenith, CBS	1937 1940
Computer	Bush, von Neumann, et al.	1940–1950
ENIAC*	University of Pennsylvania	1946
Holography	Gabor	1947
Transistor	Shockley/Bell Laboratories	1948
Silicon transistor/ semiconductor	Texas Instruments	1954
Integrated circuit	Texas Instruments	1959
Laser	Bell Laboratories	1961
Fiber optics	Bell Laboratories	1977

*Electronic numerical integrator and computer.

Figure 16-1 Transition to High Technology: Wiring a Board. *The introduction of the transistor in 1948 signaled the beginning of the microelectronics era, which led to the beginning of the information age 25 years later. (Courtesy Honeywell Information Systems, Inc.)*

successful commercial use of fiber optics, by Bell Laboratories in 1977, and the introduction of the very large scale integrated circuit (VLSI) in 1979.

Mainframe Computers

Long before the electronics connection, the beginning of the modern computer as a laborsaving office/clerical device took form. Some credit the census tabulator designed by Hans Hollerith in 1890 as the first of many inventions leading to the present-day machine (see Table 16-1). Indeed, the present-day computer leader, IBM, got its start in the information-processing field by producing various electromechanical devices, including unit record equipment, more popularly known as punch cards. IBM also made various mechanical key punch machines, tabulators, sorters, calculators, and electric typewriters before entering the electronic computer field. IBM management, as recently as the late 1940s, in fact, saw no commer-

cial application for large-scale computers. To this day, the U.S. Census of Manufacturers classifies the computer as an office machine at the four-digit SIC code level rather than as an electronic product.

IBM Domination. Prior to the 1950s, government agencies largely financed computer innovations, especially Defense Department contracts. The commercial electronic data-processing computer industry dates to the first-generation equipment developed in the early 1950s by Remington Rand and IBM (Table 16-2). Remington Rand benefited from an early lead over IBM in the late 1940s, but poor management and weak product support allowed IBM to gain dominance in the 1950s. By 1959, IBM began marketing its second-generation equipment, which used a faster transistor technology. The IBM lead in the industry continued in the 1960s with the successful introduction of the Model 360 third-generation

Table 16-2

Evolution of the Mainframe Computer

Generation	Producer	System	Date	Technology
0	University of Pennsylvania	ENIAC	1946	Relay switches
1	Remington Rand	UNIVAC I	1951	Vacuum tubes
	IBM	701	1953	Vacuum tubes
2	IBM	7090	1959	Transistors
	IBM	1401	1959	Transistors
	CDC	1604	1960	Transistors
	Sperry Rand	LARC	1960	Transistors
3	IBM	360	1964	Hybrid
	RCA	Spectra	1964	Integrated circuits
	Burroughs	5500	1964	Integrated circuits
4	IBM	370	1970	Large-scale integration (LSI)
5	?	?	198?	Very-large-scale integration (VLSI)/artificial intelligence

machine. This dominance was reinforced by the fourth-generation Model 370, which used large-scale integration (LSI).

Today, IBM continues to dominate the large mainframe computer industry, as large-capacity machines are now called, claiming a 70 percent market share. None of the many other producers in the mainframe market have traditionally claimed more than a single-digit share of sales, and the industry has been characterized as "IBM and the six dwarfs." These dwarfs included Sperry Rand (Univac), Honeywell, Burroughs, Control Data, NCR, and DEC. In 1986, however, Burroughs and Sperry merged, creating a new company, Unisys, which jumped into second place as an IBM competitor.

With headquarters in Armonk, a suburb of New York City, IBM maintains an elaborate national network of manufacturing, administrative, and research and development facilities throughout the world. This operations complex highlights a locational feature of high-technology activity. In many ways these activities are *footloose,* not being tied to a particular location. They are generally located in areas with amenities. Locations in proximity to major research universities and areas with a large, skilled labor pool receive most attention.

If IBM facilities do not create a centralized location for the production of mainframe computers, its competition is more concentrated. Control Data and Honeywell both have headquarters facilities in Minnesota. This situation makes Minnesota and New York the primary homes to most of the U.S. mainframe producers. The new Unisys Corporation will maintain dual headquarters at the outset, in Detroit (Burroughs), and Blue Bell, Pennsylvania (Sperry).

Owing to the dominance of the IBM market share in the industry, a major tendency in the large-scale computer industry today is for "plug-compatible" machines to experience unprecedented growth. Among the major firms producing compatible equipment for the IBM or other firms' hardware or software are Am-

dahl, National Semiconductor, Intel, and Hitachi. Many firms now competing with IBM and the other industry leaders were founded by former IBM or Sperry Rand employees. Gene Amdahl, for example, was largely responsible for designing the IBM 360 before leaving the firm in 1970 to make Amdahl computers. In 1980 he founded a company called Triology to make a powerful IBM-compatible computer.

Toward a Fifth Generation. Each successive generation of mainframe computers combined much faster and greater computing power in a smaller package at lower cost. More flexibility in applications and the opportunity to use a greater variety of peripheral equipment accompanied these changes. In fact, today's mini- and microcomputers have more computing power than that of first-generation mainframe computers.

The fifth-generation supercomputer has yet to emerge—but not for a lack of effort. Fierce competition exists between the United States and Japan to develop this computer, which will use very-large-scale integration (VLSI). It will also have artificial intelligence. *Artificial intelligence* refers to the ability of a computer to program itself, listen to what it is told, and make logical inferences. It will probably have the ability to understand photographs and graphics in the same way that it does words. These capabilities will make fifth-generation computers knowledge-based rather than data-based machines.

Three major efforts are under way in the United States to develop this computer. First, the Microelectronics and Computer Technology Corporation (MCC), a consortium of 14 of the leading computer firms, works on a project in Austin, Texas. The federal government finances another project through the Pentagon's Defense Advanced Research Project Agency, and the Semiconductor Research Corporation has a project under way in North Carolina.

In Japan, MITI announced in 1981 its crash program to develop a fifth-generation supercomputer. Combined with the resources of eight leading computer firms, MITI opened the Institute for New Generation Computer Technology (ICOT) in April 1982, and funded it heavily. It is possible that this new computer will emerge in the 1980s, but the Japanese are taking a longer view, anticipating that it will not be commercially operational until the early 1990s.

Today's larger machines are Cray computers, produced in Minneapolis. They produce 100 million operations a second and use the latest in microchips, but to keep the machine from overheating, built-in refrigeration is required. To overcome this disadvantage and to boost computing power, advances on many fronts will be needed for the fifth-generation computer, ranging from improved chips to better, higher-capacity external memory storage. One promising approach to developing a superfast semiconductor is to use gallium-arsenide, which conducts electricity faster and produces less heat than silicon. It can also work at higher temperatures and may have better military capabilities, as it resists ionizing radiation better than silicon. But manufacturing and stability problems have to be overcome before it becomes a commercial success.

Minicomputers

Technological improvements have created considerable segmentation in the computer industry in the past 20 years. As more and more computing power could be packaged in a smaller machine, more specialized applications emerged. Although no firm definition exists, minicomputers generally refer to intermediate-sized machines selling for under $100,000. The initial use of minicomputers centered on industrial and laboratory applications, but today administrative and management applications, including data processing and inventory control and facilities management uses, such as building utility and energy load management capabilities, claim a large share of sales.

A recent trend in the industry involves linking several minicomputers together in a network to produce the larger computing power of a mainframe at much less cost and with greater flexibility. Such a configuration is called *distributive processing*. Minicomputers can also be used to support mainframe computers.

Significantly, a major concentration of the U.S. minicomputer manufacturing industry takes place in New England, particularly in Massachusetts in the greater Boston area. Digital, Data General, Wang, and Prime all have facilities located along Route 128 or I-495, Boston's platinum perimeters (Figure 16-2). These and other firms give Massachusetts the distinction of having the second-highest concentration of computer production facilities in the country, following the lead of California. One of three manufacturing jobs in Massachusetts is now a high-technology position.

Much of the Boston area high-technology computer production claims direct ties with the Massa-

Figure 16-2 Boston's High Technology Corridor/Region. *A major concentration of minicomputer manufacturing occurs in greater Boston. Digital, Data General, Wang, Prime and others have facilities along Route 128 or I-495, Boston's platinum perimeters. One of three manufacturing jobs in Massachusetts is now a high-technology position. (Source: Courtesy Department of Public Works, Boston.)*

chusetts Institute of Technology. An article in a 1979 issue of the *Boston Globe* indicated that 300 computer-related firms existed within 2 miles of MIT. Research and development benefits from ongoing consultation with university faculty and technical personnel. These locations provide ready access for faculty contacts as well as opportunities to take advantage of laboratory and library facilities and advantages in recruiting students for employment. Such interactions stimulate problem solving, improve technologies, and assist the testing and debugging of new products.

Unlike the mainframe computer market, no single firm dominates minicomputer output. The top five producers, for example, control only 60 percent of the market. IBM is a major minicomputer producer, having a 29 percent share of the market, but due to the wide variety of applications, other competitors are much stronger in this market, including Hewlett-Packard, Wang, DEC, and Data General, among others.

Microcomputers

As the miniaturization process continued in the 1970s, it was inevitable that a powerful desktop-size personal computer would emerge. The microcomputer represents this innovation. While specialized laboratory and engineering applications initially claimed the largest market share, by the mid-1980s the market broadened to include many business uses, including data base management, graphics, and word processing. The home market also came of age, as well as video game advances. Many applications utilize stand-alone systems, while others use the micro as a terminal for access to mainframe data bases.

Microcomputer sales expanded rapidly in the 1980s, with the value of units sold in 1984 matching that of mainframes for the first time—about $12 billion each. A wide variety of companies produce microcomputers: consumer electronics firms (Tandy/Radio Shack, Atari), mainframe and minicomputer corporations (IBM, Digital), semiconductor manufacturers (Texas Instruments), and firms specializing in microcomputer products only (Apple, Commodore).

Semiconductors

The basic memory unit of a computer is a *chip* or semiconductor. Semiconductors are not a computer per se, as they have no input or output devices, but the market for these memory units is much bigger than that for computers because they are now used in many consumer and industrial products, including watches, calculators, adding machines, cash registers, automatic tellers, robots, and automobiles, among other applications, to make such products "smart."

A chip, often called a *chip board* or *memory board*, consists of a large number of diodes, transistors, resistors, and capacitors manufactured on a surface of a tiny silicon crystal which are interconnected via conducting paths that run across a surface (Figure 16-3). This configuration creates an *integrated circuit*. Over the years the density of these integrated circuits has increased and the circuits themselves have shrunk in size. The speed of memory operations occurring on these circuits has increased and they have become more reliable. Now one finds mainframe computers with very-large-scale integrated (VLSI) circuits.

Each cell on the digital memory board has a transistor and capacitor that stores a unit or *bit* of information. These circuit boards have grown in the past few years from 16K (16,000 bits), to 32K, 64K, and now to 256K. Random-access memory (RAM) is the domi-

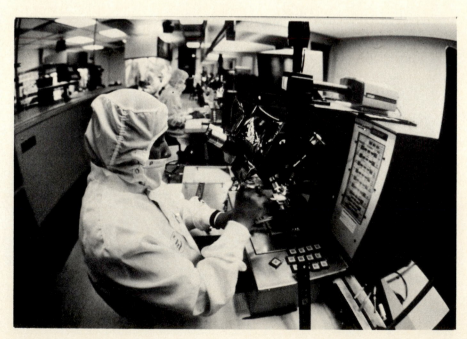

Figure 16-3 Integrated Circuit Production. *An integrated circuit consists of a large number of diodes, transistors, resistors, and capacitors manufactured on a surface of a tiny silicon crystal interconnected via conducting paths that run across the surface. Often production occurs in a clean room to assure quality control in an atmosphere free of dust and contamination. Note protective clothing, cap, and mask worn by technician and complex testing equipment. (Source: Motorola, Inc.)*

nant integrated-circuit memory on such a board, meaning that any part of the stored information can be accessed, read, and/or manipulated.

The production of semiconductors typically occurs in independent firms, with much of the output coming from small emerging firms. Fairchild Camera and Instruments, Texas Instruments, and Motorola each claimed an early lead, later joined by firms such as National Semiconductor, Intel, and Mostek. Some production now occurs in-house by established computer and telecommunications firms such as IBM, AT&T, and Hewlett-Packard. In recent years, several independent firms have been acquired by major domestic and foreign corporations to give them more control of the industry. Fairchild was acquired by Schlumberger (West Germany) in 1979 and Signetics by N. V. Philips (Netherlands) in 1975. United Technologies purchased Mostek in 1980 and General Electric bought Intersil in 1981. The market for semiconductors expanded as this consolidation occurred. World annual output of semiconductors exceeded $10 billion for the first time in the early 1980s, with the primary concentration of the industry in the United States in the Silicon Valley of California (see the next section).

Computers claim half the final market for semiconductors in the United States. Many companies, such as Texas Instruments, which accounts for 20 percent of U.S. semiconductor production by independent or "merchant" firms, integrated their operations vertically in recent years. By extending their control backward to silicon wafer purification and production, they lessened uncertainty over quality control. At the other end of the industry, firms have entered consumer markets by adding forward linkages with sales of digital watches, calculators, and computers. Some firms, such as National Semiconductor, failed in the consumer products field and have dropped out. The largest captive or in-house integrated-circuit manufacturer, IBM, accounts for over half of the total U.S. production, followed by Western Electric, General Motors–Delco, Hewlett-Packard, and Honeywell.

In Europe, a few vertically integrated electronics giants, including Philips, Siemens, and IBM, dominate the semiconductor industry (Table 16-3). European computer production nevertheless remains weak, with most of the semiconductor products directed to the consumer products market. As mentioned earlier, these companies bought into U.S. technology in the early 1980s, which helps them keep abreast of innovations. Philips, the largest European firm, owns production facilities in the Netherlands (headquarters location), France, West Germany and Switzerland. The second-largest producer, Siemens, runs facilities in West Germany and Austria.

The Japanese semiconductor industry maintains a very strong presence in the market, reaching parity with the United States in several areas in the early 1980s. By the late 1980s the Japanese share is expected to surpass that of the United States. It appears that the newest technology innovations may increas-

Table 16-3

Europe's Top 10 Electronics Firms (billions of U.S. dollars)

Rank	Company	Country of Origin	Sales
1	N. V. Philips	Netherlands	11.1
2	IBM	United States	9.4
3	Siemens	West Germany	8.7
4	ITT	United States	8.6
5	Thomas-Brandt	France	6.5
6	General Electric Co., Ltd.	Great Britain	3.8
7	AEG-Telefunken	West Germany	2.9
8	L.M. Ericsson	Sweden	2.0
9	Compagnie Générale d'Electricité	France	1.9
10	Rank Xerox	United States	1.9

Source: Keith Jones, "European Electronics Marked by Many Changes," Electronic Business, July 1982, p. 82.

ingly come from Japan, not the United States, including custom-made logic chips and microcontrollers that run robots.

The Japanese share of U.S. sales of 16K RAM units stood at 40 percent of the U.S. market in the early 1980s. For the more powerful 64K RAM, the Japanese share of the world market was 70 percent in 1980. In 1985, only two of the three companies producing the latest-generation microchip having 256K memory, Mostek and Micron Technology, produced chips in the United States. Texas Instruments imported their product from Japan. Six firms dominate the Japanese semiconductor field, including Nippon Electric Company (NEC), the leader; Hitachi, a major soga sosha; Matsushita, a leading consumer products supplier (Panasonic); and Fujitsu, the leading Japanese computer manufacturer.

Peripherals

A large share of computer products, known as *peripherals*, serve as auxiliary devices that provide computer support services. Peripherals range from data entry to storage and output machines. Keyboards, terminals, digitizers, and cathode ray tube (CRT) monitors are examples of input devices. Memory units include tape, disk, and drum storage and floppy and hard disks. Output devices include printers, CRT monitors, and plotters.

Large integrated computer firms produce many of these products, while other companies specialize in one or more products. The Silicon Valley claims a large share of the independent firms. More and more peripheral equipment marketed in the United States is also imported from Japan and other East Asian countries.

Silicon Valley

Santa Clara County, California, popularly known as *Silicon Valley*, birthed and nurtured the semiconductor industry in the United States (Figure 16-4). The valley,

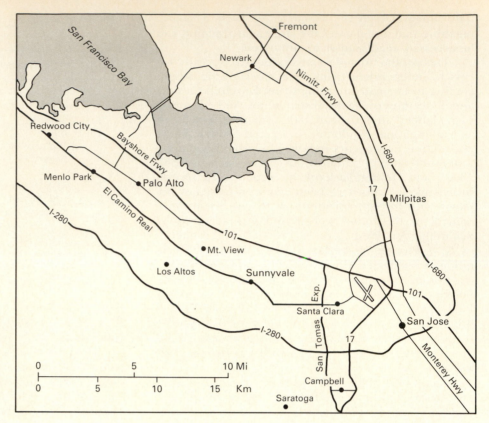

Figure 16-4 Silicon Valley at the Southern end of San Francisco Bay. *Santa Clara County, California, popularly known as a Silicon Valley, birthed and nurtured the semiconductor industry and is now the foremost high technology region in the world. The valley area includes 13 cities, including Palo Alto, the home of Stanford University, the catalyst for this growth and San Jose, the largest city in the country with a specialized high-technology work force.*

stretching over 25 miles and including 13 cities, lies in an area south of San Francisco. It now claims a large share of the headquarters, decision-making, and innovative research activity in the semiconductor industry in an area that was once rich farmland filled with prune orchards. The county grew by 1 million in population in the 1940–1970 period and by 350,000 in employment during the initial explosive-growth phase of the computer industry.

The roots of the Silicon Valley electronics boom date to World War II spin-offs from military contracts held by Stanford University in Palo Alto for weaponry research. AT&T and Bell Labs acquired a patent on the solid-state transistor in 1948, marking the beginning of this remarkable growth process. In 1955, William Schockley, one of the three Bell Labs inventors, left the firm and moved to Palo Alto, founding the Schockley Transistor Company. Another firm, Fairchild Semiconductor, spun off in 1957. By 1965, there were 10 firms, and 50 by 1970.

The Santa Clara Valley is best known today as the headquarters or regional hub of such firms as Hewlett-Packard, Intel, and Lockheed Aerospace, which each employ 20,000 or more workers (Table 16-4). In addition to computers, firms in this region produce peripheral equipment such as disk and tape drives, circuits/chips, communication equipment such as radar and microwave, and military/defense products.

Even though established firms held early patents in the industry, change occurred so rapidly in the field that only new, small operations could adjust to chang-

ing markets fast enough to incorporate innovations and introduce products rapidly. By the mid-1960s, new, rapidly growing firms overtook the traditional leaders in electrical/electronic applications such as General Electric, Sylvania, RCA, and Westinghouse.

Many communities in addition to Palo Alto shared in the Silicon Valley computer boom, including Sunnyvale, home of Atari and Amdahl; Mountain View; Cupertino, home of Apple; and Santa Clara (Table 16-4). San Jose alone has over 150,000 employment in electronics, making it the largest city in the country with a specialized high-technology work force. Over 1000 firms now operate in the valley with more than 10 employees and nearly 2000 have fewer than 10 workers. Only 25 firms employ more than 2000 persons, given the small scale and decentralized nature of the business, which is fiercely competitive. Individuals with new ideas have ready access to *venture capital* in the area and are always in demand as investors seek new growth opportunities.

Venture capital refers to funds earmarked for investment in high-risk businesses. Often, this money is specifically tagged for smaller startup firms that specialize in new product development or new technologies. In the United States this money comes from individuals, high-tech mutual funds, small business investment companies, investment banks, and subsidiaries of major corporations. Annual venture capital investments approached $3 billion in the United States in 1983.

The term *cubistic campus* applies to the small, mod-

Table 16-4

Silicon Valley Electronics Manufacturing Companies
with at Least 1000 Employees, 1984

Firm	Total Employment	City	Primary Product
		Computers	
Hewlett-Packard	17,000	Palo Alto	Computers
	3,000	Sunnyvale	Microcomputers
IBM	14,000	San Jose	Computer systems, memories
Atari	4,500	San Jose	Microcomputers
	1,200	Milpitas	Video games
Apple Computer	4,000	Cupertino	Microcomputers
	1,300	Sunnyvale	Computers
Four-Phase Systems	4,000	Cupertino	Computers' data processing equipment
Amdahl Corp.	2,500	Santa Clara	High-performance computers
		Peripherals	
Intel	20,000	Santa Clara	Memory, microprocessor electrical systems
Memorex	6,000	Santa Clara	Information storage
Shugart	3,800	Sunnyvale	Computer peripherals, disk drives
Dysan	2,000	Santa Clara	Computer memories
		Integrated Circuits/Semiconductors	
National Semiconductor	8,000	Santa Clara	Semiconductors, microcomputers
Intel Magnetics	5,000	Santa Clara	Electrical R&D, magnetics
Advanced Micro Devices	4,500	Sunnyvale	Integrated circuits
Signetics Corp.	4,000	Sunnyvale	Integrated circuits
Tandem Computers	2,200	Cupertino	Printed circuit boards
ESL/TRW	2,200	Sunnyvale	Electric systems
American Micro Systems	2,000	Santa Clara	Integrated circuits
		Communication/Microwave/Radar	
Rolm	6,200	Santa Clara	Business communication systems
GTE Sylvania	5,400	Mountain View	Electrical detection equipment, military defense
Varian Associates	4,500	Palo Alto	Microwave, light sensing
Fairchild Camera	4,300	Mountain View	LSI products group
Applied Technology/Litton	2,000	Sunnyvale	Radar warning system
Fairchild Camera	1,000	San Jose	ATE group
		Military/Aerospace/Nuclear	
Lockheed Missiles & Space	20,000	Sunnyvale	Missiles and space craft
FMC Corp.	6,000	Mountain View	Military defense
General Electric	5,000	San Jose	Nuclear power reactors
Ford Aerospace & Comm.	4,000	Palo Alto	Aerospace equipment
Westinghouse	2,800	Sunnyvale	Marine turbines
Ford Aerospace & Comm.	1,750	Mountain View	Aerospace equipment
General Electric	1,200	Sunnyvale	Nuclear power reactors

Source: Richard E. Schmieder, *Rich's Business Guide to Santa Clara County's Silicon Valley* (Palo Alto, Calif.: Business Directories, Inc., 1984).

ern, boxy single-story buildings which look more like offices than factories that now cover the Silicon Valley landscape (Figure 16-5). These facilities have gradually taken over the valley area, which formerly served as an agricultural center. Remnants of that landscape in the form of orchards, greenhouses, spur rail lines, and seed companies still exist, but the new heavily landscaped business properties with liberal setbacks and ample off-street parking interspersed with resi-

dential communities and commercial support services now spread throughout the valley such that cities have grown together, forming a loosely knit, low-density urban landscape. Much of the architecture is uninspiring, but newer buildings have more expression, especially in signing and logos (Figure 16-6).

The influx of highly skilled and paid employees to the region in the past 20 years raised housing prices and labor costs dramatically. Assembly operations in

Figure 16-5 The Cubistic Campus. *Small, modern, boxy single-story buildings which look more like offices than factories cover the Silicon Valley landscape. The new heavily landscaped business properties with liberal setbacks and earth berms to screen parking areas frame the buildings with their colorful signs, logos, and entry lobbies. Apple Computer facility in Cupertino is shown here. (T.A.H.)*

Figure 16-6 High Tech Fit and Finish. *The attractive three-dimensional sign for the Advanced Micro Devices firm, producer of integrated circuits, shown here at its headquarters in Sunnyvale, California dramatically illustrates the high profile of high technology firms in the Silicon Valley (T.A.H.)*

the industry began moving away as this uneconomical cost structure for production-line workers emerged. To take advantage of lower-cost labor and housing markets elsewhere, many Silicon Valley–based firms shifted production operations to southern California, especially the San Diego area, or to other major western urban centers, such as Phoenix and Tucson, Arizona; Austin, Texas; Colorado Springs, Colorado; or Salt Lake City, Utah. High-technology operations are also expanding in the Pacific northwest in Idaho, Washington, and Oregon. We now hear the terms Silicon Shores, Silicon Desert, Silicon Prairie, Silicon Forest, and Silicon Mountain, in these areas. Offshore operations in southeast Asia have also increased.

As decentralization of production proceeded, the Silicon Valley has matured into an elite residential and control center, especially its western foothills and northern perimeter. Moving southward in the county, one encounters progressively lower socioeconomic levels and a labor force mix more heavily dominated by production processes. A large Hispanic minority population also resides in the southern portion of the county. Traditional rural agricultural interests remain stronger to the south as well.

The rapid expansion of the Santa Clara County population in the 1970s created a severe housing shortage, a situation further exacerbated by a burgeoning no-growth movement. Restrictive land-use policies such as low-density residential zoning and re-zoning of large tracts of land from residential to industrial use limited the potential for population growth. These policies promoted inflationary price increases for housing. Many jobs remained unfilled in the region in the early 1980s as housing costs became prohibitive for prospective workers and turnover levels increased due to growing dissatisfaction with the quality of life in the region. This situation, in turn, encouraged firms to relocate operations outside the region and led to a reassessment of the restrictive policies. But by mid-decade a slowdown in computer sales led to a weakening of the market for chips and the job situation became much softer in the region.

Other regions in the United States also merit attention in the high-technology arena, owing to a concentration of activity and potential for future expansion (Table 16-5). Route 128, a beltway 10 miles from the center of Boston, has developed an international reputation in the past twenty-five years as a center for advanced technology (Figure 16-2). In the 1980s the area began to diversify away from as much dependence on computers making it less vulnerable to the computer slump that hit the Silicon Valley. In addition to computer firms such as Data General, Digital, Wang, and Prime the area also houses Raytheon, a major military contractor, Polaroid, and several biotechnology and artificial intelligence firms. The area around M.I.T.'s Artificial Intelligence Laboratory is now often referred to as "A.I. alley." Diversification along Route 128 has also brought offices of financial institutions, insurance companies, and other support service firms. In the past decade high technology de-

Table 16-5

Selected High Technology Complexes in the United States

Name(s)	Research University Nearby
Silicon Valley	Stanford
Route 128 I-495 "A.I. Alley" (Boston)	MIT
Research Triangle	Univ. of N.C., N.C. State, Duke
"Silicon Prairie" (Dallas-Austin)	University of Texas at Austin
"Bionic Valley" (Salt Lake City)	University of Utah
Route 1 "Silicon Valley East"	Princeton
Space Coast (Orlando)	Florida Tech University
Peachtree Corners "technology alley" (Atlanta)	Georgia Tech
"Silicon Desert" (Phoenix)	Arizona State

Source: modified after E. M. Rogers and J. K. Larsen, *Silicon Valley Fever.* New York: Basic Books, 1984.

velopment leaped another 15 miles outward to the I-495 beltway where cheaper land costs have enticed firms to expand production facilities as the more mature Route 128 market becomes involved in marketing and administrative control activity.

The Research Triangle of North Carolina represents another major technology complex, named for its location in a triangle between three university cities that form its corners: Durham (Duke University), Chapel Hill (University of North Carolina), and Raleigh (North Carolina State University). In addition to research and development subsidiaries of major corporations such as IBM, several nonprofit firms also conduct research at the center, including the National Institute of Environmental Health Sciences, the Environmental Protection Agency, and the National Center for Health Statistics Laboratory (Figure 16-7). In 1984 the 6000-acre complex included over 50 research and high-tech organizations and 40 commercial service support facilities.

The Princeton, New Jersey, corridor astride Route 1 between New York and Philadelphia may become the next Silicon Valley in the United States, due to its selection as a center to develop a new computer in the mid-1980s and the major corporate commitments to the area including the new research facility of Merrill Lynch. Telecommunications advances provided by fiber optics access place the area on the information mainline and the resources made available by Princeton University and the state of New Jersey promise rapid growth in technology-related activities.

Computers and the Asian NICs

The initial exposure of Asian NICs to computer production came from American firms seeking lower-production-cost locations for the assembly of com-

Figure 16-7 Research Triangle Park, North Carolina. *Named for its location in a triangle between three university cities that form its corners—Durham (Duke University), Chapel Hill (University of North Carolina), and Raleigh (North Carolina State University)—this complex houses research and development subsidiaries of major corporations, federal government agencies, and nonprofit firms. The photograph here indicates the names of firms associated with a major cooperative computer research initiative at the park. (T.A.H.)*

puter core memory component parts in the 1960s. By the end of the decade, core memories were replaced by printed circuit boards, but the seed had been planted. Today, Taiwan, South Korea, Hong Kong, and Singapore produce many types of peripheral equipment (terminals, monitors, disk drives, keyboards, printers) as well as central processing units and other components, such as plastic parts, bearings, and motors. Most of the product is for export. IBM, for example, buys components for its personal computer from several areas—monitors from Taiwan, keyboards and floppy disks from Singapore, and printers in Japan.

Some of the Asian firms are wholly owned subsidiaries of American, Japanese, or European multinationals; others are indigenous companies, the latter primarily located in Taiwan and South Korea. Despite the rapid growth and increasing sophistication of the industry in these areas, component assembly rather than product development remains the primary emphasis. Developing new technology continues to be the province of Japan and the United States, as more routine production functions shift overseas to these areas. The computer industry remains weak in several of these countries as one or more significant sectors of the industry are missing. Hong Kong and Singapore, for example, do not produce CRTs.

South Korea is gaining rapidly as a computer producer, owing to the commitment to research and development promoted by the government and the large relative size of its domestic market. The government-sponsored Korean Institute of Electronics Technology (KIET) funds work on semiconductors and computers. The major producers in South Korea are divisions of large corporations that have the resources to fund long-term research and development and efficient production, and to make marketing commitments. In Taiwan, the dominance of smaller family

corporations removes this advantage. In other areas, such as Singapore, production is mainly an assembly operation controlled by foreign multinationals.

Military Electronics

About one-fourth of the electronic equipment produced in the United States goes to the military. That market for U.S. products is three times the size of the consumer market. The leading segment of the military electronics market occurs in the area of communications/navigation, followed by data processing, electronic warfare, and radar. The five leading firms in military electronics sales in 1981 were Raytheon, Texas Instruments, RCA, Litton, and Sperry.

Software

Computer software, the newest computer-related service industry, grew rapidly in the 1980s. From a $6 billion industry in the United States in 1982, projections indicate that it will grow to a $30 billion enterprise in 1987. Software includes programs, procedures, rules, and associated documentation involved in the operation of a data-processing system. Whereas early software products were custom made for the particular needs of a user, much of the emphasis today, especially microcomputer applications, is placed on standardized packages for sale in the marketplace.

Many firms specialize in producing software only. Like hardware equipment, software development has passed through several evolutionary phases. Newer packages often integrate data base management, graphics, and word-processing capabilities, such as Lotus 1-2-3 or Symphony, whereas early versions were very specialized single-purpose programs. Major software houses today are typically independent, highly innovative ventures, whereas much of the early product development was produced in-house by hardware manufacturers. American computer service firms are the most active internationally in this field. They dominate sales in Europe, for example. The Japanese software industry has been handicapped by the complexity of the symbols used in the Japanese language but made great strides in the 1980s.

BIOTECHNOLOGY

Some high-technology experts believe that the next major thrust in the postindustrial economy in the late 1980s and the 1990s will come from *biotechnology.* Biotechnology refers to "the application of scientific and engineering principles to the processing of materials by biological agents to provide goods and services." This advance may have an impact similar to the electronics revolution of the 1960s and 1970s. "Biotechnology will do to agriculture, food and beverages, pharmaceuticals, industrial chemicals, energy production, and quality control, what microelectronics did to the efficient use of energy, manpower, plant and equipment, by replacing mechanical processes with computer aided control and processing systems."[1]

Although biotechnology is not a new field, its potential impact is now much more sweeping, spreading far beyond its roots with agriculture and food processing. Those areas continue to claim their share of breakthroughs, but attention has also shifted to pharmaceuticals, energy, basic chemicals, and mining applications. In the food and agriculture fields, work on bacteria, yeasts, algae, and enzymes has produced new vaccines, vitamins, antibiotics, and new energy sources, such as methane gas for fuel. Microbiological engineering research now provides advances in gene recombination, cell fusion, artificial body parts, and new miracle drugs.

The United States and Japan lead in biotechnology research. In 1979, 7 of 11 new antibiotics introduced internationally came from Japan, and although Japan ranks second to the United States in new drugs introduced, it does lead in interferon, artificial blood, and amino acid research. Interferon, a protective protein agent, most widely heralded as a cancer antidote, may have many other applications—to combat hepatitis, herpes, influenza, the common cold, and childhood diseases. Japanese biotechnology research has benefited from an emphasis on fermentation that grew out of work on food product development, including the bean curd, tofu, and soya sauce. At the same time, a weakness exists in physical chemistry research in Japan which may limit future breakthroughs.

Additional innovations in miracle drugs and industrial processes lie on the horizon. Within a decade drugs may be on the market to control AIDS, cure rheumatoid arthritis, cut blood cholesterol levels and alleviate cardiovascular diseases, prevent allergies and ulcers, and yield improved pain killers. Use of new petroleum–eating bacteria promises to combat pollution created by oil spills. Cracking hydrogen by splitting water with biological or solar techniques may provide an unlimited and cheap energy source before the end of the century.

As with electronics, the current U.S. lead in biotechnology has been sustained by the plentiful supply of venture-capital firms that underwrite costly product development. Close university and industrial collaboration also help the biotechnology field. Most of the activity is concentrated in California. In Japan, the government has been especially influential. In 1981, MITI identified this field as a national research priority. The industry lags in Europe because of a lack of transnational financial and scientific commitments.

ELECTRICAL MACHINERY: THE ROAD TO ROBOTS

Perhaps the original high-technology field was the electrical machinery industry. The computer industry may or may not have its roots in this field, due to an early association with office tabulating equipment, but certainly a portion of today's communications hardware industry—electronics, consumer appliances, and military hardware industries, not to mention robots—has its origin in this field. Some classifications distinguish between light electrical machinery, which includes consumer appliances, a highly market-oriented activity, and heavy electrical machinery, that portion of the field producing motors, generators, batteries, locomotives, reactors, and so on. The latter activities are traditionally located in older manufacturing centers in the northeast United States and middle west (i.e., in Philadelphia, Pittsburgh, Schenectady, and Milwaukee) near the steel industry, a skilled labor pool, and the markets for the product—mainly other manufacturers.

Larger urban centers extending from Boston, New York, Philadelphia, and Baltimore traditionally produced a wide variety of heavy machinery goods (Figure 16-8). The Chicago–Milwaukee industrial complex and other highly industrialized portions of Illinois, Michigan, and Ohio also benefited from associations with this industry. The automobile industry placed heavy demands on the electrical machinery industry for component parts, controls, and engines. The farm machinery production industry, located primarily in the middle west, further solidified the concentration in that region. California, particularly the Los Angeles and San Francisco regions, has also played an important role in this industry.

The growth of the communications industry after World War II occurred mainly in the northeastern manufacturing core area as a spin-off of the electrical machinery industry. In addition to telephone and television apparatus, this industry incorporates such specialized items as highway traffic signals, burglar

---

[1]Bruna Teso, "The Promise of Biotechnology.... and Some Constraints," *OECD Observer,* No. 118 (Sept. 1982), p. 4.

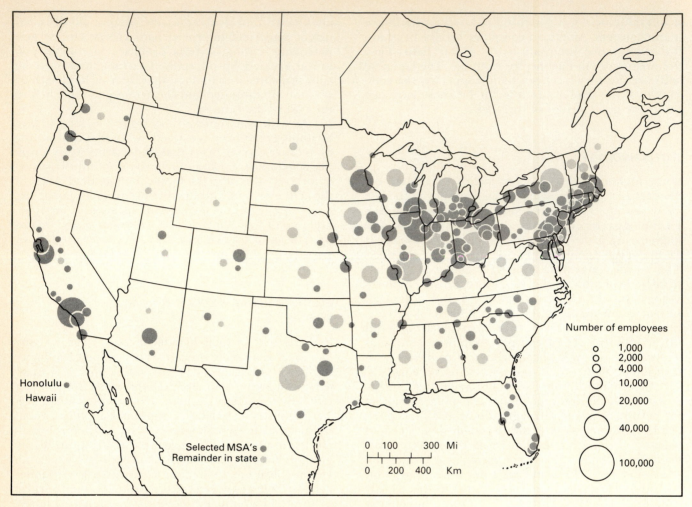

Figure 16-8 Machinery Industry in the United States. *Large urban centers in the northeast corridor extending from Baltimore to Philadelphia, and Boston as well as highly industrialized portions of upstate New York (Schenectady, Syracuse, Rochester, and Buffalo), Ohio, the lower Great Lakes (Cleveland, Detroit, Chicago, and Milwaukee) have benefited from associations with this industry. The San Francisco and Los Angeles areas have also played an important role in this industry.*

alarms, and laser systems. Mass-produced consumer communications products such as television monitors have long since left the region for lower-cost production areas, but output of their offspring in the form of sophisticated military and consumer telecommunications equipment such as satellite earth stations and microwave and fiber optics transmission equipment can still be found in that core region as well as the headquarters of many major computer corporations, as discussed previously.

Metalworking Machinery

Machines that turn, cut, drill, bore, plane, grind, mill, shear, or press metal, collectively known as the machine-tool industry, comprise another segment of the electrical machinery industry. Automobile manufacturers are the biggest users of metalworking machinery, giving Detroit a prominent role in the industry. Chicago, Milwaukee, Cincinnati, Pittsburgh, and Cleveland are also prominent in this field (Figure 16-

9). Again, the tie with steel, the availability of skilled labor, and local markets for the machines account for the localized pattern.

In many ways the strength of the machine-tool industry in a country provides a useful indicator of its level of economic development. Since manufacturers use these machines to produce other goods, local production typically evolves after other types of manufacturing are well established. Specialized high-technology applications are so critical to machinery innovations today that they rarely occur outside the United States, Japan, and western Europe.

Flexible Manufacturing Systems

Automation began restructuring the machine-tool industry in the 1950s. The first stage in this change involved *numerical control,* wherein instruments read coded numbers that gave automated instructions. A second stage involved CAD/CAM (computer-assisted design and computer-assisted manufacture) innova-

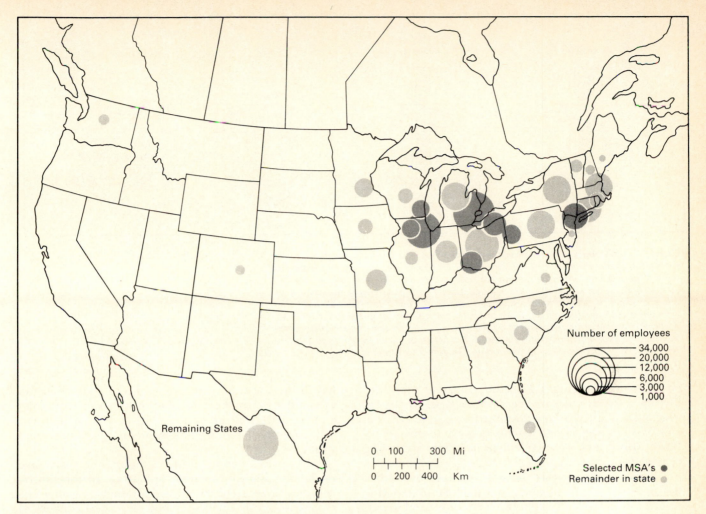

Figure 16-9 Metalworking Machinery Industry in the United States. *Automobile manufacturers are the biggest users of metalworking machinery, giving Detroit a prominent role in the industry. Chicago, Milwaukee, Cincinnati, Pittsburgh, and Cleveland are also important centers. Note the close correspondence of this industry to the manufacturing belt in the northeastern United States and the more concentrated pattern than that exhibited in Figure 16-8.*

tions. Essentially, this involves replacing the drafting table with a cathode ray tube (CRT) monitor for design work and then using computer-based instructions to drive the production process. A third stage occurs with the introduction of integrated *flexible manufacturing systems* (FMSs). This step involves introducing computer-controlled machine centers that perform many tasks, unlike the predecessor technology, in which each machine specialized in performing a single operation. This innovation allowed for more diversity in manufactured products on the assembly line. The Japanese became the world's first mass producer of computerized flexible machine tools in the 1970s.

The strategic implications for the manufacturer [of FMS] are truly staggering. Under hard automation, the greatest economies were realized only at the most massive scales. But flexible automation makes similar economies available at a wide range of scales. A flexible automation system can turn out a small batch or even a single copy of a product as efficiently

as a production line designed to turn out a million identical items. Enthusiasts of flexible automation refer to this capability as 'economy of scope.' ...

The manufacturer will be able to meet a far greater array of market needs, including quick-changing ones—even the needs of markets the company is not in now [with flexible manufacturing]. He can keep up with changing fashion in the market place—or set them himself by updating his product or launching a new one. He has many more options for building a new plant: FMS frees manufacturers from the tyranny of large-scale investments in hard automation, allowing construction of smaller plants closer to markets. ... Flexible manufacturing is the ultimate entrepreneurial system.[2]

The introduction of this flexibility accommodated the Japanese preference for *group technology* during the manufacturing process. This approach involves grouping similar parts together in families for

[2]Gene Bylinsky, "The Race to the Automatic Factory," *Fortune,* February 21, 1983, p. 54.

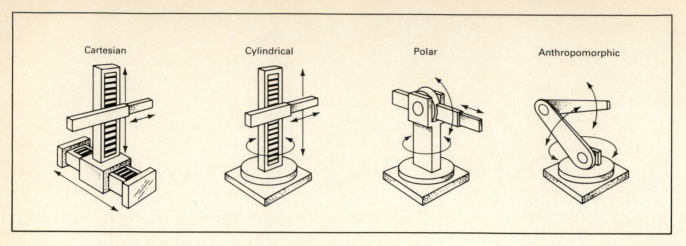

Figure 16-10 Robots Revolutionize the Factory Floor. *Automation began restructuring the machine-tool industry in the 1950s. Originally, robotics were only used to perform hazardous tasks but are now widely used for a variety of routine assembly operations, and are especially adaptable to applications in the metalworking industry. First generation robots such as those on the left side of the diagram were little more than mechanical arms. More sophisticated contemporary robots feature anthropomorphic arms that have the flexibility of a human arm because they both bend and swivel. For the future, robots with a sense of vision, touch, and sensory feedback will supplant those in use today. (Source:* The Economist, *August 29, 1981, p. 73.)*

easier manufacturing. By placing different machines together, rather than segregation by type, as is popular in the United States, manufacturers gain efficiencies. With this approach machines move to the factory floor, where they are needed, rather than occupying remote locations.

Robotics also assists the FMS process by coordinating the delivery of parts and performing routine operations. Originally, robotics were only used to perform hazardous tasks such as spot welding, spray painting, or lifting heavy parts. Since robots lower overhead costs by working longer hours, using less ventilation, and requiring less lighting, they are widely used today for assembly operations. The primary use is for automobiles and home appliances, highlighting the fact that robots seem to have the most applications in the metalworking industry.

To promote the introduction of robots, the Japanese created an incentive program, Japan Robot Lease (JAROL), in 1980, under the sponsorship of MITI and 24 robot manufacturing firms. An accelerated depreciation program encouraged the entry of medium-sized firms as robot users. This program symbolizes the aggressive stance of the Japanese in adopting robotics. In 1981 the Japanese used about 14,000 programmable robots in industry, compared with about one-third that number in the United States.

The term *robot* comes from the Czech word "robota" or serf. Early robots date to the immediate post–World War II period. In 1946, George Deval developed a "general-purpose playback device for controlling machines,"[3] Deval later sold this and other patents to Consolidated Diesel Corp. (Condec), which

led to the formation of Unimation, a robot division of Condec. Software-controlled robots were not commercialized until the early 1970s. In 1974 Cincinnati Milacron marketed the first minicomputer-controlled robot.

The primary motivation for using robots is to save on labor costs for machine operators. In the future, robots will be used to "batch produce" various machine products. They will feed pieces to clusters of automatic machines in "workcells" forming "closed-loop" manufacturing systems controlled by computers. By the turn of the century the majority of manufacturing machine operators could be displaced by such operations. As market uncertainty increases and product life cycles shorten, further incentives occur to rely on robots.

First-generation robots were little more than *mechanical arms,* often called dumb servants. In their most elementary form, a series of switches that regulated a fixed sequence of operations controlled these robots. Each operation tripped a switch sending a signal to begin another task. Changing processes involved replacing or dismantling the machine and repositioning switches. The majority of the robots at work today fall in this category.

More sophisticated robots have *reprogramable computer-driven controls* that make switching from one task to another much simpler, providing a boost to batch-manufacturing operations. Robotic arms are also becoming more sophisticated, moving away from simple cartesian, cylindrical, or polar operation to anthropomorphic arms that have the flexibility of the human arm because they both bend and swivel (Figure 16-10). For the future, users envision robots with a sense of vision, touch, and sensory feedback mechanisms.

[3]Robert Ayres and Steve Miller, "Industrial Robots on the Line," *Technology Review* (May–June 1982), 36.

The Japanese lead the world in flexible manufacturing systems and robot installations. The Japanese look at manufacturing in a holistic sense rather than as a cost accounting process, so popular in the United States. Japanese labor and management also accept new technology more willingly than do their U.S. counterparts. The Japanese also lead in machine-tool product sales. Half the machine tools sold in the United States are of Japanese origin and most U.S. producers now use Japanese technology (see Table 16-6). Indeed, several leading U.S. robot manufacturers use Japanese suppliers: Cincinnati Milacron (Dai Nichi Kiko), General Electric (Hitachi), IBM (Sankyo Seiki), and Westinghouse (Mitsubishi and Komatsu).

The United States still claims a lead in software development and may become more competitive in the future with *smart robots*. General Motors, for example, made great strides in this area in the 1980s, with three major acquisitions. In 1982, GMF Robotics joined the firm, followed in 1984 by the purchase of Electronic Data Systems, the world's largest data-processing firm, and in 1985 by the purchase of Hughes Aircraft for $5 billion. The market for these smart robots will be much larger, as they may be able to perform work as diverse as quality control on the production line, and smell fumes and scents, providing applications in fire detection and prison surveillance through detecting human beings by scent.

TELECOMMUNICATIONS

As with other high-technology fields, the telecommunications area experienced dramatic changes in the 1980s. Traditionally, telecommunications has been associated with the telephone, which used primarily voice transmission. Today, digital capabilities have broadened the potential use of the telephone to include data, facsimile copy, and video transmission. Indeed, a telephone connected with a *modem* can be used as a computer terminal. A modem is a device that permits computers to communicate over telephone lines.

Videotex services provide another dimension in telecommunication. Videotex refers to electronic communication through a computer via telephone line that allows one to conduct transactions and share information interactively. Examples include financial transactions, home shopping, access to electronic newspaper and library services, and security surveillance systems. Videotex allows for two-way communication and using information from one or more data banks. Often, a television serves as the terminal. The quality of videotex services improved with the introduction of cable TV networks. Videodisks, cassettes, and diskettes are also used for videotex but do not permit interactive transactions. An international business information system now used in Canada *Nova Tex*, provides business and technical data to users in the business and technical community in a videotex format. The data base has information on trade regulations, financial statistics, agriculture, energy, and news briefs. Recent innovations include the use of upgraded animated graphics with the interactive systems.

The role of the home television set changes from an entertainment medium to that of a central communications hub when linked to videotex services. The switch from analog to digital television signals, as used with the telephone, will provide the television more flexibility, including use as an interactive telecommunications terminal. When linked to a printer, the production of hard copies of screen images will be possible.

Fiber Optics

If telephone and television advances are important, perhaps the most promising development in the telecommunications field is the blending of digital data transmission techniques with *fiber optics*. Fiber optics refers to the transmission of light along a glass channel. Messages are sent as digital pulses. Since these channels can carry a broad band of waves and frequencies, their capacity is enormous. They can handle vast quantities of data compared with the capacity of conventional copper cables. By way of analogy, if a traditional telephone line is likened to a garden hose, a fiber optics line becomes a sewer pipe in terms of volume of flow handled.

Glass filaments as thin as human hair replace traditional wiring in a fiber optics network (Figure 16-11). These lines are cheap to construct, as they are basically silicon, and will probably entirely supplant copper cables by the turn of the century because they carry 250 times the capacity of copper wire.

While fiber optics characteristics have been known for decades, efficient use of the medium awaited proper cladding of the channel to ensure good reflectances. Dust-free coupling devices and an ability to send signals around curves were necessary before commercial applications became feasible in the late 1970s.

In 1977 in Chicago, Bell Laboratories installed for the first time a light-guide system that connected two telephone system central offices. By 1980 AT&T had

Table 16-6

Japanese Suppliers for U.S. Robot Manufacturers

U.S. Manufacturing Firm	Japanese Supplier
Admiral Equipment	Yaskawa Electric
Advanced Robotics	Nachi-Fujikoshi
Automatix	Hitachi
Cincinnati Milacron	Dai Nichi Kiko
GCA	Dai Nichi Kiko
General Electric	Hitachi
General Motors	Fujitsu/Fanuc
Hobart	Yaskawa Electric
IBM	Sankyo Seiki
Machine Intelligence	Yaskawa
Westinghouse	Mitsubishi & Komatsu

Source: Electronic Business, December 1982, p. 36.

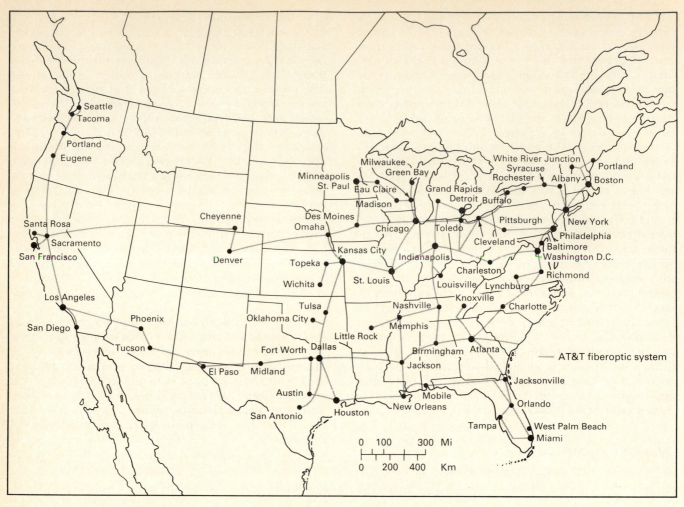

Figure 16-11 AT&T Fiber Optics Long-Haul System. *The extent of the fiber optics network in the U.S. expanded rapidly in the mid-1980s. Nearly all major cities are now connected by the AT&T network. Over fifteen other firms also provide inter-city fiber optic connections. (Data from Kessler Marketing Intelligence, Newport, Rhode Island.)*

installed about 4000 miles of optical fibers. This total zoomed upward to 200,000 miles in 1983. According to projections, within a decade the entire country will be blanketed by fiber optics networks.

Light-guide circuits may eventually replace electronic circuits in computers because of their speed advantage over electronics. An optical computer may also be on the horizon. Even communication satellites may be displaced in the future as the cost of telecommunication via fiber optics decreases.

In Japan, MITI and 11 companies, including Nippon Telegraph and Telephone (NTT), have created the optoelectronics Joint Research Laboratory to perfect fiber optics technology. AT&T, Bell Laboratories, and a host of other companies in the United States, including Corning Glass, have similar goals. Japanese visionaries see a pivotal role for fiber optics in automating the factory. Fiber optics "will perform remote-sensing chores—measuring temperatures and pressures, detecting flaws through optical inspection, and gathering data on production trends. The objective, says Koichi Murakami, a MITI administrator, is to link a company's factories with headquarters so the manufacturing operations can be run from one central location. The Japanese . . . believe fiber optics can make smokestack industries more competitive."[4]

SUMMARY AND CONCLUSION

High-technology economic activities came of age in the 1980s, replacing fanciful science fiction objects with practical, efficient, and affordable products that revolutionized the workplace and the home environment. The microcomputer and the robot best exemplify this new era, but fiber optics and biotechnology advances provided comparable breakthroughs in the fields of telecommunications and medicine. Competition between highly developed countries, especially the United States and Japan, stimulated the development of new products stemming from these advances even as more complex long-term commitments to re-

[4]"Fiber Optics: The Big Move in Communications and Beyond," *Business Week*, (May 21, 1984), p. 182.

search and development became more critical to reach new frontiers in science and technology. The large-scale efforts expended to develop artificial intelligence in Japan and the United States, heretofore unsuccessful, provide an example of the difficulty in accomplishing such an elusive goal.

In the United States, IBM dominates the main-frame computer field and has a major presence in the mini- and microcomputer markets. A major share of the minicomputer competition comes from several producers in the Boston region. The microcomputer industry is more fragmented, and producers have a more diverse background, including semiconductor manufacturers, consumer electronics firms, and those specializing in microcomputer, minicomputer, and mainframe markets. The Silicon Valley area of California has a major presence in the U.S. semiconductor industry and, indeed, represents the largest concentration of firms producing computers, peripherals, semiconductors, integrated circuits, communication, and aerospace equipment in the world. Over 1000 firms now operate in these industries in Silicon Valley.

The most competition for computer markets for U.S. producers comes from Japan. Research and development efforts are also the most sophisticated in these two areas. Government assistance to this industry has promoted its success in both countries. U.S. leadership in this area has been greatly assisted by the intense level of entrepreneurship in the country and the ready availability of venture capital. The latter two factors help explain the relative weakness of the computer industry in western Europe, because they do not exist to such a degree in that region. Third-world computer production largely emanates from branches of U.S.-dominated multinationals, but indigenous production does exist, especially in South Korea and Taiwan.

As with computers, the Japanese and Americans dominate the biotechnology field. In the electrical machinery and robotics field the Japanese excel in the production of flexible manufacturing systems (FMSs). The market for industrial robots has now expanded from hazardous applications to a broad array of uses in the metalworking industry. In the future robots will allow more flexibility in the manufacturing process as well as displacing much of the manual labor formerly handled by machine operators.

Advances have occurred in the telecommunications industry with the introduction of digital transmission applications for the telephone and television industries. Fiber optics transmission capabilities have also revolutionized the industry. Again the United States and Japan are pioneering these innovations.

As the high-technology industries discussed here provide a new technological base for a large segment of industrial economies, they also provide a means to revitalize older industries as well as create new ones. Most of the innovations are occurring in highly developed countries, but benefits are also occurring in NICs and other third-world areas as production filters down to them and they are able to assimilate the more efficient production and telecommunication advances that robotics, fiber optics, and biotechnology innovations provide. A reordering of the competitiveness of various regions and countries is also occurring as a part of the process, with the United States and Japan gaining and western Europe and the eastern bloc lagging in this transition to information-based economies.

SUGGESTIONS FOR FURTHER READING

CASTELLS, M., "High Technology, Space and Society." *Urban Affairs Annual Review,* No. 28. Beverly Hills: Sage Publications, 1985.

FORRESTER, T., *The Microelectronics Revolution.* Oxford: B. Blackwell, 1980.

FISHER, FRANKLIN, ET AL., *IBM and the U.S. Data Processing Industry: An Economic History.* New York: Praeger, 1983.

HALL, PETER, and ANN MARKUSEN, *Silicon Landscapes.* Boston: Allen & Unwin, 1985.

HANSON, DIRK, *The New Alchemists: Silicon Valley and the Microelectronics Revolution.* Boston: Little, Brown, 1982.

HAZEWINDUS, NICO, *The U.S. Microelectronics Industry.* New York: Pergamon Press, 1982.

JOINT ECONOMIC COMITTEE, *Location of High Technology Firms and Regional Economic Development.* Washington, D.C.: U.S.G.P.O., 1982.

KUHN, SARAH, *Computer Manufacturing in New England.* Cambridge, Mass.: Joint Center for Urban Studies of MIT and Harvard University Press, 1982.

McLEAN, MICK, *The Japanese Electronics Challenge.* New York: St. Martin's Press, 1982.

MARKUSEN, ANN, PETER HALL, and AMY GLASMEIER, *High Tech America.* Boston: Allen & Unwin, 1986.

O'NEILL, GERARD, *The Technology Edge.* New York: Simon and Schuster, 1983.

ORGANIZATION FOR ECONOMIC CO-OPERATION AND DEVELOPMENT, *The Semiconductor Industry.* Paris: OECD, 1985.

REES, JOHN, ed., *Technology, Regions, and Policy.* Totowa, N.J.: Roman & Littlefield, 1986.

REES, JOHN, ET AL., *New Technology in the American Machinery Industry: Trends and Implications.* Washington, D.C.: Joint Economic Committee, 1984.

ROGERS, E. M., and J. K. LARSEN, *Silicon Valley Fever.* New York: Basic Books, 1984.

SOMA, JOHN T., *The Computer Industry.* Lexington, Mass.: Lexington Books, 1976.

STANBACK, T. and T. NOYELLE, *Economic Transformation in American Cities.* Totowa, N.J.: Rowman & Allanheld, 1984.

THWAITES, A. T., and R. P. OAKEY, *The Regional Economic Impact of Technological Change.* London: F. Pinter, 1985.

UDELSON, JOSEPH, ed., *The Great Television Race.* University, Ala.: University of Alabama Press, 1982.

WILSON, JOHN W., *The New Venturers.* Reading, Mass.: Addison-Wesley, 1985.

WILSON, ROBERT, ET AL., *Innovation, Competition, and Government Policy in the Semiconductor Industry.* Lexington, Mass.: Lexington Books, 1980.

17 Cities as Service Centers

A principal function of the city is to provide services for its citizens, transients, and the surrounding hinterland. These activities embrace a variety of service and management functions, including retailing, banking, wholesaling, regional and national headquarters office activity, government, and various other personal and professional services. Recall discussion of tertiary, quaternary, and quinary services in Chapter 1. Unlike agriculture and manufacturing activity, these types of work cluster disproportionately in larger urban centers. Such activity, especially retail trade, can also be found in centers at each level of the urban hierarchy. These activities that agglomerate in urban centers, often labeled *trade centers* or *service centers,* can be accounted for using central place theory. We examine this theory here before proceeding to study each type of activity in more detail in Chapter 18.

Central place theory provides the integrating framework around which we explain the location of service activity. The term *central place* will be used here interchangeably with trade center or service center. A hierarchy of these service centers exist on the landscape, ranging in size from a rural hamlet to a large metropolitan center. At a certain point in this continuum of centers one can distinguish a city. A city is a settlement with a large concentration of people with a distinctive organization, life-style, and livelihood. Specialized land uses and a variety of social, economic, and political institutions allocate, coordinate, and manage the resources found in the city, making them very complex systems. Cities are the centers of power and control in our society. Although many are much smaller, it is often useful to think of a city as a center having at least 50,000 population. Cities of this size have distinctive residential, commercial, and public open-space areas and are large enough to strongly influence surrounding rural areas. This influence of the city takes many forms: commuting by rural citizens to the city to work, travel by urban residents to the country for recreation, and more subtle but equally influential impacts from newspaper circulation and the penetration of the electronic media airwaves into the surrounding hinterland.

Another arbitrary distinction one observes along this size continuum is that point when a transition occurs between urban and rural settlements. "The point at which a community is properly described as urban rather than rural is therefore not easily determined, and countries around the world use different population sizes to describe what is urban and what is not. In the United States it has generally been a population of 2500 persons or more that has qualified a center as urban; in Canada and Australia the threshold has been set at 1000 persons; while in Denmark and Sweden the cutoff has been even lower, at 200 persons. In Austria and Belgium, by contrast, a center must have 5000 inhabitants to qualify as urban, and in Japan only those with populations over 20,000 are considered urban!"[1] Various urban definitions em-

[1] Truman A. Hartshorn, *Interpreting the City: An Urban Geography,* 2nd ed. (New York: Wiley, 1988).

An *urban place* constitutes a basic urban population building block for census purposes. A settlement of 2500 persons, the urban place is typically an incorporated legal entity. Unincorporated centers of 2500, called *census-designated places,* defined by the census, provide a means of identifying the urban population in unincorporated areas.

A legal or corporate city having 50,000 or more population is known in census terminology as a *central city.* The central city provides the nucleus for metropolitan area definitions of the city. In 1930, acknowledging that cities had grown well beyond the city limits area, the census recognized metropolitan districts for the first time. By 1950 the Census Bureau had refined its metropolitan designation and established the standard metropolitan area (SMA) concept. The Bureau identified 168 such areas at that time. Further refinements led to the use of the term SMSA, standard metropolitan statistical area, by 1959. Changes again occurred following the 1980 census, when the nomenclature became MSA, metropolitan statistical area.

The changing numbers, definitions, and acronyms for metropolitan areas can be confusing, but the guidelines by necessity must reflect changing conditions. In 1983, for example, MSAs were subdivided into 23 CMAs and 78 PMSAs. A CMSA is a consolidated metropolitan area of more than 1 million population, so-called because it includes more than one large metropolitan area. Each of the major parts of the CMSA is known as a PMSA, primary metropolitan area.

For an area to be designated a metropolitan statistical area there must be a city of at least 50,000 population or an urbanized area of at least 50,000 with a total metropolitan population of 100,000. MSAs are defined in terms of county units, beginning with the county where the central city is located. Additional counties are added if they are integrated with the county or counties already included in the region. This integration is based on the presence of a large urban population and/or other social and economic ties, defined on the basis of commuting patterns, non-agricultural employment levels and characteristics, and population densities.

A shortcoming of the MSA definitions occurs from the arbitrary inclusion of whole counties as the definitional unit, which often leads to the incorporation of large quantities of rural territory. On the other hand, it can be argued that this rural area is a functional part of the metropolitan region serving a recreation and second home area for urban residents, not to mention its possible role as the breadbasket for the region.

Urbanized Area

The *urbanized area* (UA) definition of the city includes only the built-up areas in need of urban services provided by municipal police, fire, sewer, water, and solid-waste agencies. The UA includes the central city and surrounding suburbs but not sparsely settled fringe areas. That area includes (1) contiguous incorporated places; (2) incorporated places with fewer than 2500 inhabitants, provided that each has a closely settled area of 100 dwelling units or more; (3) adjacent unincorporated areas with a population density of 1000 or more inhabitants per square mile; and (4) other adjacent areas with a lower population density that serve to smooth the boundary.

Census Tract

A *census tract* is a multiblock region into which metropolitan areas are divided for census data reporting purposes. Tracts are designed to be relatively uniform with respect to population characteristics, economic status, and living conditions. The average tract has about 4000 residents. The most comprehensive source of population, housing, and socioeconomic data for the city is the tract level.

---

ployed by the U.S. Bureau of the Census appear in the accompanying box.

In Chapter 2 we discussed the rapid increase in size of world cities, particularly those in the third world. Rather than size, however, it is often useful to discuss these cities in terms of the number and variety of *functions* or activities that exist in them. Central place theory deals with this aspect of city life. It is particularly appropriate in an *economic geography* textbook to discuss those activities in cities that make the city the nerve center for commercial economies.

In this chapter we discuss the location and size of service centers using a central place theory framework. According to this theory, there should be a few larger centers and a larger number of smaller centers on the landscape. Of course, the settlement history of the area, the local economy, country size, technology, and environmental factors such as climate, soils, and landforms all affect the distribution and size of cities, but generalizations are possible. Before discussing the conceptual basis of central place theory, it will be useful to review the locational dynamics associated with the world's largest cities.

WORLD CITY PATTERNS

At the world scale four major urban regions can be identified using the clustering of millionaire cities as a basis for comparison (see Appendix). A list

Table 17-1

Largest Metropolitan Areas of the World, 1985
(millions of people)

Rank	Metropolitan Area	Estimated Population
1	Tokyo–Yokohama	Over 15
2	Mexico City	
3	New York	
4	Osaka–Kobe–Kyoto	
5	São Paulo	10–15
6	Seoul	
7	Moscow	
8	Calcutta	
9	Buenos Aires	
10	Bombay	
11	London	
12	Los Angeles	
13	Cairo	
14	Rio de Janeiro	5–10
15	Paris	
16	Shanghai	
17	Delhi–New Delhi	
18	Jakarta	
19	Manila	
20	Chicago	
21	Bangkok	
22	Tehran	
23	Beijing	
24	Karachi	
25	Leningrad	
26	T'aipei	
27	Istanbul	
28	Bogota	
29	Philadelphia	
30	Lima	
31	Madras	

Source: 1986 Rand McNally Commercial Atlas and Marketing Guide (Chicago: Rand McNally, 1986).

of those 31 metropolitan areas exceeding 5 million population appears in Table 17-1. The most extensive urban concentration occurs on the continent of Europe, including the Mediterranean basin. That area encompasses about 50 cities exceeding 1 million population. A second major urban concentration exists in eastern North America, extending from Montreal to Minneapolis on the north and to Miami and Houston on the south. A secondary band of cities on the west coast of North America extends from Vancouver southward to San Francisco, Los Angeles, San Diego, and inland, including Mexico City in south central Mexico. The third major urban population realm, in eastern Asia, includes cities in Japan, Korea, and Manchuria, southward to Indonesia. Several of the world's largest and fastest-growing cities exist in this area. The final concentration occurs in south Asia, including India and Pakistan.

Elsewhere in the world the urban pattern appears as a series of isolated nodes or discontinuous belts, as in South America, where the urban population concentration occurs on the periphery of the continent.

Urban clusters exhibit an irregular pattern in Africa as well as Australia, where in the latter instance the largest urban centers occur on the southeast coast.

World urban population growth accelerated rapidly during the present century. Over half the population in highly developed areas lived in urban areas by 1950, while only about 10 percent was urban in the third world. By the turn of the twenty-first century nearly 80 percent of the population of developed areas will live in urban areas, while about 25 percent of the third-world population will be urban. Nevertheless, the number of urban residents in the third world will exceed that in developed areas because their population base will be much larger. The third-world population may reach 5 billion persons by that time, and 1.4 billion in developed areas, yielding a world total of approximately 6.4 billion in the year 2000.

Today, four major urban complexes at the world scale exceed 15 million in population (Table 17-2). The New York metropolitan area presently leads in size and Mexico City places second but is much faster growing. Two Japanese urban regions, Tokyo–Yokohama and Osaka–Kobe–Kyoto, complete the grouping of the largest urban realms. To this category one can add another nine cities between 10 and 15 million persons in size. Only three of these cities occur in highly developed areas or centrally planned areas, with two falling in newly industrializing countries (Brazil and South Korea) and four in other third-world areas. The large share of these centers in the third world reflects the rapid urbanization process that will likely propel a third-world city (Mexico City) to the top of the list in population size by the year 2000.

All 13 cities in these top two size groupings, except Mexico City and Moscow, have coastal locations or are located on major rivers with nearby coastal access. Typically, these cities have a major port function, are major transportation hubs, and serve as political and economic centers for their countries. In the case of New York, the city has the added distinction of being the financial capital of the world. The 18 ad-

Table 17-2

Size Distributions of Millionaire Cities,* 1985
(millions of people)

Size Category	Number of Cities
Over 15	4
10–15	9
5–10	18
3–5	33
2–3	45
1.5–2	46
1–1.5	102
Total	257

Source: Commercial Atlas and Marketing Guide (Chicago: Rand McNally, 1986).
*Cities with more than 1 million population.

Table 17-3

Millionaire Cities,* 1870–1985

Date	Number of Cities
1870	7
1900	20
1920	30
1939	57
1951	95
1964	140
1979	213
1985	257

Source: Richard L. Forstall and Victor Jones, ''Selected Demographic, Economic and Governmental Aspects of the World's Major Metropolitan Areas,'' in Simon R. Miles, ed., *Metropolitan Problems* (Methuen: London, 1970), p. 11; and *Commercial Atlas and Marketing Guide* (Chicago: Rand McNally, various years).

*Cities with more than 1 million population.

ditional cities with 5 to 10 million population have the same relative distribution worldwide, with nearly half occurring in the third world.

The recency of the ascendancy of millionaire cities is vividly demonstrated in Table 17-3. In 1930 there were only 30 cities in the world having more than 1 million persons. That number grew to over 90 in 1950, 213 in 1979, and 257 in 1985. In the 1964–1985 period alone, over 100 cities passed the 1 million population threshold. This phenomenal growth of large cities has occurred disproportionately in the third

world and has led to a tremendous strain on local resources to cope with such explosive growth.

Whereas the economic base of cities was once closely tied to manufacturing activity, especially during the nineteenth century, a shift to service activity as a support base has occurred during the present century. In fact, the largest cities now have the largest share of service employment. The dramatic shift of employment to services has paralleled the growth of office activity and the transition to the information age. In the United States the dominance of service employment is particularly apparent in sunbelt cities, which now lead in urban growth in the country. Older industrial cities of the north central states, long the bastion of the blue-collar industrial worker, have suffered dramatically in recent years, as symbolized by the closed industrial mills throughout the rustbelt as popularized by the media. Not all older northeastern cities have suffered as much as Detroit, Cleveland, Youngstown, and Buffalo in recent years, as many have successfully transformed their support to a postindustrial base such as Boston, Rochester, and Minneapolis. Third-world cities are also heavily service oriented, as we discuss in Chapter 18.

MEGALOPOLIS

The urban corridor extending over 600 miles from Boston to Washington, D.C. in the United States, comprised of more than 30 metropolitan areas, called *megalopolis,* represents the most distinctive urban complex in the world (Figure 17-1). That area possesses the largest and wealthiest urban population in the

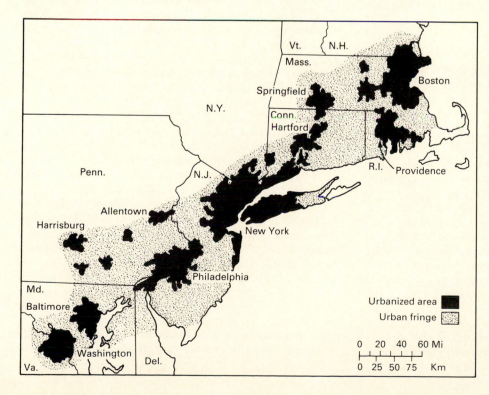

Figure 17-1 Megalopolis. *The urban corridor extending over 600 miles from Boston to Washington, D. C. in the United States represents the most distinctive metropolitan complex in the world. The more than 30 metropolitan areas comprising this region are closely tied together with interstate highways, commercial airline service, and rail passenger/freight service. The area possesses the largest and wealthiest urban population in the country and serves as the economic hinge of the national economy.*

United States and has had a long and distinctive history in its role as the economic hinge of the country. The coalescence of these urban centers is by no means unique, as similar urban realms occur in Japan and Europe. The Tokaido megalopolis, for example, extends from Tokyo–Yokohama to Osaka–Kobe–Kyoto. In Europe, the Ruhr industrial area has contributed another such region in the Essen–Dortmund–Duisburg case.

The local economies of each metropolitan area in a given megalopolis typically possess a distinctive economic base. Individually, each of the centers may serve as a port, capital, manufacturing center, headquarters city, distribution center, and/or cultural mecca. Each is also closely linked. In the case of the U.S. megalopolis, shuttle airline service and intercity metroliner passenger rail service, not to mention interstate highway linkages, closely tie the areas together physically. In Japan toll expressways and the high-speed rail service provided by the *Shinkansen* bullet train tie centers at both ends of the Tokaido megalopolis together functionally.

RANK-SIZE RULE

As can be observed in Table 17-1, there are usually many more smaller than larger urban centers. The *rank-size rule* helps account for this hierarchical arrangement of cities. This generalization occurs as a consequence of the regularity in the size of a city based on its rank in the urban hierarchy of a country. According to this guideline, one can predict the size of a given city by knowing its rank and the size of the largest city according to the following relationship:

$$P_n = P_1 \times R^{n-1}$$

where
P_n = population size of city being studied
P_1 = population size of largest city
R_n = rank of city being studied
R^{n-1} = rank of city to -1 power or $1/R_n$

In this example the second largest city in a country should be one-half the size of the largest city, the third city one-third the largest, and so on. At the world scale and in many countries this relationship works very well. Typically, the rule works better in larger countries such as the United States or the Soviet Union, in countries with a long urban history, and in areas that have a more complex social and economic system. Exceptions occur where *primate city* distributions prevail, as in the case of France or Mexico, where one city claims one-third or more of the total population of a country. Another exception exists when one or more of the various city-size groupings are missing, called an *intermediate*-city distribution, as in the case of Canada, where no dominant center occurs at the top. In the case of Australia, smaller cities are absent; most of the population occurs in a few large centers.

CENTRAL PLACE THEORY

Central place theory, developed by the German geographer Walter Christaller earlier this century, based on observations concerning settlement patterns and functions in Bavaria, provides an economic interpretation of the size, spacing, and functional activities found in cities. This theory does not apply to manufacturing or other specialized activity, but to those functions that occur in central places in response to the needs of the hinterland market, such as retail goods, banking, and professional services. In other words, these activities occur in towns of various sizes due to the size of the market or trade area served. Christaller made two assumptions about the landscape wherein ideal conditions for the central place system prevail: (1) an even topography such as a uniform plain with no interruptions created by physical features such as mountains or rivers, and (2) an economy based on providing goods and services to the surrounding population and not on the production of primary or secondary products.

When conditions of Christaller's theory are met, service centers of any particular size classification should be evenly distributed on the landscape. The classic laboratory to study central place landscapes has become the U.S. middle west, but the patterns can be observed throughout the world in agricultural areas (Figure 17-2). The smallest of these centers, *hamlets*, would be most closely spaced. In areas settled prior to the automobile, hamlets occur much more frequently than they would if settlement happened today, due to the reduced travel time needed to reach the service center from the surrounding countryside. Assuming that a person would need to reach the center in one hour's travel time (by foot or horse), and that one could travel 3.5 kilometers in an hour, these centers would typically be spaced no more than 7 kilometers apart. Because this theory postulates that service centers tend to be located in the center of the area served, the label *central place theory* applies.

Since contiguous circular trade areas would have voids and/or overlapping boundaries, Christaller determined that hexagons would be more efficient shapes to incorporate all the surrounding space in one unique trade area and to avoid the problem for each size of community of overlap or underlap. Therefore, trade areas for each size of community appear as hexagons instead of circles (Figure 17-3).

The hierarchy of trade centers builds on this hexagonal framework, with the smallest centers appearing each 7 kilometers. A second level of settlement, called a *village*, evolves in the center of an area of several hamlets. These villages provide more specialized services, drawing support from portions of several of the surrounding lower-order hamlet trade areas. The village would not only be spaced farther apart than the hamlet but also be larger in size in population and functions as well as fewer in number. The village trade area would be bounded by six hamlets (Figure 17-4).

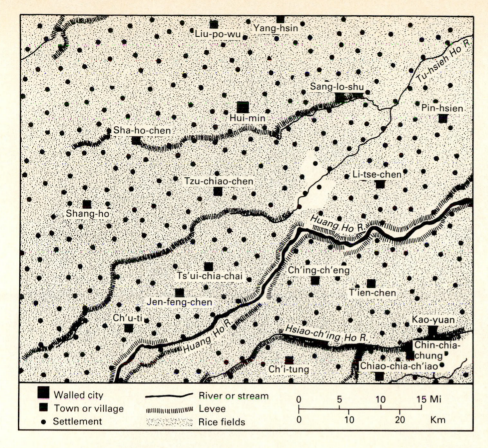

Figure 17-2 Settlement Pattern on this North China Plain. *When conditions of Christaller's theory are met, as in this instance, service centers of a particular size are evenly spaced. Note the spacing and frequency of higher order centers in relation to the lowest order settlements.*

By the same process, a third-order service center, the *town*, would emerge with a larger hexagonal trade area, bounded in this case by six villages. Fourth-order centers would be *cities*, and fifth-order centers, *regional capitals*.

Over the years the types of functions available at each level of central place have changed as life-styles and economies have changed (Table 17-4). Sometimes we refer to the *order* of the center when discussing the

types of activities in each. Lower-order centers (hamlets, villages) generally offer basic convenience goods, whereas higher-order centers provide more specialized shopper's goods. For example, the hamlet is typically limited to one or two of the following: grocery store, gasoline station, grain elevator, church. The town, on the other hand, will offer more specialized shopper's goods, such as a women's apparel store, bank, liquor store, catalog mail-order store, hardware

Figure 17-3 Alternate Trade Area Shapes. *Christaller chose the hexagon as the ideal trade area shape because it closely approximates the qualities of a cricle (all areas on boundary are equal distance from the center) and avoids a problem of underlap or overlap as shown.*

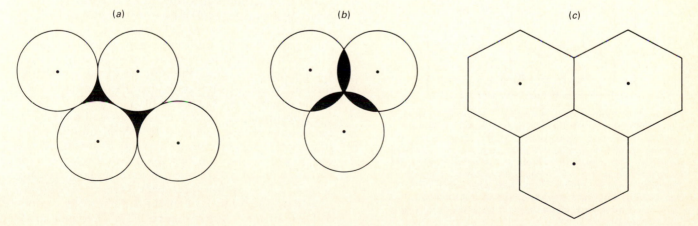

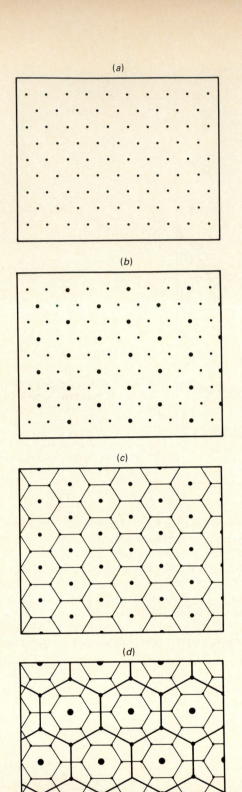

(a)

(b)

(c)

(d)

0 2 Mi

0 2 Km

Figure 17-4 Central Place Hierarchy. *When conditions of Christaller's theory are met, an even spacing of service centers occurs with higher order centers farther apart. The spacing of hamlets, the smallest of these centers, is shown in diagram (a), with both hamlets and villages shown in (b). Hexagonal trade areas for villages are shown in diagram (c). Note that the corners of the trade area occur coincidental with the locations of the six surounding hamlets. A three-level hierarchy with trade areas for towns and villages occurs in (d).*

Table 17-4

Examples of Functions in Central Places of Varying Sizes

Hamlet-level functions	Town-level functions (cont.)
Gasoline service station	Dentist
Grocery store	Insurance office
Church	Veterinarian
Grain elevator/feed store	Auto-parts store
Village-level functions	City-level functions
Barber shop	Shoe store
Beauty shop	Jewelry store
Restaurant	Florist
Bar	Hospital
Hardware store	Medical specialist
Farm equipment dealer	Hotel/motel
Bank	Newspaper
	Specialty restaurant
Town-level functions	Sporting goods store
Furniture store	Department store
Clothing store	Certified public accountant
Automobile dealer	Travel agency
Lawyer's office	Fast-food restaurants
Funeral home	Plumbing supply
Doctor	County government

store, and health clinic. Note also that the higher-order center, such as the village or town, will also have lower-order functions, as it also serves as a market center for basic convenience goods for nearby consumers (Figure 17-5).

Implicit in the distinction between orders of goods is the notion of *range of a good,* which refers to the distance that consumers are willing to travel to purchase that good or service. Higher-order goods have longer ranges. For example, consumers are willing to travel further to purchase clothing and jewelry than they are to buy groceries or travel to the post office.

Higher-order goods also have a larger *threshold.* "Threshold" refers to the size of the market required to support a particular good or service. A grocery store can thrive in a hamlet, as it needs only a few hundred customers, but a jewelry store would probably need a trade area incorporating several thousand potential customers in order to be a workable enterprise. At the upper end of the spectrum we sometimes say that a city needs 50,000 persons to justify a bus transit system and 1 million persons to justify a heavy-rail rapid-transit system. Once a threshold is reached for a good, say a bank, the number of these activities often increases more rapidly than the population size of the community (Figure 17-6).

Consumer travel preferences dovetail nicely with the notions of range of good, threshold, and the order of a good. By constructing *desire lines* that connect the origin and destination of trips by rural residents for various types of goods, one can observe the increasing length of the trip for specialized higher-order goods, as well as the absence of trips for higher-order goods to lower-order centers (Figure 17-7). The trade areas

Level of center	Types of goods and services					
	Shoppers ◄────				────► Convenience	
Metropolitan center	X	X	X	X	X	X
Regional city		X	X	X	X	X
City			X	X	X	X
Town				X	X	X
Village					X	X
Hamlet						X

X indicates that a center of the specified level provides goods and services of this order.

Figure 17-5 Center Hierarchy and Goods and Services Offered. *Lower order centers (hamlets, villages) offer basic convenience goods, whereas higher order centers provide more specialized shoppers goods. Note that higher order centers provide both lower and higher order services.*

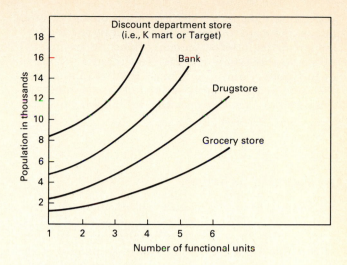

Figure 17-6 Threshold Size. *The size of the market required for a particular good or service increases for more specialized functions. Once a threshold for a particular activity is reached, say a bank, the number of these activities often increases faster than the population of the market area as the upward curving line on the graph suggests. For example, 5,000 persons may be required to support a bank, while 6,000 persons can support two banks and 8,000 will support three banks as indicated here.*

of these lower-order centers become *nested* within the area served by the higher-order center. The areal extent over which a community draws its consumers refers to the *economic reach* of that center.

By knowing the size of a community, one can usually predict the types of activities available in it. But over the years, improvements in transportation, particularly the increased accessibility provided by the automobile, has led to households making more trips and longer trips. The number and variety of activities available in lower-order centers has declined due to these transportation changes. Consumers can now bypass lower-order centers and drive to the next higher-order community in the same time that it formerly took to travel to the lower-order community. The *dying village* scenario reflects this change. This situation occurs as the retailing function of hamlets and villages has declined over the years even though the residen-

tial population of the community and its surrounding trade area may have increased.

DISPERSED CITIES

Because of historical and/or entrepreneurial factors, service centers of a given size sometimes offer functions that appear too specialized for a center of that size. An example might be the presence of an automobile dealership in a village or a major league professional sports team in a small city. A cluster of communities each offer an activity of a different type that is too specialized for that size of service center. In such cases the *dispersed city* concept may explain the size, spacing, and functions in these areas. In effect, these settlements function as a single larger-order center for a particular good or service, even though

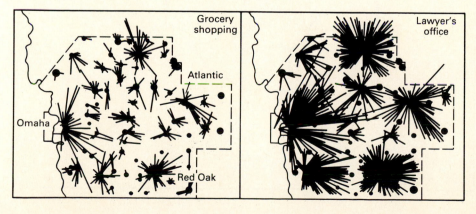

Figure 17-7 Consumer Travel Behavior. *Consumer travel preferences dove tail nicely with the notions of range of good, threshold, and the order of a good. By constructing desire lines that connect the origin and destination of trips by rural residents for various types of goods, one can observe the increasing length of trips for specialized higher-order goods. Note the distribution and length of trips for grocery shopping compared with that of travel to a lawyer's office in southwest Iowa shown here. (Redrafted from Berry, Geography of Market Centers and Retail Distribution, Prentice-Hall, 1967, p.11)*

activities are dispersed spatially. Consumers from the larger region patronize each community for the particular service offered. In many cases this means that no single area dominates and none offers a full range of functions found in a typical city that would be of the same order of the combined group. Instances of this situation have been cited in Pennsylvania, Illinois, and elsewhere.

URBAN RETAIL HIERARCHY

A counterpart to the central place urban hierarchy also occurs within metropolitan areas (Table 17-5). The metropolitan retail hierarchy vividly demonstrates this similarity. Since urban populations are more dense than those in rural areas, the areal extent of the trade area is much smaller in urban than in rural settings (Figure 17-8). The isolated Mom and Pop neighborhood grocery store or the chain limited-service convenience store replicates the hamlet in terms of function and size of population served. A second-order urban center is the *neighborhood shopping center,* typically a planned center anchored by a grocery store. Its counterpart in rural areas is a village. The neighborhood center offers a variety of convenience goods,

Table 17-5

Comparison of Retail Hierarchies in Urban and Rural Areas

Rural Hierarchy	Metropolitan Hierarchy
Hamlet	Isolated grocery store
Village	Neighborhood shopping center
Town	Community shopping center
City	Regional shopping center
Regional capital	Superregional shopping center

such as a drugstore, barber and beauty shop, coffee shop, dry cleaner/laundry, or a liquor store.

The third-level center in the metropolitan retail landscape is the *community shopping center.* Typically anchored by a junior department store (Woolworth's) or a discount variety store (K mart), the community shopping center draws more widely and offers more shopper's goods, such as apparel stores, a photography store, a restaurant, and a branch bank. In addition, the community center would offer neighborhood-level functions.

The fourth level of metropolitan retail agglomeration, the *regional shopping center,* claims one or more major department stores and dozens of other shop-

Figure 17-8 Comparison of Trade Areas in Urban and Rural Areas. *Since population densities are higher in urban than rural areas, the areal extent of the trade area is smaller in the former case as shown here. Note that the shape of trade areas conforms to the spacing of centers identified here as dots. The trade area is stretched out in the direction of lower settlement densities.*

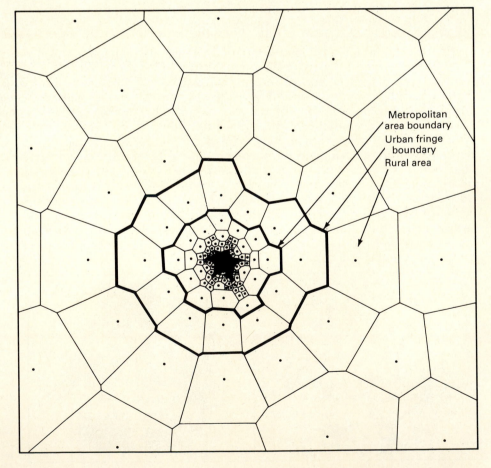

Metropolitan area boundary
Urban fringe boundary
Rural area

per's goods shops. The rural area counterpart, the city, would offer similar types of shopping opportunities. Finally, the *superregional shopping center,* with three or more department stores and up to 100 other specialty shops, replicates the functions of a regional capital. The role of the regional mall in suburban business centers is reviewed in Chapter 18.

METROPOLITAN HIERARCHIES

The discussion of city sizes this far has considered five levels of centers from the hamlet to the city. Notwithstanding this breakdown, the entire hierarchy of centers in a nation should include additional levels to distinguish among larger metropolitan centers. For our purposes we identify three levels of metropolitan centers, beginning with the largest, the *international metropolis.* The international metropolis, typically serves as the functional, corporate, and political headquarters of a nation. In the case of the United States, New York provides most of these roles, as London does for Great Britain, and Tokyo in Japan. Toronto and Montreal share this function in Canada. These metropolitan areas are the banking, business, and cultural meccas of their nations and have strong international ties. The intensity of activity in these centers is reflected by such disparate indicators as the number of corporate headquarters, the size of the financial community, and the number of air passengers handled by their airports, including international flights. The international metropolis can be labeled a *first-order* city and is typically by far the largest city in the nation, as would be expected in the context of our earlier discussion of the rank-size rule.

A *national metropolis* typifies a group of *second-order* urban centers. Typically, these centers have 2 million residents or more and dominate a major portion of the country in terms of banking linkages, wholesaling activity, and the distribution function, giving them strong national significance. A *third-order regional metropolis,* having more than 1 million population, mainly serves a regional headquarters and distribution center.

TRADE AREAS

A service center, regardless of size, casts a sphere of influence over the area from which it draws its customers to buy goods and services. This region, called a *trade area,* should ideally extend outward from the service center equally in all directions. Competition with other centers, transportation accessibility, and consumer preferences can all modify the shape and extent of this area, but based on the *principle of least effort,* consumers should travel to the nearest of the same-size communities for a particular service. The principle of least effort concept was the focus of a book by Zipf in 1949, when he explained that persons

seek the shortest distance and nearest alternative when choosing among routes and destinations. This being the case, the trade area boundary between two communities of the same size should be halfway between them. One could observe actual consumer behavior to confirm this generalization by interviewing residents concerning their travel behavior.

Determining the trade area boundary between places unequal in size is more complex but nevertheless straightforward. Using a modified *gravity model* approach, one can determine this boundary. The gravity model, so called because of its similarity with Newton's law of gravitation, postulates that the potential power of attraction between two bodies increases with the product of their masses and decreases with distance between them. In this case mass refers to the population size of the service center and distance to the miles or kilometers separating the centers.

The gravitational attraction of two places is sometimes expressed as an example of *interaction theory.* Interaction theory accounts for the strength of the economic connections between two places, which varies positively according to their size and negatively according to the intervening distance. The larger the populations of the two places, the greater their economic interaction, and the greater the distance between them, the less the interaction. In other words, population and distance can be taken into account simultaneously in accounting for levels of movement using this gravity model approach. For example, suppose that there are three cities, X (with 20,000 people), Y (with 10,000), and Z (with 30,000), located as in Figure 17-9, with Z 50 miles from Y and 100 miles from X. How would the amount of business contacts between Y and Z compare to that between X and Z? Interaction theory suggests an answer: that the amount of business will vary directly with the product of the two populations and inversely with the distance between them.

The theory is expressed in the following formula:

$$i = \frac{P_1 P_2}{d}$$

where
i = interaction
P_1 = population of one of the places
P_2 = population of the other place
d = distance between them

Accordingly, the index of business between Y and Z would be computed as follows:

$$\frac{10,000 \cdot 30,000}{50} = \frac{300,000,000}{50} = 6,000,000$$

The index number between X and Z is computed as follows:

$$\frac{20,000 \cdot 30,000}{100} = \frac{600,000,000}{100} = 6,000,000$$

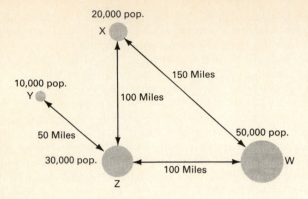

Figure 17-9 Interaction Levels Between Centers. *The strength of economic connections between two places depends on their size and distance apart. In this case the level of interaction between y and z is the same as that between x and z even though x is larger than y, because it is also farther away.*

Thus the interaction between Y and Z would tend to be equally as strong as that between X and Y. Even though X is bigger than Y, both share similar ties with Z because Y is also twice as far away from Z as is X.

Other evidence in support of this theory shows that the interaction theory roughly predicts the number of telephone messages, bus passengers, airline traffic, and other types of interaction that generally occur between pairs of cities.

Breaking-Point Theory

A modification of the interaction theory occurs with the *breaking-point theory,* which is an endeavor to provide a way to predict the location of the boundary line separating trade areas around two cities unequal in size. Referring to Figure 17-9 again, where would the boundary line run separating the trade area of city Y from that of city Z? Would it be halfway between the two cities, or closer to one of them? If the latter, how close would it be?

The breaking-point theory predicts the location of this line by means of the following formula:

distance of breaking point (BP) from the smaller trade center } equals

{ *Numerator:* distance between the two trade centers

Denominator: 1 plus the square root of { population of larger city *divided by* population of smaller city }

In symbols, the ratio is

$$BP = \frac{d}{1 + \sqrt{Pop\ Z/Pop\ Y}}$$

From Figure 17-9 the appropriate values can be inserted in the formula as follows:

$$BP = \frac{50}{1 + \sqrt{\dfrac{30,000}{10,000}}} = \frac{50}{1 + \sqrt{3}}$$

$$BP = \frac{50}{1 + 1.73} = \frac{50}{2.73} = 18.3 \text{ miles (from X)}$$

In reality, of course, the position of a trade area boundary is influenced by the forces of distance and size of trade centers (as considered in the interaction theories just cited) but also by transportation routes, political boundaries, and so on.

To determine the trade area boundary of a city for a particular type of good or service, one would have to determine what appropriate competing cities to include in the comparison and then calculate a series of breaking-point distances between successive pairs of cities and connect the breaking points. Figure 17-10 shows such a circumstance involving the expected newspaper circulation trade area of the *Atlanta Constitution* in Georgia. Note that only major regional centers in the southeast surrounding Atlanta are involved in the calculation and that straight-line distances between centers were modified somewhat to reflect interstate highway mileage.

Law of Retail Trade Gravitation

A second modification of interaction theory is W. J. Reilly's *law of retail trade gravitation,* an attempt to provide a way to predict the volume of retail trade patronage that a city's residents will give to other cities. For example, in Figure 17-9, what are the relative volumes of patronage that inhabitants of city X will give to businesses in city Z and to those in city W?

Reilly's theory can be stated as follows:

$$\frac{\text{volume of X's patronage to Z}}{\text{volume of X's patronage to W}} =$$

$$\frac{\text{population of Z}}{\text{population of W}} \cdot \left(\frac{\text{distance X–Z}}{\text{distance X–W}}\right)^2$$

From Figure 17-9, we can insert values in the formula above; thus

$$= \frac{30,000}{50,000} \cdot \left(\frac{150}{150}\right)^2$$

$$= \frac{3}{5} \cdot \left(\frac{10}{15}\right) = \frac{3}{5} \cdot 666 = \frac{1.32}{5}$$

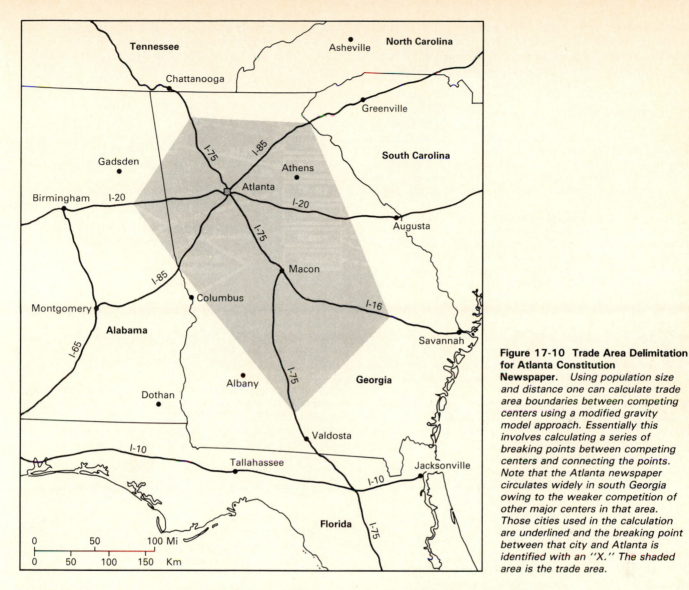

Figure 17-10 Trade Area Delimitation for Atlanta Constitution Newspaper. *Using population size and distance one can calculate trade area boundaries between competing centers using a modified gravity model approach. Essentially this involves calculating a series of breaking points between competing centers and connecting the points. Note that the Atlanta newspaper circulates widely in south Georgia owing to the weaker competition of other major centers in that area. Those cities used in the calculation are underlined and the breaking point between that city and Atlanta is identified with an ''X.'' The shaded area is the trade area.*

Thus for every $1.32 of goods that X's inhabitants purchase in Z, they will tend to purchase 5 dollars' worth in W. The principle can be summed up in this way: The degree to which the residents of a city patronize another city is directly proportional to the other city's population and inversely proportional to the square of the distance between them.

While most urban residents patronize retailers in their own community whenever possible, central place theory tells us that residents must travel to larger communities for more specialized goods. Using the law of retail gravitation approach presented here one can estimate the share of purchases residents will make in communities unequal in size based on their relative population size and distance apart.

SUMMARY AND CONCLUSION

In this chapter we have reviewed the overarching theory and principles that account for the location and function of cities as service centers. Central place theory provides the primary basis for explaining hierarchy of these centers, their spacing, and the type of functions (i.e., retail and service activities) one finds in them. The rank size-rule concept provides additional insight into the hierarchical order of cities. World city patterns generally conform to these principles.

The retail sales mix and volume in a given center reflects the position of that center in the central place hierarchy. Trade or market areas for centers also reflect their rank. The interaction or gravity model described in Chapter 1 and reviewed again here assists in the delimitation of trade areas, especially in determining boundaries between competing centers.

Large concentrations of urban populations occur in Europe, North America, East and South Asia. These large concentrations occur in both developed and developing areas, reflecting the robust nature of the world urbanization process. The megalopolis concept, first used to delimit the urban corridor in the northeastern United States extending from Boston to Washington, D.C., for example, has also been used to describe urban agglomerations in Europe and Japan. In Chapter 19 we will explore comparative world urbanization processes in more detail, following a discussion of the commercial structure of the city in Chapter 18, including a more in-depth review of retail activity.

SUGGESTIONS FOR FURTHER READING

AGNEW, JOHN, ET AL., *The City in Cultural Context*. Boston, Mass.: Allen & Unwin, 1984.

BASKIN, C. W., *Central Places in Southern Germany*. Englewood Cliffs, N.J.: Prentice-Hall, 1966. Translation of Walter Christaller, *Die zentralen Orte in Suddeutschland*. Jena, Fischer, 1933.

BEAVON, KEITH S., *Central Place Theory: A Reinterpretation*. London: Longman, 1977.

BERRY, BRIAN J. L., *Geography of Market Centers and Retail Distribution*. Englewood Cliffs, N.J.: Prentice-Hall, 1967.

EYRE, J. D., *Nagoya: The Changing Geography of a Japanese Regional Metropolis*. Chapel Hill, N.C.: University of North Carolina, 1982.

GOTTMANN, JEAN, *Megalopolis*. New York: The Twentieth Century Fund, 1961.

HALL, PETER G., *London 2000*. 2nd ed. New York: Praeger, 1969.

HALL, PETER G., and DENNIS HAY, *Growth Centres in the European Urban System*. Berkeley, CA: University of California Press, 1980.

HARTSHORN, TRUMAN A., *Interpreting the City: An Urban Geography*. 2nd ed. New York: John Wiley & Sons, 1988.

HAYNES, KINGSLEY, and A. STEWART FATHERINGHAM, *Gravity and Spatial Interaction Models*. Beverly Hills: Sage Publications, 1984.

HERBERT, DAVID T., and COLIN J. THOMAS, *Urban Geography: A First Approach*. New York: John Wiley & Sons, 1982.

JACOBS, JANE, *Cities and the Wealth of Nations*. New York: Random House, 1984.

KING, LESLIE J., *Central Place Theory*. Beverly Hills: Sage Publications, 1984.

McGEE, TERENCE G., *The Southeast Asian City: A Social Geography of the Private Cities in Southeast Asia*. New York: Praeger, 1967.

McGEE, TERENCE G., *The Urbanization Process in the Third World: Explorations in Search of a Theory*. London: G. Bell & Sons, 1971.

NEUTZE, GRAEME MAX, *Urban Development in Australia: A Descriptive Analysis*. London: Allen & Unwin, 1977.

NOBLE, ALLEN G., and ASHOK K. DUTT, *Indian Urbanization and Planning: Vehicles of Modernization*. New Delhi, India: Tata-McGraw-Hill, 1977.

PINCH, STEVEN, *Cities and Services*. London: Routledge and Kegan Paul Inc., 1985.

PRED, ALLAN, *City-Systems in Advanced Economies*, New York: Wiley, 1977.

PRED, ALLEN, *Urban Growth and City Systems in the United States 1840–1860*. Cambridge, Mass.: Harvard University Press, 1980.

REPS, JOHN W., *Cities of the American West: A History of Frontier Urban Planning*. Princeton, N.J.: Princeton University Press, 1979.

ZIPF, GEORGE K., *Human Behavior and the Principle of Least Effort*. Cambridge: Addison-Wesley, 1949.

18 Commercial Activity in the City

As discussed in Chapter 17, the service economy and cities are synonymous. As the nerve center of commercial economies, the city provides a home for a vast array of tertiary activities which feature the fastest-growing segments of modern economies today (Figure 18-1). Generally, the number and variety of these services increases with city size, especially office-related functions which agglomerate to take advantage of other supporting services found in larger metropolitan areas. Examples of these services include better nonstop air service to other large centers, enhanced computer and technical support service opportunities, and financial services. The transition to a knowledge-based information society has had a tremendous impact on these office activities, creating totally new fields of work in recent years, not to mention a revolution in work patterns in established businesses.

In third-world cities the nature of the service economy is much different, but nevertheless just as integral to the functioning of the city. In fact, the majority of the labor force is often engaged in informal bazaar-type service activities in third-world cities (Figure 18-2). This activity involves individuals or family groups working as peddlers, vendors, street hawkers, artisans, shoemakers, or tailors, together with many other low-skill activities. This service or trade sector is very different from that in developed areas, where more of these activities involve formal market transactions in retail outlets or professional office settings. The lack of a middle-class merchant and clerical class

in third-world areas means that retail trade and service activity is handled by self-employed entrepreneurs through an informal, often underground, economy.

Rarely do cities, even in the third world, host service activities exclusively. Often, manufacturing activity or another specialized support base also provides employment opportunities. In some situations most of the citizenry earns a livelihood from the primary and secondary activities discussed earlier. It is uncommon, however, not to find at least a modicum of service employment in any city, even if it is limited to the retail or education sector.

In this chapter we discuss the location and character of service activities within the city, placing emphasis on retail and office activity, but also mentioning the wholesale/distribution function and hotel activity. The focus is on the North American experience, with some mention of western Europe, Japan, and the third world to give proper recognition to the importance of historical, cultural, and financial factors in influencing the diversity in the form and function of the city.

HISTORICAL TRADITIONS

Not only do cities date to antiquity as settlement forms but so do the roots of many contemporary service-related functions and land uses still found in them. Indeed, some ancient settlements have survived the mil-

Figure 18-1 Las Colinas Suburban Downtown. *Located midway between Dallas and Fort Worth, in Irving this master planned 12,000 acre development contains over 20 million square feet of office space, in addition to several hotels, restaurants, and retail centers. It will also house 50,000 residents upon completion in the 1990s. The lavishly landscaped and amenity-laden Las Colinas complex is surrounded by water features and has received several design awards. Note downtown Dallas skyline in the distance at top left and the Texas Stadium in Arlington at top center. (Photo courtesy of LANDIS AERIAL PHOTO, INC.)*

Figure 18-2 Informal Economy in Aurangabad, India. *Open markets provide the primary means of exchanging goods in many third world areas. Clothing items and produce are displayed here in the street market. (Courtesy Eugene Loring)*

Figure 18-3 The Ancient Walled City of Rome and the Vatican. *This perspective of Rome at the time of St. Leo I, about 450 a.d., shows the walled city and major landmarks including the coliseum and St. Peters basilica in the Vatican located across the Tiber River to lower left. Note that a single bridge crossed river at that time. The wall around the Vatican still stands, but urban development now surrounds the entire region (see also Figure 18-9). (Courtesy Vatican Bookstore)*

lennia, flourishing today as strongly as ever. In Europe, the tradition of London as a major trade center, having origins as a Roman town, and Rome itself, demonstrate the long history of European cities (Figure 18-3).

Middle Eastern and Chinese cities, although not as well known to Americans, continue to provide strong ties to the past, physically and culturally. Sophisticated architectural techniques used by the Chinese to build imposing buildings, for example, date back 7000 years. Traditional city forms with buildings placed around a courtyard can still be observed in Chinese cities today (Figure 18-4.). Impressive government buildings in the center of the city typically faced south on a central north-south axis. Shops, factories, and markets lined streets radiating outward from this central axis, together with residential blocks built around courtyards.

The ancient Greek and Roman city also exhibited many characteristics familiar to Americans today. The *agora*, originally an open-market area in the city center, and later a mixed-use area, is the precursor of the modern central business district. Greek cities also gave us the formal *grid* street pattern. We often think of Roman cities for their elaborate public buildings and land-use allocation patterns that demonstrated a hierarchical social order. The camptown or *castrum* found throughout the Roman Empire, which extended northward to England and the Rhine Valley, best demonstrated this hierarchical order through the careful placement of administrative and religious facilities (Figure 18-5). Much later, the Spanish mission towns brought this hierarchical ordering concept to the New World. As with their European forebears, these towns featured a plaza in the center, surrounded by a grid street system and a cathedral on the plaza. Houses typically faced inward around courtyards. Notwithstanding these and other formal planned traditions brought by colonists, most American cities followed a more pragmatic plan as their guiding principle.

One example of a less-structured city form is that provided by the *medieval bastide*. Bastides, built by lords and bishops in the Middle Ages in England, Spain, and France, were placed in previously unsettled areas that had strong potential for agricultural production and trade. The bastide was established as an economic place, and even though the typical city had a formal street pattern, there was no hierarchical land assignment as in the Roman tradition.

The laissez-faire *land speculation* format that created early U.S. cities may have roots in the medieval bastide. Regardless of origin, the speculator city form represents the most influential and typically American urban settlement type. According to the land speculation principle, economics, not an elaborate social order or formal planning, guides the development process. The distinctive American city therefore has been the businessman's city, whereby market forces allocate land uses.

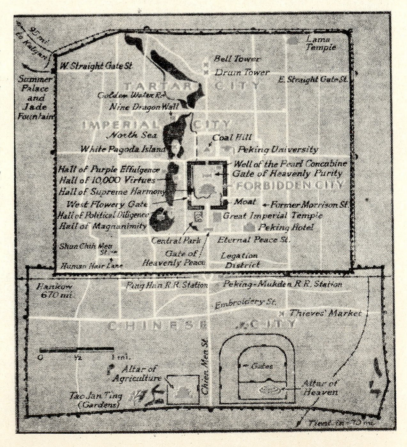

Figure 18-4 Beijing in the 1950s. *Beijing has experienced several periods of rebuilding in its 2500 year history at the hands of various emperors and invaders. It still retains the geometric flair of Emperor Yung Lo who followed the advice of his astrologer to make the city in the likeness of a monster. The resulting geometric form resembled a puzzle box with a series of walled cities inside one another. The Tartar City, 4 miles in diameter, contained the Imperial City which in turn enclosed the Forbidden City. (New York Public Library Picture Collection)*

Figure 18-5 Ancient and Modern Corfu. *Located on an island in the Ionian Sea off the Coast of Greece, this area was settled by the Corinthians in 700 BC. Later a Roman and Venetian possession, the city numbers about 50,000 persons today. (New York Public Library Picture Collection)*

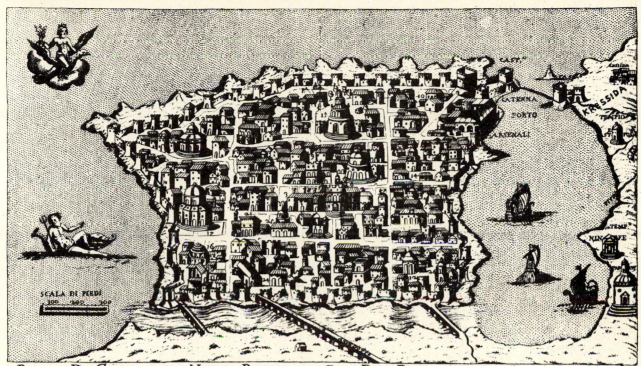

PIANTA DI CHERSSOPOLI HORA PALEOPOLI, CON DOI PORTI ANTICA CITTA DI CORFV

1 Pallaggio Reggio	3 Zecce	6 Stat. di German	9 Temp. di Gioue	12 Di Apollo	15 Di Bacco	18 Di Eolo	21 Di Cerere	24 Di Alcinoo	27 Porta Itaclia
2 Tribuna di Giud.	4 Fonte Mpal	7 Stat. di Seuero	10 Di Nettuno	13 Di Mercurio	16 Di Corcira	19 Di Agreste	22 Di Fortuna	25 Di Belerofronte	28 Porta Iasson
...dici Basilica	5 Stat. d'Alles.	8 Statua M.Aurelio	11 Di Giuna	14 Di Marte	17 Di Hercole	20 Di Cibele	23 Di Vittoria	26 Di Perseo	29 Porta Fimi.

CORFOU ANTIQUE, DU TEMPS DES CORINTHIENS

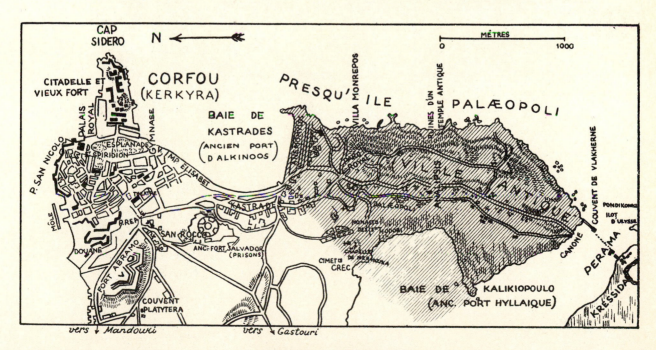

CORFOU MODERNE

The earliest American cities, located in a linear band on the east coast, comprised a series of unplanned port cities. These cities served as points of entry for imported goods from Europe and the Caribbean, as embarkation points for agricultural exports, and as centers of commerce and trade among the colonies. Later, they aided the colonial drive to self-sufficiency through the expansion of manufacturing activity while growing rapidly in an irregular organic manner, as evidenced by the street pattern of Boston or Providence, during this colonial period (Figure 18-6).

CHANGING PHYSICAL STRUCTURES

Whether discussing the ancient Chinese city, the historic European city, or early American cities, one observes a common dimension—compact size. Until the advent of the horse-drawn streetcar and later the trolley in the middle and late 1800s, settlements for thousands of years were primarily walking cities and very compact physically by contemporary American standards. Home and work were closely knit together, often under the same roof, and rarely farther than a block or two apart. Most families were self-sufficient

Figure 18-6 Layout of Early Boston.
The original site for Boston was a small land peninsula jutting into marshy coastal waters. Landfills have since greatly enlarged the city in all directions. Note the location of Boston Common, which is now the oldest public park in the U.S. Beacon Hill, on which elegant homes, were built, overlooked the harbor. (New York Public Library)

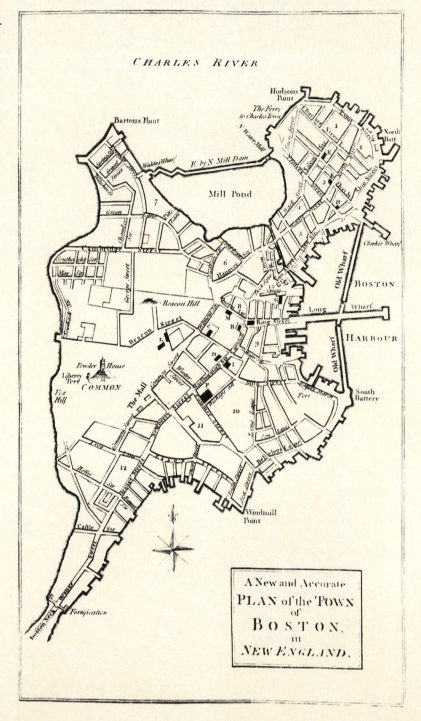

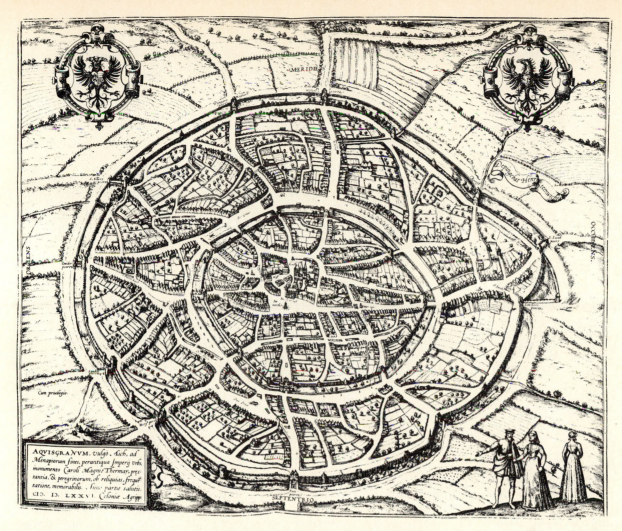

Figure 18-7 Aix-la-Chapelle in the Sixteenth Century. *Now the West German city of Aachen, located near the Belgian border, this city dates to Roman times. It was the second capital of Charlemagne's empire after 768. A town hall was built in 1353 on the ruins of Charlemagne's Palace. (New York Public Library Picture Collection)*

in that era, and domestic handicraft work provided a major source of employment. Conceptually, it is often useful to think of these cities as circular in form (Figure 18-7).

Many cities were walled cities in ancient and medieval times, and population densities were highest at or just beyond the walls, where the underprivileged lower classes resided. Farmers often lived in this area, concentrated on the edges of towns and villages, rather than dispersed in nuclear units on the rural landscape as is common in the western world today. Living on the edge of the city provided much easier daily access to their farmland.

MECHANIZED TRANSPORTATION

By the late nineteenth century the advent of mechanized transportation dramatically expanded urban accessibility, opening up new corridors for develop-

ment. As with other transportation innovations to follow, railroads, streetcars, and trolleys provided greater opportunities for residential and business development. These American city transportation modes as a group led to a city with a more pronounced radial form (Figure 18-8) due to improved accessibility along emerging transportation corridors. Nearly all large American cities boasted extensive trolley line networks that truly provided cheap mass transportation at the turn of the present century, stimulating considerable radial growth.

The streetcar era experienced a premature decline with the advent of the mass marketing of automobiles in the 1920s and the introduction of the motor bus. The automobile literally created a city on wheels, opening up vast areas between the tracks, so to speak, for settlement (Figure 18-8). Development expanded outward from the city center at an unprecedented rate after World War II, due to the flexibility of movement provided by the automobile. Former sat-

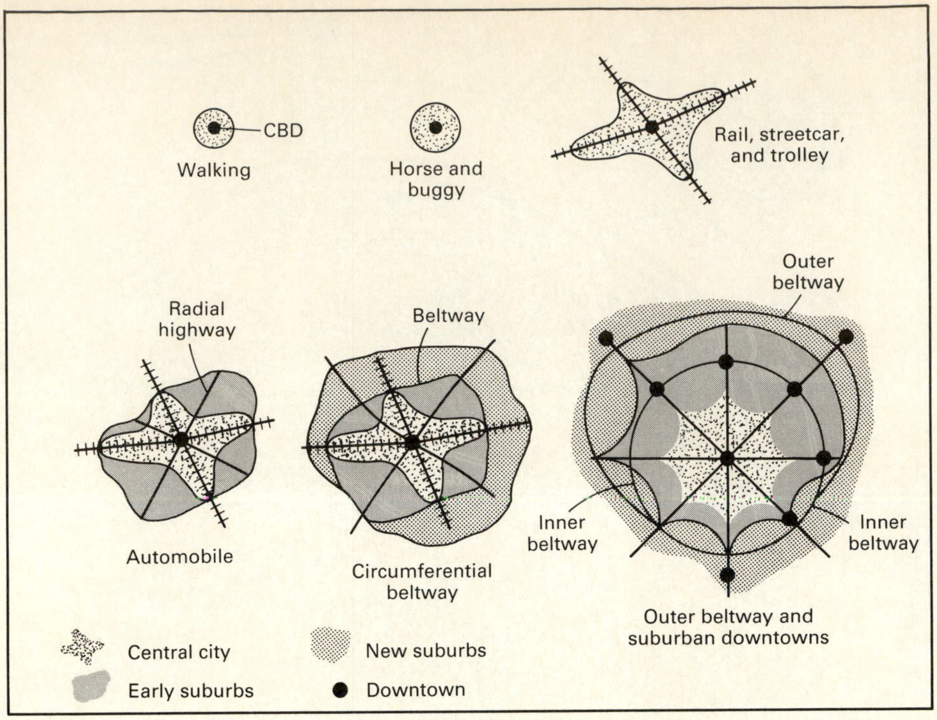

Figure 18-8 Transportation and the Evolution of Urban Form. *The city evolved from a compact entity into a more linear, elongated form with the advent of mechanized transportation in the nineteenth century as demonstrated here. The improved accessibility provided by radial transit lines associated with the rail, streetcar, and trolley era provided the means for this change. Later, the automobile era further expanded the urban frontier. The interstate highway program in the 1960s and 1970s led to the development of both radial and circumferential freeways which in turn led to the development of suburban downtowns, outer beltways, and the full blown polycentric metropolis in the 1980s.*

ellite communities became more directly integrated into the urban fabric as urban settlements became metropolitan in scale in dozens of urban regions.

By way of contrast, the European city continued to depend more heavily on public transportation in the post–World War II era. The European central city also retained its compact walking scale. Periodic large-scale public works renovations, including the development of broad boulevards, have allowed automobile incursions, but the traditional imprint remains largely intact in the European central city, as in contemporary Rome or Amsterdam (Figure 18-9).

By the 1960s the addition of extensive limited-access freeways in American cities linked downtown areas with the metropolitan fringe as well as providing better circumferential access, leading to an even more intense wave of urban development and activity decentralization (Figure 18-8). Most recently, the beltways and radial freeways have become major commercial development generators (Figure 18-10). Suburbs in European and Japanese cities have also witnessed rapid development but with a greater share of movement dependent on public transportation via commuter rail systems. In many older European cities this public transportation accessibility advantage has spawned a high-density, high-rise suburban band of housing surrounding the central city, together with tracts of single-family units and garden apartments. Commercial activity decentralization has followed as in the United States, making the suburbs in Europe and the United States more similar than their central-city counterparts.

EMERGENCE OF POLYCENTRIC CITY

Not surprisingly, the rapid waves of urban development since World War II have turned the American city inside-out and are having similar, if lesser, impacts in areas such as Europe and Japan. Replacing a former urban region that focused on a single downtown central business district, we now have a *polycentric city* with many downtowns. Although the old adage that "all roads lead downtown" may still be true, the "downtowns" are many in number today and no longer located exclusively in the traditional center of the city.

Also, the urban region continues to evolve. Whereas we mentioned earlier that all commercial activity was once located near the city center, this is no longer the case. The breakaway of economic activities paralleled, if not followed the rapid expansion of the suburban residential population after World War II. Indeed, new job formations and construction activity in the suburbs now far exceeds that of the central city in most U.S. metropolitan areas. Today, it is no longer simply a case of decentralization of activity away from the center city, because much suburban growth now occurs from expansion of activities already located in the suburbs and from new businesses such as high-technology activity, which is almost exclusively suburban in nature. In Europe, the central business district was never as clearly defined, as retailing existed at the street level in residential blocks, but today it too has decentralized to suburban shopping malls, emphasizing automobile accessibility.

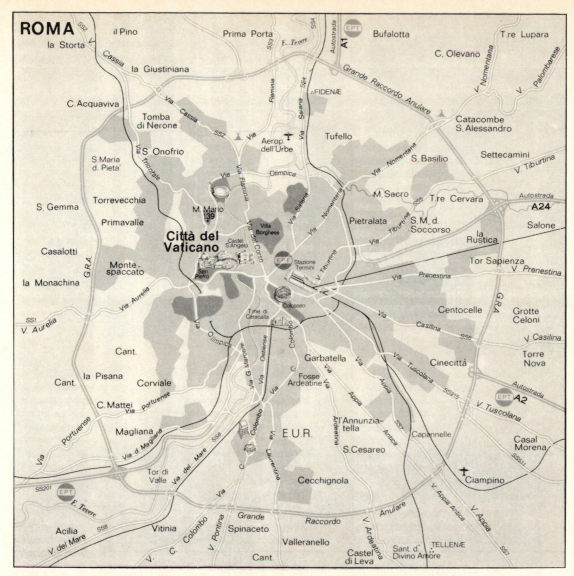

Figure 18-9 Contemporary Rome. *While contemporary Rome is a sprawling metropolis with a suburban beltway, land uses in the historic core of the city exhibit a strong imprint from the past, and it remains a lively, if congested district. Note location of the Vatican (left center) and the Coliseum (right center) in relation to the city form shown in Figure 18-3. (Roma, Ente, Provinciale per il Tourismo, 1986, Italian Tourism Bureau)*

STAGES OF SUBURBAN DEVELOPMENT

Suburban growth, although nearly as old a phenomenon as the city itself, has occurred for centuries as cities grew, but it took on new meaning and intensity following World War II, first in America, then in Europe and other parts of the world. The pent-up demand for new housing at that time, coupled with a desire to leave the problems and costs of city living behind, came at a time that new transportation facilities and other infrastructure improvements converged to make living outside the political limits of the city, but within commuting distance, a feasible alternative for the middle class.

Stage I: Bedroom Community

The first stage of this process, called *bedroom community* formation, typically involved a mass exodus of middle-class families from the city to unincorporated suburban areas in pursuit of a single-family detached home on a relatively large lot beginning in the 1950s (Figure 18-11). Residents typically continued to find employment in the city. Commuting distances to work increased together with the expansion of radial freeways and arterials to serve work-trip needs as this bedroom community phenomenon intensified. Sometimes this form has been called a *mercantilistic suburb*, referring to the fact that residents imported their live-

Figure 18-10 Suburban Freeway Office Corridor. *Radial and circumferential highways have become the preferred locations for massive agglomerations of high-rise offices and associated commercial activity. The visibility and prestige of these locations nicely complements their accessibility advantage. These office buildings face the East-West Tollway in the Chicago suburb of Oak Brook. (T.A.H.)*

Figure 18-11 Tract Housing in Levittown. *This mass-produced complex of 17,000 modest-priced houses was built on 12,000 acres of Long Island potato fields, 1947–1951, by William Levitt. Noted for its complete dependence on automobiles this pace-setting complex included schools, playgrounds, and shopping centers but no office or industrial activity. (Courtesy Nassau County Museum Reference Library)*

lihood (income) from the city but did not contribute to its tax base. Low-order services in the form of grocery stores, gas stations, and fast-food restaurants quickly sprung up to serve the mushrooming suburban market, foreshadowing many changes yet to come.

Stage II: Independence

Fast on the heels of residential growth came jobs and higher-order retail activity to suburban areas. In the 1960s industrial and office parks sprung up in suburban settings in response to the increased accessibility offered by the expanding freeway system. Industrial parks increasingly catered to light industry and the distribution/warehouse function, both of which actively sought imageable sites in landscaped parks (Figure 18-12).

Expansion of industrial parks came at the same time as manufacturers began seeking alternative locations for businesses located in deteriorating multistory, congested, railroad corridor locations in the central city (Figure 18-13). Associated tenement hous-

ing, vandalism, crime, and poor automobile and truck accessibility commonly plagued these older industrial sites. A change in preference from vertical to horizontal manufacturing processes occurred as manufacturers introduced new production techniques.

The shift away from railway dependence to trucking as a mode of transportation also required more docking and parking space than was typically available in older central-city locations. The so-called *truck–auto–freeway* trilogy concept captures the importance of the interstate highway network in promoting suburban industrial expansion, the attraction of improved employee access, and the addition of highly visible and imageable settings for firms striving to enhance public awareness. These changes occurred at a time when a vast expansion occurred in light industry itself, together with a cessation of growth of the belching smokestack industries often associated with traditional manufacturing. Again, trucks served growing light-industry firms better, due to the greater flexibility provided and lesser dependence on bulky raw materials. Associated warehouse and distribution facili-

Figure 18-12 Traditional Rail Served Industrial Park. *This classic perspective of an industrial park in Denver dates to the 1960s and shows the tie of the traditional industrial park to the railroad. The single-story high-bay buildings are typical industrial park facilities. (Courtesy Upland Industries)*

Figure 18-13 Abandoned Central City Industrial District in Philadelphia. *This abandoned manufacturing plant in the inner city of Philadelphia typifies the plight of many older smokestack industrial districts in American cities. (T.A.H.)*

Figure 18-14 Overnite Transportation Company, Memphis Terminal. *Headquartered in Richmond, Va., Overnite's fleet of 4,000 tractors and 9,500 trailers operate from 117 terminals in the U.S. Paper products, textiles, tobacco products, and food items comprise the primary traffic base. A merger of the firm with the Union Pacific Corporation was completed in 1987.*

ties geared to trucking services also gravitated to these industrial park settings (Figure 18-14).

Suburban office park facilities initially catered to firms having an emphasis on intensive clerical operations, involving routine paper-processing tasks such as the back offices of banks and insurance companies, as well as regional sales offices, which preferred the intercity highway accessibility provided by outlying locations. These types of operations experienced savings from lower rental rates in suburban offices and a high level of worker satisfaction with the lower-density work environment.

Back-office operations mainly depend on internalized transactions, and telephone, postal, and electronic data transmission techniques readily provide any needed external contact. Sites in the suburbs therefore provided attractive work locations for employees without sacrificing efficiency for the employer (Figure 18-15). Often, suburban expansion involved splitting up office operations, with the clerical functions breaking away from middle- and upper-level management and research and development functions, which clung to downtown locations.

Complementing the office and industrial function, often in adjacent physical settings, developers built *regional shopping centers,* with department stores and other specialized shopping goods outlets offering an alternative to downtown shopping for the suburbanite (Figure 18-16). The role and impact of the regional shopping center are discussed in a later section.

Stage III: Catalytic Growth

The development of suburban landscapes continued to evolve in the 1970s and 1980s with the expansion of high-income housing and the growing tendency for more specialized office functions to migrate to emerging suburban business centers with expressway exposure. In some cases these centers took the form of corridors with linear belts of high-rise offices and hotels lining expressways. In other cases clusters emerged as rings of offices encircled by regional malls (Figure 18-17). A third type of complex, the large-scale mixed-use center, provides retail, office, and hotel facilities in an integrated town center complex.

Figure 18-15 Executive Park in Atlanta. *This pioneering office park dating to 1964 lies adjacent to the I-85 radial highway corridor in suburban northeast Atlanta. Midrise office buildings in a heavily landscaped campus setting house many regional and national headquarters activities. A Marriott Courtyard hotel is under construction in the park at left center in this 1987 photograph (Courtesy EQUITEC Properties)*

Figure 18-16 Southdale Center in Minneapolis. *When it opened in 1956, this regional mall became the first totally enclosed shopping center in the U.S. It was also one of the first centers to include two competing department stores as anchors. In this instance Victor Gruen built the shopping center for Dayton's, the largest retailer in the twin cities, but also included space for Donaldson's, the major competitor. This design quickly became the norm for all shopping centers. This aerial view of the mall dates to 1958. (Courtesy of Minnesota Historical Society)*

Figure 18-17 Lenox Square Regional Mall Center, Atlanta. *As the suburban commercial landscape matures during its catalytic phase of growth more specialized office activity typically creates a high-rise ring around the regional mall. In this case, office and hotel activity emerged as a major growth factor around Lenox Square in Atlanta in the 1980s, over two decades after the mall was first built. Located in the affluent Buckhead neighborhood this area has now become one of several downtowns in Atlanta. A station on the north line of the rail transit system occurs at the right center of photograph, Lenox Square lies in the center, and Peachtree Road cuts across the left side from top to bottom. (Courtesy Dillon-Reynolds Aerial Photography, Inc.).*

Whatever form these centers assume, the occupant/tenant profile typically includes middle- and upper-management personnel overseeing regional and national headquarters firms. As the business center skyline began to offer more recognition and prestige for these centers, they also began competing more directly with the downtown for the most specialized office functions, such as mortgage banking, corporate legal offices, and accounting services, once thought to be immovable bastions of downtown enterprise. Luxury hotels also joined the litany of businesses seeking a growing share of the suburban market, together with additional specialized retail outlets, complementing the office function.

Stage IV: High Rise–High Tech

As the process of business center differentiation continued, suburban areas became increasingly recognized in the 1980s for their prestigious buildings designed by nationally reknown architects. In addition to the increasing use of postmodern architecture, more color, and decked parking, leasors demanded larger blocks of space and it increasingly became obvious that suburban centers in many cities would equal or surpass downtown areas in office activity as they had earlier begun dominating retail sales (Figure 18-18). Houston is perhaps the quintessential example of this type of development, with suburban city Post Oak downtown boasting the largest suburban office building in America, the 65-story Transco Tower (Figure 18-19).

The high-technology stage is also known for growing research and development activity related to the postindustrial information age, involving biotechnology, electronics, and telecommunications among other activities, as discussed in Chapter 16. Typically, this R&D activity occurs at more remote suburban locations than the setting of a typical suburban downtown (Figure 18-20). R&D activities gravitate to low-profile buildings in heavily landscaped and buffered settings. Often the building itself, called *high-tech flex*, incorporates many functions—reception, office, production, engineering, warehousing, and distribution functions under one roof—and must be of a design that can be readily modified as the needs of the firm change. Single-tenant users predominate in these one- and two-story facilities, often having a mezzanine level for offices, an elaborate entry lobby, and a so-called *clean room* for specialized, sensitive technical work. Extensive advanced telecommunications and computer cabling services are frequently required with access to rooftop satellite telecommunication transmission facilities often necessary. These firms are price and competition conscious and prefer individualized work environments for their highly motivated and specialized work force, which zealously covets an environment free of distractions from within (i.e., management) or externally (i.e., competitors). A strong sense of corporate identity places great stock in signage,

Figure 18-18 Buckhead Plaza mixed-use center, Atlanta. *This post-modern office building with a copper roof and green tinted windows is the first phase of a major mixed-use development. The building design incorporates many set-backs which multiply opportunities for desirable corner offices. Various design features distinguish post-modern buildings including arches, pediments, and roofs that resemble "hats." (T.A.H.)*

proper setbacks, and a classy exterior skin for these facilities, described in the aggregate as a *cubistic campus* in Chapter 16.

LOCATION OF SUBURBAN BUSINESS CENTERS

During the 1960s and 1970s a stereotype of suburban business landscapes emerged featuring *sprawl* and *visual blight* as prominent characteristics. This image came from the rapid expansion of strips of low-order automobile-related services (gas stations, repair shops, dealerships), fast-food franchises, strip shopping centers, and various low-order consumer services strung along urban arterial highways and freeway frontage roads. These strips, complete with endless signs and parking lots, continue to flourish and multiply and are discussed in a later section.

Figure 18-19 Transco Tower in City Post Oak, Houston. *This 65-story office tower is the tallest office building in suburban America. Connected to the Galleria Shopping Mall, this structure provides a visual anchor for the City Post Oak suburban downtown. Note the Neiman-Marcus Department store in foreground. (T.A.H.)*

Figure 18-20 Boeing 737-300 Jetliner Flight Management System. *The Boeing Company digital systems laboratory tests components prior to installation in the aircraft. The unit mounted on shelf at far left is the flight management computer. Left to right on the table are two processors for the autopilot, an integrated mode control panel, autothrottle computer, and, on the far side, the control display unit for the flight management computer. (Boeing Photo)*

Whereas strip retailing is rather ubiquitous on the suburban landscape, located in all parts of the metropolitan area, higher-order business centers, reflected by stages III and IV in the evolution of suburban landscapes, gravitate predominantly to higher-income sides of the metropolitan area. American cities are notorious for their sectoral socioeconomic spatial structure. That is, one can usually differentiate distinctive subareas of socioeconomic status based on occupation, education, and income in various directions away from the city center. Several models of city growth that provide support for this contention are discussed in greater detail in Chapter 19.

The higher-income side of Houston, for example, lies to the west; in the case of Denver, the south/southwest encompasses the high-income sector; the northside offers a similar desirable location in Atlanta. In these higher-income sectors one would expect to find a disproportionate concentration of the major suburban business centers. The explanation for this phenomenon has its roots in the pattern of higher-income residential neighborhoods. Business managers who make the location decision for their firms prefer work locations near their residential base that have the same or similar amenities. Once this process is set in motion, considerable "follow the leader" or snowballing of development also occurs. As one area becomes fully developed, a leapfrogging of development farther from the city center typically occurs, as has recently been demonstrated in the three cities just mentioned—Houston, Atlanta, and Denver (Figure 18-21).

Typically, the finest shopping, hotel, and restaurant facilities in an urban area are found in these suburban environments along with the concentrations of work activity. The highest order suburban business center forms include several types of corridors and clusters as shown in Table 18-1. The Suburban Freeway Corridor is typically represented by a string of high-rise office buildings fronting a freeway. The retail strip corridor offers a plethora of goods and services from fast food to furniture and motor vehicle sales. The high technology corridor provides a nexus for research and development activity in 1- and 2-story dispersed buildings, described elsewhere as a cubistic campus.

The clusters include several types of mixed use centers that have regional malls, offices, hotels, restaurants, and cultural centers as focal points. Some are evolving from the ground-up as centrally planned new town centers. A diagram that includes one or more of each of these types of centers appears

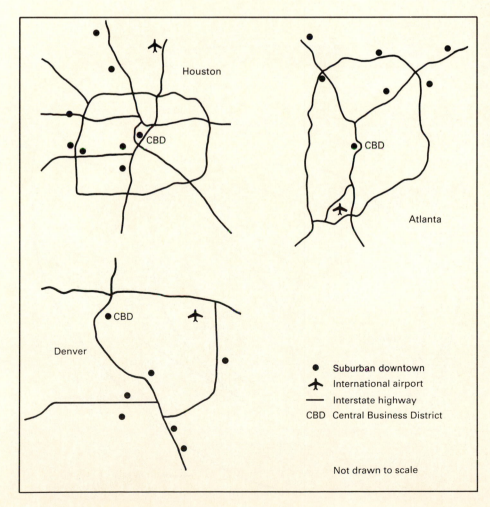

Figure 18-21 Contrasting Directional Growth Patterns in Selected Metropolitan Areas. *Suburban downtowns typically develop more intensively on the higher income sides of metropolitan areas—to the west in the case of Houston, to the southeast and southwest in the case of Denver, and the north in Atlanta.*

● Suburban downtown
✈ International airport
— Interstate highway
CBD Central Business District

Not drawn to scale

Table 18-1

High Order Suburban Business Center Forms

Form	Examples
Corridors	
Suburban Freeway Corridor	West Houston I-10 Corridor - Houston
	I-494 Corridor - Minneapolis
Retail Strip Corridor	Roswell Road - Atlanta
	Memorial Drive - Houston
High Technology Corridor	Peachtree Corners - Atlanta
	I-495 Corridor - Boston
Clusters	
Regional Mall Center	Tyson's Corners - Washington, D.C.
	Galleria - Houston
	King of Prussia - Philadelphia
Diversified Office Center	Greenway Plaza - Houston
	Perimeter Center-Ga. 400 - Atlanta
New Town Center	North Park Town Center - Atlanta
	Galleria - Dallas
	South Coast Metro - Costa Mesa
Old Town Center	Bethesda - Washington, D.C.
	Decatur - Atlanta
	Stanford - New York City
Specialty Commercial Center	O'Hare Int'l Airport - Chicago
	NASA/Clear Lake - Houston

in Figure 18-22. Note that development predominantly occurs on the north side of this hypothetical metropolitan area, due to the concentration of the higher income neighborhoods on that side of the urban region. Development also occurs in other parts of metropolitan areas, but typically is restricted to retail activity except for special circumstances, as in the case of airport locations or other specialized markets. The slower pace of business activity formation in suburban districts not perceived to be as desirable by developers has become an important development issue in many cities. There are other problems associated with these higher-order developments as well, including access to suburban workplaces by lesser-skilled central-city blue-collar service workers.

An important issue associated with the relative stagnation of central cities compared to the burgeoning growth of suburban activity is access to jobs among America's central-city blue-collar workers. Neither personal nor public transportation are typically available to carry workers to those locations while numerous service work opportunities remain unfilled. To overcome this housing and transportation spatial mismatch, fast-food and hotel employers often have to provide shuttle bus services or private-sector-financed day-labor buses to provide the linkage. Wage rates are also bid up in these labor-starved growth areas.

Figure 18-22 Model of High Order Metropolitan Commercial Corridors and Clusters. *Specialized retail, office, and hotel activity can be found in several combinations in a wide variety of settings, including the traditional CBD, regional mall centers, old town centers, diversified office centers, retail strip centers, new town centers, specialty centers, suburban freeway corridors, and high technology corridors. Note the disproportionate clustering on one side of the region and the orientation to the arterial highway and freeway system.*

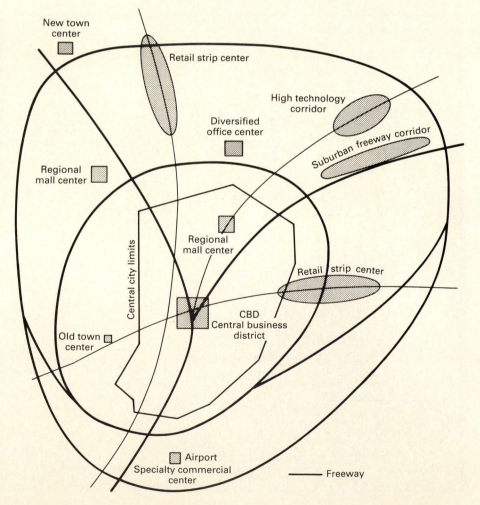

CHANGING ROLE OF CENTRAL BUSINESS DISTRICT

As discussed earlier, the central business district (CBD) once housed virtually all commercial activity in the city. In some special cases in North America today, most notably in Chicago, New York, and San Francisco in the United States and Toronto and Montreal in Canada, the downtown remains not only the premier office and financial center of the region but also the cultural and entertainment nexus for the region. In other, more typical cases the downtown thrives but has lost considerable clout to the burgeoning suburban centers just discussed (Figure 18-23). Sunbelt cities in particular exhibit the latter tendency. In still other instances the downtown has declined considerably as a major business center, as in the case of Phoenix, or one has never existed, as in the case of Orange County in California or Nassau/Suffolk on Long Island, where new business centers have emerged relatively recently without a traditional CBD.

The basis of the decentralization of economic activity away from the downtown central business district can be traced to the growing dependence on the personal automobile as a mode of transportation over the past 50 years. As recently as the immediate post–World War II era, the downtown still commanded a monopoly over retail activity, hotel space, offices, movie theaters, and a host of related services. But once the process of breakaways began it has successively affected nearly every type of function.

First came residential dispersion, typically followed by the exodus of lower-order population-serving neighborhood activity, including grocery stores, taverns, movie theaters, and then church and school closures. The suburban regional shopping mall emerged as critical threshold markets and changing merchandizing techniques allowed in the 1960s. In time this trend led to the closure of many higher-order downtown retail outlets, including the remaining department stores and many specialty shops (Figure 18-24).

Despite these and other losses, the resilience and adaptability of the downtown should not be underestimated. As retailing declined, many downtowns experienced a net expansion of office space as a new

Figure 18-23 Houston Central Business District. *The impressive Houston skyline results from the mix of several tall and distinguished buildings including, from left, the Texas Commerce Tower with 75 stories, the gable-like profile of the Pennzoil building, the multiple spires of the 56-story Republic Bank Center, and the rounded face of the 71-story Allied Bank Plaza. The Houston central business district contains 37 million square feet of office space. (Courtesy Houston Chamber of Commerce)*

Figure 18-24 Blighted Retail Block in downtown Detroit. *Retail sales in downtown areas of American cities today rarely account for more than 2 percent of the metropolitan total. Some downtowns have no department stores and no weekend or evening shopping hours. The decline of retailing is shown dramatically here with the boarded up storefronts, including bars and adult book stores, in downtown Detroit. (T.A.H.)*

wave of office construction enhanced the office market. Part of this expansion emanated from the growth of government offices, but the greatest share occurred in the private sector. But as the 1970s unfolded, a slowdown in downtown office expansion occurred in all but the strongest CBD markets, such as New York, Chicago, and San Francisco. Vacancies in existing space also increased as more firms chose suburban alternatives.

Some cities adjusted to the changes wrought by declining office activity by creating a new downtown support base, such as promoting hotel and convention activity or tourism. This trend led to increased competition among cities in building new conference facilities. Such an emphasis has occurred in Dallas, Atlanta, New York, Chicago, and San Francisco, among others (Figure 18-25). Hotels and entertainment facilities, however, do not employ as many persons per square foot as office facilities or require as sophisticated a work force as the activities they typically replace.

Employment Shares

Rarely does the CBD today in any particular city account for more than 20 percent of metropolitan employment. A survey of several cities suggests that an average of 15 percent of metropolitan employment occurred in downtown areas in larger areas (Table 18-2). All indications show that this share continues to erode in the 1980s except in a few cases where downtown locations remain unusually attractive, as in New York. In the latter instance foreign investment and the strong attachment to the financial services industry in Manhattan partly explain the continued vigor of the area as an employment center.

Revitalization Issues

The downtown remains in nearly all instances the leading single employment center in metropolitan areas, despite decentralization. Downtown areas having the most favorable performances as retail and fi-

Figure 18-25 J.K. Javits Convention Complex in New York City. *Designed by I.M. Pei and named after the late Senator from New York, this mammoth new crystal palace complex of nearly 1 million square feet reflects the Manhattan skyline from its Hudson River location. The 15-story lobby is large enough to hold the Statue of Liberty. (New York Convention & Visitors Bureau)*

nancial centers and a solid record of employment stability also have a middle-class residential population base in or relatively close to the downtown area and their metropolitan areas typically have a strong post-industrial service-related employment base. New York and Chicago, of course, are major headquarter cities and fall in a class by themselves. Nevertheless, they also have a strong middle-class residential base downtown and a strong service economy. Industrial cities such as Akron, Cleveland, Detroit, and Milwaukee generally have weaker downtowns, as do many sun-belt cities, which have grown rapidly in the post–World War II era, when suburban areas had a competitive advantage for service-related employment. In this category are included San Jose, San Diego, Phoenix, and Orlando, among others.

One factor inhibiting downtown revitalization, and a problem facing the development of a successful hotel/convention support base, is that the downtown in many cases no longer functions on a 24 hour-a-day basis. Years of residential decentralization have created a 9-to-5 downtown, leaving city streets deserted during evening and weekend periods. Whereas freeways were originally designed as main streets to bring more people and activity to the downtown area, they had the opposite effect—that is, encouraging decentralization of activity and residential areas.

Some observers have likened the downtown situation to a neutron-bomb syndrome wherein there are many physical structures but no people on the streets. Indeed, many renewal projects in the past two decades have intensified rather than ameliorated the problem. Downtown demolitions have typically exceeded new construction in many cities, such as Atlanta, giving the downtown fabric an increasingly pock-marked aspect in terms of continuous develop-

ment at the street level. Demolitions have left vacant parcels and idle parking lots interspersed among islands of development (Figure 18-26).

New megastructures have also contributed to the problems facing downtown areas, even though highly touted as their salvation a decade ago. It seems that these inward-looking facilities have walled themselves off from the downtown, effectively turning their back on it. Belatedly, many cities have placed more emphasis on renovating older buildings rather than destroying them and on ensuring that newer developments face toward the street, the downtown lifeline, enriching the pedestrian experience.

Table 18-2

Central Business District Employment Shares in Selected Metropolitan Areas

Metropolitan Area	Percent CBD Share
Atlanta	10 (1980)
Denver	14 (1972)
Washington, D.C.	24 (1973)
Indianapolis	16 (1973)
Louisville	12 (1975)
Baltimore	10 (1970)
Rochester	11 (1973)
Dayton	9 (1975)
Columbus	18 (1974)
Pittsburgh	14 (1975)
Salt Lake City	8 (1973)
Seattle	17 (1973)
Boston	21 (1972)
Cleveland	15 (1970)
Milwaukee	9 (1972)

Figure 18-26 Nodal Development in the Atlanta Central Business District. *Several loosely connected concentrations of activity occur in downtown Atlanta as is typical in cities that have experienced years of decentralization and selective commercial redevelopment. The Peachtree Street spine provides an alignment for several of these nodes, including Peachtree Center at the bottom. A convention/sports complex lies to the left, and a government center to the south. Note the vast quantities of parking facilities between and among the centers (Courtesy Dillon-Reynolds Aerial Photography, Inc.)*

The recently completed Horton Plaza complex in San Diego is an example of a massive new downtown project that seeks to overcome mistakes made earlier in other downtown renewal projects (Figure 18-27). That mixed-use project, with an emphasis on retailing, spans an area of 11 acres. Playing off themes from the mission heritage of the city, including bell towers and pastel colors, this complex opens outward to celebrate the street. Several older structures, including theaters, have been integrated into the facility, which includes offices, a hotel, several department stores, a cultural center, and restaurants.

Many cities have attempted to revive downtown retailing by turning main streets into pedestrian malls—with mixed results. In some cases this has involved widening sidewalks and narrowing streets, restricting automobile use. In other cases the mall becomes a transit way, as with Nicolet Mall in Minneapolis or the 16th Street Mall in Denver (Figure 18-28). Banning all vehicular traffic presents yet another approach.

The pedestrian mall concept has been most popular in medium-sized to smaller cities in the size range 50,000 to 250,000, where downtown retailing has remained relatively stronger than in larger metropolitan areas. These cities also have the most to lose because their downtowns are virtually totally dependent on retailing, as few other functions occur in their CBDs. Some enclosed downtown malls, virtually identical to their suburban counterparts, occur in several larger cities, such as the Gallery in Philadelphia, Water Tower Place in Chicago, and Tandy Center in Fort Worth. The largest downtown mall in the world, Eaton Centre in Toronto, has three levels of shopping, anchored on either end by department stores and office towers, and is surrounded by high-rise parking decks (Figure 18-29). The center, which parallels Yonge Street, has access to the subway system which runs under the street, via an elaborate network of underground passageways lined with retail outlets as well as access to office buildings and department stores. A light-rail system running on Dundas and Queen

Figure 18-27 Horton Plaza in San Diego. *This innovative downtown mixed use center has successfully revived a formerly decayed 7-block downtown area. City and suburban dwellers alike support its 4 department stores, 135 specialty shops and restaurants, two performing arts centers, and farmers market. The ecclectic post-modern architectural style incorporates locally prominent pastel colors and a mix of old and new construction featuring domes, arches, and bell towers inspired by Italian, Spanish, and Indian themes. (Courtesy Horton Plaza)*

Figure 18-28 Transit Mall in Denver. *Located on the major shopping corridor in downtown Denver, this 16th Street bus transit mall connects central bus depots at either end of the mall which serve as major service hubs for the transit system. The downtown Denver office/financial district parallels this corridor on 17th Street. (Courtesy Regional Transit District, Denver)*

Figure 18-29 Eaton Centre in Toronto. *When opened in the late 1970s Eaton Centre had the distinction of being the largest downtown retail mall in the world. Linking two department stores on either end is a three-story glass-vaulted mall. The upper floors behind the solid walls offer enclosed parking. Note the Canada geese suspended in the public space. (T.A.H.)*

Corporate headquarters traditionally choose locations in the dominant national metropolitan center: New York in the case of the United States, London in the British case, and Paris in France. Center-city locations offered the prime locational setting for these office operations, given the accessibility advantage of downtown, its vast office space inventory, its labor pool, its network of support services, and the opportunities for maximum face-to-face contacts with other professionals. Headquarters offices also clustered disproportionately in the downtown area of a single metropolitan area, Manhattan in the U.S. case, due to strong ties to financial markets and to other corporate headquarters (Figure 18-31). Large metropolitan centers also provided unparalleled cultural opportunities and enhanced opportunities for international contacts.

As with all economic activity, decentralization began affecting corporate headquarters locations in the 1960s and 1970s, especially in the United States. This occurred in two ways. First, regional centers that had gained prominence began competing more aggressively for headquarter functions, and second, suburbs in both the fringes of traditional headquarter cities such as New York and those in the regional centers became more competitive. As an example of the dramatic impact of this decentralization, consider that New York City controlled $106 billion in assets in the finance and utilities industries in 1957, including banking and life insurance, or "1.2 times the assets controlled by the next 10 metropolitan centers."[1] By 1970 the value of the assets controlled by New York had increased to $700 billion, but the next 10 centers had now surpassed New York in assets.

This relative loss of control phenomena by New York City can also be demonstrated by examining the number of Fortune 500 corporations located in the city at various points in time. In 1968, for example, there were 138 Fortune 500 firms headquartered in New York City, but only 73 in 1980 and 65 in 1984. The decline is somewhat tempered by the fact that New York's suburbs captured about 75 percent of these losses. In 1984, for example, New York suburbs claimed 62 Fortune 500 companies, meaning that about half the headquarters in the greater New York region had city locations and half, suburban settings.

The largest share of suburban headquarters in the New York region occurs in southern Connecticut (31 firms), Stamford with nine and Greenwich with six (1984 figures). The New Jersey suburbs added another 23 headquarter firms, together with eight on Long Island and other nearby New York State suburbs. The concentration of activity in southern Connecticut now makes that suburban area the second-most-important

Streets serves the north and south ends of the Eaton Centre complex, respectively. The high quality of the transit system, the vast array of cultural opportunities available downtown, and the strong residential character of the immediate area all contribute to the continued health of the Toronto CBD.

In some cities retailing has been promoted by renovating historic areas, which become the homes of specialty shops and boutiques, as in the case of Larimer Square in Denver, or festival marketplaces, as in the case of Faneuil Hall in Boston, Harbor Place in Baltimore (Figure 18-30), and the South Street Seaport in New York. The latter three areas represent very successful attempts to revitalize former rundown areas by the Rouse Corporation. In many ways these projects attempt to recreate the Roman agora by emphasizing pedestrian strolling as well as shopping.

[1]R. Keith Semple and Alan G. Phipps. "The Spatial Evaluation of Corporate Headquarters within an Urban System," *Urban Geography,* 3 (1982), p. 273.

Figure 18-30 Harbor Place in Baltimore. *This festival market consisting of several 2-story retail/specialty store pavilions has successfully revitalized the Baltimore waterfront. The Pratt Street pavilion is shown here with the U.S.F. Constellation tied up in the harbor at right. The American Cities Corporation, a subsidiary of the Rouse Corporation, has developed similar projects in many other cities. (T.A.H.)*

Figure 18-31 Mid-town Manhattan. *The mid-town Manhattan skyline came of age in the 1930s with the completion of the Chrysler Building (top right with spire) and the world's most famous office tower the Empire State Building (foreground). The 1,250 foot 102 story Empire State Building is now overshadowed by the World Trade Center towers in lower Manhattan but it still symbolizes the contribution of the twentieth century skyscraper to building technology. (New York Convention & Visitors Bureau)*

Figure 18-32 General Foods Corporate Headquarters. *Located in suburban New York in White Plains, this stark white architectural masterpiece fronts a reflecting lake. Ranked 38th in size on the* Fortune 500 *list of leading manufacturers in 1985, the firm was acquired later in the year by Philip Morris which ranked 12th in size in the country in 1986, following the merger. (T.A.H.)*

headquarters location in the country after Manhattan itself. These outlying headquarters are frequently located in rather isolated wooded settings, but nevertheless in close proximity to elegant high-income residential districts, as in the case of General Foods in Rye, New York, and AT&T in Basking Ridge, New Jersey (Figure 18-32).

In a metropolitan context, the New York area (127 firms) retains the distinction as the overwhelming corporate headquarters leader in the United States, followed by Chicago (49), San Francisco (16), Pittsburgh (14), Los Angeles (12), Boston (11), and Houston and Dallas (10 each). In nearly each instance about half of these headquarters claim suburban locations. In greater San Francisco 8 of the 16 firms are located in the Silicon Valley. In some instances firms headquartered in second order centers have relocated their corporate headquarters from New York (such as the movement of Shell to Houston, and American Airlines to Dallas), but locally based industries are also increasingly a part of the equation, such as Texas Instruments (electronics) in Dallas, and Tenneco (petroleum) in Houston. Several other sunbelt cities have

also made recent strides in attracting and spawning Fortune 500 firms, including Coca-Cola and Georgia-Pacific in Atlanta and Manville in Denver (Figure 18-33).

The decentralization of headquarters activity away from the dominant national center (New York) appears to be a part of a spatial reorganization process leading to a more dispersed pattern of decision making as the country matures.[2] This decentralization process is more advanced in the United States than in Canada, Europe, or Japan, where decision making remains very centralized in the largest urban centers.

Notwithstanding the balance in number of headquarters offices between city and suburb in U.S. metropolitan areas, examination of the overall office market reveals that with the notable exception of New York and a few other areas, the bulk of office activity now occurs in suburban settings in U.S. cities, as well as the overwhelming majority of new construction activity. Regional offices of firms, and professional business services, including data processing, management

[2]Ibid., p. 278.

services, and sales and marketing firms, now largely prefer suburban settings. Higher-order corporate services can also be found in suburban settings, including investment and brokerage services, corporate legal firms, and advertising and public relations organizations. The downtown remains a center for government and quasi-government office work and a major financial center in the largest cities, but in the majority of cases it has lost in relative importance in all types of office activity.

METROPOLITAN RETAIL STRUCTURE

As discussed in Chapter 17, the metropolitan retail landscape in cities exhibits a hierarchy of centers. In addition, linear corridors or ribbons of retail-dominated activity line many arterial highways and bypass routes. Retail centers and corridors can be planned or unplanned, but the latter are mainly unplanned in the sense that they represent an aggregate landscape created by the decisions of many independent business interests. This is not to say that these activities are owned and managed by small family-based companies because most are controlled by large national and international corporations and operated as franchise operations.

We focus our discussion here on two types of business activities: the centers and the corridors. As also mentioned earlier, the metropolitan commercial landscape today is a product of the automobile era and has evolved to its present form relatively recently—in the past 30 years. Retailing in the suburbs today is only a part of the complex commercial structure as it is in the downtown area. Smaller cities and metropolitan areas do not offer as many levels in the retail center hierarchy nor as many establishments in each functional category, in keeping with central place theory principles, such as threshold size and the order of goods, discussed in Chapter 17. Each retail center has a trade area proportional in size to its degree of specialization and square footage, with lower-order centers such as neighborhood and community centers nested within the sphere of influence of larger centers. Lower-order centers are much more numerous than are higher-order functions (Figure 18-34). The individual retail grocery store unit at the lowest end of the hierarchy is not shown in Figure 18-33, but it would be the most frequently occurring activity and have the smallest trade area.

Figure 18-33 Manville Corporate Headquarters in suburban Denver. *The Manville corporate headquarters is located in the foothills of the Rocky Mountains in the suburbs of southwest Denver. The striking building design is well integrated into the topography of the area creating an imposing setting when approaching the facility by car. (Courtesy Manville Corporation)*

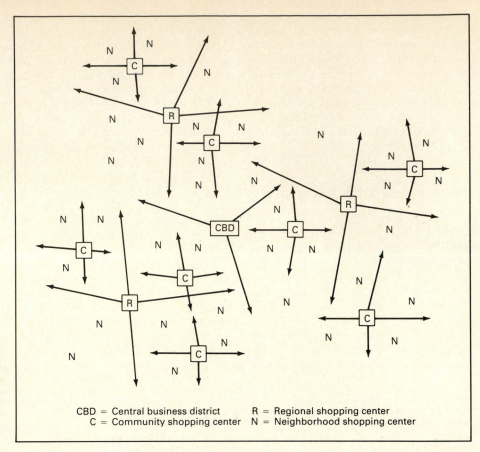

Figure 18-34 Retail Trade areas for a Hierarchy of Centers. *The CBD functions as one of several regional shopping centers in U.S. metropolitan areas. Community centers are more numerous, but command smaller trade areas. The neighborhood center is most ubiquitous in a metropolitan area, serving a small market area.*

CBD = Central business district   R = Regional shopping center
    C = Community shopping center   N = Neighborhood shopping center

SHOPPING CENTER HIERARCHY

Clusters of stores in shopping centers can contain as few as six or as many as 150 or more retail outlets. Superregional centers at the high end of the scale frequently have more than 125 stores and occur almost exclusively in metropolitan areas of 1 million or more persons. Regional shopping centers can thrive in cities as small as 50,000 persons. At one time the central business district was the single most specialized center in the hierarchy as its trade area encompassed the entire urban region. But after World War II, with the decentralization of population and retailing opportunities, suburban department stores and specialty shops opened in regional shopping centers, first in the largest cities in the mid-1950s and by the early 1960s in smaller metropolitan centers.

Today, the central business district functions as one of many regional shopping centers in the city and no longer serves at the top of the retail hierarchy. In essence it serves as the regional shopping center for the central city, frequently an area of lower income, hence the sharply reduced number and variety of retail opportunities in the area compared with the situation 25 or 30 years ago.

Convenience Center

At the lowest level of the retail hierarchy the convenience center frequently occurs as an individual gro-

cery store, drugstore, tavern, or gas station. The grocery store function formerly occurred as a family-run "mom and pop" grocery, but this type of outlet is rapidly being replaced by franchised chain-store operations open 24 hours a day. They are often operated by immigrant families or unskilled workers paid the minimum wage. These minimarts offer a variety of frequently demanded household items in a limited range of choices (Figure 18-35). The staples might include bread, milk, soft drinks, beer, magazines, basic drugstore needs, and gasoline. Originally, consumers on foot frequented these centers, but automobile access is now preferred. The trade area of this center encompasses only a few city blocks, depending on population densities, and many serve as few as 1000 persons.

Neighborhood Center

The neighborhood center offers a wider choice of low-order goods and services and is typically anchored by a full-line grocery store or supermarket that also claims the largest floor area in the center (Figure 18-36). Serving a population of 7000 to 15,000 persons, these centers offer a variety of services for the family—barbershop, beauty shop, drugstore, dry cleaner, branch bank, liquor store, and an ice cream store, among other activities. Changes in retailing approaches by major chain stores have led to the development of huge supermarkets, called superstores and

Figure 18-35 Seven-Eleven mini-market. *As the name of this chain of stores implies, the convenience store mini-market offers extended hours for the shopper. Typically located on an arterial highway such stores have small trade areas and offer basic grocery items. Some stores provide self-service gasoline pumps, snack bars, and video tape rentals (note sign). (T.A.H.)*

Figure 18-36 Neighborhood Shopping Center. *Anchored by a grocery store, the neighborhood shopping center also offers 8-10 other functions typically including a drug store, branch bank, barber/beauty shop, and laundromat/dry cleaners, as is the case here. (T.A.H.)*

warehouse stores in the United States and *hypermarkets* in Europe, that offer a much wider range of consumer services today—all under one roof and management umbrella. Newer supermarkets in the United States and Europe often offer services as diverse as a specialized drugstore counter, branch bank operation, liquor stand, photo shop, health food counter, bakery, and meat deli. These stores can exceed 100,000 square feet of space in comparison to the standard supermarket of 50,000 square feet a few years back. Nevertheless, the function of the neighborhood center remains the same—the provision of low-order consumer goods to the immediate market.

Large corporations dominate the retail food market business in the United States today. The leaders, Kroger and Safeway, operate thousands of stores in dozens of states, as well as manufacturing plants, warehouses, and enormous truck fleets. Several other major food retailers operating in the United States today, once independent, now function as subsidiaries of European-based corporations. A&P, for example, is owned by Tengelman from West Germany and Grand Union by Cavenham, a British firm. European firms have also invested in nonfood retailing in the United States, such as the British-American Tobacco Co. (BAT Industries) acquisition of Gimbel Brothers, owners of Saks Fifth Avenue, in the 1970s. This transnationalization process has also led American retailers to invest overseas, and the trend accelerated in the 1980s.

Community Center

A transition to offering shoppers goods occurs with the step up to the community shopping center. In addition to neighborhood-level goods and services, the community shopping center provides a junior department store, such as a Woolworths or Walgrens, or a discount department store such as K mart or Target. About one-third of all shopping centers fall in this category and they can range up to 300,000 to 500,000 square feet in size. The junior department or discount store differs from a major department store in the quality and price of goods offered, which generally emphasize lower-quality lines, but the trend today is

Figure 18-37 Community Shopping Center. *The community shopping center is most often a planned open-air linear or L-shaped center with a junior department store or discount store anchor, K mart in this case. A grocery store also typically occupies considerable space in the mall. Alternatively, multiple screen movie theatres can also serve as anchors. The Winn Dixie grocery store in this mall (at left) has recently been replaced by a family style cafeteria. (T.A.H.)*

Figure 18-38 Regional Mall Food Court. *The food court has become a prominent function of the regional mall shopping center in recent years as part of a strategy to lengthen the time a shopper spends in the mall and broaden the appeal to the entire family. (T.A.H.)*

to offer more and more quality name-brand merchandise at lower prices by not offering as much personal service to the customer.

Typically, these centers have two or more supermarkets, women's- and men's-wear stores, a cafeteria or restaurant, a jewelry store, bookstore, gift shop, shoe store, appliance store, and a multiscreen theater complex, among others. The community shopping center today is most often a planned open-air linear or L-shaped center with anchor stores interspersed among smaller shops (Figure 18-37). Covered walkways typically join the stores, which front large parking lots, providing plentiful offstreet surface parking. Many older centers built 20 to 30 years ago remain strong community shopping centers, while others have required renovation and expansion. Changing consumer tastes and retail merchandising strategies today require specialized smaller stores rather than the larger units originally constructed. In some cases these centers were transformed into factory outlet centers or off-price centers offering high-quality merchandise at a discount price in a minimal-overhead cost environment offering few customer services.

Regional Center

At the regional shopping center level, low-order convenience goods give way almost exclusively to higher-order shopping goods. Early regional centers in the 1950s may have had only one department store anchor, but the trend soon became established for these centers to offer two major department stores. These centers ranged in size up to 1 million square feet of space. Typically, the trade area includes about 50,000 to 200,000 persons. Travel times to regional centers are longer than for other centers but rarely more than 15 minutes for the typical metropolitan shopper. Res-

idents normally patronize the nearest regional shopping center exclusively rather than splitting their purchases among several centers. In turn, the shopping opportunities available are essentially the same at each regional center, modified somewhat by the buying power of consumers in the trade area and their tastes and preferences. These centers are typically located adjacent to freeway interchanges, which promotes access and visibility. As they provide higher-order goods and services, consumers are willing to travel longer distances to these centers in classic central place theory fashion.

The most frequently occurring functions found in the regional shopping center include clothing stores, shoe stores, jewelry stores, gift stores, and bookstores. It is also popular today to include a food court in the shopping center, comprised of a wide variety of popular fast-food products (Figure 18-38). Many nonretail service and cultural functions are also provided by the regional shopping center, almost exclusively in an enclosed air-conditioned mall environment. Medical and dental offices, travel agencies, libraries, auditoriums, and theaters occupy permanent quarters, while many promotional events, such as auto shows, craft shows, art exhibitions, and flower shows, serve to identify the mall locally as the town center for the community. An unprecedented cultural feature of the Southcoast Plaza regional mall in Orange County, California, for example, is the construction of a major center for the performing arts, which opened in 1985 (Figure 18-39).

In an enlightening book on the topic, *The Malling of America*, Kowinski describes the mall as special space "achieved by enclosure, protection, and control. Those are its secrets, the keys to the kingdom, the whole mall game."[3] Not only does the mall provide an

[3]William Kowinski, *The Malling of America: An Inside Look at the Great Consumer Paradise* (New York: William Morrow, 1985), p. 61.

Figure 18-39 Southcoast Metro in Orange County California. *This area houses the single most complete concentration of urban business, retail, cultural, and residential activity in suburban america today. A portion of an urban sculpture garden and park near the Orange County Performing Arts center shown here in foreground. (T.A.H.)*

efficient selling environment but also an escape from everyday routines. "The mall environment is itself a magic theater—trees grow out of the tiled floor! Plants flourish without sun or rain. . . ."[4] "You can buy anything from diamonds to yogurt in them, go to church or college, register to vote, give blood, bet, score, jog and meditate in them, and in some you can get a motel room, apartment, or condominium—and live there."[5]

Superregional Center

Larger regional centers, in the form of superregional centers, appeared on the landscape in the 1970s. These centers typically have three or more department stores and more than 1 million square feet of retail space—some even approach twice this amount. The largest regional malls in the country fall in this category, such as Woodfield Mall at Schaumburg, Illinois, in the Chicago area near O'Hare airport and

[4]Ibid., p. 62.
[5]William Kowinski, "The Malling of America," *New Times*, 10 (May 1978), p. 33.

the King of Prussia business center in suburban Philadelphia (Figure 18-40).

Superregional centers offer the same functions as regional centers, but with more duplication in each type of specialty store. A shopper may find 10 or more shops offering a particular type of merchandise, such as shoes or leisure wear. Again, the superregional mall serves as the social nexus of the community, being a popular meeting place for the teenager, family shopper, and elderly patron alike. Lavish landscaping complete with trees, plants, fountains, and sitting spaces gives the mall a restful setting amidst the rush of shoppers.

RIBBON CORRIDORS

Retail corridors, stretching along traditional shopping streets, arterial, and bypass highways, serve as locations for a wide variety of functions—automobile-related services, including new car dealerships; fast-food stores and restaurants; and space-intensive stores, such as lumber and furniture outlets. Although no precise figures exist on sales-level comparisons be-

Figure 18-40 The Court and Plaza at King of Prussia. *Shown here at upper left is the original mall at King of Prussia (the Plaza) anchored by three department stores. The recent expansion to lower right (the Court) includes three more department store anchors. Along with dozens of other specialty stores the mall complex now offers 2.47 million square feet of retail space. The Pennsylvania Turnpike is at right and Route 202 to the left.*

tween strip developments and the centers as a group, strip centers account for about one-third of total metropolitan area sales.

One way to distinguish functions along a retail ribbon is to classify recurring groups of activities from the city-center outward. Using this technique, one finds many duplications of retail activities, such as fast-food franchises, gas stations, auto parts, tire, brake, and repair shops. Other strip functions have rather unique and distinctive locational characteristics. Hospitals, nursing homes, funeral parlors, and florist shops often gravitate in close proximity to areas with good accessibility, such as that provided by the intersections of major arterial streets. Printing shops and wholesale distributorships with ties to the city center but also needing suburban access also gravitate to these corridors.

Automotive dealerships, once clustered downtown near former high-income residential areas, largely abandoned the central city in the 1970s, only to locate along major arterial strips in the suburbs, frequently relatively close together to promote comparison shopping (Figure 18-41). Indeed, the concept of the automobile dealership mall housing several

businesses has appeared in some markets. Outlying areas also attract space-intensive users such as furniture stores, garden supply and landscape nurseries, mobile-home sales, do-it-yourself lumberyards, and home supply centers. Recreation and amusement centers, golf courses, and sporting centers can also be found in peripheral areas. Farthest out on these strips on the rural/urban fringe are junkyard dealers, construction companies, and other activities that support the building trades, such as irrigation equipment suppliers, swimming pool companies, and plumbing and electrical system suppliers.

From a planning perspective retail strips pose many problems. Owing to dynamic urbanization processes these centers often deteriorate rapidly as customers are lured to newer facilities. Recycling older strip areas is difficult, as former markets may not exist. Often, lower uses, such as flea markets or farmer's produce markets, replace retail outlets, along with continued high vacancy levels in many stores (Figure 18-42).

Visual blight provides another problem facing ribbon developments, which arises due to haphazard activity arrangements, inconsistent designs, frequent

Figure 18-41 Automobile Row. *Automobile dealerships traditionally located in or adjacent to the central business district. In the past 15 years decentralization has led to a new automobile rows on major arterials in suburban settings. In the short half-mile section shown here one can select from over 12 different brands of new automobiles, domestic and foreign, as well as trucks and used cars. Automobile dealerships also cluster at regional mall sites and along by-pass highways in smaller cities. (T.A.H.)*

vacancies, poor sign controls, unregulated curb cuts, traffic congestion, and overhead wires, among other problems. More effective zoning and planning guidelines could mitigate many of these problems, including traffic, design, and density issues.

TRADE AREA ANALYSIS

Retail trade areas of shopping centers can be delimited using a modification of the retail gravity model described in Chapter 17 to measure trade areas around cities. In the earlier instance, population and distance became the critical measures, but in this instance the size of the shopping center in square feet is typically substituted for population, and travel times in minutes replaces the distance measure. A method developed by Huff uses this approach in expressing the attractiveness of a center to a consumer based on probabilities of reaching a particular center in relation to others rather than developing a fixed breaking

point as presented in the earlier analysis. "Shopping center offerings are an expression of the center's mass [its attractive force]. The fundamental argument is that the larger the number of items carried by the shopping center, the greater will be the consumer's expectation of a successful shopping trip and willingness to travel to that center. Travel time is inversely related to a shopping center's utility. The farther the customer lives from the center the less likely a trip will be made to the center."[6] The advantage of using probabilities to describe shopping behavior is appealing because it more closely approximates the uncertainty and element of choice facing a consumer than a rigid interpretation would permit. Closer to the center, probabilities of a consumer choosing that center to shop increase correspondingly, as shown in Figure 18-43. But the element of choice remains. The Huff formulation is therefore a behavioral model.

[6]Truman A. Hartshorn, *Interpreting the City: An Urban Geography,* 2nd ed. (New York: Wiley, 1988).

Figure 18-42 Blighted Retail Center. *This formerly thriving community shopping center reflects the life cycle of the community. As income declined in the neighborhood following residential succession, stores moved away leaving behind a high rate of vacancies and less intense uses, such as flea markets, used car sales, a bingo parlor, and a weekend night club/ballroom. (T.A.H.)*

WHOLESALE ACTIVITY

Unlike retailing activity, wholesaling does not occur in every community in direct proportion to its population, but rather is dependent on the size, function, and location of the city. Wholesaling represents an intermediate service between the production of a product and its consumption. It frequently involves breaking bulk lots into smaller units, extending credit, and distributing the product to its consumer(s).

At one time itinerant merchants, manufacturers' agents, brokers, and factors traveled widely peddling wares and establishing distribution channels. Other

Figure 18-43 Probabilities of Shopping at a Regional Mall. *Using the Huff model one can develop probabilities of a consumer traveling to competing centers. Probabilities of a successful trip to a particular center increase as distance to the center decreases. The Probabilities of residents shopping at three competing malls are shown here. (From Hartshorn, Interpreting the City, Wiley, 1980, p. 356; after David L. Huff, "A Probabilistic Analysis of Shopping Center Trade Areas," Land Economics, 39 (1963), p. 81–90.)*

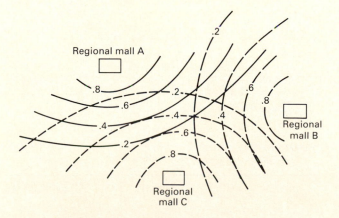

early wholesalers were the middlemen who assisted with the importing and exporting of goods at waterfront locations in port cities. Customhouses, commodity exchanges, and brokerage functions evolved in such areas. Financial and legal institutions also emerged in connection with this activity, as happened in the case of Wall Street in lower Manhattan, where a financial district sprang up to serve the needs of shippers.

As the interior of the United States was settled in the nineteenth century, wholesaling became identified with successive waves of frontier gateway centers to the west, such as Chicago, St. Louis, Kansas City, Dallas, Denver, and Salt Lake City (Figure 18-44). As a group, cities with strong wholesale activity today serve as distribution centers in agricultural areas (e.g., Lubbock, Texas), or as major regional service centers (e.g., Atlanta, Denver), but New York City remains the best example of a wholesale center. Due to its financial and headquarters role, as well as its role as the leading urban market in the country, it has been described as a *summit of convergence,* and an *arbiter market,* which establishes consumption preferences as the cultural capital of the country.[7]

Within the city, wholesaling activity moved from the waterfront to areas adjacent to rail terminals in the early 1800s, only to move again in the 1920s and thereafter, when the motor truck began supplanting the rail as a locational factor. This decentralization process accelerated in the 1930s with the advent of rural free delivery provided by the postal service. Parcels could now be shipped by mail, further loosening the tie to the railroad industry. Rapid growth of mail-order houses such as Montgomery Ward and Sears, Roebuck also dates to that era.

[7]James E. Vance, Jr., *The Merchant's World: A Geography of Wholesaling* (Englewood Cliffs, N.J.: Prentice-Hall, 1970).

Figure 18-44 Infomart in Dallas. *Opened in 1985 this educational and marketing facility features permanent product demonstration facilities from more than 60 computer hardware, software, telecommunications, and consulting companies—the largest such concentration in the world. This eight-story 1.5 million square foot building was inspired by London's Crystal Palace, the 19th century permanent exhibition hall of the Industrial Revolution. (Courtesy INFORMAT)*

Today, wholesaling involves an ever-greater variety of activities and functions. Merchandise and trade marts in major cities house both permanent and seasonal shows at which professional buyers make selections for their clients—the retailers. As business products become more complex, specialized sales agencies and sales representatives have emerged to demonstrate and sell the products. Some company representatives now operate from their homes and/or cars, creating a new type of work environment. Arterial streets also have a large share of the wholesaling business, as they provide locations for automotive parts distributors, building trades suppliers, and salvage and scrap material handlers. Finally, the stock supplier provides yet another function. Such firms cater to specialized needs of the business, such as providing vending machine services (candy, cigarettes, coffee, soft drinks), printing and office supplies, film and camera services, and recreation and amusement services.

HOTEL ACTIVITY

As with retail facilities, the majority of lodging facilities (motels, hotels) in the United States are relatively small, with over half having fewer than 25 rooms. Not-withstanding this statistic, about one-half the hotels in the country have more than 150 rooms. By far most of the facilities are located in cities, although one-third have highway locations, primarily along interstate highways or in resort areas (Table 18-3).

Traditionally, hotels were located near rail passenger stations in the center city. It was not until after World War II that motor hotels oriented to the personal automobile began appearing, in both downtown and suburban locations. Until the current generation of hotel investment activity began in the 1960s,

Table 18-3

Location of Hotels in the United States, 1984

Location	Percent of Facilities	Percent of Rooms
Downtown	19	26
Suburbs	24	21
Highway	37	27
Resort	14	17
Subtotal	94	91

Source: Urban Land Institute, *Development Review and Outlook, 1984–1985.*

Figure 18-45 Hilton Hotel, Suburban Peachtreee Corners High Technology Corridor, Atlanta. *This post-modern hotel, located in the midst of Atlanta's high technology market 20 miles north of the downtown, offers facilities for sales meetings, conferences, and trade shows. (T.A.H.)*

the hotel stock in the United States was becoming very aged, due to meager additions to the inventory earlier in the century. Of course, roadside cabins and, later, family-owned motels began catering to travelers and businessmen much earlier, but it was not until national accrediting associations became active in the 1950s, together with newly formed national chains, that the motel industry became a major player in the lodging industry. The interstate highway system promoted this growth, and by the 1960s the inventory of motel rooms exceeded the hotel facilities for the first time. As this growth occurred, the demand for more services and deluxe accommodations also expanded. The motor hotel concept, combining the advantages of the hotel (fine restaurants, convention facilities, personal services) and motel (free parking, informality, privacy) in one package, had a great impact on suburban markets in this period. Locations in office parks and suburban business centers became rather standard in this period, as the business/commercial sector accounted for the largest market share of the hotel industry (Figure 18-45). Businesses use hotels for regular sales meetings, conferences, and trade shows, as well as accommodations for visiting out-of-town corporate personnel.

The convention business has become a major growth industry for many cities, particularly downtown areas. "Over 20,000 conventions are held in the United States each year, but only one-third of them are international, national, or regional in scope. The appeal of the convention trade to hotels is that it generates lucrative demand for many support services (banquets, luncheons, cocktail parties, exhibits, entertainment, catering)."[8] The largest trade shows are restricted to the largest urban markets with 10,000 or more hotel rooms with easy access to the convention facility itself, which is increasingly a mammoth publicly run facility with over a half-million square feet of exhibit space. The leading convention cities in the United States are New York, Chicago, Atlanta, Washington, D.C., and Dallas.

As with the retail industry, hotel markets today are becoming increasingly segmented. National chains target their market by offering three or more types of facilities to the customer. For example, Holiday Inn now offers limited service economy facilities to the business traveler under the name of Hampton Inn. Marriott offers a Courtyard chain in the same market. Both chains provide full service hotels under their corporate logo, and in the upscale luxury grouping Holiday Inn began marketing the Crown Plaza in the 1980s, and Ramada Inns, the Renaissance. These firms also operate worldwide, a trend also associated with their European and Japanese counterparts.

Spartan discount budget hotels, such as Econo-Hotel, Red Roof, Motel 6, and Days Inn, have also en-

[8]Hartshorn, op. cit., p. 381.

joyed considerable expansion in the past decade, catering to family and itinerant traveler groups. Other market niches are served by the all-suite hotel, such as Guest Quarters and Embassy Suites, or the longer-term residence hotel which caters to business persons away from home for extended periods and relocating families. Resort markets also offer time-sharing accommodations that can be purchased by many individuals/businesses for a specific number of days a year on a sequential basis. These facilities can be either a condominium or a hotel property, depending on the services offered and market considerations.

SUMMARY AND CONCLUSION

An understanding of the changing spatial structure of the modern metropolitan area is essential to comprehend the location of commercial activities. In the nineteenth-century city nearly all commercial activity occurred in the city center, but transportation improvements, especially those associated with the automobile, led to a steady decentralization process in the twentieth century. In the early stages of suburban growth, central locations still provided most of the employment and retail opportunities, but with the coming of suburban office and industrial parks and regional malls in the 1950s and 1960s, suburban areas became independent work and living centers.

In the past 25 years the growth of suburban office, retail, and hotel activity accelerated with the emergence of high-income residential areas and the development of superregional shopping centers. Much of this activity now occurs in suburban downtowns boasting impressive skylines that have increasingly congregated in high-income suburban areas. These centers occur in many forms: (1) as clusters focused on regional malls, (2) as corridors of development

fronting expressways, or (3) as urbanizing mixed-use town centers either developed from scratch or built around an older urban core. Strip retailing activity along major arterials has expanded as fast-food, automobile sales, and other space-intensive users dependent on consumer automobile access have multiplied. Wholesale and distribution functions round out the suburban work environment.

As suburban areas became stronger work centers, downtown areas lost their former dominance, but they typically remain the leading single metropolitan area employment center. Headquarters activity, regional offices, and retailing have decentralized strongly in the majority of American cities, with suburban areas now accounting for at least half of the employment related to each of these categories. Fortunately, new activity has emerged downtown to take up the slack created by decades of decentralization. The most prominent of these activities has been the expansion of hotel/convention functions. Those cities with modern service-oriented economies, as compared with cities with a traditional strong manufacturing activity emphasis, generally have stronger downtown employment concentrations. Those cities with a close-in middle-class housing market also generally have stronger downtowns in terms of retailing and cultural activities.

Emphasis in this chapter has focused on commercial activity patterns in the U.S. city, with particular attention placed on the impact of various transportation modes as a city-shaping force, and anecdotal references to the European situation. In Chapter 19 we discuss characteristics and changes in the broader form of the city, including the residential overlay using several conceptual models. U.S. experiences are given prominent coverage, but a more systematic treatment of the third-world, European, Japanese, and Soviet situation is also included.

SUGGESTIONS FOR FURTHER READING

BATEMAN, MICHAEL, *Office Development: A Geographical Analysis.* New York: St. Martin's Press, 1985.

BERRY, BRIAN J. L., *Geography of Market Centers and Retail Distribution.* Englewood Cliffs, N.J.: Prentice-Hall, 1967.

BERRY, BRIAN J. L., *Comparative Urbanization: Divergent Paths in the Twentieth Century.* New York: St. Martin's Press, 1981.

CERVERO, ROBERT, *Suburban Gridlock.* Brunswick, N.J.: Center for Urban Policy Research, Rutgers University, 1986.

CHRISTALLER, W., *Die Zentralen Orte in Suddeutschland, Jera: Guslav Fischer Verlag, 1933.* [See also translation in English: C. W. Baskin, *Central Places in Southern Germany* (Englewood Cliffs, N.J.: Prentice-Hall, 1966).]

CLAY, GRADY, *Close-up: How to Read the American City.* New York: Praeger, 1973.

DANIELS, PETER, *Service Industries: A Geographical Appraisal.* New York: Methuen, 1986.

HARTSHORN, TRUMAN A., *Interpreting the City: An Urban Geography,* 2nd ed. New York: John Wiley and Sons, 1988.

HARTSHORN, TRUMAN A., and PETER O. MULLER, *Suburban Business Centers: Employment Implications,* Final Report Prepared for U.S. Department of Commerce, Economic Development Administration, 1986.

HARVEY, DAVID, *The Urbanization of Capital.* Baltimore, Md.: The Johns Hopkins University Press, 1985.

HOROWITZ, RICHARD, *The Strip: An American Place.* Lincoln, Neb.: University of Nebraska Press, 1985.

JACKSON, KENNETH T., *Crabgrass Frontier.* New York: Oxford University Press, 1985.

JACKSON, RICHARD H., *Land Use in America.* New York: Wiley, 1981.

KOWINSKI, WILLIAM S., *The Malling of America: An Inside Look at the Great Consumer Paradise.* New York: William Morrow, 1985.

MULLER, PETER O., *Contemporary Suburban America.* Englewood Cliffs, N.J.: Prentice-Hall, 1981.

SAWERS, LARRY, and WILLIAM K. TABB, eds., *Sunbelt/Snowbelt: Urban Development and Regional Restructuring.* New York: Oxford University Press, 1984.

VANCE, JAMES E., JR., *This Scene of Man: The Role and Structure of the City in the Geography of Western Civilization.* New York: Harper's College Press, 1977.

WARNER, SAM BASS, JR., *Streetcar Suburbs: The Process of Growth in Boston, 1820–1900.* Cambridge, Mass.: Harvard University and the MIT Press, 1962.

19 Comparative Urban Structure

Just as their economies vary in emphasis, cities also exhibit considerable variation in their internal spatial structure. Many historical, cultural, physical, technological, and economic factors explain these variations. Even cities that share common institutions and histories provide considerable contrasts, but the most pronounced differences occur among cities in the developed, developing, and centrally planned economies.

In this chapter we discuss the contrasting patterns and processes that create these similarities and differences, beginning with the North American experience. Discussion of third-world, European, Japanese, and Soviet cities follows.

NORTH AMERICAN URBAN AREAS

U.S. Urban Areas

The present form of the U.S. city evolved over the past two centuries in an unplanned, yet systematic manner largely influenced by the land speculation principle discussed in Chapter 18. Accessibility and topographic considerations have traditionally played an important role in this growth process. The importance of transportation as a city-shaping force and the transformation of the city from a monocentric to a polycentric structure was discussed in Chapter 18. This transition has evolved to its highest form in the United States, but evidence of this process also exists

in Japan, western Europe, and Canada, among other areas. In this discussion we focus on the evolving pattern of residential and industrial areas of the city as well as the polycentric commercial pattern in the context of several relevant models. It should be mentioned at the outset, however, that the traditional models of the city have all posited a single commercial core, a notion that can no longer be supported in the United States. Nevertheless, it is important to review these models, as they still apply with respect to an understanding of residential and industrial structure.

Land Values and Land Uses. Although one could argue that land uses largely determine land values in market economies, it could also be argued that land values determine land uses. Regardless of the perspective one takes, there is no question that the best accessibility situation in terms of proximity to the urban region traditionally occurred in the downtown areas, as discussed in Chapter 18. Accessibility advantages led to the most intense level of commercial development in the downtown area and to the coining of the phrases *100 percent corner,* or *peak land value intersection,* to refer to the highest land values at the major crossroads location (Figure 19-1). Due to high land values, such locations commanded intense development in high-rise buildings occupied by commercial tenants that could afford to pay high rents. Retail activities that benefited from strong exposure to pedes-

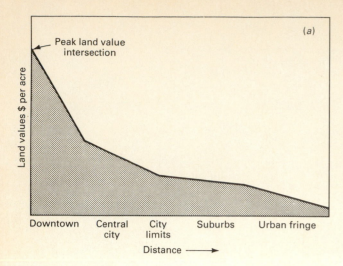

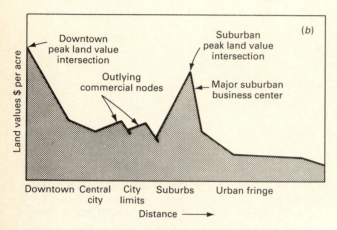

Figure 19-1 Changing Urban Land Values with Distance in U.S. cities. *The traditional land value profile (a) indicates very high prices occur at the primary downtown street intersection, with values falling dramatically short distances away, followed by a zone of gradual decline in the suburbs and urban fringe. A more contemporary interpretation (b) shows a similar situation in the downtown area, but much higher values in outlying suburban downtowns and other outlying commercial nodes, with values declining again on the urban fringe.*

trian traffic typically chose such choice locations, including jewelry stores and other specialty shops. Banks, offices, hotels, and department stores sought nearby sites as cities developed more specialized commercial core areas in the mid-nineteenth century. Land values fell dramatically short distances away from prime downtown locations. It was not uncommon for downtown values, which might exceed $1 million an acre near the peak land value intersection in a metropolitan area with a population of 1 million persons or more, to fall to one-third that level within a few blocks.

Values in the center have remained high, but in recent years they have increased dramatically in outlying areas, with good access. Values now rise to levels comparable with downtown prices at major intersections of beltways and circumferential expressways at suburban business center locations. Developers and real estate specialists remind us of the importance of accessibility in determining land values by saying that the worth of a property is based on three factors: *location, location, location.*

Using a metropolitan land value gradient as just discussed, one can construct a model of urban structure based on an ability to pay with increasing distance from commercial business centers, similar to that discussed earlier in Chapter 5 for agriculture and refined for this purpose by Alonso and others.[1] In the present model there are only three types of uses: commercial, residential, and industrial. This gradient sloped downward to the right from the city center outward in the traditional city as in Figure 19-2a, such that commercial activity captured close-in locations, followed by residential and industrial activity. As with the agricultural situation, this bidding process created rings of economic activity with the least competitive use located in fringe areas. Industrial areas are located on the edge of the region, as they are space-extensive and cannot afford high-priced land. Over time, transportation improvements and suburban growth have led to a flattening out of the bid-rent curves with distance from the city center due to higher bid-rent levels of residential and commercial uses in outlying areas.

Today, outlying commercial centers dramatically impact the form of this model, essentially replacing the former single gradient with multiple downward-sloping profiles (Figure 19-2b). The intensity of uses and values still generally decreases with distance from these commercial nodes, but in some cases commercial nodes at major suburban radial and beltway locations may even create values and rents equal to or higher than those downtown. Surrounding each of these commercial nodes are residential areas, with industrial areas relegated primarily to the periphery of the region, although special needs (i.e., waterfront locations, rail access, etc.) may dictate close-in locations. While industrial uses compete the least favorably for high-priced land, they nevertheless require metropolitan locations to take advantage of urban transportation, labor markets, and financial services.

Population Density. The population density profile exhibited in metropolitan areas exhibits a similar pattern to the land-value gradient and has experienced comparable changes over time. Although many approximations of this relationship have been postulated, perhaps the most appealing is the nega-

[1]William Alonso, "A Theory of the Urban Land Market," *Papers and Proceedings,* Regional Science Association, 6 (1960), 149–157; Edgar Hoover, *The Location of Economic Activity* (New York: McGraw-Hill, 1948).

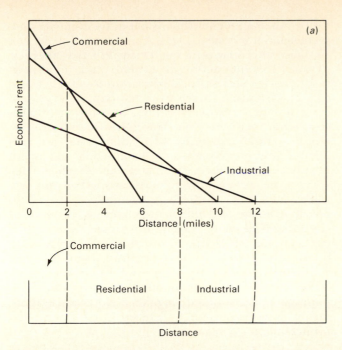 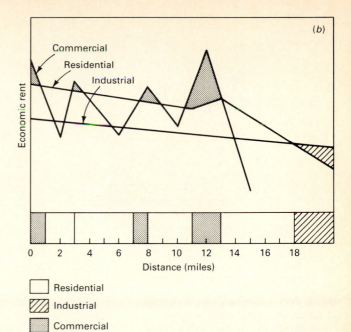

Figure 19-2 Rent gradients and Urban Land Uses in U.S. metropolitan areas. *The traditional model of urban structure (a) indicates that commercial activity captures close-in locations followed by residential and industrial uses at greater distances from the center. The contemporary form of this model (b) reflects the impact of outlying commercial centers. Surrounding major suburban centers one finds residential areas, with industrial uses on the periphery. Note that some suburban centers generate values and rents equal to or higher than those downtown with the contemporary model.*

tive exponential formulation (Figure 19-3).[2] The major differences between the land value and population density profiles include the contrast in the very high land values in the downtown and relatively lesser population densities due to commercial uses and the relatively steeper population density gradient elsewhere. This distinction has weakened over time, as differences between the higher-density characteristics of the older central city housing stock have decreased in relation to suburban densities.

Population densities in American cities generally peak a few miles from the city center. Densities are typically lower in the downtown core, rising gradually with distance until a peak is reached a few miles from the city center. The decreased densities close in, labeled a *crater effect,* occur due to the absence of housing in many commercial areas, primarily as a result of urban renewal programs and the clearing of slum residential properties in the past 25 years. Unlike the situation in Europe and elsewhere in the world, Americans generally do not live in residential units above street-level commercial uses except in the largest and oldest cities, such as the situation in New York and Chicago.

Over time, population density profiles have flattened out as close-in densities have fallen and outly-

ing areas have increased. Particularly pronounced have been the density increases around perimeter beltways and radial expressway corridors as higher-density apartment and townhouse units have expanded to meet growing market needs.

The actual pattern of land values and densities is enormously more complicated than these land-value and population-density models imply, but the simplicity of these models more than offsets these weaknesses. Tastes and preferences for housing, life-styles, mobility, and the preference for separation and isolation can have profound impacts on population distribution and density patterns. The field of *new urban economics* has attempted to introduce some of this complexity into models by including the public sector and other *externalities* as factors in explaining land values and densities.[3]

Externalities refer to factors such as physical, environmental, social, or political phenomena that may play a part in explaining variations. Some externalities may act as "hidden mechanisms" that affect distributions or serve to subsidize one area or group of people in relation to others. Governmental policies or actions using public funds, for example, can generate such advantages or disadvantages. Zoning, land-use controls, and tax policies are examples of public-sector actions that impact population patterns.

[2]Richard Muth, "The Variations of Population Density and Its Components in South Chicago," *Papers and Proceedings,* Regional Science Association, 15 (1965), 173–183.

[3]Harry W. Richardson, *Urban Economics* (Hinsdale, Ill.: Dryden Press, 1978).

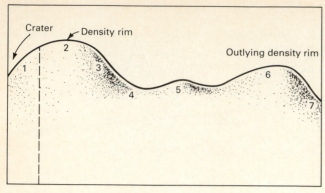

1. High-rise middle-income housing and low-income public housing
2. Walk-up apartments/row houses
3. Detached single family units—moderate income
4. Middle-income single family units
5. Garden apartments/condominiums
6. Midrise apartments/condominiums/town houses—high income
7. Large lot upper-income units

Figure 19-3 Population density profile in a U.S. metropolitan area of 2 million persons. *Note that population densities are lower in the city center, rising gradually with distance until a peak is reached a few miles from the city center. Densities then increase again in the suburbs creating a second density rim in areas with more apartments, condominiums, town houses, cluster houses, etc., near beltways and suburban downtowns. Densities fall again in large lot middle and high income neighborhoods on the urban fringe.*

Urban Growth Models. Sociologists, economists, and geographers have each been credited with creating urban growth models, including the *concentric zone, sector,* and *multiple nuclei* models, respectively. Each model provides insights into city-building processes and lends further credence to the notion that no one principle or perspective has yet been developed to account for the complexity exhibited by city structure. Most recently, in fact, sociologists, geographers, and others have assimilated ideas from each of the disciplinary-based models mentioned above, forming a more general statement about urban growth with the *social area analysis* approach.

Concentric-Zone Model. Chicago school sociologists, centered on the work of Burgess, developed the concentric-zone model of urban growth in the early 1900s (Figure 19-4).[4] Inspired by the new field of *human ecology,* with its roots in plant ecology and social Darwinism, this model used several concepts derived from the biological community to account for patterns of specialized land use, with an emphasis placed on residential locations. These concepts included *competition, invasion and succession, dominance, segregation,* and *specialization.* The methodological underpinning of the concentric-zone model is based on the notion that like plants and animals, people in cities develop a sense of territory and sort themselves out in rela-

[4]E. Burgess, "The Growth of the City: An Introduction to a Research Project," *Publications,* American Sociological Society, 18 (1924), 85–97.

tively uniform neighborhoods based on ethnicity and socioeconomic status.

The concentric-zone model suggests that as cities expand, the interaction of people and their economic, social, and political organizations create rings of urban growth outward from the city center which contain a specialized central business district (CBD) commercial area. Around the CBD occurred a *zone of transition,* intermediate between commercial and residential areas, encompassing wholesaling and light-manufacturing activity. This area has also been described as a *gray zone,* owing to its association with factories, slum areas, and ethnic communities living in boarding houses and tenement properties. In this area conversions of older housing to commercial uses occurred as the CBD grew. Frequently, this land is held by speculators and absentee landlords, with housing considered a temporary or interim use. A third zone housed the lower-income blue-collar working-class population, surrounded rings of middle- and upper-income housing, originally labeled a commuter bedroom community zone.

Figure 19-4 Concentric Zone Model of Urban Structure. *This model indicates that rings of relatively uniform land uses emerge away from the central business district including a zone of transition into lower income residential areas, followed by successively higher income neighborhoods at greater distances. (Redrawn from Harris and Ullman, "The Nature of Cities", in H. Mayer and C. Kohn [eds.], Readings in Urban Geography, University of Chicago Press, 1959, p. 281. Copyright © 1959 by the University of Chicago)*

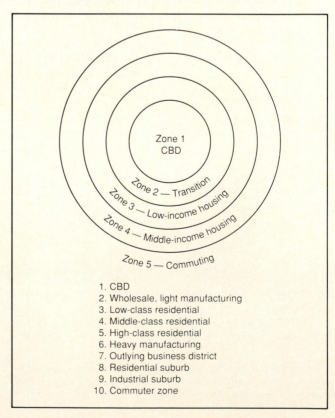

1. CBD
2. Wholesale, light manufacturing
3. Low-class residential
4. Middle-class residential
5. High-class residential
6. Heavy manufacturing
7. Outlying business district
8. Residential suburb
9. Industrial suburb
10. Commuter zone

This model correctly posits the tendency in American cities for lower-income persons to live closer to the city center and higher-income persons farther out. Given the higher land values toward the city center, one might sense a contradiction with this arrangement. It is generally explained by the fact that closer-in residents are more likely to live at higher densities, be renters, and have fewer locational options due to greater dependence on public transportation than higher-income persons who can afford to live at lower densities, on larger lots in higher-priced houses, as well as afford to commute longer distances to work by automobile. The model also correctly predicts that as households climb the socioeconomic ladder and experience life-cycle changes they undertake a series of outward moves across one or more rings and are replaced by new waves of lower-socioeconomic groups, also experiencing upward mobility. This chain of moves creates considerable change in the ethnic and cultural identity of neighborhoods over time.

The concentric-zone model fails to account properly for the expansion of commercial and industrial areas in outlying sections of the city. It also overlooks the impact of radial highway and rail transportation routes in creating specialized land-use corridors. Nevertheless, the model remains the most popular conceptualization of American urban structure.

Sector Model. In the 1930s the economist Hoyt suggested an alternative model of urban growth to overcome weaknesses of the zonal model.[5] The *sectoral model* he developed described growth as occurring in the context of wedges of similar activity, including residential, industrial, and commercial sectors (Figure 19-5). The influence of transportation corridors away from the central business district, such as rail lines and river valleys, on the pattern of industrial activity provided a basis for an industrial corridor according to Hoyt. Frequently, lower-income housing for blue-collar workers occurred near industrial districts in low-lying basins and valleys, creating another sector. High-income housing which gravitated to higher, rolling, or wooded ground on one side of the city, rather than forming a belt around the city, created another wedge or cluster, according to this perspective.

Hoyt suggested that residential areas, like industrial zones, expanded along transportation routes, whether trolley lines or highways, with the growth of various land-use groups occurring at the periphery of existing districts along existing development axes rather than as successive waves of rings. In this model a high-income wedge could evolve from the core of the city outward to the fringe following a ridge of high ground, along a commuter rail corridor such as the Main Line in Philadelphia, or along a prestigious tree-lined boulevard such as Peachtree Street in Atlanta.

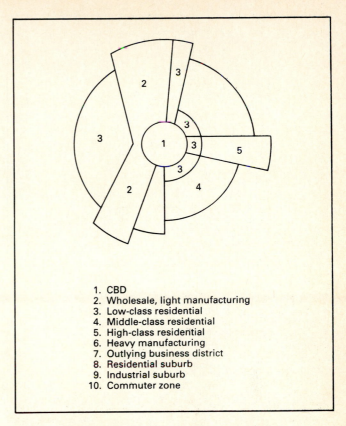

1. CBD
2. Wholesale, light manufacturing
3. Low-class residential
4. Middle-class residential
5. High-class residential
6. Heavy manufacturing
7. Outlying business district
8. Residential suburb
9. Industrial suburb
10. Commuter zone

Figure 19-5 Sector Model of Urban Structure. *The influence of transportation corridors creates axes of uniform development according to this conceptualization. These wedges of activity may include industrial areas in low lying river valleys, middle class residential communities along major arterials, and upper income neighborhoods on high, rolling, or wooded ground. (Source: see Fig. 19-4)*

The appeal of the sector model as a conceptual device to understand urban growth lies in its emphasis on transportation routes in forming development axes. These wedges of growth are generally much more pronounced in newer, rapidly growing sunbelt cities, while the influence of railroad corridors and waterfront locations remains most pronounced in older industrial cities in the northeastern and north central states. A problem with this model is that it does not properly account for outlying commercial centers by placing too much emphasis on the central business district. It also does not adequately account for residential succession and urban mobility.

Multiple-Nuclei Model. In the mid-1940s two geographers, Harris and Ullman, suggested that cities developed around many centers, not one, and developed the multiple-nuclei model to accommodate this conceptualization (Figure 19-6).[6] The model recog-

[5]Homer Hoyt, *The Structure and Growth of Residential Neighborhoods in American Cities* (Washington, D.C.: Federal Housing Administration, 1939).

[6]Chauncy D. Harris and Edward L. Ullman, "The Nature of Cities," *Annals, American Academy of Political and Social Science*, 240, No. 2 (1945), 7–17.

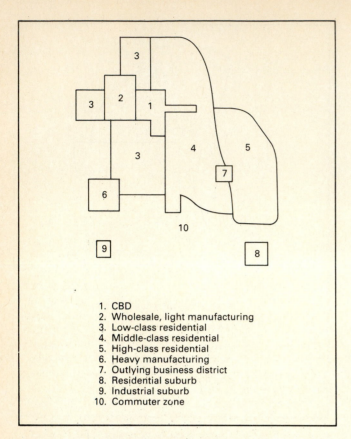

1. CBD
2. Wholesale, light manufacturing
3. Low-class residential
4. Middle-class residential
5. High-class residential
6. Heavy manufacturing
7. Outlying business district
8. Residential suburb
9. Industrial suburb
10. Commuter zone

Figure 19-6 Multiple Nuclei Model of Urban Structure. *This model recognizes that cities can develop around many centers and that different functions have varying locational and accessibility requirements. These specialized focal points for development might include an airport, university, cultural center or commercial district. (Source: see Fig. 19-4)*

nizes that different functions have varying locational and accessibility requirements. It provides for the separation of activities detrimental to one another. The model also acknowledges a role for historical inertia in maintaining an ongoing identity for existing specialized areas as they are assimilated into the urban complex.

The Harris and Ullman formulation accommodates the existence of multicentered cities that spawn specialized focal points for development, such as a university, cultural center, government complex, or airport, by correctly recognizing that many land uses generate development opportunities, including residential districts and associated commercial activities. As with the sectoral model, the multiple-nuclei model also provides for the separation of residential and industrial districts, including the expansion of middle- and high-income housing on one side of the city. The model is a bit dated, however, in that it does not reflect the profound impact of radial and circumferential beltways and large-scale suburban business centers that have literally turned urban structure inside-out in recent years.

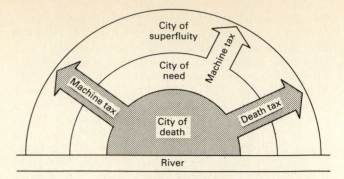

Figure 19-7 Exploitive Model of Urban Structure. *This model divides the city into three functional areas based on the ownership of resources and ability to pay. The city of death is the poorest inner city area, surrounded successively by the city of need, and the city of superfluity. (From William Bunge, "Detroit Humanly Viewed," Ronald Abler et al., eds.,* Human Geography in a Shrinking World. *Duxbury Press, 1975, p. 153).*

Exploitive Model. An adaptation of the concentric-zone model developed by Bunge, based on research in Detroit but having general applicability elsewhere, is the *exploitive model* of urban structure (Figure 19-7). This model divides the city into three areas based on the ownership of resources and ability to pay, with exploitation occurring because of the flows of money from inner-city needy areas to affluent suburban sections.

Bunge labeled the three areas in the exploitive model as (1) the *city of death,* (2) the *city of need,* and (3) the *city of superfluity.* The city of death is the poor inner-city area that is exploited by the rest of the city through the exaction of a "machine tax" which results from wage payments below the workers' worth. The poor residents in this area also pay a "death tax," which involves the payment of higher prices for food, housing, insurance, and other services than occurs in other parts of the city. This area suffers additionally from a lack of city services and amenities and is described as a slum area in the model.

The city of need occupies an intermediate location just beyond the city of death. It is inhabited by the blue-collar working class, "the 'hard hats,' the solid union members of Middle America."[7] This area is also exploited by the suburban-based business interests and the politicians. The outer ring, the city of superfluity, is the home of the elite entrepreneurs and managers, who live a life of leisure and mass consumption subsidized by the exaction of payments from the other groups. Even though small in number and not conspicuous, this affluent group controls the allocation of resources according to this model. While the model provides insight into the problems of the inner-city poor, especially their housing situation, critics would argue that the economic system itself is not the root of the problem, but the lack of skills and training to-

[7]William Bunge, "Detroit Humanly Viewed: The American Urban Present," in Ronald Abler et al., eds., *Human Geography in a Shrinking World* (North Scituate, Mass.: Duxbury Press, 1957), p. 158.

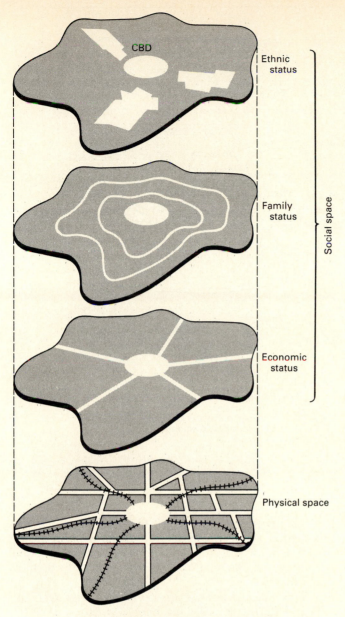

Figure 19-8 Social Area Analysis Model. *This model incorporates aspects of the concentric ring, sector, and multiple nuclei models, allowing for the simultaneous operation of several patterns and processes. This perspective helps one comprehend the behavioral underpinning of uniform socioeconomic neighborhoods. (Redrafted from Robert A. Murdie,* Factorial Ecology of Metropolitan Chicago, 1951– 1961. Department of Geography Research Paper, # 16, 1969, *p. 9. Copyright by the University of Chicago, Department of Geography)*

gether with the absence of job opportunities in lower-income communities.

Social Area Analysis. The social area analysis model of urban structure incorporates certain aspects of the ring, sector, and multiple nuclei models in a more comprehensive statement of urban form.[8] Es-

[8]Eshref Shevky and Wendell Bell, *Social Area Analysis: Theory, Illustrative Application, and Computational Procedures* (Stanford, Calif.: Stanford University Press, 1955).

sentially, this approach involves delineating uniform subareas in the city using three independent indices: (1) *family status,* (2) *socioeconomic status,* and (3) *ethnic status* (Figure 19-8). By mapping metropolitan areas according to their characteristics in each of these three areas, researchers have found that family status indicators such as population and age variables, housing ownership and age characteristics, family size and marital status, and other life-cycle characteristics exhibit a zonal or ring pattern with older, smaller, renter households located closer to the city center, and owner-occupied, newer, larger, family households farther from the city center in suburban settings. Socioeconomic status variations are measured by occupation, education, income, and house-value figures, which in the aggregate demonstrate a sectoral pattern. Third, the ethnic status indicator identifies with minority-group segregation patterns, which have been characterized as portraying a multiple-nuclei tendency. Some researchers have also characterized minority patterns as exhibiting a reverse sector tendency (Figure 19-9). See the next section for an elaboration of the minority experience in U.S. cities.

The advantage of the social area analysis perspective is that it allows for several different patterns and processes to operate in the city simultaneously. It also has a behavioral component that is useful in that it shows how individuals sort themselves out in urban space in relatively uniform life-style subareas that creates an urban mosaic not unlike a patchwork quilt. This process produces an amalgam of relatively uniform residential enclaves or neighborhoods having similar life-styles but at the same time considerable variation from neighborhood to neighborhood (Figure 19-10). Social area analysis is much weaker, explaining commercial and industrial structure and in accounting for changes in urban structure over time.

Figure 19-9 Reverse Sector Black Residential Space Model. *The reverse sector model suggests lower class minority communities will form near the central business district, with working and middle class areas farther away. Growth of black residential space in the city typically occurs along an axis created by the existing space. (Redrafted from Harold M. Rose,* The Black Ghetto. *©McGraw-Hill Book Company, 1971)*

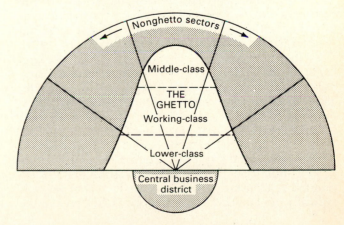

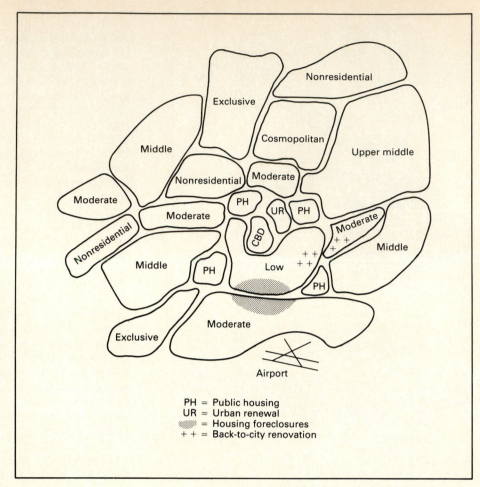

Figure 19-10 Generalized map of uniform residential enclaves in a metropolitan area. *An urban mosaic not unlike a patchwork quilt is created by families sorting themselves out according to life-style and socioeconomic status. In this instance distinctions are made between moderate, middle, upper middle, exclusive (high), and cosmopolitan neighborhoods. The latter refers to areas catering to highly educated professionals and intellectuals where a variety of religious, ethnic, and political persuasions exist, as often occurs in neighborhoods in close proximity to a university. (Source: Hartshorn,* Interpreting the City, *Wiley, 1980, p. 258)*

PH = Public housing
UR = Urban renewal
░░ = Housing foreclosures
+ + = Back-to-city renovation

Minority Residential Space. Minority residents, typically European immigrants seeking jobs in factories, traditionally took up residence in close-in neighborhoods in U.S. cities in the nineteenth century and with time, succeeding generations experienced rapid upward mobility and assimilation into the urban fabric as they dispersed outward. Beginning in the 1920s a new urbanizing group rapidly relocated to northern industrial cities—the southern rural black. This group faced much more difficulty in entering the mainstream.

Residential segregation essentially led to the emergence of two cities with separate housing markets: one white, one black. Decades after residential discrimination was declared illegal in the early 1960s, U.S. cities remain largely segregated racially. The largest black population concentrations occur in the largest cities in the country, including New York and Chicago. Several large metropolitan areas have black majorities in their central cities, including Detroit, Washington, D.C., Baltimore, New Orleans, and Atlanta. Nationally, 60 percent of the black population lives in the central city and 14 percent in adjacent suburbs. Middle-class black residential space has also expanded significantly in these cities in recent years. It typically lies adjacent to the majority black area in the central city (Figure 19-11). Within this separate black residential space, patterns of residential neighborhoods exhibit ring tendencies, in terms of socioeconomic status, with affluent areas generally farther from the city center.

The most recent expansion of minority residential growth in U.S. cities has been associated with the expansion of Hispanic communities. Largely Caribbean (Cuban, Puerto Rican, Haitian) or Mexican in origin, this group now occupies considerable close-in residential space in a growing number of American cities, but assimilation and decentralization appears to be occurring as well. Asian communities (Korean, Vietnamese, Chinese) also comprise growing minority space in U.S. cities. In the mid-1980s Los Angeles became the largest port of entry for immigrants into the United States, owing to the strength of the flow from Pacific rim nations. Previously, east coast cities such as New York and Miami processed the most immigrants.

Canadian Cities

In comparison with the United States, where the national space-economy is relatively completely settled and developed, large portions of Canada remain

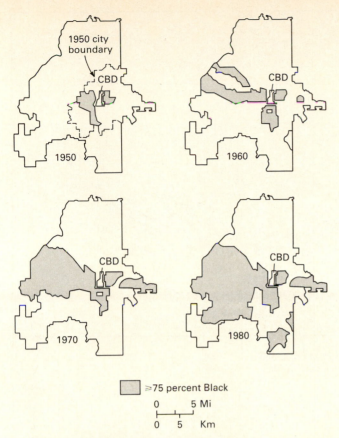

Figure 19-11 Black residential expansion in Atlanta, 1950-1980. *The expansion of black residential space in the city of Atlanta 1950-1980, generally conforms to the reverse sector model. By 1980 considerable expansion of this space had occurred in suburbs to the southwest and east of the city which are not shown here.*

≥75 percent Black

0   5 Mi

0   5   Km

there are also two Canadas, the Ontario/Quebec heartland and the lesser-developed western and eastern hinterland.

Political and Financial Inducements. Notwithstanding the contrasting national space-economy circumstances and ethnic traditions, one might expect Canadian and American cities to be quite similar, given their proximity and common cultural and economic experiences. Although this assessment is partially true, it is also misleading and superficial, because significant differences do occur between U.S. and Canadian cities. Several of these differences can be traced to differing federal government roles and distinctive taxation policies, but the ring, sector, and multiple-nuclei models, as in the United States, generally apply to urban structure itself. The role of planning also differs significantly between the United States and Canada. Much more authority is held at the municipal level in granting approvals for building in Canada than in the United States.

One traditional role of government in Canada has been to serve as an investment catalyst due to the lack of capital in many lesser-developed parts of the country. Vast distances must be overcome to integrate the economy. Transportation and utility subsidies assist private-sector initiatives to overcome these problems. Crown corporations, which are major companies partially or totally owned by governmental agencies, have also been created to assist economic development.

Despite the strong presence of government and public planning in the economic development arena, the national government in Canada is generally less involved in urban programs than in the United States. There is no national highway building program in Canada comparable to the federal interstate system in the United States. There is less direct federal-aid funding for cities in Canada than the United States, such as money for urban renewal or block grants for infrastructure improvements. Provincial governments, on the other hand, are more involved in urban programs

sparsely populated and undeveloped. Most large Canadian cities, for example, occur in a narrow band just north of the U.S. border (Figure 19-12). Canadian urban structure does not exhibit a complete urban hierarchy, nor is the economy itself as mature and independent as that in the United States. Many provinces, in fact, remain very resource dependent: the Maritimes rely on fishing; the western prairie provinces on agriculture, minerals, and energy; and the Pacific coast on timber resources and port activity.

No one city dominates Canada as in the case of the United States (Table 19-1). Toronto and Montreal, each with about 3 million population, dominate their hinterlands, focused on Ontario and Quebec, respectively, and serve as major national cities. Vancouver serves a similar function on the west coast, although it is less than one-half the size. In many ways, the Toronto and Montreal metropolitan areas symbolize the presence of two Canadas: one anglo, oriented to Toronto, and one francophone, with allegiance to Montreal. Two official languages (French and English) exist in Canada and a greater sense of interprovincial rivalry occurs than at the state level in the United States. From an economic development perspective

Table 19-1

Leading Canadian Metropolitan Areas, 1986
(thousands of people)

Rank	Name	Population
1	Toronto	2,999
2	Montreal	2,828
3	Vancouver	1,268
4	Ottawa-Hull	718
5	Edmonton	657
6	Calgary	593
7	Winnipeg	585
8	Quebec City	576
9	Hamilton	542

Source: 1986 Rand McNally Commercial Atlas and Marketing Guide 117th ed. (Chicago: Rand McNally, 1986).

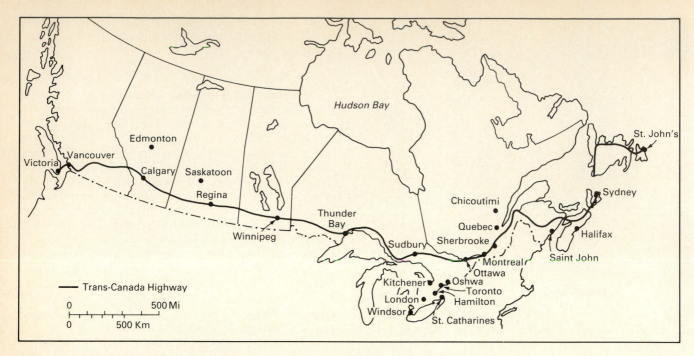

Figure 19-12 Leading Canadian Metropolitan Areas. *Most large Canadian cities occur in a narrow band just north of the U.S. border, many of which are connected by the Trans-Canada Highway.*

than are the states in the United States. Another important difference between U.S. and Canadian cities lies in the area of taxation and mortgage-interest payment deductions. In the United States, interest paid on home mortgages is tax deductible according to current practices, creating a powerful home-ownership subsidy. The lack of such an incentive in Canada makes housing relatively more expensive and has led to relatively higher residential densities.

Given these contrasting policies, one might expect subtle differences in urban structure between U.S. and Canadian cities, which is indeed the case. Canadian cities are generally more compact and higher-density centers than their U.S. counterparts. Fewer inducements exist for urban sprawl, given the taxation and urban planning situation, and the more restrained automobile expressway program means that downtown areas remain stronger relative to suburban commercial areas. Retail activity is much stronger in downtown areas in Canada than in the United States (Figure 19-13). A major exception to this generalization occurs in the case of the West Edmonton Mall in suburban Edmonton, Alberta, reputed to be the largest mall in the world, which accounts for up to 25 percent of retail sales in the Edmonton metropolitan area. Boasting a covered indoor amusement park complete with submarines in a lake and a roller coaster, this shopping complex has become a major tourist attraction for residents in the northwestern U.S. and Canadian west.

Greater dependence on public transportation also distinguishes mobility in Canada in comparison with

the United States. The larger metropolitan centers of Toronto and Montreal provide an integrated heavy-rail, commuter rail, light-rail, and bus transit system that provides much more service than is available in comparable U.S. cities. Edmonton, Calgary, and Vancouver also offer both bus and light-rail service. Despite the presence of public transportation, dependence on the private automobile is very high in Canada, and the lack of a comprehensive expressway building program has created significant mobility problems in urban areas comparable to those in U.S. cities.

Downtown skylines in both Canadian and American cities appear similar from a distance, but closer scrutiny reveals significant differences in urban structure. Large Canadian developers such as Cadillac-Fairview and Trizec have done an excellent job in introducing new mixed-use residential and commercial centers into the downtown market. The restrictive planning environment in the Canadian city has also improved the skills of developers and lenders in putting such projects together. Historic preservation also receives a high priority, and downtown areas remain more attractive for the pedestrian, with more continuous built-up space. Several larger Canadian cities have developed elaborate downtown shopping malls underground, which benefit not only the pedestrian but also the CBD retail market. For these reasons one does not find the same level of central-city decay in Canadian cities as in the United States.

Canadian cities house large ethnic population groups. One estimate, for example, is that nearly one-

Figure 19-13 Copley Place in Downtown Boston. *Retail activity has made a comeback in some U.S. downtown areas such as Boston and Chicago which have a nearby middle class residential market. This photograph shows the Copley Place mixed-use complex built in downtown Boston in the early 1980s over the Massachusetts Turnpike and three railroad lines. It includes a high-rise hotel on either end, and a 2-level shopping gallery that links the project together. Glass pedestrian bridges cross streets in the area linking the center to the Back Bay. (T.A.H.)*

third of Toronto residents were born outside the country. These minority groups include many Europeans, Chinese, and Middle Eastern households, but significantly fewer blacks than in U.S. cities.

THIRD-WORLD URBANIZATION

Most third-world cities, despite long settlement histories, remained small by contemporary standards until the post–World War II era. The largest of these cities today, however, are not those with long traditions but settlements dating to the colonial era associated with European domination in the eighteenth and nineteenth centuries. Peculiar circumstances associated with the colonial era have also given these cities a legacy and structure quite different from the experience shared by North American and European cities. The mercantilist and colonial eras in fact distorted indigenous urbanization processes in the third world by disrupting local economies. Coastal *entrepôt* port cities, such as Calcutta, Shanghai, Jakarta, Lagos, Lima, and Buenos Aires, for example, grew rapidly as a result of brisk import–export activity in the colonial period.

This foreign-driven transformation process severely undercut the natural evolution of urban growth and technology development in the third world and led to cities with an imbalanced economic structure. As European-directed trade flowed into and out of these centers, the cities became very dependent on the largesse of foreign business interests. Local traders became passive collectors as a part of this process, and domestic institutions gradually weakened.

The imprint from the colonial era remains in third-world cities today, decades after independence following World War II. In country after country, pri-

mate cities, claiming one-third or more of their nation's population, remain anachronisms in the local context. They have failed to develop diversified industrial or postindustrial economies associated with cities in the developed world and have become havens for immigrating unskilled rural families.

One characteristic of these cities, a direct carry-over from colonial days, is the presence of a small but powerful elite class that controls the economy and maintains strong international ties. The vast majority of the urban population remains poor and unskilled, having strong emotional and cultural ties to their recent rural past. This situation occurs in Latin America, Africa, the Middle East, and Asia alike.

From a spatial perspective, the third-world city often takes the form of a reverse concentric-ring model, owing to the presence of the elite business/administrative group in the city center adjacent to the central business district and progressively lower-income groups located farther from the city center. A large middle class, as in developed areas, is conspicuously absent. Often, substandard *squatter housing* forms the outermost ring of the city.

The Latin American Experience

Latin American cities differ somewhat from their counterparts in Asia and Africa. The Spanish Law of the Indies guidelines strongly influenced the form of the colonial Latin American city.[9] Inspired by the Roman tradition discussed in Chapter 18, with a formal grid pattern and central plaza, land uses in the European-dominated Latin American city demonstrated a strict hierarchical social order. These cities were im-

---

[9]The Spanish "Laws of the Indies" building guidelines were issued in the name of King Philip II of Spain in 1537.

portant trade and administrative centers, an activity that occurred close to the central plaza.

As elsewhere in the third world, very little industrialization occurred in Latin American cities in the colonial era and very few opportunities existed to support the growth of a middle class. Instead, cities housed elites near the center and unskilled low-income groups elsewhere in areas notorious for their lack of public services. Typically, only the elite residential sector had access to the full range of urban services, such as paved and lighted streets, water, sewage, police and fire protection, and public parks. The central business district and industrial sector usually developed in close proximity to the elite spine, as shown in Figure 19-14. Around these central areas one observed a series of zones of lower-quality housing.

The elite residential sector located along the spine presents a good example of the Hoyt sector model discussed earlier in the U.S. context. Typically, new growth of exclusive residences occurred toward the fringe of this area and middle-class residents moved into adjacent closer-in neighborhoods. Outward from the central spine occur rings of residential areas, showing an inverse relationship to that demonstrated in North America. Rather than increasing in value and quality with distance from the center, housing decreased in quality. The *zone of maturity* encompasses an area of modest houses built by residents which were upgraded over time, creating a relatively uniform residential environment. Typically, this area houses older residents, is relatively stable, and has developed a full complement of city services over time.

The *zone of in situ accretion* represents a newer residential area having a wider variety of house styles and quality levels. Typically, it is a mixed area in terms of services since streets may not all be paved and electricity not available. Some squatter shacks may be found among better-quality units. In general, the area is being upgraded, however, as more permanent housing is built and renovations occur, such as adding a second story to the house, which may have had origins as a squatter shack.

The *squatter settlement zone* typically surrounds the city. Practically no city services exist in this area. Housing is built in a makeshift fashion based on the availability of materials, such as scrap galvanized-metal roofing strips, cardboard, and discarded lumber. Rather than an attitude of despair, residents in these areas often have an upbeat macho reputation in the city because they have taken the initiative to solve their housing problems and provide for themselves. The root of the problem is not that the people are lazy or shiftless but that so many unskilled rural peasants are moving to the city so rapidly that there are neither jobs, housing opportunities, nor resources to build housing for their needs in either the public or private sector. Thus the poverty present on the fringe of the Latin American city provides a great contrast to the affluence of the North American city found in the United States and Canada.

The recent experience of Mexico City provides a

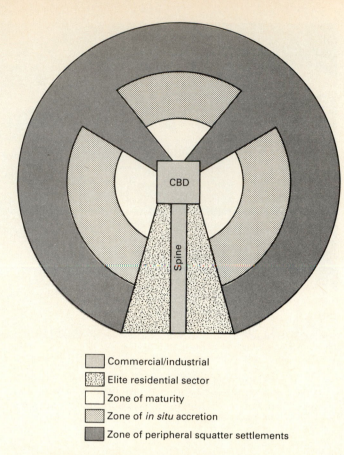

Commercial/industrial

Elite residential sector

Zone of maturity

Zone of *in situ* accretion

Zone of peripheral squatter settlements

Figure 19-14 Latin American City Structure Model. *An elite residential sector typically surrounds a fashionable boulevard, labeled the spine, leading from the CBD. Outward from the central spine occur rings of residential areas with lower quality areas at greater distances. The* zone of maturity *is a middle class area while the* zone in situ accretion *refers to newer areas with less consistent quality and incomplete services. The* squatter settlement *zone occupies peripheral areas and undesirable land closer to the city center such as steep slopes. [Redrafted from E. Griffin and L. Ford,* Geographical Review, *Vol 70 (1980), with the permission of the American Geographical Society].*

graphic example of problems associated with mushrooming growth, largely in the form of squatter housing. So great has been the growth of Mexico City—a process that will probably continue—that many observers predict that it will be the world's largest city by the turn of the century, rising from a 15 million population in 1980 to 31 million. "By 1950, Mexico City was the fifteenth-largest city in the world, with a population of 3 million. Thirty years later its population had swelled fivefold, to make it the third-largest city in the world. Within 20 years, more people will live in Mexico City than have ever lived in any city before."[10] Others have said that Mexico City is the archetype of third-world urbanization problems.

The contradictions of Mexico City run legion. On

[10]Clifford D. May, "Mexico City: Omens of Apocalypse," *Geo Magazine* (1981), 31.

Paseo de la Reforma, an eight-lane, tree-lined boulevard running 2 miles from the city core, the Zo'calo, westward to Chapultepec Castle and Park, one finds fancy homes, high-class shopping opportunities, embassies, and museums, not unlike the Champs Elysées in Paris. This is the classic manifestation of the elite spine described in the model above (Figure 19-15). On the other hand, there are the shantytown suburbs such as the Santa Fe squatter settlement surrounding the city dump (Figure 19-16).

Chronic air pollution occurs in Mexico City as a result of the spewing of pollutants from industrial smokestacks and motor vehicle exhausts in an oxygen-poor environment due to the high elevation of the city, which exceeds 7000 feet. Frequent air stagnation further exacerbates the problem, as the city lies in a valley between two mountains. Sanitation services are poor—over 2 million residents have no running water; 3 million residents have no sewage facilities; and an estimated 30,000 children die each year from chemical and biological poisons. Corruption, inflation, and inadequate planning add to the litany of problems, but life goes on, along with uncontrolled growth.

The African and Middle Eastern Situation

Africa, the least urbanized of the third-world regions, also provides examples of primate cities with roots in the colonial period. Unlike the Latin American situation, where colonial implants occurred primarily along the coast, in Africa they were also established in the interior, owing to the need to gain access to mineral resources. But as elsewhere, a small, affluent elite and a majority of poor people remains a legacy of the colonial era.

The Middle East provides yet another example of third-world poverty and a traditional preindustrial economy. Exceptions to the prevailing trends occur in the oil-rich states discussed in earlier chapters, but the typical circumstance remains—an impoverished preindustrial city. European colonial involvement was less influential in the Middle East than elsewhere and

Figure 19-15 Paseo de la Reforma, Mexico City. *This eight-lane highway spine when viewed from a distance from the wooded Chapultepec Park, shown here in the foreground, appears as a high-rise corridor. The Polanco neighborhood seen here includes hotels, luxury condominiums, and embassies, and the area has a long history as an elite residential community. The tallest building is the El Presidenté Hotel, with the new Nikko Hotel to its left. In foreground at bottom is the Museum of National History. (Courtesy Mexico Ministry of Tourism)*

Figure 19-16 Santa Fe Squatter Settlement. *Located on the outskirts of Mexico City at a dump site, this squatter settlement typifies the poor quality housing in similar districts of other third world cities. (T.A.H.)*

a different cultural/religious heritage provides the Middle Eastern city with a unique character.

Due to the overriding impact of Islamic religious dogma, which treats all individuals equally, with a lesser role provided for civil authority, Middle Eastern cities have an egalitarian flair. They have been described as irregular and formless, with a strong emphasis on individuality as evidenced by the privacy afforded individual houses. A permanent market, referred to as a *bazaar,* typically provides the commercial core of the Islamic city, housing a variety of stalls for vendors (Figure 19-17). Other distinctive features of the Middle Eastern city include shrines, public baths, and mosques. Various ethnic groups occupy separate neighborhoods called *quarters,* but very little sorting out by socioeconomic status occurs, due to the egalitarian principle mentioned earlier. Houses typically face inward courtyard style, with no windows toward the street, bringing to mind the overriding concern to protect family privacy and shield women from the public. Overall, Middle Eastern cities have changed very little in the twentieth century, maintaining an appearance centuries old, but modern high rise structures do provide a modern skyline for the city center.

EUROPEAN CITIES

Even in the face of industrialization, increasing affluence, and continued growth, European cities reflect a strong imprint from the past (Table 19-2). Artifacts in the form of gates, walls, cathedrals, canals, castles, palaces, and narrow streets abound (Figure 19-18). A twentieth-century passion with historic/conservation planning continues the tradition of the past in the contemporary city. A pedestrian scale still pervades the historic urban core, and cities have a compact, if high-density, aspect. Of course, the automobile has been accommodated to some extent, and newer suburban areas rely on the motor car very heavily, but physical constraints such as narrow alleys and streets, the lack of offstreet parking, and greater dependence on integrated public transportation systems (Figure 19-19) mean that the automobile has not had as much structural impact on the European city as it has had in America (Figure 19-20). Furthermore, there is no such thing as an expressway through the city center in Europe.

Europeans have been much more innovative in establishing car-free areas and zones in which the car is more heavily disciplined than in the United States.

Figure 19-17 Street Market in the Fes, Morocco Medina. *Artisans sell their wares in small stalls throughout this ancient section of Fes. Selling goods in this manner is common throughout north African and Middle Eastern cities. (Courtesy of Sanford H. Bederman)*

The *woonerf* developed in the Netherlands, for example, has gained wide acceptance in Europe as one approach. Woonerf, literally translated, means "residential precinct." In the present context it refers to a public open space used in an integrated way by pe-

Table 19-2

Leading Western European Metropolitan Areas, 1985 (thousands of people)

Rank	Metropolitan Area	Population
1	London	11,100
2	Paris	9,450
3	Essen (the Ruhr)	5,125
4	Madrid	4,515
5	Barcelona	3,975
6	Berlin	3,775 *
6	Milan	3,775
8	Roma	3,115
9	Athens	3,027
10	Manchester	2,775
11	Naples	2,765
12	Birmingham	2,675
13	Brussels	2,400
14	Lisbon	1,950
15	Munich	1,940
16	Stuttgart	1,935
17	Frankfurt am Main	1,880
18	Vienna	1,875
19	Cologne	1,815
20	Amsterdam	1,810
21	Glasgow	1,800
22	Turin	1,600
23	Leeds	1,540
24	Liverpool	1,525

Source: 1985 Rand McNally Commercial Atlas and Marketing Guide, 116th ed. (Chicago: Rand McNally, 1985).
*East Berlin, 1.1; West Berlin, 2.0.

destrians, bicycles, cars, and children (Figure 19-21). In such areas pedestrians and bicycles receive priority, while cars are restrained. Many design elements can be employed in various combinations, including curved and broken road segments involving shifts of axis, using planters, bumps, bollards, and one lane for two-lane traffic. Parking is highly regulated in these areas, to assist visual sight lines by drivers in areas where children are playing.

Notwithstanding the woonerf technique, problems associated with cars abound in the European city, related to noise, pollution, and congestion problems. In their favor, however, European cars are smaller than their American counterparts, especially in southern Europe, greatly aiding navigation, but as has been discussed earlier, fuel costs are also much higher. Bicycles, mopeds, and motorcycles are, of course, more popular than in the United States, due to cost and maneuverability considerations, not to mention the shorter distances often traveled.

In such an environment it goes without saying that the central city is much stronger than its U.S. counterpart. As in the case of third-world cities, housing closer to the center city is typically held by high-income groups, with lower-income families on the fringe (Figure 19-22). Often mid- and high-rise housing exists on the edge of the city, a situation totally absent in the United States (Figure 19-23). The high demand for moderate-income housing in the fringe, the shortage of suitably zoned land, and the greater propensity for Europeans to be renters rather than owners explains this situation. Often, an industrial belt separates the central city from the suburbs, and outlying areas also take on a higher-density form than in the United States. Stronger dependence on rail commuting, tighter zoning controls, and the high cost of land explain this situation.

Figure 19-18 Medieval City Gate in Munich. *The fountain in the foreground frames the entrance to the Karlstor Gate which dates to the 14th century. Beyond the gate lies the pedestrian-oriented Neuhauser Strasse and the old city of Munich with its historic churches and cultural and political landmarks. (T.A.H.)*

Figure 19-19 Transit in Vienna, Austria. *Most large European cities offer an integrated transit system, incorporating buses, light rail, and inter-city rail services. A light rail or streetcar vehicle is shown here. (T.A.H.)*

Figure 19-20 Coping with the Automobile in Amsterdam. *Heavy dependence on the automobile in older European cities causes many problems with parking, traffic flow, and pedestrian safety, especially in residential areas. In this Amsterdam street scene parking occurs along the canal partially blocking the street. (Courtesy of Richard Pillsbury)*

Figure 19-21 Woonerf Automobile Restraint Concept in Netherlands. *In order to minimize automobile conflicts in residential neighborhoods, innovative car restraint concepts have been introduced to restrict movements. This large circular barrier, for example, prohibits automobile access to the street, while permitting easy access to pedestrians, bicycles, and mopeds. (Courtesy of Richard Pillsbury)*

Figure 19-22 Piazza Navona residential area in Rome. *High-income residents have traditionally lived in the city center in Europe. This popular plaza with many historic landmarks was once the site of a stadium which gives it its shape. The Plaza is ringed by high-income residential units. Note Basilica and Bernini fountain in the distance. (T.A.H.)*

Figure 19-23 Mid-Rise Housing on Periphery, Amoter Netherlands. *It is typical to find multiple-story housing on the periphery of European cities. The high demand for moderate income housing on the fringe, the shortage of suitably zoned land, and the propensity for Europeans to be renters rather than owners explains this situation. (Courtesy of Richard Pillsbury)*

A higher proportion of *social housing* also exists in Europe than in the United States. This term refers to public or subsidized housing. A trend for older units in the city center to be occupied either by wealthy foreign nationals or poor ethnic *guest workers* (immigrants) has strengthened in the past decade (Figure 19-24). The largest flows of immigrants to European countries has traditionally been associated with persons leaving former colonial possessions. In recent years, however, the flow from eastern and southern Europe and northern Africa to western Europe has grown in significance. The largest flows to these cities today emanate from Turkey, Bulgaria, Yugoslavia, and North Africa.

These foreign workers as originally envisioned would stay only temporarily, but today it looks increasingly like they may become permanent residents. Although they do face discrimination and economic constraints, immigrants frequently remain segregated spatially, due to personal preference.

Traditionally, commercial activity took place on the ground floors of midrise residential housing blocks in European cities. A single, physically distinctive downtown business district commercial area such as occurs in the United States never developed in Europe. Rather, several mixed shopping/residential streets evolved. In some cases former residential units on upper floors have been converted to business use, as in Amsterdam or Vienna, and the structures now serve as office centers. In the 1980s, decentralization of office activity to suburban areas in modern highrise blocks has also intensified in Amsterdam, Paris, Frankfurt, Munich, and Vienna, among others. The new La Défense suburban office/commercial core in Paris is a case in point (Figure 19-25). In some ways European cities can be thought of as more Japanese than American in form, due to the mixing of residential and commercial uses and the presence of higher-income groups toward the city center. European and Japanese cities also have higher proportions of highrise housing in their suburbs than their U.S. counterparts.

JAPANESE CITIES

Urban traditions in Japan, heavily influenced by Chinese principles, began centuries ago with the development of castle cities, many of which have evolved over the milennia into world-class centers (Table 19-3). The Chinese contribution to Japanese urban design included a rectangular city plan and grid street pattern. The administrative and ceremonial functions in castle cities occurred in elaborate centrally located shrines, temples, and castles. One of these centers,

Table 19-3

Leading Japanese Metropolitan Areas, 1986 (thousands of people)

Rank	Metropolitan Area	Population
1	Tokyo	26,200
2	Osaka	15,900
3	Nagoya	4,625
4	Fukuoka	1,575
5	Hiroshima	1,525
6	Kitakyushu	1,515
7	Sapporo	1,450

Source: 1986 Rand McNally Commercial Atlas and Marketing Guide, 116th ed. (Chicago: Rand McNally, 1986).

Figure 19-24 Guest Quarters Housing in Vienna. *Many central-city residential blocks in Vienna have attracted immigrant guest workers in recent years as the Viennese move to the suburbs. This residential complex is built around a typical courtyard. (T.A.H.)*

Kyoto, remained the political capital of the country until the Emperor Meiji returned to power in 1868. At that time, often called the beginning of the modern era, the capital was moved to Tokyo.

Today, two major urban complexes dominate Japan—Tokyo and Osaka. Each, with a waterfront location, commands a large urbanized hinterland—numbering nearly 30 million in the case of Tokyo and about 16 million in the case of Osaka (Table 19-3). Osaka, the economic capital of the country, became the center of textile production following industrialization and later, a diversified machinery manufacturing complex. Nearby Kobe served as the major port city for Osaka, a role similar to that provided by Yokohama for the Tokyo region.

Japanese cities developed dual downtown areas in the late 1800s with the addition of a commercial node around the railroad station to the more traditional commercial/residential functions tied to the palace/castle compound. Even today large prospering underground retail shopping complexes can be found in all major cities at railroad stations (Figure 19-26). Outward from the city center, which is served by a radial street and railway service, development becomes more ad hoc and random in character.

Japanese cities appear very congested and confusing to the outsider. Jumbled land uses, heavy automobile and truck traffic on streets, limited parking opportunities, and bustling pedestrian activity on the sidewalks all contribute to the confusing scenario (Figure 19-27). Many mom-and-pop retail shops, especially food stores and cafes, occupy ground-level sites in two- or three-story buildings with upstairs living quarters in traditional sections of the city. These functions are rapidly giving way to mid- and high-rise buildings—showing a great contrast between the old and the new (Figure 19-28).

Population densities are very high in Japanese cities, with a strong residential function found in the city center, unlike the American situation. This gives the city a more active profile throughout the day and night. Much commuting typically occurs by subway or commuter rail. Tokyo has an elaborate network of 15 subway lines.

Suburban development in Japanese cities is heavily influenced by rail access. Modern high-rise apartment blocks create very high densities in these outlying areas as well (Figure 19-29). Single-family detached residential units are often preferred by middle-class families, but their extraordinary cost, up to

Figure 19-25 La defense suburban downtown, Paris. *The intense office activity concentration in this suburban downtown is supported by excellent public transportation services. A multiple-level platform complex separates pedestrian, vehicular, and rail service with the pedestrian mall shown here on the top level. (T.A.H.)*

Figure 19-26 Underground Shopping Mall, Kyoto, Japan. *This shopping complex in the city center at the rail station is typical for station areas. Attendant with sign is advertising a massage parlor. (T.A.H.)*

Table 19-4

Leading Soviet Metropolitan Areas, 1986
(thousands of people; estimated)

Rank	Metropolitan Area	Population
1	Moscow	12,400
2	Leningrad	5,550
3	Kiev	2,635
4	Tashkent	2,165
5	Donetsk	2,140
6	Gorkiy	1,880
7	Baku	1,880
8	Kharkov	1,825

Source: 1985 Rand McNally Commercial Atlas and Marketing Guide, 116th ed. (Chicago: Rand McNally, 1986).

urbanscape, expressways, often financed as tollroads, must be elevated, often positioned over canals or existing streets or placed in tunnels (Figure 19-30). In Tokyo one moat around the Imperial Palace grounds has been drained for such an expressway. Parking is also a big problem. Tokyo car buyers, in fact, are required to show a parking permit before they can purchase a vehicle. Despite these problems and others associated with lagging city services such as sewage treatment, the Japanese city is a colorful and affluent place. Billboards and signs create striking visual impacts, and the pace of activity impresses the visitor. Despite the densities and the enormity of the city in size, youngsters can travel by themselves—a commentary on the Japanese life-style, one characterized by strong motivation, self-reliance, and freedom from crime.

SOVIET CITIES

The typical Soviet city has also shown considerable growth in recent decades, especially larger metropolitan areas, which are preferred areas for the citizenry (Table 19-4). Citizen choices favor larger cities, due to better quality-of-life opportunities with respect to the provision of personal and professional services, such as health care and cultural opportunities, largely lacking in smaller centers. As a result, large cities have grown rapidly, even when official policies have discouraged the process. "Considerable land use reorganization and regimentation has been undertaken in recent decades to reduce socioeconomic inequities, creating more homogeneity in lifestyles, architecture, and urban design than in the Western world."[11] The overriding goal of the *socialist city* is, in fact, to create a classless society.

The central area of the socialist city, as demonstrated in the Soviet Union, reveals more space devoted to administrative, cultural, and educational functions around a large open square than its western

[11]Truman A. Hartshorn, *Interpreting the City: An Urban Geography,* 2nd ed. (New York: Wiley, 1988).

three or four times as much as comparable single-family housing in America, make this dream largely prohibitive in large metropolitan areas.

Fully 80 percent of Japanese urbanites consider themselves middle class. Evidence of poverty and substandard housing is rare indeed in the Japanese city. Despite heavy damage during World War II, complete rebuilding occurred in a few years and waves of redevelopment continued. High-rise office towers became more visible in multiple locations in cities, creating a polycentric commercial form. The Sinjuku and World Trade center sections are two such nodes in Tokyo, in addition to intense commercial developments near the palace grounds and central railroad station. Improved building technology and strict building codes now allow structures to reach 60 stories in height or more, a format prohibited in the past due to earthquake hazards in Tokyo.

The influence of the automobile led to the development of an urban expressway system in Japanese cities in the 1970s. Due to the fully built nature of the

Figure 19-27 Complex Land Uses in Tokyo. *The intense street level activity, including shoe shine stalls, heavy automobile traffic, and retail uses and colorful signs characterize commercial areas in downtown Tokyo. Digital sign on building monitors environmental conditions. (T.A.H.)*

Figure 19-28 Contrasts in Development density, Tokyo. *Tokyo has been rebuilt twice since World War II. It appears as a modern metropolis with the newest buildings occupying a much larger footprint than those being replaced. Red Cross headquarters building shown in center surrounded by mixed housing and commercial buildings. (T.A.H.)*

Figure 19-29 High Rise Suburban Housing, Tokyo. *Large apartment blocks in the suburbs of Tokyo are typically well-served by rail commuter lines. Bicycles are heavily used to access stations and for short shopping and social trips. (T.A.H.)*

Figure 19-30 Urban Expressway, Tokyo. *Superimposing expressways on the high-density land uses in Tokyo has required innovative design solutions. Many expressways are elevated over canals and rivers as is this 1000 meter harbor-front route. (T.A.H.)*

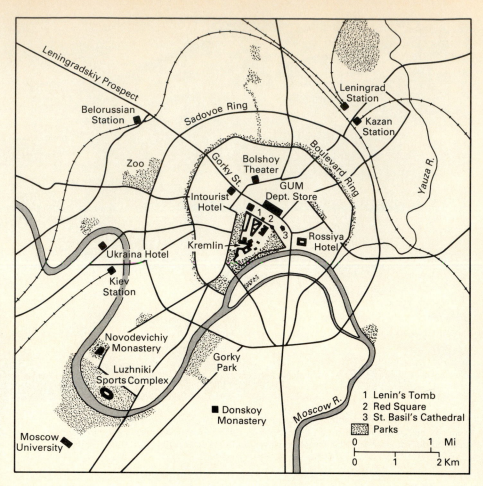

Figure 19-31 Land Use in Central Moscow. *The central area of Moscow reflects a greater quantity of space devoted to administrative, cultural, and educational functions, along with large open squares and parks than its western counterparts, which have greater concentrations of commercial activity in the center. Many of these uses date to prerevolutionary czarist period. Note the ring roads and locations of rail stations away from core area. (Redrafted from Brunn and Williams, Cities of the World, Harper & Row, 1983, p. 138; after Moscow, Verlitz, 1976, pp. 32–33)*

counterpart, which would have more commercial enterprises, such as hotels, banks, retail outlets, and office buildings (Figure 19-31). Public gatherings, parades, and ceremonial functions take place on the open square. The concentration of cultural and administrative activities in the city center, as in the case of Moscow parks, museums, theaters, gardens, and so on, is largely a legacy of the past, dating to prerevolutionary czarist days. The same origin characterizes various highly political institutions, such as Red Square and the Kremlin.

The internal structure of Soviet cities is guided primarily by centrally conceived plans, not by land values or economics, but by a plan to allocate resources equally. These plans place great priority on decentralizing activity, whether it be bureaucratic administrative functions, housing, or cultural activities, giving the city a distinctive patchwork structure. Often vast areas are unused or underutilized because of errors in judging space needs. In such a system there is no pressure to use land intensively or to use vertical space, but a recent tendency has been to move away from five-story walk-up apartment blocks to 15- to 20-story residential structures.

The provision of additional housing has been a particularly big problem facing the Soviet city, and acute housing shortages remain a big issue. Massive apartment blocks, typically comprised of small, poorly

equipped units, have replaced lower-density housing, but their construction and design, with an emphasis on prefabrication, has been notoriously poor. The ponderous architectural style of these buildings has also been criticized.

Not being part of a consumer-oriented society, the Soviet city devotes far less space to retail and commercial structures than does its western counterpart. The *micro-rayon* planning concept seeks to create relatively self-sufficient residential districts of 10,000 to 15,000 residents on about 100 acres, car-free and self-contained, by providing most needed services internally. Therefore, these areas provide their own schools, parks, stores, and recreation facilities.

Soviet urban planning suffers from several institutional and economic problems. First, it is not clear who governs the Soviet city, and as a result many gaps exist between the ideal and reality. Considerable competition and conflict also occurs among ministries in implementing plans. Beginning with the Stalinist period in the 1920s, industrial development received priority and no comparable program existed to expand urban services or make infrastructure improvements. Severe housing shortages resulted, which continues to the present. Not only are available units very small, but waiting lists are long. Fully 20 percent of the urban population lives communally sharing facilities with others, and 5 percent of the urban population lives in urban dormitory facilities. The housing shortage is exacerbated by a continuing food shortage, especially meat and dairy products, as discussed in Chapters 5 and 6.

Labor shortages also exist in cities, as the population is aging, and productivity remains low due to lack of motivation, high absenteeism, and chronic alcoholism. This labor shortage is occurring at the same time as growth caps limit expansion in major cities. An associated problem with Soviet industrial planning involves the emphasis on labor-intensive operations rather than mechanization, which would decrease the need for more labor.

Some observers feel that industrial ministries have destroyed the city in the Soviet Union. Control lies at the ministry, not metropolitan, level for the provision of housing and services. In relation to the party and enterprises, cities remain weak. Ministries operate them as company towns. Unfortunately, ministries often share in the provision of various functions such as electricity, and not all parts of the city have comparable services.

> In the game of bureaucratic politics some bureaucracies are more powerful and successful than others. This also applies to the triangular struggle over who controls Soviet cities: the managers of enterprises, the first party secretary of the city, and the chairman of the executive committee of the city soviet are all party members. Most participate in an interlocking directorate, the city party bureau. But not all are equal. Because of the institutions they represent, representatives pursue different interests and carry different responsibilities. They are also backed by and responsible to higher authorities whose political and economic influences play an important role.[12]

Just as governance is a problem, implementing overall comprehensive urban plans, referred to as the *general plan*, is similarly weak in the Soviet Union:

> Soviet cities are handicapped by overlapping and ill-coordinated planning. The USSR State Planning Committee's Council for the Utilization of Productive Forces draws up long-range territorial plans; the USSR State Construction Committee designs borough layouts; various state agencies compile their own economic plans; and architectural planners draw up general plans for urban development. Each of these documents deals with only one facet of urban development; not one of them treats the city as an integral unit.[13]

Notwithstanding these problems, the Soviet bureaucracy does regularly issue revised general plans. Since World War II, for example, "Novokuznetsk has had eight general plans, Volgograd six, and Kharkov three."[14]

[12]Henry W. Morton and Robert C. Stuart, *The Contemporary Soviet City*. (Armonk, N.Y.: M. E. Sharpe, 1984), p. 19.

[13]A. V. Dmitriyev, as reported in James H. Bater, *The Soviet City* (Beverly Hills, Calif.: Sage Publications, 1980).

[14]Ibid.

SUMMARY AND CONCLUSION

The strongest international contrasts in city structure occur between the developed and developing areas, but contrasts also occur in Soviet and western urban areas. Although relatively prosperous downtown areas occur in each environment, third-world and Soviet cities do not possess large middle-class suburban communities, greatly shrinking the extent of prosperous residential neighborhoods, which are so extensive in North American, European, and Japanese cities. The rather small elite residential district in third-world cities occurs along a single corridor close to the downtown. The affluent upper class in the North American context is much more likely to live at some distance from the city center. Lower-income groups in the United States and Canada generally occupy close-in locations. In Europe, higher-income residences traditionally occur nearer the city center. The widespread poverty present in the third world is demonstrated by the extensive squatter settlements in suburban areas, characterized by few or no urban services and very poor housing conditions.

Problems associated with accommodating the automobile constitute a common problem in cities in all parts of the world. In Japan, Europe, and parts of the third world the introduction of the car in a traditional pedestrian-scale environment has created enormous problems. Even in countries emphasizing public transportation, as is the case in Canada, coping with the car is difficult due to the absence of a comprehensive expressway system. Disciplining and restricting the automobile has been accomplished successfully in Europe, a trend that may be copied elsewhere in the future.

North American cities exhibit strong ring and sectoral characteristics due to the sorting out of residential communities by age, size, socioeconomic status, and ethnic characteristics. The minority residential space in these cities has been called a city within a city and demonstrates an internal neighborhood-level sorting-out process not unlike that of the greater urban region. An inverse ring model may best characterize the third-world city, due to the presence of the poor on the edge and the wealthy in the center, the opposite tendency found in North America.

Exceptions to the prevailing demographic and socioeconomic sorting process occur in all areas—undeveloped, planned, and developing economies. The Middle Eastern Islamic city, characterized by a strong religious-based emphasis on egalitarianism, exhibits much more formlessness than much of the rest of the third world. Japanese and Soviet cities also demonstrate less variation, the former because of more mixing of activities throughout and the latter due to an emphasis on a classless decentralized society.

The primate city form, dating to colonial times, provides the most conspicuous and problematic example of third-world urbanization. Rapidly growing, but lacking a modern diversified economy, poverty and despair characterize large shares of the populous in these cities.

Canadian cities differ in important ways from their U.S. counterparts, due to varying economic, governmental, and financial circumstances. The varying roles of the federal government, transportation, and taxation policies explain much of the variation. Canadian central cities are generally stronger in relative population and employment shares than their U.S. counterparts as a result of these policy differences.

Looking ahead, it appears that the largest cities in the world by the close of the present century will be third-world cities. Moreover, it appears that the problems facing these areas today will only intensify, creating even larger problems of economic development in the future. By contrast, cities in developed areas will grow much more slowly and maintain the prevailing middle-class life-style.

In summary, urban structure continues to exhibit considerable spatial variation worldwide. While similar land uses occur in all cities, the strong imprint from the past on patterns remains, and institutional forces affecting uses also vary considerably. Even the automobile, which has had a tremendous impact on cities in highly developed areas, is not a universal transportation mode in the late twentieth century, as evidenced by the continuing dependence on the bicycle in China. In an age of strong and growing global business ties, these differences may be waning, as evidenced by the increasingly similar high-rise commercial architecture styles worldwide. Nevertheless, extreme levels of poverty continue in the third world, which lacks a strong middle class. Housing styles and densities continue to exhibit strong contrasts not only between developed and developing areas but also with centrally planned economies. In Chapter 20 we explore the growing trends toward international business linkages and the mechanisms that tie nations together.

SUGGESTIONS FOR FURTHER READING

BATER, JAMES H., *The Soviet City: Ideal and Reality.* Beverly Hills, Calif.: Sage, 1980.

BERRY, BRIAN J. L., and JOHN KASARDA, *Contemporary Urban Ecology.* New York: Macmilian, 1977.

BOURNE, LARRY, *Internal Structure of the City: Readings on Urban Form, Growth and Policy.* New York: Oxford University Press, 1982.

BRUNN, STANLEY, and JACK F. WILLIAMS, *Cities of the World: World Regional Urban Development.* New York: Harper & Row, 1983.

BURTENSHAW, DAVID, M. BATEMAN, and G. J. ASHWORTH, *The City in Western Europe.* New York: Wiley, 1981.

DWYER, DENNIS J., ed., *The City in the Third World.* London: Macmillan, 1974.

FRENCH, R. A., and F. E. IAN HAMILTON, eds., *The Socialist City: Structure and Policy.* New York: Wiley, 1979.

GOLDBERG, MICHAEL A., and JOHN MERCER, *The Myth of the North American City.* Vancouver: University of British Columbia Press, 1986.

HAKIM, BESIM SELIM, *Arabic-Islamic Cities.* New York: Routledge & Kegan Paul, 1986.

HALL, PETER, *Third World Cities,* 3rd ed. New York: St. Martin's Press, 1984.

HARTSHORN, TRUMAN A., *Interpreting the City: An Urban Geography,* 2nd ed. New York: Wiley, 1988.

KORNHAUSER, DAVID, *Urban Japan: Its Foundation and Growth.* New York: Longman, 1976.

MCGEE, T. G., *The Southeast Asian City: A Social Geography of the Primate Cities of Southeast Asia.* New York: Praeger, 1967.

MILLS, EDWIN S., *Studies in the Structure of the Urban Economy.* Baltimore, Md.: Johns Hopkins University Press, 1972.

MORTON, HENRY W., and ROBERT C. STUART, *The Contemporary Soviet City.* Armonk, N.Y.: M. E. Sharpe, 1984.

MUMFORD, LEWIS, *The City in History: Its Origins, Its Transformations, and Its Prospects.* New York: Harcourt Brace & World, 1961.

POTTER, ROBERT, *Urbanization and Planning in the Third World: Spatial Perceptions and Public Participation.* New York: St. Martin's Press, 1985.

REPS, JOHN, *The Making of Urban America: A History of City Planning in the U.S.* Princeton, N.J.: Princeton University Press, 1965.

RICHARDSON, HARRY, *The New Urban Economics and Alternatives.* London: Pion, 1977.

SCOTT, IAN, *Urban Spatial Development in Mexico.* Baltimore, Md.: Johns Hopkins University Press, 1982.

SMITH, NEIL, and PETER WILLIAMS, eds., *Gentrification of the City.* Winchester, Massachusetts: Allen & Unwin, 1986.

STELTER, GILBERT and ALAN F. J. ARTIBISE, eds., *Power and Place: Canadian Urban Development in the North American Context.* Vancouver: University of British Columbia Press, 1986.

STITT, VICTOR, *Chinese cities: The Growth of the Metropolis since 1949.* New York: Oxford University Press, 1985.

VANCE, JAMES E., JR., *This Scene of Man: The Role and Structure of the City in the Geography of Western Civilization.* New York: Harper's College Press, 1977.

Geography of International Business

20

International business operations date to the mercantilist period of the 1500s or earlier. English and Dutch trading companies, among others, vastly increased the wealth of their countries from this activity. Remarkably, these two nations still retain leadership in international business in the late twentieth century. Whereas the bulk of this activity involved the trade of goods in the earlier period, today it not only involves vastly increased levels of trade but also direct foreign investment and far-flung operations of enormous multinational corporations (Figure 20-1).

Multinational corporate giants based in developed and developing countries engage in manufacturing activity as well as services and agricultural operations that account for nearly one-fourth of the world's output in the late 1980s. They dominate some industries, such as petroleum, automobiles, and mineral resources, including copper and bauxite extraction. Multinationals also frequently control the largest banks, wholesale businesses, engineering firms, and construction companies in the service sector.

In this chapter we discuss the maturing of the world system in terms of the growth of the internationalization process, together with the financial reforms that have accompanied this expansion and the countervailing forces that temper this growth. The changing mix of the headquarters locations of multinational firms and shifts in direct investment patterns are also analyzed, together with the challenges facing free trade.

INTERNATIONAL TRADE DYNAMICS

The growing internationalization process in manufacturing operations served as a major integrating theme in earlier chapters, but the scope of changing international relationships has not yet been placed in proper perspective. Here we review recent developments in world trade, then examine other international business initiatives, including direct investments, joint ventures, and licensing agreements, among others.

The dominance of developed countries in international business, in terms of the value of exported goods, can be expected given their advanced technology and financial resources. Developing countries similarly lead in the volume of exported goods, given the emphasis on raw materials flows from those areas. What is not well known is that the export of manufactured goods from developing countries, excluding oil-exporting nations, now exceeds the value of their raw material exports. The import-substitution policy followed by these nations created the basis for this recent expansion, but it has also introduced greater third-world debt burdens from the purchase of the technology and equipment required by this new economic order.

In the early 1980s, the United States and the European Economic Community (EEC) nations began treating newly industrialized countries (NICs) differently than other third-world countries, in recognition

Figure 20-1 Coca Cola delivery truck in Dar es Salaam, Tanzania. *Coca Cola is distributed widely in Africa. This delivery is being made in downtown Dar es Salaam. Note advertising slogan in Swahili. (Courtesy Sanford H. Bederman).*

of their maturing manufacturing sector, by excluding them from a general system of import preferences, a policy that had encouraged the movement of manufactured goods from developing nations to developed areas. In other words, the newly industrializing nations began behaving more like developed countries than developing ones and were treated as such.

Other signs of the maturing of the world system, in terms of international trade indicators, relate to the sheer growth in the value of world trade, the growing dependence of some nations on exports for income (such as Canada and Japan), and the declining balance-of-trade problems facing many countries. In the 1970s alone, the worldwide annual value of exports increased from less than $500 billion to nearly $2 trillion. Most of these exports came from developed countries, increasing their interdependence on one another. Several European states derive more than one-fourth of their gross national product (GNP) from exports, paced by the 50 percent share for the Netherlands (Table 20-1). Switzerland, Sweden, Great Britain, Italy, and West Germany all derive more than 20 percent of their GNP from exports. By contrast, the U.S. share of GNP from exports is less than 10 percent.

In the early 1980s, international trade levels actually declined, but they rebounded by 1984 as the negative affects of the 1980–1982 worldwide recession waned and the dampening effect of the high value of the U.S. dollar decreased. The U.S. dollar increased in value at the beginning of the decade, due to the influx of foreign investment attracted by high interest

Table 20-1

Dependence on Foreign Trade as Percent of Gross National Product, 1983

Netherlands	50
Sweden	31
Switzerland	26
West Germany	26
Canada	24
Great Britain	21
Italy	21
France	18
Japan	13
United States	6

Source: John Hein, *World Trade: Reversing the Decline,* Report 82, (New York: The Conference Board, 1984) p. 3.

rates. The growing balance-of-trade deficit facing many nations in the 1970s, especially acute in developing areas following the oil boycott in 1974, continued to be a problem in the 1980s, and inflation rates in these areas remained at very high levels. In Latin America, for example, inflation rates zoomed upward from 73.3 percent in 1982 to 185.3 percent in 1985. Many developing countries curtailed imports in this period to cope with growing debt burdens associated with earlier purchases of western technology and higher energy import costs. In the mid-1980s even energy-exporting countries such as Brazil and Venezuela suffered setbacks as oil prices rolled back.

Despite problems associated with inflation early in the decade, which was followed by volatility in the relative strengths of major world currencies, including a decline in the value of the U.S. dollar relative to Japanese and European currencies in 1986 and 1987, direct foreign investment by many countries accelerated in the 1980s. This also meant that economic adjustments in one country increasingly had more direct impacts on other areas. "No matter how the structure of the world economy is analyzed, its various component parts are functioning more and more like parts of a machine. . . . If one gear slows down, others also slacken."[1] As a result, it became increasingly difficult for one area to gain relative strength vis-à-vis another.

This interdependence situation is no better illustrated than by the dilemma facing both developed and developing countries as they groped for solutions to the problems arising from the decline of traditional industries in developed countries. As pressures mounted for more tariff and nontariff barriers to cushion these industries from within, threats of retaliatory measures increased by overseas competitors. The recession in the early 1980s exacerbated the problem as economic activity slowed worldwide. One author invoked the bicycle analogy to shed light on the problem: When economies slow down, as with bicycles, balancing the load becomes more difficult.

Just as trade levels have intensified among highly developed nations, so has the level of intrafirm trade controlled by multinational corporations, such as that between parent companies and affiliates, and among affiliates themselves. Nearly one-fourth of U.S. foreign trade, for example, occurs as intrafirm transactions of U.S.-controlled multinational firms. When multinationals from all countries are included, about 40 percent of U.S. imports and exports occur as intrafirm transactions. These companies can take advantage of economies of scale, as well as marketing and financial expertise, giving them maximum advantage in the marketplace when conducting this trade. Indeed, multinational firms can create their own demand for products and significantly influence consumption patterns given their large presence in many world markets simultaneously.

[1]John Hein, *Major Forces in the World Economy: Concerns for International Business* (New York: The Conference Board, 1981) p. 7.

While multinationals influence the overall framework of trade, the development levels of the country and *spatial interaction* principles also play a role in accounting for the patterns of trade. Accordingly, the highest levels of trade typically occur among advanced nations relatively close to one another, a situation that exists among western European countries and between the United States and Canada. Canada, for example, is the largest trading partner with the United States. As a region, however, the Pacific basin ranks as the largest U.S. trade partner, displacing western Europe as the leader in the 1980s. A combination of the strength of the Japanese economy and the expansion of NIC exports explains this recent development.

The United States traditionally enjoyed a trade surplus, meaning that the value of exports exceeded imports. In the past decade large deficits replaced earlier surpluses, rising to an annual deficit rate in excess of $100 billion in the mid-1980s. The United States, in fact, posted a trade deficit with most of its major trade partners in the mid-1980s, paced by the $37 billion differential with Japan in 1984, $20 billion with Canada and $11 billion with Taiwan (Figure 20-2). The large deficit in trade between western European countries and Japan also merits attention.

The large annual budgetary deficit of the United States in the 1980s, in excess of $200 billion at mid-decade, contributed to the trade deficit by keeping interest rates high, which in turn attracted more foreign investments, and promoted an overvalued dollar,

Figure 20-2 Directions of U.S. Import-Export Trade, 1985. *The United States posted a large trade deficit with most of its trade partners in the mid-1980s, including Japan, Canada, and western Europe. While imports from western Europe are the largest, the deficit with Japan is the highest.*

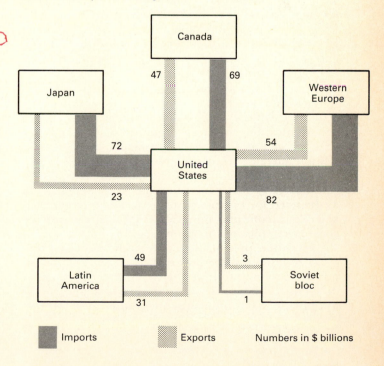

which in turn constrained exports and promoted imports. The lion's share of the U.S. deficit occurred with manufactured goods, but the traditional U.S. surplus in services also dropped in the 1980s. By 1986 the strength of the dollar declined against the Japanese yen and interest rates plummeted in the United States and inflation rates remained low. This adjustment caused imported goods to become more expensive relative to domestic output, but did little in the short term to counteract the massive national deficit.

More elaborate strategies to finance international trade emerged in the 1980s, focused particularly on alternatives to traditional sales arrangements for goods and services sold to third-world countries and centrally planned economies by developed areas. *Countertrade,* or noncash sales arrangements, provide such a mechanism. Countertrade can involve *barter* deals, wherein products of equal value are swapped; buy-back agreements; and/or offset trade. Examples of barter arrangements would include trading coffee, tea, or cotton for farm machinery. *Buy-back agreements* refer to a policy wherein sellers agree to buy products from the purchasing country as a condition of making a sale. *Offset trade* refers to agreements to assist foreign buyers in selling their production as a condition of selling technology or machinery to them. Trading bauxite for automobiles, or as in the case of Nigeria, exchanging oil for sugar provide illustrations. An example of U.S. countertrade with the Soviet Union involves a soft-drink manufacturer trading bottling equipment for vodka.

CHANGING FORMS OF INTERNATIONAL BUSINESS

The major change in international business activity in the post–World War II era involved the shift away from an emphasis on trade as international business in favor of direct foreign investments, joint ventures, and a multitude of other types of operations conducted by multinational firms. In accounting for this expansion, one must examine many factors related to the firm and its internal operations, the nature of the economic framework in which it operates, changes in communications and transportation technology over time, as well as shifting raw materials, market, technology, and finance considerations affecting the particular good or service. As firms expand, many forces can lead to growing international involvement, such as the need for raw materials or access to larger markets. Petroleum and mineral resource firms provide classic examples of this process.

From the product-life-cycle model we learn that firms become involved abroad in stages, first by exporting product, followed later by direct investment in production facilities in other markets, and finally by moving labor-intensive operations to lower-cost producing areas. Often, the most successful of these transitions occur in oligopolistic markets where few producers compete. A Marxist interpretation suggests that international involvement occurs as a result of the growth of a few monopolistic firms eager to expand their markets. The *geobusiness model* of international business assists an understanding of the factors associated with the expansion of this activity (see the accompanying box).

Improvements in transportation, communication, and management techniques, especially those associated with air transportation, the telephone, and computer applications, have vastly increased the ability of firms to operate effectively in a broader international context. At the same time, the world economy itself has grown remarkably in the second half of the twentieth century, creating many more opportunities for firm expansion, in both the developing and developed economies.

Today, about 10,000 firms worldwide have direct investments outside their headquarters country, along with 90,000 foreign affiliates. The sales level of this activity is even more impressive. Multinational firms sold about $3 trillion of goods annually in the mid-1980s, with $1 trillion or one-third the total produced outside their home country and another one-third exported from the home country. Frequently, as in the case of the United States and Japan, exports of multinationals account for one-half of a country's total exports and 80 percent of all direct foreign investment. A very few large firms, perhaps 5 percent, conduct over three-fourths of this activity. Most multinationals are privately held, but state enterprises partially or totally controlled by government agencies, frequently occur in Europe in the petroleum, steel, and automobile industries. State-controlled or state-managed enterprises also exist in Canada and the third world, not to mention their dominance in centrally planned economies.

Multinational Ownership Trends

As expected, the overwhelming majority of multinational firms operate from headquarters facilities in developed countries (95 percent), but the share based in developing areas continues to expand. The United States has the largest share (21 percent) and has had the longest history of this involvement in modern times, defined here as the post–World War II era. U.S. firms initially invested in Canada, western Europe, and the Middle East, but later became involved in import-substitution manufacturing expansion in the third world. As a region, western Europe–based multinationals have grown the most rapidly in the past two decades. Much of this investment came to the United States. A perception of an antibusiness bias at home partially motivated this expansion. Of the world's largest multinationals about 14 percent each are headquartered in West Germany and the United

The *geobusiness* model developed by Robock and Simmonds explains international business activity on the basis of the characteristics of (1) the product, (2) the market, (3) the firm, (4) the competition, and (5) restrictions or incentives provided by home and/or host countries in an international framework.* These factors are grouped into three clusters of variables: (1) conditioning, (2) motivation, and (3) control. We discuss each of these in turn.

Conditioning Variables

Conditioning variables include (1) characteristics of the product, (2) the situation in the home and host country, and (3) international constraints. The product items include technology access, production process characteristics, management and marketing arrangements, among others. Home-country factors include experience levels with overseas operations and the relative size of the domestic market. In Switzerland and Japan, for example, the small domestic market provided an early incentive for international investments, whereas the large size of the U.S. market has discouraged such activity. International variables include financial system characteristics and the trade framework. The post–World War II international financial system, for example, encouraged decades of expansion, followed by contraction in the 1970s. Producer associations (e.g., OPEC) and common markets have similarly stimulated or retarded trade.

Motivation Variables

As before, motivation variables include firm characteristics, the nature of the competition, and international political and financial considerations. Reflecting a wide range of goals that motivate investment decisions, some businesses operate so as to avoid risks, whereas others aim to intimidate their competition. Market-seeker firms typically pursue horizontal integration, whereas vertical integration is a goal of resource seekers. Technology seekers, on the other hand, often buy foreign companies to gain access to technology.

Control Variables

National political controls and policies frequently guide international investment decisions. The United

*Stefan Robock and Kenneth Simmonds, *International Business and Multinational Enterprises*, 3rd ed. (Homewood, Ill.: Richard D. Irwin, 1983), pp. 50–56.

States, for example, restricts the export of technology that may jeopardize national security. Home governments can encourage overseas expansion by providing political risk insurance, subsidizing investment loans, and providing tax credits or payments. Host governments can similarly become active in encouraging and/or discouraging foreign business activity. By requiring local content in products, mandating a share of local ownership, or restricting investments, host governments can manage business priorities and the "foreignness" of firms.

The geobusiness model posits that business transactions cross international boundaries when such an arrangement is mutually beneficial to all parties concerned. Firms choose markets to enter and develop operational strategies based on raw material sources, markets, labor and capital, as well as research and development resources and opportunities. These operations occur under the watchful eye of host-country and home-country governmental policies. Although the factors affecting international operations are largely the same as those in domestic business, they are more complex and varied. Many more choices are available and a much greater understanding of currency fluctuations and the uncertainties of international finance are required for global operations. Firms can opt to engage in traditional trade arrangements as an initial step in internationalizing operations, and many firms make this their only strategy to broaden involvement beyond the domestic market. Other firms gradually expand operations to include production in foreign countries, and later complete internationalization of their business. We have noted examples of each of these behaviors in earlier chapters.

Using the geobusiness model as a prototype, one can account for the early Japanese drive to gain access to foreign sources of raw material and energy, and later, as home-country labor costs increased and overseas import restrictions loomed ahead, the shift of production operations to foreign sites. These factors are the conditioning and motivation variables influencing these policies. At the same time, control variables protected Japanese producers from import competition at home. Not only can one observe the simultaneous influence of conditioning, motivation, and control variables in the Japanese economy at a given time by using the geobusiness framework, but also the unfolding of changing strategies over time as circumstances dictate. A similar analysis could be made of policies in other countries.

Table 20-2

Twenty-five Leading Multinational Firms, 1981 (billions of U.S. dollars)

Rank	Company	Industry	Headquarters Nation	Worldwide Sales	Sales of Overseas Subsidiary as Percent of Worldwide Sales
1	Exxon	Oil	United States	113	74
2	Royal Dutch/Shell	Oil	Netherlands/Great Britain	82	60
3	Mobil	Oil	United States	69	65
4	General Motors	Autos	United States	63	25
5	Texaco	Oil	United States	58	67
6	BP	Oil	United States	52	63
7	Standard/California	Oil	United States	45	53
8	Ford	Autos	United States	38	49
9	Standard (Indiana)	Oil	United States	31	18
10	ENI	Oil	Italy	31	19
11	Gulf	Oil	United States	30	38
12	IBM	Computers	United States	29	48
13	Atlantic Richfield	Oil	United States	28	9
14	General Electric	Electrical	United States	28	23
15	Unilever	Food	Great Britain/Netherlands	24	39
16	Dupont	Chemicals	United States	23	17
17	Total	Oil	France	23	59
18	Veba	Misc.	West Germany	22	—
19	Fiat	Autos	Italy	19	—
20	Elfaquitaine	Oil	France	19	26
21	Renault	Autos	France	19	45
22	BAT	Tobacco	Great Britain	17	78
23	ITT	Electrical	United States	17	47
24	Nissan	Autos	Japan	17	—
25	Philips	Electrical	Netherlands	17	93

Source: J. M. Stopford and J. D. Dunning, *Multinationals: Company Performance and Global Trends* (London: Macmillan, 1983).

Kingdom, 7 percent in Switzerland, and 6 percent each in France and the Netherlands (Table 20-2). Only 6 percent of these large firms are based in Japan and 4 percent in Canada.

The U.S. share of multinational firm headquarters has declined relatively in recent years, but it still accounts for one-half of the top 25 multinationals worldwide. All but four of these large U.S. firms belong to the oil or automobile industry group. At the same time, the sales strength of West German, Swiss, and Japanese firms continues to expand. Outside the highly developed world, Hong Kong and Singapore lead as multinational centers.

Japanese firms initially concentrated on the export of textiles and consumer electronics, followed later by steel and automobiles. The soga sosha trading companies initially developed extensive networks of overseas raw material sources, followed by the development of branch offices and overseas production operations. The gigantic operations these firms control today also offer a wide range of services, including investment and trade financing, technology transfer, and service contracts to assist affiliates. As discussed earlier, the cooperation of business and government in Japan in working out long-term equity investments in overseas operations gives the Japanese increasing clout in international business as well as providing more stability for the domestic economy.

Japanese multinationals show a disproportionate concentration in the wholesale trade sector, owing to the strength of the soga soshas, as discussed in Chapter 6. Japanese firms control over 40 percent of the worldwide foreign investment in wholesale trade. Japanese multinationals are also disproportionately concentrated in banking. In addition to New York, international banking focuses on London, Tokyo, Hong Kong, and Frankfurt (Table 20-3 and Figure 20-3).

Although international operations generally run smoothly in host countries, nationalization pressures and taxation issues often surface. Problems of nationalization in several countries very dependent on mineral resource mining were discussed in Chapter 10. Third-world countries have been very effective in recent years in gaining concessions from these firms. These concessions have involved the provision of increased economic development assistance for developing countries and more access to western technology. Developing countries have also restricted foreign investments in some sectors of the economy which they prefer to reserve for themselves, such as banking, insurance, public utilities, and the airline industry.

Screening of investments assists developing coun-

Table 20-3

Leading Commercial Banks Outside the United States, 1984
(billions of U.S. dollars)

Rank	Bank	Headquarters Nation	Assets	Number of Employees (thousands)
1	Dai-Ichi Kangyo	Japan	125	22
2	Fuji	Japan	116	16
3	Sumitomo	Japan	114	15
4	Mitsubishi	Japan	111	16
5	Sanuva	Japan	102	16
6	Banque Nationale de Paris	France	98	60
7	Caisse Nationale de Credit Agricole	France	92	74
8	Credit Lyonnais	France	90	55
9	Societé Générale	France	87	44
10	Barclay's	Great Britain	85	126
11	Bank of Tokyo	Japan	84	15
12	Norinchukin	Japan	84	3
13	Industrial Bank of Japan	Japan	83	6
14	National Westminster	Great Britain	83	90
15	Tokai	Japan	78	14
16	Mitsui	Japan	75	12
17	Deutsche	West Germany	74	48
18	Midland	Great Britain	71	81
19	Long-Term Credit Bank of Japan	Japan	70	4
20	Royal Bank of Canada	Canada	67	38
21	Mitsubishi Trust and Banking	Japan	63	7
22	Hong Kong and Shanghai Banking	Hong Kong	62	45
23	Taiyo Kobe	Japan	61	16
24	Sumitomo Trust and Banking	Japan	59	6
25	Bank of Montreal	Canada	58	34

Source: *Fortune*, August 19, 1985.

tries meet development objectives and has generally led to a more restrictive regulatory environment in developing countries than in developed countries for multinational firms. Protecting against balance-of-payments deficits provides some of the motivation for these safeguards, but worries about past abuses, excessive political influence, and fears of oligopoly power also occur.[2]

Third-world multinationals traditionally carved out a unique market niche for their firms, with scaled-down, labor-intensive activities, typically focused on the textile, apparel, construction, and electronic components fabrication fields. These firms produce standardized products and compete mainly on the basis of price.[3] In recent years, however much more diversification has occurred. Third-world countries often take charge of externally controlled multinational firms in stages. By gaining access to service contracts, engaging in joint ventures, and placing more nationals in managerial positions, business expertise has increased in these countries. They typically continue to rely on the multinationals for specialized international marketing, distribution, and technical services.

Other problems also exist for multinationals in their home and host countries. In the taxation field, both host and home countries have increasingly lobbied to tax multinationals on the basis of world sales totals or assets rather than solely on the basis of local business levels. This approach, often referred to as a *unitary tax,* causes great concern to many firms and has played a role in the location decision-making process, as areas implementing such taxes are perceived as antibusiness.

Multinational Activity by Sector

Multinational firm activity in the manufacturing sector occurs most intensively in high-technology fields requiring heavy research and development investments, such as pharmaceuticals, chemicals, computers, robots, electrical machinery, and electronics (Figure 20-4). Second, concentrations also occur in capital-intensive mass-produced items such as automobiles, tires, and home appliances. Mass-produced nondurable consumer goods products also attract

[2] William Dymza, "Trends in Multinational Business and Global Environments: A Perspective," *Journal of International Business Studies,* 15 (1984), 25–46.

[3] Louis T. Wells, *Third World Multinationals* (Cambridge, Mass.: MIT Press, 1983).

Figure 20-3 Banking and Finance in Frankfurt. *Dating to Charlemagne's time, Frankfurt has always been a center for commerce and trade in Europe. After World War II the city benefited from the shift of financial institutions away from Berlin. Banking activity occurs in many downtown high-rise office buildings, giving Frankfurt a modern skyline. (T.A.H.)*

many multinationals, including the cigarette, soft-drink, and toiletries industries.

In the service field, multinationals operate large finance, insurance, wholesale, real estate, hotel, airline, engineering, management, and construction firms. The importance of services in international business is not well understood or appreciated, but the trade of this activity has grown more rapidly than other sectors in recent years, exceeding $650 billion in 1980.

Service investments are often less appreciated by host countries than manufacturing involvement because they are perceived erroneously as being less productive. There are also many crucial barriers facing service firms seeking entry to foreign markets despite reforms, such as the lack of standardization in the telecommunication and data-processing fields. Nationalization often limits the involvement of multinationals in the insurance field in developing coun-

tries. Foreign accounting firms cannot bid on large official contracts in France, and in Mexico foreign drivers and trucking firms cannot operate their equipment. Some countries, such as Norway, Sweden, and Portugal, ban foreign banks completely, while in Denmark only EEC member countries can operate, and in Canada only 8 percent of total financial transactions can be handled by foreign-owned banks. The United States is not without restrictions either. One example dates to the Jones Act in the 1930s, which forbids foreign vessels from transporting freight among coastal U.S. ports. Only U.S.-based ships can carry goods between U.S. port cities.

Multinational activity in the agriculture field has been discussed in Chapters 5 to 7, but it deserves mention here that agriculture-based multinationals are as dominant in some areas as they are in the manufacturing and service trades. Examples of agricultural fields in which multinational operators dominate include many subtropical plantation crops, such as coffee, tea, bananas, pineapple, and sugar, as well as tobacco.

INTERNATIONAL BUSINESS STRATEGIES

The faster pace of economic change in terms of product life cycles and product innovations, together with the need for greater flexibility in resource allocations associated with heightened competition facing the multinational firm from both the developed and developing country realms, accelerated the urgency for firms to adopt new management styles in the 1980s. Firms also began to develop external strategies to replace the self-sufficient or "go it alone" attitudes of the past. Not only did this adjustment involve the need to respond creatively to the shift of traditional industries from the north (highly developed countries) to the south (developing countries), but also to substitute knowledge/service-intensive activities for capital-intensive activities.[4] Economic and social problems, such as recycling old plants and job training, associated with these shifts also had to be addressed. Development officials generally agree that policies which facilitate these changes will probably be more successful in the long run than short-sighted protectionist strategies attempting to preserve the status quo.

Managers in the future, it is said, must be more cosmopolitan, flexible, and responsive to global economic changes and yet operate in a framework of long-term strategic planning. The greater fluidity in the flows of international capital, for example, will force greater conformity in the economic structure of countries and at the same time require management to cope with "political, economic, technological, and competitive discontinuities in the global environ-

[4]Franklin R. Root, "Some Trends in the World Economy and Their Implications for International Business Strategy," *Journal of International Business Studies,* 15 (1984), 21.

Figure 20-4 Sandoz Multinational Chemical Firm, Bazel, Switzerland. *Many large chemical firms are headquartered in Switzerland, as is Sandoz, on the banks of the Rhine River. (Courtesy Hofmann-LaRoche)*

ment.... Cooperative ventures among MNC's to share scarce corporate resources (such as R&D, new product development, and manufacturing plant) and to gain market access with mutual 'market swaps' will become common responses to the new global industrial map."[5] This global work environment will also be more dynamic and "surprise intensive."

In the past American firms have been less inclined than their global competition to adopt long-term strategic goals, favoring instead the pursuit of short-term profits. In their favor, however, they have been keenly responsive to market shifts, and there is no reason to doubt their ability to make other adjustments as suggested here. One observer has noted that there is even less room for complacency regarding the needs of new high-technology industries, which are global from the start and footloose due to their high value per unit of weight and ability to withstand considerable transportation costs. Notwithstanding these characteristics, highly developed nations have the initial advantage in the development of high-technology activity, due to its technology-in-

[5]Ibid., p. 23.

tensive nature and their specialized market applications for such products.

Foreign Investment in the United States

Foreign investment in the United States has had a long history, due primarily to long-standing English and Canadian involvement. In 1984, however, a dramatic turnaround occurred, because for the first time Japanese investments became the most numerous, totaling 87 initiatives in a single year out of a total of 325 foreign investments (Table 20-4). In large measure this expansion could be accounted for by the Japanese anticipation of impending automobile import quotas in the United States and the desire to avoid such an impediment to exports by establishing production facilities in the United States, which would be exempt from any such future trade restrictions. Automobile-parts suppliers followed suit in order to remain major providers for these automobile plants and also to open up new markets as suppliers to domestically owned firms. The decisions of Honda, Toyota, Mitsubishi, and Nis-

Table 20-4

Leading Foreign Nations Investing in the United States, 1984

Rank	Nation	Number of Investments	Percent of World Total
1	Japan	87	27
2	Great Britain	56	17
3	West Germany	46	14
4	Canada	35	11
5	France	22	7
6	Netherlands	14	4
	Subtotal	260	80
	World total	325	100

Source: David Bauer, "Foreign Investment in the United States—the Year of the Japanese," *World Business Perspectives,* Report 87 (New York: The Conference Board, 1985).

Table 20-5

Foreign Investments in the United States, by Sector, 1984

Rank	Sector		Percent of U.S. Total
1	Manufacturing		35
2	Services		21
	Real estate	9	
	Banking	6	
	Insurance	6	
3	Petroleum		16
4	Wholesale trade		15
	Subtotal		87

Source: David Bauer, "Foreign Investment in the United States—The Year of the Japanese," *World Business Perspectives,* Report 87 (New York: The Conference Board, 1985).

san, among others, to expand U.S. production has been discussed in Chapter 15. The Japanese also bought into U.S. firms, forming joint ventures at an increasing rate in the early 1980s, including investments in associated steel and electronics industries.

Although Japanese investments in the United States have increased in number, they traditionally have been smaller in value than those of their European and Canadian counterparts. European investors frequently invest in existing businesses rather than start with new plants, partially explaining this differential. BASF, the chemical firm headquartered in West Germany, for example, bought Inmont in 1984, and Nestlé of Switzerland bought Scovil in 1984. In addition to big transactions in the chemicals field, these investors made big electrical machinery purchases in 1984. European and Canadian investors also made large banking and real estate purchases in the United States in the mid-1980s, especially in office buildings and shopping centers.

Foreign direct investments in U.S. businesses (Table 20-5) are less concentrated in the manufacturing field (35 percent) than in the past, largely due to expanded interest in services (21 percent), wholesale trade (15 percent), and petroleum (16 percent). Third-world countries have invested in the United States primarily in textile, electronics, and construction activities. A list of leading states receiving foreign investments appears in Table 20-6. Although New York is clearly the favorite of many foreign investors, California and several southern states also rank high, due largely to the clustering of Japanese investments in these areas.

FINANCIAL REFORMS

Along with transportation and communication reforms, financial institution innovations have been the most influential forces in promoting international business advances in the twentieth century. These advances include the creation of the international monetary system, the abandonment of the gold standard, and the switch from a fixed to a floating currency exchange system. These changes paralleled tremendous increases in the level of international banking transactions in the post–World War II era. In 1960, for example, eight U.S. banks operated 131 branches overseas, a number that increased to 151 banks with 1150 foreign offices in 1980. Banks based in other countries expanded their operations even more intensely, such that U.S. banks placed only two entries in the list of the top 10 international bank institutions in 1979. The leading commercial banks in the United States and their asset totals appear in Table 20-7.

U.S.-based banks assumed a growing share of the loans extended to developing areas to finance third-world development in this period. Several developing countries, on the brink of default, renegotiated these loans in the mid-1980s as the full impact of higher energy prices, inflation, recession, and lesser trade levels took their toll. At the same time, new players entered the lending market, primarily OPEC-based fi-

Table 20-6

Leading States for Foreign Investments in the United States, 1984

Rank	State	Number of Investments
1	New York	49
2	California	33
3	North Carolina	20
4	Georgia	17
5	Texas	16
6	New Jersey	15
6	Illinois	15
8	Pennsylvania	14
9	Tennessee	13

Source: David Bauer, "Foreign Investment in the United States—The Year of the Japanese," *World Business Perspectives,* Report 87 (New York: The Conference Board, 1985).

Table 20-7

Top 10 U.S. Commercial Banks, 1984
(billions of dollars)

Rank	Company	Headquarters	Assets
1	Citicorp	New York	151
2	Bank America	San Francisco	118
3	Chase Manhattan	New York	87
4	Manufacturers Hanover	New York	76
5	J. P. Morgan & Co.	New York	64
6	Chemical New York	New York	52
7	Security Pacific	Los Angeles	46
7	First Interstate Bank Corp	Los Angeles	46
9	Bankers Trust New York	New York	45
10	First Chicago	Chicago	40

Source: Fortune, June 10, 1985.

nancial institutions, eager to invest their surplus currency. The exposure of U.S. banks in third-world finance is reflected by the fact that they held $320 billion of the $360 billion Latin American debt in 1984.

International Monetary System

A very different basis for trade occurred until World War II than we are familiar with today. National currencies were valued according to a *gold standard* which linked them to a specific weight in pure gold. This system served the primary trading partners of the time—western European countries and the United States—quite well. The shocks of the worldwide recession in the 1930s and World War II, however, necessitated the reforms introduced in 1944. As an outgrowth of the Bretton Woods conference held that year, a *gold exchange* standard became the official exchange-rate system. The value of a country's currency, expressed in terms of gold, was established by this system. To maintain a steady value for the currency, gold was bought or sold or the currency devalued. The U.S. dollar had a special role in this system, as it was used as a standard of value against such to assign the worth of other currencies. The Bretton Woods Conference also created the International Monetary Fund (IMF) to assist orderly exchange arrangements, as well as establish the World Bank.

This gold exchange-rate system worked well at first, but stability problems increased in the 1960s as the strength of the European and Japanese economies increased and the Vietnam buildup led to greater U.S. dollar outflows. It became increasingly difficult to correct imbalances in the system during the 1970s with the pressure of rapidly escalating oil prices and the demands created by expanding economies in developing countries. These problems led to the adoption of a *managed float system* in 1976 to replace the fixed-exchange approach.

With the floating system, the relative price of currencies depended on daily supply and demand forces for that currency. Currencies of many smaller countries were tied to their major trading partner, typically meaning the U.S. dollar. European country currencies floated together against the rest of the world after 1979. The major untied currencies in addition to the U.S. dollar included the Japanese yen, the British pound, Swiss franc, and Canadian dollar. Centrally planned country currencies were not included in the system.

The *Eurocurrency* market represents another innovation in the post–World War II era. This market, using *Eurodollars,* provided a market for the trade of U.S. dollars outside U.S. national regulations and government restrictions. The origin of this system lies with the European dollar deposits controlled by the Soviet Union after World War II and the holdings of U.S. dollars in Europe by OPEC countries in the 1970s. That the dollar remained the single preferred investment currency in the period testifies to its continuing importance as an international standard of exchange.

In the late 1970s, West German and French leaders proposed the formation of a European Monetary System to establish a stable market system based in Europe. As originally envisioned, the system would consist of an exchange-rate mechanism and a European Monetary Fund. To date, the exchange-rate mechanism has been implemented and a common *European currency unit* (ECU) established for EEC members. That currency has become very important in international lending markets and may form the basis of common European currency in the future.

FREE-TRADE INITIATIVES

In addition to monetary reforms, world leaders saw the need for eliminating tariffs and other barriers to world trade at the close of World War II in order to promote economic recovery. In the 1930s, for example, U.S. tariffs were very high, in the 60 percent range, effectively keeping out imports and a contributing factor to the severity of the depression. A tariff is a tax or duty levied on a good or service when it crosses a national boundary, either as an import or export, most often the former. Nontariff barriers, which also impede trade, are discussed later in the chapter.

The GATT System

The post–World War II trade barrier reduction process became a reality with the creation of the General Agreement on Trade and Tariff (GATT) group in 1947. From that time until the mid-1970s great strides occurred in reducing trade restrictions among nations, especially with manufactured goods. Highly developed countries such as the United States took the lead in this process. The growing world market provided ample opportunities for both developing and developed areas to benefit from such a process. Accordingly, emphasis in negotiating tariff and nontariff restrictions centered on reciprocal and mutually advantageous measures. Challenges to this prevailing liberal trade philosophy gained momentum, however, following the Arab Oil Embargo in 1974, which, along with higher inflation rates, caused many nations to question unfettered trade agreements. Energy-poor third-world areas felt that they could not compete as strongly without some protection, and developed areas feared for the future of their traditional industries, which faced growing third-world competition. At the close of the 1970s, a worldwide recession further reduced trade opportunities as demand for products contracted.

Fantastic advances occurred in liberalizing trade policies between 1947 and 1967 despite later setbacks. Extended GATT negotiations occurred in six rounds during this period. Tariff rates on imported raw materials fell to 2 percent as a result of these initiatives, and various manufactured products faced rates of only 6 to 10 percent.

A seventh negotiating session, the Tokyo Round, occurred from 1977 to 1979 involving 99 countries. By that time pressures for protection began increasing again and nontariff barriers became a big issue. Nevertheless, several innovations were adopted that continued the philosophy of fewer barriers, including (1) simplification of import licensing, (2) a resolution calling for no discrimination in government procurement, and (3) a policy to enhance opportunities for foreign firms to secure large government orders. Moreover, tariffs themselves were projected to decline by another one-third by the late 1980s.

Earlier challenges to the GATT framework gained credibility in the 1980s, and growing numbers of countries accelerated the introduction of bilateral agreements outside the GATT umbrella. Some critics called this process protectionism à la carte. These protectionist pressures increased in both the developing and developed countries, including the United States, as the dollar became stronger relative to other currencies. Declining industries in developed areas faced more foreign competition, fledgling firms in developing areas sought relief, and worldwide recession and higher energy prices caused more uncertainties (Figure 20-5). Even though there was a mechanism in GATT to give other countries relief under a provision allowing for temporary protection of critically threatened industries, challenges continued to mount. Flaws in the GATT system added to the problem, including the fact that its emphasis on trade did not cover situations involving direct international investment. Also, the fact that multinationals increasingly shifted goods around as internal corporate transfers, outside normal markets, meant that they avoided GATT conditions. Moreover, GATT guidelines did not apply to the most rapidly growing segments of developed world economies—the service sector—not to mention the avoidance of a comprehensive agricultural policy. Finally, GATT had no enforcement powers, making it increasingly difficult to function as challenges mounted.

A special meeting of GATT was called in 1985 in Geneva at the behest of the United States to consider incorporating service activities, including transport, computer, banking, and insurance services, in the system, but the plea fell on deaf ears. Few developing countries favor such a move because their economies do not depend on professional services as heavily as the United States and they feared greater foreign penetration into these sectors of their economies, as well as deals whereby they would be forced to trade manufactured goods for services.

Regional Trade Organizations

Despite the fact that the GATT system appeared increasingly stalemated in the mid-1980s, several initiatives occurred outside the system. The GATT system, for example, encouraged countries to form regional trade agreements to decrease trade barriers, as has occurred in several instances. Regional organizations typically take one of three forms:[6]

1. *Free trade area:* no barriers to trade among members, but restrictions on trade with nonmembers
2. *Customs unions:* no barriers to trade among members, but a common tariff applied to nonmember trade
3. *Common market:* no trade barriers among members, a common external tariff, and no restrictions on the movement of labor and capital among members

The most successful example of an active trade association is the European Common Market (EEC). Established in 1958, based on the Treaty of Rome, with six members, this organization grew to 10 members in the early 1980s, and to 12 in 1986. This group of countries creates an entity comparable in population size and gross national product (GNP) to the United States. All internal customs among members were abolished in 1968 and common external tariffs have consistently fallen. The one area receiving protection, agriculture, creates both internal problems within the organization and problems with major world trading partners, as discussed in Chapters 5 and 6. Considerable progress in expanding cooperative ventures among members has occurred over the years, including the development of a common antitrust law, patent law, transport policy, and the European Monetary System (EMS). A vision that this organization might one day mature to assume political responsibilities remains unfulfilled, but the creation of a European Parliament with limited authority in 1979 anticipated such a development.

Two other trade organizations also exist in Europe: (1) the European Free Trade Association (EFTA),

---

[6]Stefan Robock and Kenneth Simmonds, *International Business and Multinational Enterprises* (Homewood, Ill.: Richard D. Irwin, 1983), p. 148.

Figure 20-5 Mounting Protectionist Sentiment in the U.S. *As manufacturers in the U.S. faced stiffer foreign competition and began layoffs and plant closures, labor unrest mounted and pleas for limits on imported goods increased. These union workers are taking their case to company management. (Courtesy Fletcher Drake)*

and (2) the Council of Mutual Economic Cooperation (COMECON). The EFTA organization, founded in 1960 with seven member nations, lost two members following the withdrawal of Great Britain and Denmark to join the EEC, only to return to seven as Finland and Iceland joined. Austria, Norway, Sweden, Portugal, and Switzerland constitute the continuing member group. No internal tariff barriers exist among members, but each has its own external tariff.

The COMECON group consists of the centrally planned economies of eastern Europe (Bulgaria, Czechoslovakia, East Germany, Hungary, Poland, and Romania) and the Soviet Union. Since market mechanisms do not allocate goods in these economies, no tariffs or subsidies as such exist. National economic plans determine trade policies such that prices and costs in particular industries are not as important as

overall planning goals. It should also be mentioned that trading companies totally separate from producing entities, the state enterprises, take responsibility for imports and exports in centrally planned economies. External trade involving COMECON countries is typically tied to prevailing world prices negotiated on a bilateral basis because currencies of COMECON members are not freely convertible.

Developing countries in various regions have generally been less successful than in other areas in establishing trade organizations. The ASEAN group in southeast Asia is a notable exception, due to its continuing strength. Nevertheless, developing countries have successfully established international commodity agreements to assist resource-based activity trade in world markets. The instability in the demand, supply, and price for many resource-based products they produce has encouraged the creation of buffer arrangements to smooth out destabilizing tendencies. In the tin industry, for example, a commodity association provides for a fund manager to oversee the trade of product within preestablished minimum and maximum price ranges. Producer associations have also been utilized, of which OPEC is the most visible. Such associations are most effective when a few member states control a majority of the product, which is itself in strong demand.

Nontariff Barriers

If established tariffs traditionally provided the focus of trade barrier reduction negotiations and the source of international tension, less visible but just as significant nontariff restrictions assumed greater importance in the 1980s. These barriers come in many forms, summarized here in five categories:[7]

1. *Government involvement:* takes many forms, including subsidies and discriminatory practices, such as local content requirements and prohibitions
2. *Customs and entry procedures:* strict requirements concerning product documentation, health, safety, and classification requirements
3. *Standards:* conditions based on performance levels, quality, packaging, and labeling and testing guidelines
4. *Quotas and licenses:* restrict trade in services, such as accounting, architecture, and legal matters
5. *Import deposits, credit restrictions, documentation, and ownership requirements*

The scope of each of these nontariff barriers can range from a specific restriction aimed at a particular country or product to a global discrimination policy. Quotas are the most common type of nontariff barrier. The advantage of a quota is that it is specific, definite, and can easily be negotiated up or down or withdrawn.

Developing countries frequently use nontariff barriers to protect domestic currency reserves and to cushion fledgling industries from foreign competition. Among developed nations, the Japanese have been singled out by many U.S. firms for their reluctance to remove many domestic nontariff barriers that work against the penetration of American goods into their markets (Table 20-8). These barriers also extend into the service sector. U.S. manufactured goods face a variety of nontariff barriers in the international marketplace.

One example of a disadvantage that U.S. goods face from foreign competitors involves taxes incurred in their development which cannot be rescinded in the United States, as is the case with the European Economic Community (EEC), which refunds taxes collected by the government to manufacturers when they

[7]Ibid., p. 138.

Table 20-8

Examples of Japanese Tariff and Nontariff Barriers to U.S. Goods

Product	Barrier
Communication satellites	Special frequencies, desire to develop domestic industry
Telephone switching equipment	Special performance standards
Fiber optics	Restrictive testing procedures, cumbersome licensing procedures
Medical devices	Restrictive testing requirements
Machine tools	"Buy Japanese" policy, preferential tax laws
Particleboard	13.1% tariff
Kraft paper	9.3% tariff
Tobacco	Government monopoly over prices and distribution
Food and food products	Quotas
Coal	Preference given to suppliers in which Japan has equity investments

Source: John S., McClenahan and Michael McAbee, "Congress Declares War on Japan," *Industry Week,* April 1985, p. 17.

cross international boundaries. Part of this policy difference relates to variations in the taxes themselves. In the United States a direct tax on goods is levied, whereas in the EEC an indirect value-added tax (VAT) exists. The EEC not only refunds to producers the value-added tax on exports, but levies it on corresponding imports, creating an advantage for exports and a deterrent to imports.

Other types of problems facing U.S. firms occur in the financial environment in which they must operate and the restrictive antitrust laws which govern U.S. firms abroad, including the operations of all affiliates. Generally, this antitrust policy governing U.S. firms is more restrictive than that in other highly developed countries and sometimes weakens the competitive position of U.S.-based firms. In the financial arena, the structure of the banking system in the United States in comparison to its competitors in Japan, for example, creates disadvantages. In Japan, the central Bank of Japan reports to the Ministry of Finance, whereas its U.S. counterpart, the Federal Reserve, is more independent, making it more difficult for industry to get favorable treatment. The large number of business firms in the United States in a particular field in comparison to the situation in Japan, where a few enormous firms may control an industry, creates additional contrasts. A final example of a nontariff barrier, and certainly one not confined to the United States, relates to government restrictions on the sale of defense products and high-technology goods for fear that they might fall into enemy hands or enhance the transfer of technology to unfriendly nations.

FUTURE PROSPECTS

Evidence of future growth of economic internationalism continued to accumulate in the late 1980s (Figure 20-6). International trade, for example, is expected to continue growing faster than domestic trade, as will foreign investments. Short-term setbacks that continued to surface during the 1980s also suggest that this process may occur at a slower, more deliberate rate in coming years.

In the third world, the high debt burdens, continuing poverty, and the prospect that more controls could dampen growth of exports to developed areas indicate that economic development advances will probably be more difficult to achieve in the future. Moreover, the drought in Africa in the 1980s and uncertainty everywhere about energy prices and supplies created additional concerns. Indeed, the term *lost decade* became a more and more realistic assessment of prospects in developing areas when referring to the 1980s as these setbacks unfolded. Growing pressures in highly developed economies for protectionism to shelter declining industries also cast a pall over prospects for increased international cooperation. In-

deed, a growing share of international transactions fell in the category of *managed trade,* due to the fact that controls affected nearly half of all activity in the late 1980s.

Offsetting lingering problems facing international manufacturing activity, service employment mushroomed in developed areas, approaching two-thirds of all employment in the United States and about half of the gross national product, with shares nearly as large in other developed economies. This expansion fueled urban growth and accompanied the transition to the postindustrial information age. Although restrictions limited the export of services somewhat, international initiatives were taken to reduce these barriers. More sophisticated and comprehensive techniques can be expected in the future to finance development, and the U.S. dollar will probably remain the dominant international currency. A global securities market may also evolve to ease investment transactions.

There were other bright spots as well in the 1980s. Prospects looked good for sharing resources of the sea in future decades. Pacific-rim nations appeared to

Figure 20-6 American Fast Food Franchises in Japan. *American fast-food franchises have been very successful worldwide as have soft drink franchises. Japanese families, for example, eat western style meals several times a week. (T.A.H.)*

be on the threshold of accelerated expansion, including China, South Korea, Taiwan, Singapore, and Hong Kong. Several new disciplines, such as genetic engineering and biotechnology, harnessed with ongoing high-technology initiatives, offered the prospect of a new technology revolution that would benefit humankind immensely and provide the basis for considerable employment growth in the 1990s and beyond. Finally, it appeared that global marketing was coming of age and faced a bright future. Improvements in communication and transportation, as well as accelerated levels of travel, suggested that products designed from the beginning for a global market, a process well under way in the automobile and electronics industries, would gain momentum.

SUMMARY AND CONCLUSION

Evidence of the maturing of international business operations comes from many facets of the global system that evolved rapidly in the post–World War II era but whose roots date back to the mercantilist period and the industrial revolution. First, business operations expanded in the developing, developed, and centrally planned economies, in the form of foreign investment, trade, and the dispersal of activity that created a broader geographic base of operations in all major realms. These trends created more interdependencies among nations and increased their exposure to external political pressures and economic conditions. The problems facing developed countries with declining industries, for example, became problems for developing countries attempting to nurture these same businesses.

In the 1980s, multinationals based in all countries increasingly adopted strategies involving long-term joint-venture agreements with their affiliate partners, creating more stability in the world system. Following the lead of Japanese-based firms, American, European, and third-world companies engaged in these initiatives to decrease uncertainty and risk and to moderate stiffer international competition. Other forms of international corporate arrangements having similar goals included international licensing agreements, management contracts, franchising, and buy-back arrangements. Trade among affiliate firms also accounted for a greater share of world trade in this period. A new management style accompanied these changes as well, one based on improved interpersonal relations and shorter strategic planning horizons to provide the firm with more flexibility to meet changes in the global marketplace. Even with the greater use of joint-venture arrangements, multinational firms typically maintained close control of operations at the top-management level by maintaining exclusive access to product and process technology, the use of product trademarks, and through marketing expertise.

Financial reforms, including the creation of the international monetary system, the abandonment of the gold standard, and the transition from a fixed to a floating exchange system greatly aided this internationalization process. Accompanying these changes was the continuing role of the U.S. dollar as the central world currency and the development of new dollar markets, such as the Eurodollar currency system. These changes also led to expanded roles for private financial institutions as facilitators of economic development in developing areas. In turn, third-world nations accumulated unprecedented debt levels, largely held by U.S. banks. A strong U.S. dollar, sharply higher energy prices, and worldwide recession in the early 1980s greatly aggravated this debt problem, as did continued high inflation rates in developing areas later in the decade. Highly developed areas also faced more uncertainty as their economies confronted the need for dramatic restructuring in order to compete in the postindustrial information age.

The growth of service activities provided the greatest source of change in developed nation economies in the 1970s and 1980s. Not only did banking activity expand, but so did the insurance, real estate, and wholesale institutions, along with retailing, telecommunications, and computer industries. Foreign investment in these areas became more intense, including vast new flows of funds into the United States, placed by Japanese and European investors.

During the first 20 years following World War II the general trend was for barriers to trade to fall, as all areas enjoyed economic expansion. But when conditions changed following the Arab oil boycott in 1974, which increased energy prices dramatically, more concern surfaced for protecting domestic business, even if it meant restricting imports. The one agency that served as a fulcrum for the discussion of these changes was the GATT system, created in 1947 and expanded several times in following decades, to oversee the emergence of a free-trade environment among nations. That organization has been very successful in reducing trade barriers in the manufacturing sector, but as dissatisfaction increased from greater external competition, many countries invoked bilateral agreements to limit trade in specific goods in the 1980s. Generally speaking, however, developed countries remained committed to the free-trade concept, developing countries preferred more protection for their infant industries, and centrally planned economies desired improved trade relationships with both groups. A lingering weakness of the GATT system occurs with its lack of jurisdiction over service activities.

Nontariff barriers such as quotas also became a big issue in international trade in the 1980s, with a large part of the discussion focused on barriers to the entry of foreign goods into Japan, as especially voiced by the United States. Again, the strong U.S. dollar and the increasing jeopardy facing declining domestic industries in the United States added intensity to the argument for reforms. By 1986 the U.S. dollar began declining against foreign currencies placing U.S. produced goods in a better competitive posture abroad.

Prices of Japanese and western European goods destined for U.S. markets increased. The Japanese were also cited by the U.S. government in 1987 for unfair trade practices in the semiconductor chip field. At that time the U.S. government imposed a 100 percent tariff on imports of selected Japanese television sets, personal computers, and power tools.

For the future, business leaders remain very optimistic, as they anticipate worldwide economic expansion, even if at a slower pace. The prospect of new economies opening up around knowledge/service-intensive research frontiers such as biotechnology and artificial intelligence is partially responsible for this positive outlook, but so is the prospect of the expansion anticipated among Pacific-rim nations, harvesting the wealth of the sea, and the prospects for the global marketing of goods and services engineered and produced from inception for the world market. Further improvements in transportation and telecommunications will also allow a greater level of sharing of information, creating a world with a better-informed public, in the likeness of a true global village.

In summary, the geography of international business has evolved into a study of heightened global interdependencies and more transnational ownership levels, at the same time as more segmentation of traditional activity patterns has occurred between north and south, a process that will continue to intensify in the future. The developing world (including China), for example, may account for one-fourth of the value added by manufacture in the year 2000, up from one-eighth in 1970, while the value-added share occurring in highly developed countries could fall from two-thirds to one-half.[8] New forms of knowledge/service-intensive activities may be more footloose and volatile than those they replace, leading to dramatic changes in the location of economic activity in the future, but it is likely that the highly developed nations will remain in control of the process and that their financial and technological resources will continue to shape the world of tomorrow.

[8]Organization for Economic Cooperation and Development, *Facing the Future.* (Paris: OECD, 1979), p. 88.

SUGGESTIONS FOR FURTHER READING

BUSINESS INTERNATIONAL CORPORATION. *New Directions in Multinational Corporate Organization.* New York: BIC, 1981.

CLARK, DAVID, *Post-Industrial America: A Geographical Perspective.* New York: Methuen, 1985.

CLARK, GORDON L., et al., *Regional Dynamics.* Winchester, Mass.: Allen and Unwin, 1986.

CLINE, WILLIAM R., *Exports of Manufacturers from Developing Countries: Performance and Prospects for Market Access.* Washington, D.C.: The Brookings Institution, 1984.

DIEBOLD, WILLIAM, *Industrial Policy as an International Issue.* New York: McGraw-Hill, 1980.

DUNNING, JOHN J., *International Production and the Multinational Enterprise.* London: Allen & Unwin, 1981.

GRUB, PHILLIP., ed., *The Multinational Enterprise in Transition,* 2nd ed. Princeton, N.J.: Darwin Press, 1984.

GRUNWALD, JOSEPH, and KENNETH FLAMM. *The Global Factory: Foreign Assembly in International Trade.* Washington, D.C.: The Brookings Institution, 1985.

HEIN, JOHN, *Major Forces in the World Economy: Concerns for International Business.* New York: The Conference Board, 1981.

INGRAM, JAMES C., *International Economics.* New York: Wiley, 1983.

KORTH, C. M., *International Business: Environment and Management.* Englewood Cliffs, N.J.: Prentice-Hall, 1985.

KRASNER, STEPHEN D., *Structural Conflict: The Third World Against Global Liberalism.* Los Angeles: University of California Press, 1985.

LALL, SANJAYA, *The New Multinationals.* New York: Wiley, 1983.

LEONTIDADES, JAMES C., *Multinational Corporate Strategy: Planning for World Markets.* Lexington, Mass.: Lexington Books, 1985.

LEVITT, THEODORE, *The Marketing Imagination.* New York: Free Press, 1983.

OMAN, CHARLES, *New Forms of International Investment.* Paris: Development Centre, Organization for Economic Cooperation and Development, 1984.

ROBOCK, STEFAN, and KENNETH SIMMONDS, *International Business and Multinational Enterprises.* Homewood, Ill.: Richard D. Irwin, 1983.

ROOT, FRANKLIN R., *Foreign Market Entry Strategies.* New York: AMACOM, 1982.

STOPFORD, JOHN M., and JOHN D. DUNNING, *Multinationals: Company Performance and Global Trends.* London: Macmillan, 1983.

TSOUKALIS, LOUKAS, ed., *The Political Economy of International Money: In Search of A New Order.* Beverly Hills: Sage Publications, 1985.

VERNON, RAYMOND, *Sovereignty at Bay.* New York: Basic Books, 1971.

VERNON, RAYMOND, *Storm over the Multinationals.* Cambridge, Mass.: Harvard University Press, 1977.

WELLS, LOUIS T., *Third World Multinationals.* Cambridge, Mass.: MIT Press, 1983.

Appendices

Standard Industrial Classification (SIC) Hierarchy For Selected
Product Groupings

	1 Digit	2 Digit
Division A	0 Agriculture, forestry, fisheries	
Division B	1 Mining	
Division C	1 Construction	
Division D	2 Manufacturing (nondurable)	
		20 Food products
		21 Tobacco manufactures
		22 Textile mill products
		23 Apparel products
		24 Lumber and wood products
		25 Furniture and fixtures
		26 Paper products
		27 Printing, publishing industries
		28 Chemicals
		29 Petroleum refining
Division D	3 Manufacturing (durable)	
		30 Rubber and plastics
		31 Leather
		32 Stone, clay, glass, concrete
		33 Primary metals
		34 Fabricated metals
		35 Machinery, except electrical
		36 Electrical and electronic machinery
		37 Transportation equipment
		38 Instruments, photographic goods, optical goods, watches, and clocks
		39 Miscellaneous
Division E	4 Transportation, communication, utilities (electric, gas sanitary services)	
Division F	5 Wholesale trade	
Division G	5 Retail trade	
Division H	6 Finance, Insurance, and real estate	
Division I	7 Services (personal, business)	
Division I	8 Services (professional, educational)	
Division J	9 Public administration (federal, state, and local government agencies)	
Division K	Nonclassifiable establishments	

3 Digit	4 Digit
201 Meat products	
	2011 Meat-packing plants
	2012
	2013 Sausages and other prepared meat products
	2016 Poultry dressing plants
	2017 Poultry and egg processing
202 Dairy products (Remainder of 20 series and 21 to 34 series omitted here)	
	2021 Creamery butter
	2022 Cheese
	2023 Condensed and evaporated milk
	2024 Ice cream and frozen desserts
	2026 Fluid milk
351 Engines and turbines	[4-digit breakdown omitted here]
352 Farm and garden machinery	
353 Construction and related machinery	
354 Metalworking machinery	
355 Special industry machinery	
356 General industrial machinery	
357 Office and computing machines	
358 Refrigeration and service machinery	
359 Miscellaneous machinery	
361 Electric distribution equipment	[4-digit breakdown omitted here]
362 Electrical industrial apparatus	
363 Household appliances	
364 Electric lighting and wiring equipment	
365 Radio and TV receiving equipment	
366 Communication and equipment	
367 Electronic components and accessories	
368 Miscellaneous electrical equipment and supplies	
371 Motor vehicles and equipment	
	3711 Motor vehicles and passenger car bodies
	3712
	3713 Truck and bus bodies
	3714 Motor vehicle parts and accessories
	3715 Truck trailers
	[4-digit breakdown omitted here]
372 Aircraft and parts	
373 Ship and boat building	
374 Railroad equipment	
375 Motorcyles, bicycles, and parts	
376 Guided missiles and space vehicles and parts	
379 Miscellaneous transportation equipment (Remainder of 38 and 39 series omitted here)	

Source: Modified after Office of Management and Budget, *Standard Industrial Classification Manual,* USGPO, Washington, D.C., 1972.

Metropolitan Areas with Population of One Million or More

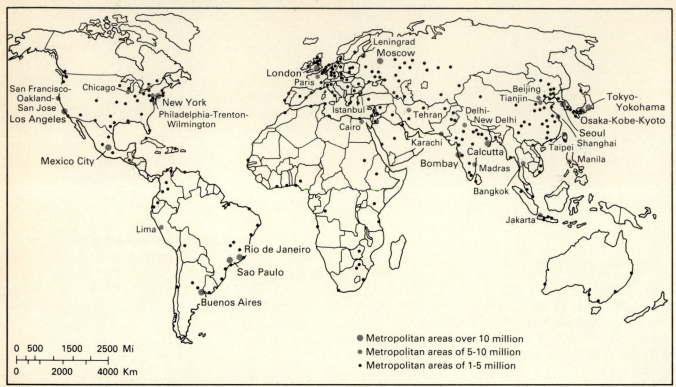

Metropolitan Areas of One Million or More Population: 1987. *The distribution of large metropolitan areas at the world scale shows a concentration in western Europe, and east Asia with secondary clusters occurring in North America and south Asia. Conspicuous voids occur in Africa and Australia. A large number of centers in the 1- to 5-million population category occur in China, India, Soviet Union, Europe, and the United States. Please also note that large metropolitan areas typically have coastal locations. Moscow and Mexico City are significant exceptions to this generalization. (Source: Estimates by Richard L. Forstall for Rand McNally & Company)*

Index